# MULTIDIMENSIONAL
# DIGITAL SIGNAL PROCESSING

**PRENTICE-HALL SIGNAL PROCESSING SERIES**

*Alan V. Oppenheim, Editor*

ANDREWS and HUNT    *Digital Image Restoration*
BRIGHAM    *The Fast Fourier Transform*
BURDIC    *Underwater Acoustic System Analysis*
CASTLEMAN    *Digital Image Processing*
CROCHIERE and RABINER    *Multirate Digital Signal Processing*
DUDGEON and MERSEREAU    *Multidimensional Digital Signal Processing*
HAMMING    *Digital Filters, 2e*
HAYKIN, ED.    *Array Signal Processing*
LEA, ED.    *Trends in Speech Recognition*
LIM, ED.    *Speech Enhancement*
MCCLELLAN and RADER    *Number Theory in Digital Signal Processing*
OPPENHEIM, ED.    *Applications of Digital Signal Processing*
OPPENHEIM, WILLSKY, with YOUNG    *Signals and Systems*
OPPENHEIM and SCHAFER    *Digital Signal Processing*
RABINER and GOLD    *Theory and Applications of Digital Signal Processing*
RABINER and SCHAFER    *Digital Processing of Speech Signals*
ROBINSON and TREITEL    *Geophysical Signal Analysis*
TRIBOLET    *Seismic Applications of Homomorphic Signal Processing*

# MULTIDIMENSIONAL
# DIGITAL SIGNAL PROCESSING

**DAN E. DUDGEON**
*Lincoln Laboratory, M.I.T.*

**RUSSELL M. MERSEREAU**
*School of Electrical Engineering*
*Georgia Institute of Technology*

*PRENTICE-HALL, INC.*, Englewood Cliffs, New Jersey   07632

*Library of Congress Cataloging in Publication Data*

Dudgeon, Dan E.
    Multidimensional digital signal processing.

    (Prentice-Hall signal processing series)
    Includes bibliographical references and index.
    1.  Signal processing—Digital techniques.
I. Mersereau, Russell M.   II. Title.   III. Series.
TK5102.5.D83 1984      621.38′042      83-3135
ISBN 0-13-604959-1

Editorial/production supervision and
    interior design: *Ellen Denning*
Manufacturing buyer: *Anthony Caruso*

Printed in the United States of America

10  9  8  7  6  5  4  3  2  1

ISBN 0-13-604959-1

PRENTICE-HALL INTERNATIONAL, INC., *London*
PRENTICE-HALL OF AUSTRALIA PTY. LIMITED, *Sydney*
EDITORA PRENTICE-HALL DO BRASIL, Ltda., *Rio de Janeiro*
PRENTICE-HALL CANADA INC., *Toronto*
PRENTICE-HALL OF INDIA PRIVATE LIMITED, *New Delhi*
PRENTICE-HALL OF JAPAN, INC., *Tokyo*
PRENTICE-HALL OF SOUTHEAST ASIA PTE. LTD., *Singapore*
WHITEHALL BOOKS LIMITED, *Wellington, New Zealand*

To Judy and Lindsay and to Martha, Adam, and David

# CONTENTS

# 7. INVERSE PROBLEMS                                                    348

# PREFACE

This book is the result of a suggestion by Professor Alan V. Oppenheim, who was our doctoral thesis supervisor at M.I.T. and who serves as the series editor for the Prentice-Hall series of books on signal processing, that we should write a senior- or first-year-graduate-level textbook on multidimensional digital signal processing. It is intended to be used in a one-semester course which would follow a basic course in digital signal processing using a text such as *Digital Signal Processing* by Oppenheim and Schafer (Prentice-Hall, 1975).

This text provides the student with a basic background of multidimensional signal processing theory with an emphasis on the differences and similarities between the one-dimensional and multidimensional cases. We have endeavored to write a text that will develop the student's intuition and motivation for this field without boring him or her with lengthy formal derivations, theorems, and proofs. Mathematical formality has its place, of course, but we feel that it should spring from an intuitive understanding of how things work, not the other way around. We hope the more mathematically inclined readers will understand, and benefit from, our informal approach.

There are several good books on the topic of digital image processing already available, so we have not attempted to duplicate the bulk of that subject matter in our book. Instead, we have tried to develop the theory of multidimensional signal processing, which not only serves as the foundation for image processing but also has applicability to other areas, such as array processing (e.g., radar, sonar, seismic signal processing, and radio astronomy).

In this book we assume that the reader has knowledge of one-dimensional digital signal processing theory, including linear shift-invariant systems, the discrete Fourier transform (DFT), the fast Fourier transform (FFT), linear filtering, the $z$-transform, stability, and power spectrum estimation. These concepts are not reviewed explicitly, but they are introduced in the two-dimensional context in a straightforward and rudimentary way.

Chapter 1 introduces the basic concepts of multidimensional signals and systems, focusing in particular on two-dimensional signals and linear shift-invariant (LSI) systems. The notion of the impulse response is introduced as one way to characterize LSI systems. The multidimensional Fourier transform is defined and used to compute the frequency response of two-dimensional LSI systems. Strategies for sampling two-dimensional continuous signals are also discussed.

In Chapter 2, the multidimensional discrete Fourier transform is introduced and algorithms for its efficient computation are presented in detail. The fast Fourier transform algorithm is shown to be applicable to signals sampled with arbitrary periodic sampling geometries. The close relationship between one-dimensional and multidimensional DFTs is also discussed.

Chapter 3 focuses on the design and implementation of two-dimensional finite-extent impulse response (FIR) filters. The direct, frequency-domain, and block convolution implementations of these filters are discussed, and design algorithms, including the window method, optimal methods, and the transformation method, are presented.

In Chapters 4 and 5, we examine infinite-extent impulse response (IIR) filters which can be represented by two-dimensional constant-coefficient difference equations. Chapter 4 lays the groundwork by introducing the concepts of the two-dimensional difference equation, the $z$-transform, stability, and the complex cepstrum. Chapter 5 follows with a discussion of implementation strategies and design techniques for two-dimensional IIR filters, including stabilization techniques.

Chapter 6 discusses the use of multidimensional signal processing in the context of processing signals received by an array of sensors. This broad application area is used as a vehicle for introducing the concepts of beamforming and power spectrum estimation. Beamforming represents a linear filtering approach to the problem of determining the strength and direction of propagating energy, while power spectrum estimation represents a modeling and parameter estimation approach to the same problem. Modern spectrum estimation techniques such as high-resolution, all-pole, and maximum entropy methods are discussed together with the more classical techniques. The significant theoretical differences between the one-dimensional and two-dimensional cases are brought out.

In Chapter 7, we discuss inverse problems in which one tries to deduce or reconstruct a signal from limited measurements and *a priori* information. Three examples are examined: deconvolution with constraints, seismic-wave migration, and signal reconstruction from projections.

In writing this text, we have tried to include topics and examples which illustrate fundamental principles of multidimensional signal processing. (Difficult or advanced material is presented in sections flagged by an asterisk.) Since we did not intend to compile an encyclopedic compendium of this still-evolving field, our book is necessarily incomplete. Some may view this as a deficiency, but our goal was to give a broad rather than deep view in the hopes of seducing students into pursuing research projects in this field. We have tried to compensate for any lack of depth by including important references to give readers an entrée into the technical literature.

We have enjoyed the encouragement and patience of many people during the five years it took us to conceive, outline, write, rewrite, and polish this manuscript. Our families were very supportive and patient; time working on the book was generally time spent away from them. Al Oppenheim has had a significant influence on this book in his role as series editor. More importantly, he and Ron Schafer have had a significant influence on the authors in their roles as teachers, supervisors, sounding boards, colleagues, models, and friends.

Over the course of our careers, we have enjoyed the opportunity of interacting with several exceptional colleagues who have made important technical contributions to the field of multidimensional digital signal processing and who have stimulated our own thinking and research in this field. In some cases, their technical contributions are presented explicitly in our book; in other cases, their influence has been more subtle. Among these colleagues, whom we also count as friends, are Professors Demetrius Paris and Monson Hayes, Dr. Mark Richards and Ms. Theresa Speake of Georgia Tech, Dr. Gary Shaw, Dr. Thomas Quatieri, and Dr. Stephen Pohlig of Lincoln Laboratory, Dr. James McClellan of Schlumberger, Professor Jae Lim of M.I.T., and Professor Don Johnson of Rice University. We also owe a measure of gratitude to our respective institutions, the Lincoln Laboratory, M.I.T., and the Georgia Institute of Technology, for providing the intellectual environment which encourages the pursuit of excellence in signal processing, as well as in other areas of engineering and science.

DAN E. DUDGEON
RUSSELL M. MERSEREAU

# INTRODUCTION

One of the by-products of the computer revolution has been the emergence of completely new fields of study. Each year, as integrated circuits have become faster, cheaper, and more compact, it has become possible to find feasible solutions to problems of ever-increasing complexity. Because it demands massive amounts of digital storage and comparable quantities of numerical computation, multidimensional digital signal processing is a problem area which has only recently begun to emerge. Despite this fact, it has already provided the solutions to important problems ranging from computer-aided tomography (CAT), a technique for combining x-ray projections from different orientations to create a three-dimensional reconstruction of a portion of the human body, to the design of passive sonar arrays and the monitoring of the earth's resources by satellite. In addition to its many glamorous and humble applications, however, multidimensional digital signal processing also possesses a firm mathematical foundation, which allows us not only to understand what has already been accomplished, but also to explore rationally new problem areas and solution methods as they arise.

Simply stated, a signal is any medium for conveying information, and signal processing is concerned with the extraction of that information. Thus ensembles of time-varying voltages, the density of silver grains on a photographic emulsion, or lists of numbers in the memory of a computer all represent examples of signals. A typical signal processing task involves the transfer of information from one signal to another. A photograph, for example, might be scanned, sampled, and stored in the memory of a computer. In this case, the information is transferred from a variable silver density, to a beam of visible light, to an electrical waveform, and finally to a sequence of numbers, which, in turn, are represented by an arrangement of magnetic domains on a computer disk. The CAT scanner is a more complex example; information about the structure of an unknown object is first transferred to a series of electromagnetic waves, which are then sampled to produce an array of numbers, which, in turn, are processed by a computational algorithm and finally displayed on the phosphor of a cathode ray tube (CRT) screen or on photographic film. The digital processing which is done cannot add to the information, but it can rearrange it so that a human observer can more readily interpret it; instead of looking at multiple shadows the observer is able to look at a cross-sectional view.

Whatever their form, signals are of interest only because of the information they contain. At the risk of overgeneralizing we might say that signal processing is concerned with two basic tasks—information rearrangement and information reduction. We have already seen two examples of information rearrangement—computer-aided tomography and image scanning. To those we could easily add other examples: image enhancement, image deblurring, spectral analysis, and so on. Information reduction is concerned with the removal of extraneous information. Someone observing radar returns is generally interested in only a few bits of information, specifically, the answer to such questions as: Is anything there? If so, what? Friend or foe? How fast is it going, and where is it headed? However, the receiver is also giving the observer information about the weather, chaff, birds, nearby buildings, noise in the receiver, and so on. The observer must separate the relevant from the irrelevant, and signal processing can help. Other examples of information-lossy signal processing operations include noise removal, parameter estimation, and feature extraction.

*Digital signal processing* is concerned with the processing of signals which can be represented as sequences of numbers and *multidimensional digital signal processing* is, more specifically, concerned with the processing of signals which can be represented as multidimensional arrays, such as sampled images or sampled time waveforms which are received simultaneously from several sensors. The restriction to digital signals permits processing with digital hardware, and it permits signal processing operators to be specified as algorithms or procedures.

The motivations for looking at digital methods hardly need to be enumerated. Digital methods are simultaneously powerful and flexible. Digital systems can be designed to be adaptive and they can be made to be easily reconfigured. Digital algorithms can be readily transported from the equipment of one manufacturer to another or they can be implemented with special-purpose digital hardware. They can be used equally well to process signals that originated as time functions or as spatial functions and they interface naturally with logical operators such as pattern classifiers. Digital signals can be stored indefinitely without error. For many applications, digital methods may be cheaper than the alternatives, and for others there may simply be no alternatives.

Is the processing of multidimensional signals that different from the processing of one-dimensional ones? At an abstract level, the answer is no. Many operations that we might want to perform on multidimensional sequences are also performed on one-dimensional ones—sampling, filtering, and transform computation, for example. At a closer level, however, we would be forced to say that multidimensional signal processing can be quite different. This is due to three factors: (1) two-dimensional problems generally involve considerably more data than one-dimensional ones; (2) the mathematics for handling multidimensional systems is less complete than the mathematics for handling one-dimensional systems; and (3) multidimensional systems have many more degrees of freedom, which give a system designer a flexibility not encountered in the one-dimensional case. Thus, while all recursive digital filters are implemented using difference equations, in the one-dimensional case these difference equations are totally ordered, whereas in the multidimensional case they are only partially ordered. Flexibility can be exploited. In the one-dimensional case, the discrete Fourier transform (DFT) can be evaluated using the fast Fourier transform (FFT) algorithm, whereas in the multidimensional case, there are a host of DFTs and each can be evaluated using a host of FFT algorithms. In the one-dimensional case, we can adjust the rate at which a bandlimited signal is sampled; in the multidimensional case, we can adjust not only the rate, but also the geometric arrangement of the samples. On the other hand, multidimensional polynomials cannot be factored, whereas one-dimensional ones can. Thus, in the multidimensional case, we cannot talk about isolated poles, zeros, and roots. Multidimensional digital signal processing can be quite different from one-dimensional digital signal processing.

In the early 1960s, many of the methods of one-dimensional digital signal processing were developed with the intention of using the digital systems to simulate analog ones. As a result, much of discrete systems theory was modeled after analog systems theory. In time, it became recognized that, while digital systems could simulate analog systems very well, they could also do much more. With this awareness and a strong push from the technology of digital hardware, the field has blossomed and many of the methods in common use today have no analog equivalents. The same trend can be observed in the development of multidimensional digital signal processing. Since there is no continuous-time or analog two-dimensional systems theory to imitate, early multidimensional systems were based on one-dimensional systems. In the late 1960s, most two-dimensional signal processing was performed using separable two-dimensional systems, which are little more than one-dimensional systems applied to two-dimensional data. In time, uniquely multidimensional algorithms were developed which correspond to logical extrapolations of one-dimensional algorithms. This period was one of frustration. The volume of data demanded by many two-dimensional applications and the absence of a factorization theorem for two-dimensional polynomials meant that many one-dimensional methods did not generalize well. Chronologically, we are now at the dawn of the age of awareness. The computer industry, by making components smaller and cheaper, has helped to solve the data volume problem and we are recognizing that, although we will always have the problem of limited mathematics, multidimensional systems also give us new freedoms. These combine to make the field both challenging and fun.

In this book we summarize many of the advances that have taken place in this exciting and rapidly growing field. The area is one that has evolved with technology. Although we do describe many applications of our material, we have tried not to make it too technology dependent, lest it become technologically obsolete. Rather, we emphasize fundamental concepts so that the reader will not only understand what has been done but will also be able to extend those methods to new applications.

To accomplish all of this, it is necessary to assume some background on the part of the reader. Specifically, we assume that the reader is familiar with one-dimensional linear systems theory and has a basic understanding of one-dimensional digital signal processing (at the level of Oppenheim and Schafer [1], Chaps. 1–6).

In this book our interest is in the processing of all signals of dimensionality greater than or equal to 2. Whereas there is a substantial difference between the theories for the processing of one- and two-dimensional signals, there seems to be little difference between the two-dimensional and higher-dimensional cases, except for the issue of computational complexity. To avoid cluttering up the discussions, equations, and figures of the book, we therefore state the majority of our results only for the two-dimensional case, which is the most prevalent one in applications. In most cases, the generalizations are straightforward, and when they are not, they will be explicitly given. In a similar spirit, we do not belabor results that are obvious generalizations of the one-dimensional case.

We hope the reader will find what we found when we first became involved in the area of multidimensional digital signal processing. It is an area to which a great deal of intuitition may be carried over from the one-dimensional causal world, and yet there are many places where the final form of a result is unexpected and its implications are surprising and counterintuitive. The way in which some one-dimensional results generalize to several dimensions can give the reader new insights into the structure of multidimensional as well as one-dimensional signal processing operations.

# REFERENCE

1. Alan V. Oppenheim and Ronald Schafer, *Digital Signal Processing* (Englewood Cliffs, N.J.: Prentice-Hall, Inc., 1975).

# 1

# MULTIDIMENSIONAL SIGNALS AND SYSTEMS

A multidimensional signal can be modeled as a function of $M$ independent variables, where $M \geq 2$. These signals may be classified as continuous, discrete, or mixed. A *continuous* signal can be modeled as a function of independent variables which range over a continuum of values. For example, the intensity $I(x, y)$ of a photographic image is a two-dimensional continuous signal. A *discrete* signal, on the other hand, can be modeled as a function defined only on a set of points, such as the set of integers. A *mixed* signal is a multidimensional signal that is modeled as a function of some continuous variables and some discrete ones. For example, an ensemble of time waveforms recorded from an array of electrical transducers is a mixed signal. The ensemble can be modeled with one continuous variable, time, and one or more discrete variables to index the transducers.

In this chapter we are concerned primarily with multidimensional discrete signals and the systems that can operate on them. Most of the properties of signals and systems that we will discuss are simple extensions of the properties of one-dimensional discrete signals and systems and therefore, most of our discussions will be brief. The reader who desires further details is referred to one of several excellent textbooks that cover the one-dimensional case [1–3]. It will become apparent, however, that many familiar one-dimensional procedures do not readily generalize to the multidimensional case and that many important issues associated with multidimensional signals and systems do not appear in the one-dimensional special case. In these cases, our treatments will necessarily be more complete.

## 1.1 TWO-DIMENSIONAL DISCRETE SIGNALS

A two-dimensional (2-D) *discrete signal* (also referred to as a *sequence* or *array*) is a function defined over the set of ordered pairs of integers. Thus

$$x = \{x(n_1, n_2), -\infty < n_1, n_2 < \infty\} \tag{1.1}$$

A single element from the sequence will be referred to as a *sample*. Thus $x(n_1, n_2)$ represents the sample of the sequence $x$ at the point $(n_1, n_2)$. Sample values can be real or complex. On occasion, if $n_1$ and $n_2$ are interpreted as variables, a reference to $x(n_1, n_2)$ will be interpreted as a reference to the entire sequence. Although this convention is abusive, it has become commonplace in the engineering literature and it should cause no confusion.

It may be helpful, on occasion, to regard the signal $x$ as the collection of its samples rather than simply as a function that is evaluated at integer values of its arguments. With this interpretation, there is no temptation to define $x$ for values of $n_1$ and $n_2$ other than integers. A 2-D sequence is depicted graphically in Figure 1.1.

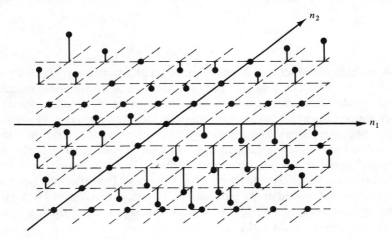

**Figure 1.1** Graphical representation of a two-dimensional sequence.

Two-dimensional sequences, as we have defined them, extend infinitely far since $n_1$ and $n_2$ may take on any integer values. In practice, however, most 2-D sequences have sample values which are known only over a finite region in the $(n_1, n_2)$-plane. For example, when a black-and-white photograph is scanned, samples are not taken beyond the edges of the photograph. Rather than restrict the domain of definition of the resulting 2-D sequence, we simply assume that the values of the samples outside the finite region are all equal to zero.

### 1.1.1 Some Special Sequences

Some sequences are sufficiently important to warrant special names and symbols. One of these is the 2-D *unit impulse*, $\delta(n_1, n_2)$, also called the *unit sample*. The unit impulse is defined by

$$\delta(n_1, n_2) = \begin{cases} 1, & n_1 = n_2 = 0 \\ 0, & \text{otherwise} \end{cases} \tag{1.2}$$

If the one-dimensional (1-D) unit impulse is defined as

$$\delta(n) = \begin{cases} 1, & n = 0 \\ 0, & n \neq 0 \end{cases} \tag{1.3}$$

then the 2-D unit impulse can be written as the product of two 1-D unit impulses.

$$\delta(n_1, n_2) = \delta(n_1)\delta(n_2) \tag{1.4}$$

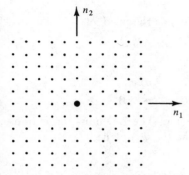

**Figure 1.2**  Two-dimensional unit sample sequence $\delta(n_1, n_2)$. The circle represents a sample of value 1. The dots represent samples of value 0.

In Figure 1.2 we show a stylized graphical representation of the 2-D unit impulse.

A 2-D *line impulse* is a sequence that is uniform in one variable and impulsive in the other. The sequences

$$x(n_1, n_2) = \delta(n_1) \tag{1.5a}$$

and

$$y(n_1, n_2) = \delta(n_2) \tag{1.5b}$$

which are represented in Figure 1.3, are examples of line impulses. In the *M*-dimensional case, we can define not only *M*-dimensional unit impulses, but also *M*-dimensional line impulses, *M*-dimensional sheet impulses, and so on, in the obvious fashion.

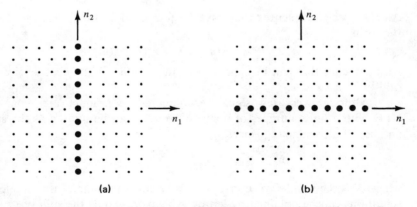

(a)    (b)

**Figure 1.3**  Two examples of line impulses. (a) $x(n_1, n_2) = \delta(n_1)$. (b) $x(n_1, n_2) = \delta(n_2)$.

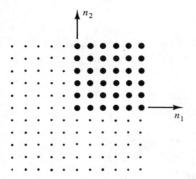

**Figure 1.4** Two-dimensional unit step sequence $u(n_1, n_2)$.

Another special sequence is the *2-D unit step* sequence $u(n_1, n_2)$, which is shown in Figure 1.4. The step is defined by

$$u(n_1, n_2) = \begin{cases} 1, & n_1 \geq 0 \quad \text{and} \quad n_2 \geq 0 \\ 0, & \text{otherwise} \end{cases} \qquad (1.6)$$

We can also interpret $u(n_1, n_2)$ as the product

$$u(n_1, n_2) = u(n_1)u(n_2) \qquad (1.7)$$

where

$$u(n) = \begin{cases} 1, & n \geq 0 \\ 0, & n < 0 \end{cases} \qquad (1.8)$$

is the one-dimensional unit step. The 2-D unit step is non-zero over one quadrant of the $(n_1, n_2)$-plane.

*Exponential sequences* are defined by

$$x(n_1, n_2) = a^{n_1}b^{n_2}, \qquad -\infty < n_1, n_2 < \infty \qquad (1.9)$$

where $a$ and $b$ are complex numbers. When $a$ and $b$ have unity magnitude, they may be written as

$$a = \exp(j\omega_1); \qquad b = \exp(j\omega_2)$$

In this case, the exponential sequence becomes the complex sinusoidal sequence

$$x(n_1, n_2) = \exp(j\omega_1 n_1 + j\omega_2 n_2)$$
$$= \cos(\omega_1 n_1 + \omega_2 n_2) + j\sin(\omega_1 n_1 + \omega_2 n_2) \qquad (1.10)$$

Exponential sequences are particularly important because, as we shall see later, they are eigenfunctions of 2-D linear shift-invariant systems.

### 1.1.2 Separable Sequences

All the special sequences that have been defined to this point can be expressed in the form

$$x(n_1, n_2) = x_1(n_1)x_2(n_2) \qquad (1.11)$$

Any sequence that can be expressed as the product of 1-D sequences in this form is said to be *separable*.

Although very few signals encountered in practice are separable, any 2-D array with a finite number of nonzero samples can be written as the sum of a finite number of separable sequences:

$$x(n_1, n_2) = \sum_{i=1}^{N} x_{i1}(n_1)x_{i2}(n_2) \qquad (1.12)$$

where $N$ is the number of nonzero rows or columns. One of the simplest such representations can be obtained by letting $x$ be expressed as the sum of its isolated rows.

This is done by choosing

$$x_{t1}(n_1) = x(n_1, i) \tag{1.13a}$$

$$x_{t2}(n_2) = \delta(n_2 - i) \tag{1.13b}$$

Other sum-of-separable decompositions are possible and, on occasion, they can be quite useful.

Separable sequences can be quite valuable when used as test inputs for evaluating and debugging experimental systems.

### 1.1.3 Finite-Extent Sequences

Finite-extent 2-D sequences are another important class of discrete signals. The modifier "finite-extent" implies that these signals are zero outside a region of finite extent (or area) in the $(n_1, n_2)$-plane. This region is called the signal's *region of support*. One typical finite-extent sequence, shown in Figure 1.5, is nonzero only inside the rectangle

$$0 \leq n_1 < N_1; \qquad 0 \leq n_2 < N_2 \tag{1.14}$$

Although rectangular and square shapes are used most often for the region of support of finite-extent sequences, it is also possible to consider regions with other shapes as well.

The alert reader will recognize that there is an ambiguity in the definition of the region of support for a 2-D finite-extent sequence. Clearly, if a sequence is zero outside a region $R$, it is also zero outside any larger region that contains $R$. By embedding a sequence with an irregularly shaped region of support into a larger rectangular region, the representation of that sequence and operations performed on it can often be simplified.

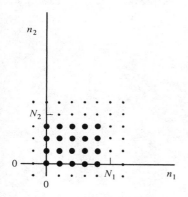

**Figure 1.5**   Finite-extent sequence with a rectangular region of support.

### 1.1.4 Periodic Sequences

Periodic discrete signals form another important class of 2-D sequences. Like its 1-D counterpart, a periodic 2-D sequence can be thought of as a waveform that repeats itself at regularly spaced intervals. Because a 2-D signal must repeat in two different directions at once, however, the formal definition of a periodic 2-D sequence is more complex than that of a periodic 1-D sequence. We shall build to the general definition with a special case.

Consider a 2-D sequence $\tilde{x}(n_1, n_2)$ which satisfies the following constraints:

$$\tilde{x}(n_1, n_2 + N_2) = \tilde{x}(n_1, n_2) \tag{1.15a}$$

$$\tilde{x}(n_1 + N_1, n_2) = \tilde{x}(n_1, n_2) \tag{1.15b}$$

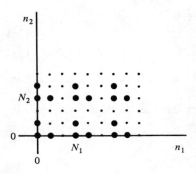

**Figure 1.6** 2-D periodic sequence with $N_1 = N_2 = 3$.

This sequence is doubly periodic; its values are repeated if the variable $n_1$ is incremented by $N_1$ or if the variable $n_2$ is incremented by $N_2$. Figure 1.6 shows a sketch of such a sequence. We shall call $N_1$ and $N_2$ the *horizontal* and *vertical periods* of $\tilde{x}$ if they are the smallest positive integer values for which equations (1.15) are true.

Only $N_1N_2$ samples of $\tilde{x}$ are independent; the remaining samples are determined by the periodicity condition. Any connected region of the $(n_1, n_2)$-plane containing exactly $N_1N_2$ samples will be called a *period* of $\tilde{x}$ if those sample values are independent. Often the most convenient shape for the period is the rectangle $\{(n_1, n_2), 0 \leq n_1 \leq N_1 - 1, 0 \leq n_2 \leq N_2 - 1\}$ but this is not the only possibility. The shaded region shown in Figure 1.7, for example, can also be used to represent one period of a periodic sequence.

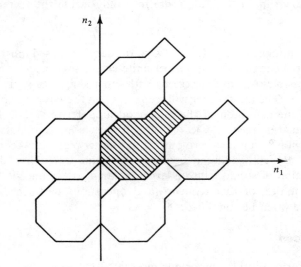

**Figure 1.7** 2-D periodic sequence with an irregularly shaped period.

Now consider a 2-D sequence $\tilde{x}(n_1, n_2)$ that satisfies the more general periodicity constraints

$$\tilde{x}(n_1 + N_{11}, n_2 + N_{21}) = \tilde{x}(n_1, n_2) \tag{1.16a}$$

$$\tilde{x}(n_1 + N_{12}, n_2 + N_{22}) = \tilde{x}(n_1, n_2) \tag{1.16b}$$

where

$$D \triangleq N_{11}N_{22} - N_{12}N_{21} \neq 0 \tag{1.17}$$

The ordered pairs $(N_{11}, N_{21})'$ and $(N_{12}, N_{22})'$ can be interpreted as vectors $\mathbf{N}_1$ and $\mathbf{N}_2$ which represent the displacements from any sample to the corresponding samples of two other periods. (The prime denotes the transposition operation, which converts the ordered pair into a column vector.) One period of such a sequence is contained in the parallelogram-shaped region whose two adjacent sides are formed by $\mathbf{N}_1$ and $\mathbf{N}_2$. We leave it to the reader to show that the number of samples in this region is $|D|$. Figure 1.8 depicts a general 2-D periodic sequence with $\mathbf{N}_1 = (7, 2)'$ and $\mathbf{N}_2 = (-2, 4)'$.

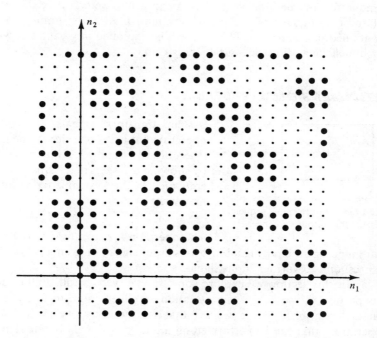

**Figure 1.8**    Periodic sequence with periodicity vectors $(7, 2)'$ and $(-2, 4)'$.

This idea of periodicity is readily generalized to $M$-dimensional signals. For notational convenience, we shall let $\mathbf{n}$ denote the ordered $M$-tuple of integer variables $(n_1, n_2, \ldots, n_M)'$. Then $\tilde{x}(\mathbf{n})$ is an *$M$-dimensional periodic sequence* if there exist $M$ linearly independent $M$-dimensional integer vectors, $\mathbf{N}_1, \ldots, \mathbf{N}_M$ such that

$$\tilde{x}(\mathbf{n} + \mathbf{N}_i) = \tilde{x}(\mathbf{n}), \qquad i = 1, \ldots, M \qquad (1.18)$$

The vectors $\mathbf{N}_i$ are called *periodicity vectors* and they can be arranged to form the columns of an $M \times M$ matrix $\mathbf{N}$ called the *periodicity matrix*.

$$\mathbf{N} = [\mathbf{N}_1 \,|\, \mathbf{N}_2 \,|\, \cdots \,|\, \mathbf{N}_M] \qquad (1.19)$$

The requirement that the periodicity vectors be linearly independent is equivalent to requiring that $\mathbf{N}$ have a nonzero determinant. In the special case that $\mathbf{N}$ is a diagonal matrix, we will say that $\tilde{x}(\mathbf{n})$ is *rectangularly periodic*. This is the special case we considered earlier.

If $\tilde{x}(\mathbf{n})$ is periodic with periodicity matrix $\mathbf{N}$, then

$$\tilde{x}(\mathbf{n} + \mathbf{Nr}) = \tilde{x}(\mathbf{n}) \tag{1.20}$$

for any integer vector $\mathbf{r}$. Using this fact, we see that if $\mathbf{P}$ is any integer matrix, then $\mathbf{NP}$ will also be a periodicity matrix for $\tilde{x}(\mathbf{n})$. Thus the periodicity matrix is not unique for any periodic sequence. As an aside, we can note that the absolute value of the determinant of the periodicity matrix gives the number of samples of $\tilde{x}(\mathbf{n})$ contained in one period. This fact will be exploited in Chapter 2, where we define an $M$-dimensional discrete Fourier transform.

## 1.2 MULTIDIMENSIONAL SYSTEMS

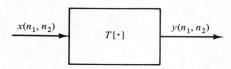

**Figure 1.9** Pictorial representation of a system. Here $x$ represents the input sequence and $y$ represents the output sequence.

Systems transform signals. Formally, a system is an operator that maps one signal (the input) into another (the output). Figure 1.9 illustrates this simple point by showing a system that maps $x$ into $y$. The operator embodied in this system is represented by $T[\cdot]$, so we may write

$$y = T[x] \tag{1.21}$$

The operator $T[\cdot]$ can represent a rule or set of rules for mapping an input signal into an output signal, or even a list of output signals that correspond to various input signals.

In this section we explore some simple, yet useful, multidimensional systems. In particular, we focus our attention on linear shift-invariant systems and their characterizations. Before we get that far, however, we shall discuss some simple operations that can be performed on multidimensional discrete signals.

### 1.2.1 Fundamental Operations on Multidimensional Signals

Signals may be combined or altered by a variety of operations. Here we describe some of the basic operations on signals which will serve as building blocks for the development of more complicated systems.

Let $w$ and $x$ represent 2-D discrete signals. These signals can be added to yield a third signal, $y$. The addition is performed sample by sample so that a particular sample value $y(n_1, n_2)$ is obtained by adding the two corresponding sample values $w(n_1, n_2)$ and $x(n_1, n_2)$.

$$y(n_1, n_2) = x(n_1, n_2) + w(n_1, n_2) \tag{1.22}$$

Two-dimensional sequences may also be multiplied by a constant to form a new sequence. If we let $c$ represent a constant, we can form the 2-D sequence $y$ from the scalar $c$ and the 2-D sequence $x$ by multiplying each sample value of $x$ by $c$.

$$y(n_1, n_2) = cx(n_1, n_2) \tag{1.23}$$

A 2-D sequence $x$ may also be linearly shifted to form a new sequence $y$. The operation of shifting simply slides the entire sequence $x$ to a new position in the $(n_1, n_2)$-plane. The sample values of $y$ are related to the sample values of $x$ by

$$y(n_1, n_2) = x(n_1 - m_1, n_2 - m_2) \qquad (1.24)$$

where $(m_1, m_2)$ is the amount of the shift. An example of shifting a 2-D sequence appears in Figure 1.10.

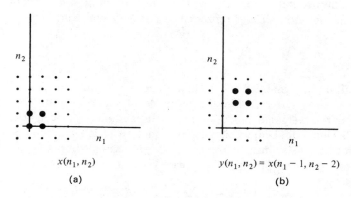

$$x(n_1, n_2)$$

(a)

$$y(n_1, n_2) = x(n_1 - 1, n_2 - 2)$$

(b)

**Figure 1.10**   Operation of shifting the 2-D sequence $x(n_1, n_2)$.

Using the fundamental operations of addition, scalar multiplication, and shifting, it is possible to decompose any 2-D sequence into a sum of weighted and shifted 2-D unit impulses.

$$x(n_1, n_2) = \sum_{k_1=-\infty}^{\infty} \sum_{k_2=-\infty}^{\infty} x(k_1, k_2)\delta(n_1 - k_1, n_2 - k_2) \qquad (1.25)$$

Here $\delta(n_1 - k_1, n_2 - k_2)$ represents a unit impulse that has been shifted so that its nonzero sample is at $(k_1, k_2)$; the values $x(k_1, k_2)$ can be interpreted as scalar multipliers for the corresponding unit impulses.

Two other fundamental operations on 2-D sequences are worth mentioning. The first, which we call a *spatially varying gain*, can be viewed as a generalization of scalar multiplication. Each sample value of a 2-D sequence $x$ is multiplied by a number $c(n_1, n_2)$ whose value depends on the position of the corresponding sample.

$$y(n_1, n_2) = c(n_1, n_2)x(n_1, n_2) \qquad (1.26)$$

The collection of numbers $c(n_1, n_2)$ may also be regarded as a 2-D sequence. Thus equation (1.26) can also be interpreted as the sample-by-sample multiplication of two sequences.

Two-dimensional sequences may also be subjected to nonlinear operators. One important type of nonlinear operator, called a *memoryless nonlinearity*, operates on each sample value of a 2-D sequence independently. For example, consider the sequence formed by squaring each sample value in a 2-D sequence $x$.

$$y(n_1, n_2) = [x(n_1, n_2)]^2 \qquad (1.27)$$

The squaring operation is a memoryless nonlinearity, since the computation of the output value at $(n_1, n_2)$ depends only on the single input value at $(n_1, n_2)$.

### 1.2.2 Linear Systems

A system is said to be *linear* if and only if it satisfies two conditions: if the input signal is the sum of two sequences, the output signal is the sum of the two corresponding output sequences, and scaling the input signal produces a scaled output signal. Therefore, if $L[\cdot]$ represents a linear system, and

$$y_1 = L[x_1]; \qquad y_2 = L[x_2]$$

then

$$ay_1 + by_2 = L[ax_1 + bx_2] \tag{1.28}$$

for all input signals $x_1$ and $x_2$ and all complex constants $a$ and $b$.

Linear systems obey the principle of superposition. The response of a linear system to a weighted sum of input signals is equal to the weighted sum of the responses to the individual input signals. In equation (1.25), an arbitrary 2-D sequence was represented as a linear combination of shifted unit impulses. If we use this sequence as the input to a 2-D discrete linear system $L[\cdot]$, we will get the output sequence

$$y(n_1, n_2) = L\left[ \sum_{k_1=-\infty}^{\infty} \sum_{k_2=-\infty}^{\infty} x(k_1, k_2)\delta(n_1 - k_1, n_2 - k_2) \right]$$

By exploiting the fact that the system is linear, this can be rewritten as

$$
\begin{aligned}
y(n_1, n_2) &= \sum_{k_1=-\infty}^{\infty} \sum_{k_2=-\infty}^{\infty} x(k_1, k_2)L[\delta(n_1 - k_1, n_2 - k_2)] \\
&= \sum_{k_1=-\infty}^{\infty} \sum_{k_2=-\infty}^{\infty} x(k_1, k_2)h_{k_1 k_2}(n_1, n_2)
\end{aligned}
\tag{1.29}
$$

where $h_{k_1 k_2}$ is the response of the system to a unit impulse located at $(k_1, k_2)$. If the spatially varying impulse response $h_{k_1 k_2}(n_1, n_2)$ is known for each $(k_1, k_2)$, the response of the linear system to *any* input can be found by superposition.

### 1.2.3 Shift-Invariant Systems

A *shift-invariant* system is one for which a shift in the input sequence implies a corresponding shift in the output sequence. If

$$y(n_1, n_2) = T[x(n_1, n_2)]$$

the system $T[\cdot]$ is shift invariant if and only if

$$T[x(n_1 - m_1, n_2 - m_2)] = y(n_1 - m_1, n_2 - m_2) \tag{1.30}$$

for all sequences $x$ and all integer shifts $(m_1, m_2)$.

Linearity and shift invariance are independent properties of a system; neither property implies the presence of the other. For example, the spatially varying gain,

$$L[x(n_1, n_2)] = c(n_1, n_2)x(n_1, n_2) \tag{1.31}$$

which multiplies the input sequence by $c(n_1, n_2)$, is linear but it is not shift invariant. On the other hand, the system

$$T[x(n_1, n_2)] = [x(n_1, n_2)]^2 \qquad (1.32)$$

is shift invariant, but it is not linear.

### 1.2.4 Linear Shift-Invariant Systems

To study multidimensional systems productively, it is necessary to restrict our investigations to certain classes of operators which have properties in common. Linear shift-invariant (LSI) discrete systems are the most frequently studied class of systems for processing discrete signals of any dimensionality. These systems are both easy to design and analyze, yet they are sufficiently powerful to solve many practical problems. The behavior of these systems can also, in many cases, be studied without regard to the specific input to the system. The class of linear shift-invariant systems is certainly not the most general class of systems which can be studied, but it does represent a good starting point.

In (1.29) we derived an expression for the output sequence of a linear system to the input $x$. If this system is also shift invariant, further simplifications can be made. The spatially varying impulse response is defined by

$$h_{k_1 k_2}(n_1, n_2) \triangleq L[\delta(n_1 - k_1, n_2 - k_2)] \qquad (1.33)$$

For the special case where $k_1 = k_2 = 0$, we have

$$h_{00}(n_1, n_2) = L[\delta(n_1, n_2)] \qquad (1.34)$$

Applying the principle of shift invariance embodied by equation (1.30), we get

$$h_{k_1 k_2}(n_1, n_2) = h_{00}(n_1 - k_1, n_2 - k_2) \qquad (1.35)$$

The spatially varying impulse response becomes a shifted replica of a spatially invariant impulse response. Defining $h(n_1, n_2) \triangleq h_{00}(n_1, n_2)$, we can then write the output sequence as

$$y(n_1, n_2) = \sum_{k_1=-\infty}^{\infty} \sum_{k_2=-\infty}^{\infty} x(k_1, k_2) h(n_1 - k_1, n_2 - k_2) \qquad (1.36)$$

This relation is known as the 2-D *convolution sum*. Conceptually, the input sequence $x(n_1, n_2)$ is decomposed into a weighted sum of shifted impulses according to equation (1.25). Each impulse is transformed by the LSI system into a shifted copy of the impulse response $h(n_1, n_2)$. Superposition of these weighted, shifted impulse responses forms the output sequence, with the weighting coefficients given by the sample values of the input sequence $x(n_1, n_2)$. Equation (1.36) implies that an LSI system is completely characterized by its spatially invariant impulse response $h(n_1, n_2)$.

If we make the substitution of variables $n_1 - k_1 = l_1$ and $n_2 - k_2 = l_2$, equation (1.36) can be written in the alternative form

$$y(n_1, n_2) = \sum_{l_1=-\infty}^{\infty} \sum_{l_2=-\infty}^{\infty} h(l_1, l_2) x(n_1 - l_1, n_2 - l_2) \qquad (1.37)$$

Thus we see that convolution is a commutative operation. As a notational device, we shall use the double asterisk (**) to denote 2-D convolution. [A single asterisk (*) will denote 1-D convolution.] Equations (1.36) and (1.37) can be written using this shorthand notation as

$$y = x ** h = h ** x \tag{1.38}$$

By using vector notation, the output sequence of an $M$-dimensional LSI system can be expressed as the $M$-dimensional convolution of the output sequence and the impulse response

$$y(\mathbf{n}) = \sum_{\mathbf{k}} x(\mathbf{k})h(\mathbf{n} - \mathbf{k}) \tag{1.39}$$

Two-dimensional convolution is not substantially different from its 1-D counterpart. As in the 1-D case, there is a computational interpretation for the convolution sum. Consider $x(k_1, k_2)$ and $h(n_1 - k_1, n_2 - k_2)$ as functions of $k_1$ and $k_2$. To generate the sequence $h(n_1 - k_1, n_2 - k_2)$ from $h(k_1, k_2)$, $h$ is first reflected about both the $k_1$ and $k_2$ axes and then translated so that the sample $h(0, 0)$ lies at the point $(n_1, n_2)$ as illustrated in Figure 1.11. The product sequence $x(k_1, k_2)h(n_1 - k_1, n_2 - k_2)$ is

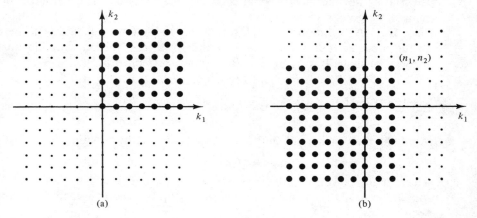

**Figure 1.11**   (a) The sequence $h(k_1, k_2)$. (b) The sequence $h(n_1 - k_1, n_2 - k_2)$ for $n_1 = 2, n_2 = 3$.

formed, and the output sample value $y(n_1, n_2)$ is computed by summing the nonzero sample values in the product sequence. As $n_1$ and $n_2$ are varied, the sequence $h(n_1 - k_1, n_2 - k_2)$ is shifted to other positions in the $(k_1, k_2)$-plane, leading to other product sequences and, consequently, other output sample values. If the alternative form of the convolution sum, equation (1.37), is used, the roles of $x(n_1, n_2)$ and $h(n_1, n_2)$ are interchanged in describing the computation.

**Example 1**

Let us consider a 2-D discrete LSI system whose output at the sample $(n_1, n_2)$ represents the accumulation of the input sample values over a region below and to the left of the point $(n_1, n_2)$. Roughly speaking, this system is one type of 2-D digital integrator; its impulse response is the 2-D unit step sequence $u(n_1, n_2)$ described in Section 1.1.1.

For the input sequence $x(n_1, n_2)$, we shall use a 2-D finite-extent sequence whose sample values are equal to 1 inside the rectangular region $0 \le n_1 < N_1; 0 \le n_2 < N_2$ and equal to 0 outside it.

To compute the output sample value $y(n_1, n_2)$ using equation (1.36), we form the product sequence $x(k_1, k_2)h(n_1 - k_1, n_2 - k_2)$. Depending on the particular value of $(n_1, n_2)$, the nonzero regions of $x(k_1, k_2)$ and $h(n_1 - k_1, n_2 - k_2)$ overlap by different amounts. We can distinguish five cases which are illustrated in Figure 1.12.

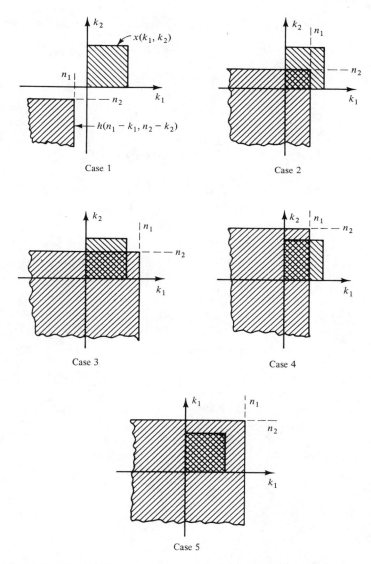

**Figure 1.12**   Convolution of a square pulse with a two-dimensional step sequence. The nonzero regions of each sequence are crosshatched. The product sequence $x(k_1, k_2)h(n_1 - k_1, n_2 - k_2)$ is nonzero only in the doubly crosshatched areas.

In this figure, the nonzero portions of each sequence are crosshatched, and the zero samples are not shown.

*Case 1.*   $n_1 < 0$ or $n_2 < 0$. Looking at Figure 1.12, we see that for these values of $(n_1, n_2)$, $h(n_1 - k_1, n_2 - k_2)$ and $x(k_1, k_2)$ do not overlap. Hence their product and the value of these samples of the convolution are zero.

*Case 2.*   $0 \leq n_1 < N_1$, $0 \leq n_2 < N_2$. Here there is partial overlap. The accumulation of the nonzero sample values in the product sequence yields

$$y(n_1, n_2) = \sum_{k_1=0}^{n_1} \sum_{k_2=0}^{n_2} 1 = (n_1 + 1)(n_2 + 1) \tag{1.40}$$

*Case 3.*   $n_1 \geq N_1, 0 \leq n_2 < N_2$. Here we can write

$$y(n_1, n_2) = \sum_{k_1=0}^{N_1-1} \sum_{k_2=0}^{n_2} 1$$
$$= N_1(n_2 + 1) \tag{1.41}$$

*Case 4.*   $0 \leq n_1 < N_1, n_2 \geq N_2$. By analogy with case 3, we have

$$y(n_1, n_2) = N_2(n_1 + 1) \tag{1.42}$$

*Case 5.*   $n_1 \geq N_1, n_2 \geq N_2$. In this final case the reflected shifted step sequence $h(n_1 - k_1, n_2 - k_2)$ completely overlaps the pulse $x(k_1, k)_2$. Thus

$$y(n_1, n_2) = N_1 N_2 \tag{1.43}$$

Therefore, the overall convolution is

$$y(n_1, n_2) = \begin{cases} 0, & n_1 < 0 \text{ or } n_2 < 0 \\ (n_1 + 1)(n_2 + 1), & 0 \leq n_1 < N_1, \quad 0 \leq n_2 < N_2 \\ N_1(n_2 + 1), & n_1 \geq N_1, \quad 0 \leq n_2 < N_2 \\ (n_1 + 1)N_2, & 0 \leq n_1 < N_1, \quad n_2 \geq N_2 \\ N_1 N_2, & n_1 \geq N_1, \quad n_2 \geq N_2 \end{cases} \tag{1.44}$$

which is shown in Figure 1.13.

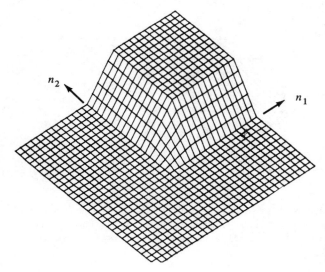

**Figure 1.13**   Convolution of the two sequences represented in Example 1.

For this particular example, it should be noted that $x$ and $h$ are both separable sequences and that their convolution is also separable, since we can write

$$y(n_1, n_2) = y_1(n_1)y_2(n_2) \qquad (1.45)$$

where

$$y_1(n_1) = \begin{cases} 0, & n_1 < 0 \\ n_1 + 1, & 0 \le n_1 < N_1 \\ N_1, & n_1 \ge N_1 \end{cases}$$

$$y_2(n_2) = \begin{cases} 0, & n_2 < 0 \\ n_2 + 1, & 0 \le n_2 < N_2 \\ N_2, & n_2 \ge N_2 \end{cases}$$

This property is true in general; the convolution of two separable sequences is always separable (see Problem 1.9).

**Example 2**

In some cases, one may only be interested in the extent of the nonzero region of the output of a convolution operation. For example, consider the convolution of the finite-extent signal $x(n_1, n_2)$ shown in Figure 1.14(a) with the finite-extent impulse

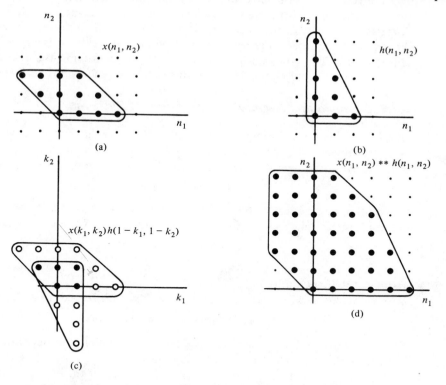

**Figure 1.14**   Pictorial representation of the convolution in Example 2. (a) The input sequence. (b) The impulse response. (c) The product sequence at $(n_1, n_2) =$ (1, 1). (d) The region of support of the convolution.

response $h(n_1, n_2)$ shown in Figure 1.14(b). [For the moment, we shall not be concerned with the values of the nonzero samples of $x(n_1, n_2)$ and $h(n_1, n_2)$.] It is obvious that the result of this convolution, which we shall call $y(n_1, n_2)$, will also be a signal of finite extent. We want to sketch the region of support for this output signal.

Proceeding as before, we form the 2-D sequence $h(n_1 - k_1, n_2 - k_2)$ as a function of $(k_1, k_2)$. Starting with $(n_1, n_2) = (0, 0)$, we slide $h(n_1 - k_1, n_2 - k_2)$ over the sequence $x(k_1, k_2)$. When the two sequences overlap, we have a (potentially) nonzero point in the output sequence $y(n_1, n_2)$. Figure 1.14(c) shows the overlap for $(n_1, n_2) = (1, 1)$ and Figure 1.14(d) shows the region of support for $y(n_1, n_2)$.

Even within this region, some samples of $y(n_1, n_2)$ may have a value of zero, because the terms in the summation on the right side of equation (1.36) may cancel each other for a particular value of $(n_1, n_2)$. In general, however, $y(n_1, n_2)$ will be nonzero in this region, and it will certainly be zero outside it.

As an exercise, the reader can compute the values of the samples of $y(n_1, n_2)$ in its region of support for the simple case where $x(n_1, n_2)$ and $h(n_1, n_2)$ are both equal to one in their respective regions of support [Figure 1.14(a) and (b)].

In this section we have presented two relatively simple examples of performing 2-D convolutions. You have undoubtedly noticed that some effort is involved in these calculations. Fortunately, we do not often perform such calculations by hand. Some familiarity with the basic operations, however, is necessary in order to write the required computer programs and to interpret their results. It is virtually impossible to perform a 2-D convolution correctly without identifying the relevant cases to consider. This should be the first step whenever a convolution is performed.

### 1.2.5 Cascade and Parallel Connections of Systems

One of the virtues of linear shift-invariant systems is the ease with which they can be analyzed when they are connected together. This is due in part to some properties of the convolution operator. We have already seen that convolution is commutative.

$$x ** h = h ** x \qquad (1.46)$$

Convolution is also associative. If the convolution of $x$ with $h$ is convolved with $g$, the result is the same as if $x$ were convolved with the convolution of $h$ and $g$.

$$(x ** h) ** g = x ** (h ** g) \qquad (1.47)$$

Because of the associative property, the parentheses can be omitted when talking about $N$-fold convolutions.

Convolution also obeys the distributive law with respect to addition.

$$x ** (h + g) = (x ** h) + (x ** g) \qquad (1.48)$$

The associative and distributive properties of the convolution operator are straightforward to demonstrate. This is left as an exercise for the reader (see Problem 1.4).

Two systems are said to be connected in *cascade* if the output of the first is the input to the second, as illustrated in Figure 1.15. If the two systems are linear and shift invariant, their cascade connection can be shown to be linear and shift invariant.

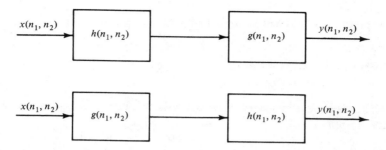

**Figure 1.15**   Each figure represents a cascade connection of two systems. If both systems are linear and shift invariant, the order of the cascade is immaterial and from an input–output point of view the two cascades above are equivalent.

If $w$ denotes the output of the first system in the cascade, it follows that

$$w = x ** h$$
$$y = w ** g = (x ** h) ** g \tag{1.49}$$

From the associative law, however, (1.49) can be rewritten as

$$y = x ** (h ** g) \tag{1.50}$$

and thus the equivalent impulse response for the cascade system is

$$h_{\text{equiv}} = h ** g \tag{1.51}$$

Going one step further and applying the commutative law, we see that the equivalent impulse response is unchanged if the order of the two systems in the cascade is reversed. Thus two cascaded LSI systems which are identical except for the order of their subsystems are equivalent; they will produce the same output if they are excited by the same input. If $N$ LSI systems are arranged as a cascade combination, the equivalent impulse response is the $N$-fold convolution of their respective impulse responses. Furthermore, these systems can be cascaded in any order without affecting the equivalent impulse response.

Figure 1.16 illustrates two systems which are connected in *parallel*. They have a common input and their individual outputs are summed to produce a single output. It can be straightforwardly shown that if these two systems are linear and shift invariant, the overall system will be linear and shift invariant. To find the equivalent impulse response, we observe that

$$y = (x ** h) + (x ** g) \tag{1.52}$$

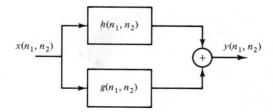

**Figure 1.16**   Parallel connection of two systems.

Applying the distributive law, it follows that

$$y = x ** (h + g)$$ (1.53)

from which it follows that

$$h_{equiv} = h + g$$ (1.54)

This rule generalizes in the obvious fashion to parallel connections of more than two LSI systems.

Sometimes it is useful to decompose an impulse response into several components, particularly if the impulse response has a finite, but oddly shaped, region of support which can be represented as a collection of smaller, more regularly shaped regions. The input sequence can be convolved with each of the component impulse responses and the final output sequence formed by summation, thus implementing the system of interest as the parallel connection of simpler systems.

### 1.2.6 Separable Systems

A *separable system* is an LSI system whose impulse response is a separable sequence. The input signals processed by a separable system and the output signals produced by it need not be separable. As with any other LSI system, the output can be computed from the input using the convolution sum, but for separable systems, the convolution sum decomposes. As we shall see in Chapters 3 and 5, this property makes these systems very efficient to implement. To see how the convolution sum decomposes, let the impulse response of the system be denoted by

$$h(n_1, n_2) = h_1(n_1)h_2(n_2)$$ (1.55)

The output of the system is then

$$y(n_1, n_2) = \sum_{k_1=-\infty}^{\infty} \sum_{k_2=-\infty}^{\infty} x(n_1 - k_1, n_2 - k_2)h_1(k_1)h_2(k_2)$$
$$= \sum_{k_1=-\infty}^{\infty} h_1(k_1) \sum_{k_2=-\infty}^{\infty} x(n_1 - k_1, n_2 - k_2)h_2(k_2)$$ (1.56)

The innermost sum represents a 2-D array of numbers. If we define

$$g(n_1, n_2) \triangleq \sum_{k_2=-\infty}^{\infty} x(n_1, n_2 - k_2)h_2(k_2)$$ (1.57)

equation (1.56) becomes

$$y(n_1, n_2) = \sum_{k_1=-\infty}^{\infty} h_1(k_1)g(n_1 - k_1, n_2)$$

The array $g(n_1, n_2)$ can be computed by performing a 1-D convolution between each column of $x$ ($n_1$ = constant) and the 1-D sequence $h_2$. The output array $y$ can then be computed by convolving each row of $g$ ($n_2$ = constant) with the 1-D sequence $h_1$. Alternatively, the row convolutions could be computed before the column convolutions; the same output signal will result in either case. The important point is that the output can be obtained as a series of 1-D convolutions.

The $M$-dimensional case is similar. A separable system can again be implemented using 1-D convolutions, but the number of such convolutions grows rapidly with the dimensionality of the signal. For example, consider the $M$-dimensional input sequence $x(n_1, n_2, \ldots, n_M)$ defined over the $N \times N \times N \times \cdots \times N$ hypercube. If this signal is convolved with a separable sequence of the form $h(n_1, n_2, \ldots, n_M) = h_1(n_1)h_2(n_2) \cdots h_M(n_M)$, then $MN^{M-1}$ 1-D convolutions are required to obtain the output sequence.

### 1.2.7  Stable Systems

As in the 1-D case, the only truly useful systems are those which are stable. It is reasonable to require, for example, that a system's output sequence should remain bounded if its input sequence is bounded. To distinguish this type of stability from others, we will say that such systems are BIBO (bounded input, bounded output) stable. For a BIBO stable system, when $|x(n_1, n_2)| \le B$, there must exist a $B'$ such that $|y(n_1, n_2)| \le B'$ for all $(n_1, n_2)$. A necessary and sufficient condition for an LSI system to be BIBO stable is that its impulse response be absolutely summable.

$$\sum_{n_1=-\infty}^{\infty} \sum_{n_2=-\infty}^{\infty} |h(n_1, n_2)| = S_1 < \infty \tag{1.58}$$

The proof of this fact is identical to the 1-D case [1].

A weaker form of stability is mean-square stability. An LSI system is mean-square stable if

$$\sum_{n_1=-\infty}^{\infty} \sum_{n_2=-\infty}^{\infty} |h(n_1, n_2)|^2 = S_2 < \infty \tag{1.59}$$

A BIBO stable system is mean-square stable, but the converse is not necessarily true. If we simply refer to a system as stable, we will be referring to a system which is BIBO stable.

The definitions above would seem to imply that multidimensional stability is very similar to 1-D stability. As we shall see in Chapter 4, this is definitely not the case. Multidimensional stability is far more difficult both to understand and to test than 1-D stability.

### 1.2.8  Regions of Support

In studying 1-D systems, we found it useful to characterize systems as causal if their outputs could not precede their inputs. Such systems were useful for processing signals whose independent variable was time, both because the constraint made physical sense and because it yielded systems that could be implemented in real time.

For most 2-D applications, the independent variables do not correspond to time, and causality is not a natural constraint for such systems. We are forced to consider generalizations of causal systems, however, when we consider system implementations.

The impulse response $h(n)$ of a causal 1-D discrete LSI system is zero for $n < 0$.

Consequently, one generalization of the concept of causality can be made by requiring that an impulse response be zero outside some region of support.

Earlier we discussed the special case of sequences whose support is a region of finite extent. Sequences that are nonzero only in one quadrant of the $(n_1, n_2)$-plane form another important special case. They are said to possess *quadrant support*. The notion of quadrant support can be generalized to include regions of support that are wedge-shaped. We will say that a sequence has *support on a wedge* if it is nonzero only at points inside (and on the edges) of a sector defined by two rays emanating from the origin, providing that the angle between the two rays is strictly less than 180 degrees. An example of a sequence with support on a wedge is shown in Figure 1.17(a).

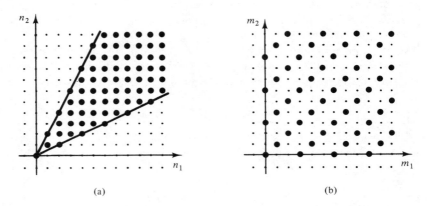

(a)                                          (b)

**Figure 1.17**  (a) Sequence with support on a wedge. (b) Sequence with support on one quadrant derived from the sequence in (a) under the linear transformation of variables with $(N_{11}, N_{21}) = (2, 1)$ and $(N_{12}, N_{22}) = (1, 2)$. Small dots imply a sample of value zero.

Any sequence with support on a wedge can be mapped into a sequence with support on the first quadrant through a linear transformation of variables [4]. For example, suppose that the vectors

$$\mathbf{N}_1 = (N_{11}, N_{21})'$$
$$\mathbf{N}_2 = (N_{12}, N_{22})'$$
(1.60)

lie along the edges of a wedge-shaped region. ($N_{11}$, $N_{21}$, $N_{12}$, and $N_{22}$ are integers.) Furthermore, assume that there are no common factors between $N_{11}$ and $N_{21}$ or between $N_{12}$ and $N_{22}$. Since $\mathbf{N}_1$ and $\mathbf{N}_2$ are not collinear

$$D = N_{11}N_{22} - N_{12}N_{21} \neq 0$$
(1.61)

The change of variables

$$m_1 = N_{22}n_1 - N_{12}n_2$$
$$m_2 = -N_{21}n_1 + N_{11}n_2$$
(1.62)

will map the wedge in question *onto* the first quadrant. Such a transformation is not unique. In this case, $\mathbf{N}_1$ is mapped onto $(D, 0)'$ and $\mathbf{N}_2$ is mapped onto $(0, D)'$. Figure 1.17(b) shows the result of mapping the wedge in Figure 1.17(a) onto the first quadrant. Because $D = 3$ for this example, not every point in the first quadrant of Figure 1.17(b) lies in the range space of the linear transformation (1.62). There will be samples in the $(m_1, m_2)$-plane to which no sample in the $(n_1, n_2)$-plane will be mapped. This phenomenon is a result of using discrete signals; we insist that an *integer* vector $(n_1, n_2)'$ be mapped into another *integer* vector $(m_1, m_2)'$. It can be shown that in order for every integer-valued order pair in the first quadrant of the $(m_1, m_2)$-plane to lie in the range space of the linear transformation, it is necessary and sufficient that $|D| = 1$.

### *1.2.9  Vector Input–Output Systems

Systems with several inputs and/or several outputs are important in some practical applications. We shall take a very brief look at these systems and how they are related to multidimensional LSI systems.

Consider a system that processes signals received by an array of sensors equally spaced along a line. The $i$th sensor provides a 1-D discrete-time signal to the system, which produces a number of 1-D discrete-time output signals. We shall denote the $i$th input signal by $x_i(n)$ and the $j$th output signal by $y_j(n)$. For simplicity, we shall assume that the system under consideration is linear and time invariant (shift invariant in the discrete variable $n$). If the $i$th input signal is a 1-D unit impulse $\delta(n)$ and all the other input signals are zero, the $j$th output signal will be the impulse response $h_{ij}(n)$. In general, of course, there will be arbitrary discrete-time signals at each input port of the system, so the $j$th output signal must be written

$$y_j(n) = \sum_i \sum_m h_{ij}(m) x_i(n - m) \tag{1.63}$$

We can relate equation (1.63) to the 2-D convolution sum discussed in Section 1.2.4 by defining the 2-D sequences

$$
\begin{aligned}
p(i, m) &\triangleq x_i(m) \\
q(j, n) &\triangleq y_j(n)
\end{aligned}
\tag{1.64}
$$

At this point we will naively assume that the sequences $p$ and $q$ are related by the 2-D convolution sum

$$q(j, n) = \sum_i \sum_m f(i, m) p(i - j, n - m) \tag{1.65}$$

which can be subjected to a change of variables to give

$$q(j, n) = \sum_i \sum_m f(i - j, m) p(i, n - m) \tag{1.66}$$

By comparing equations (1.66) and (1.63), we see that the linear time-invariant vector input/output system can be regarded as a 2-D LSI system if

$$h_{ij}(m) = f(i - j, m) \tag{1.67}$$

This requirement essentially imposes shift invariance in the variable corresponding to the input–output index. If equation (1.67) does not hold, then the vector input–output system can still be regarded as a 2-D linear system, but not a 2-D LSI system.

## 1.3 FREQUENCY-DOMAIN CHARACTERIZATION OF SIGNALS AND SYSTEMS

In the preceding section we saw that the response of a 2-D LSI system to an input signal could be obtained by convolving the input signal with the impulse response of the system. By representing the input signal as the superposition of shifted impulses, the output signal could be represented as the superposition of shifted impulse responses. Frequency-domain representations of LSI systems also exploit the superposition principle, but in this case the elemental sequences are complex sinusoids. Let us begin by considering the responses of LSI systems to sinusoidal inputs.

### 1.3.1 Frequency Response of a 2-D LSI System

Consider a 2-D LSI system with unit impulse response $h(n_1, n_2)$ and an input which is a complex sinusoid of the form

$$x(n_1, n_2) = \exp(j\omega_1 n_1 + j\omega_2 n_2) \tag{1.68}$$

where $\omega_1$ and $\omega_2$ are real numbers called the horizontal and vertical frequencies, respectively. We can determine the output signal by convolution.

$$
\begin{aligned}
y(n_1, n_2) &= \sum_{k_1=-\infty}^{\infty} \sum_{k_2=-\infty}^{\infty} \exp[j\omega_1(n_1 - k_1) + j\omega_2(n_2 - k_2)]h(k_1, k_2) \\
&= \exp(j\omega_1 n_1 + j\omega_2 n_2)\left[\sum_{k_1=-\infty}^{\infty}\sum_{k_2=-\infty}^{\infty} h(k_1, k_2)\exp(-j\omega_1 k_1 - j\omega_2 k_2)\right] \\
&= \exp(j\omega_1 n_1 + j\omega_2 n_2)H(\omega_1, \omega_2)
\end{aligned} \tag{1.69}
$$

The output signal is a complex sinusoid with the same frequencies as the input signal, but its amplitude and phase have been altered by the complex gain $H(\omega_1, \omega_2)$. This gain is called the system's *frequency response*, and it is given by

$$H(\omega_1, \omega_2) \triangleq \sum_{n_1}\sum_{n_2} h(n_1, n_2)\exp(-j\omega_1 n_1 - j\omega_2 n_2) \tag{1.70}$$

An LSI system is able to discriminate among sinusoidal signals on the basis of their frequencies. If $|H(\omega_1, \omega_2)|$ is approximately equal to one for a particular value of the ordered pair $(\omega_1, \omega_2)$, sinusoidal signals at that frequency will pass through the system without being attenuated. On the other hand, if $|H(\omega_1, \omega_2)|$ is close to zero for some $(\omega_1, \omega_2)$, sinusoids at that frequency will be rejected by the system.

It is straightforward to show that the frequency response $H(\omega_1, \omega_2)$ is periodic in both the horizontal and vertical frequency variables with a period of $2\pi$.

$$
\begin{aligned}
H(\omega_1 + 2\pi, \omega_2) &= H(\omega_1, \omega_2) \\
H(\omega_1, \omega_2 + 2\pi) &= H(\omega_1, \omega_2)
\end{aligned} \tag{1.71}
$$

The proof is left to the reader (see Problem 1.12).

**Example 3**

As a simple example, let us compute the frequency response of the system with the impulse response

$$h(n_1, n_2) = \dot{\delta}(n_1 + 1, n_2) + \delta(n_1 - 1, n_2) + \delta(n_1, n_2 + 1) + \delta(n_1, n_2 - 1) \quad (1.72)$$

This sequence is drawn in Figure 1.18(a). The frequency response is given by

$$
\begin{aligned}
H(\omega_1, \omega_2) &= \sum_{n_1=-\infty}^{\infty} \sum_{n_2=-\infty}^{\infty} h(n_1, n_2) \exp(-j\omega_1 n_1 - j\omega_2 n_2) \\
&= \sum_{n_1=-\infty}^{\infty} \sum_{n_2=-\infty}^{\infty} [\delta(n_1 + 1, n_2) + \delta(n_1 - 1, n_2) + \delta(n_1, n_2 + 1) \\
&\quad + \delta(n_1, n_2 - 1)] \cdot \exp(-j\omega_1 n_1 - j\omega_2 n_2) \\
&= e^{j\omega_1} + e^{-j\omega_1} + e^{j\omega_2} + e^{-j\omega_2} \\
&= 2(\cos\omega_1 + \cos\omega_2)
\end{aligned}
\quad (1.73)
$$

This frequency response is shown as a perspective plot in Figure 1.18(b).

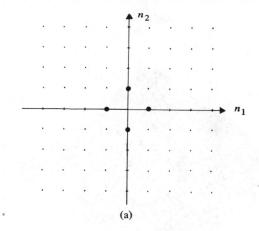

(a)

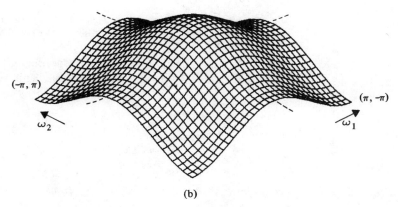

(b)

**Figure 1.18**    (a) Impulse response of Example 3. (b) Frequency response of Example 3.

**Example 4**

Consider the system whose impulse response is given by

$$h(n_1, n_2) = \begin{cases} 0.125, & n_1 = \pm1, \quad n_2 = \pm1 \\ 0.25, & n_1 = \pm1, \quad n_2 = 0 \\ 0.25, & n_1 = 0, \quad n_2 = \pm1 \\ 0.5, & n_1 = n_2 = 0 \\ 0, & \text{otherwise} \end{cases} \quad (1.74)$$

Applying the definition of the frequency response, we get

$$\begin{aligned} H(\omega_1, \omega_2) &= \sum_{n_1=-\infty}^{\infty} \sum_{n_2=-\infty}^{\infty} h(n_1, n_2) \exp{(-j\omega_1 n_1 - j\omega_2 n_2)} \\ &= 0.5 + 0.25(e^{-j\omega_1} + e^{j\omega_1} + e^{-j\omega_2} + e^{j\omega_2}) \\ &\quad + 0.125(e^{-j\omega_1}e^{-j\omega_2} + e^{-j\omega_1}e^{j\omega_2} + e^{j\omega_1}e^{-j\omega_2} + e^{j\omega_1}e^{j\omega_2}) \\ &= 0.5(1 + \cos{\omega_1})(1 + \cos{\omega_2}) \end{aligned} \quad (1.75)$$

which is shown in Figure 1.19. This is an example of a simple lowpass filter. The gain of the filter is approximately 2 near the origin, and it is approximately 0 when either $\omega_1 \approx \pm\pi$ or $\omega_2 \approx \pm\pi$.

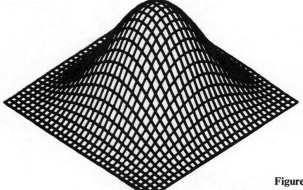

**Figure 1.19** Frequency response of the simple lowpass filter of Example 4.

The system in Example 4 has a separable impulse response, and from (1.75) we see that its frequency response is also a separable function. This result is true in general. If

$$h(n_1, n_2) = f(n_1)g(n_2) \quad (1.76)$$

then

$$H(\omega_1, \omega_2) = F(\omega_1)G(\omega_2)$$

where

$$F(\omega_1) = \sum_{n_1} f(n_1) \exp{(-j\omega_1 n_1)} \quad \text{and} \quad G(\omega_2) = \sum_{n_2} g(n_2) \exp{(-j\omega_2 n_2)} \quad (1.77)$$

The proof is left for the reader in Problem 1.13.

If the input sequence to an $M$-dimensional LSI system is a complex sinusoid of the form

$$x(n_1, n_2, \ldots, n_M) = \prod_{i=1}^{M} \exp\left(j\omega_i n_i\right) \qquad (1.78)$$

its output is also a complex sinusoid of the same form multiplied by a complex gain. Using vector notation, we can rewrite equation (1.78) as

$$x(\mathbf{n}) = \exp\left(j\boldsymbol{\omega}'\mathbf{n}\right) \qquad (1.79)$$

where $\mathbf{n} = (n_1, n_2, \ldots, n_M)'$ and $\boldsymbol{\omega} = (\omega_1, \omega_2, \ldots, \omega_M)'$. The output of the $M$-dimensional LSI system is given by

$$y(\mathbf{n}) = H(\boldsymbol{\omega}) \exp\left(j\boldsymbol{\omega}'\mathbf{n}\right) \qquad (1.80)$$

where the $M$-dimensional frequency response $H(\boldsymbol{\omega})$ is given by

$$H(\boldsymbol{\omega}) = \sum_{\mathbf{n}} h(\mathbf{n}) \exp\left(-j\boldsymbol{\omega}'\mathbf{n}\right) \qquad (1.81)$$

### 1.3.2 Determining the Impulse Response from the Frequency Response

The frequency response of a discrete LSI system is generally a continuous 2-D periodic function which can be expressed as a linear combination of harmonically related complex sinusoids, as demonstrated by the definition of $H(\omega_1, \omega_2)$ in equation (1.70). This relation not only defines $H(\omega_1, \omega_2)$, it also serves as a 2-D Fourier series expansion of $H(\omega_1, \omega_2)$. The coefficients of the expansion are the values of the impulse response samples $h(n_1, n_2)$. It should not be surprising, therefore, that the impulse response of an LSI system can be obtained from the frequency response.

The inverse relationship can be derived by multiplying both sides of equation (1.70) by a complex sinusoid and integrating over a square in the frequency plane. In detail, we form

$$\frac{1}{4\pi^2} \int_{-\pi}^{\pi} \int_{-\pi}^{\pi} H(\omega_1, \omega_2) \exp\left(j\omega_1 k_1 + j\omega_2 k_2\right) d\omega_1 \, d\omega_2$$

$$= \frac{1}{4\pi^2} \int_{-\pi}^{\pi} \int_{-\pi}^{\pi} \sum_{n_1} \sum_{n_2} h(n_1, n_2) \exp\left(-j\omega_1 n_1 - j\omega_2 n_2\right)$$

$$\cdot \exp\left(j\omega_1 k_1 + j\omega_2 k_2\right) d\omega_1 \, d\omega_2 \qquad (1.82)$$

$$= \sum_{n_1} \sum_{n_2} h(n_1, n_2) \left[ \frac{1}{2\pi} \int_{-\pi}^{\pi} \exp\left[-j\omega_1(n_1 - k_1)\right] d\omega_1 \right]$$

$$\cdot \left[ \frac{1}{2\pi} \int_{-\pi}^{\pi} \exp\left[-j\omega_2(n_2 - k_2)\right] d\omega_2 \right]$$

It is straightforward to demonstrate that

$$\frac{1}{2\pi} \int_{-\pi}^{\pi} \exp\left[-j\omega(n - k)\right] d\omega = \delta(n - k) \qquad (1.83)$$

so that the right side of equation (1.82) becomes, upon evaluation of the double sum, simply $h(k_1, k_2)$. This, then, provides a means of evaluating the value of the impulse response at $(k_1, k_2)$.

Restating this result in terms of the more familiar integer variables $(n_1, n_2)$, we get

$$h(n_1, n_2) = \frac{1}{4\pi^2} \int_{-\pi}^{\pi} \int_{-\pi}^{\pi} H(\omega_1, \omega_2) \exp(j\omega_1 n_1 + j\omega_2 n_2)\, d\omega_1\, d\omega_2 \qquad (1.84)$$

The area of integration for the double integral in (1.84) extends over exactly one period of $H(\omega_1, \omega_2)$. Although we chose to use the period centered at the origin in writing these expressions, in fact any period of $H(\omega_1, \omega_2)$ could have been used.

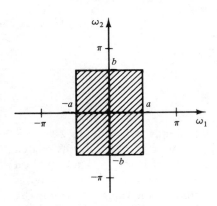

**Figure 1.20** Frequency response of the ideal rectangular lowpass filter used for Example 5.

**Example 5**

Let us use this result to find the impulse response of the ideal lowpass filter specified by the frequency response

$$H(\omega_1, \omega_2) = \begin{cases} 1, & |\omega_1| \leq a < \pi, \quad |\omega_2| \leq b < \pi \\ 0, & \text{otherwise} \end{cases} \qquad (1.85)$$

which is shown in Figure 1.20. This example is quite simple because it is a separable system. Thus

$$\begin{aligned} h(n_1, n_2) &= \frac{1}{4\pi^2} \int_{-a}^{a} \int_{-b}^{b} \exp(j\omega_1 n_1 + j\omega_2 n_2)\, d\omega_2\, d\omega_1 \\ &= \frac{1}{2\pi} \int_{-a}^{a} \exp(j\omega_1 n_1)\, d\omega_1 \\ &\quad \frac{1}{2\pi} \int_{-b}^{b} \exp(j\omega_2 n_2)\, d\omega_2 \\ &= \frac{\sin an_1}{\pi n_1} \frac{\sin bn_2}{\pi n_2} \end{aligned} \qquad (1.86)$$

**Example 6**

As a slightly more complex example, consider the problem of determining the impulse response of the ideal circular lowpass filter given by

$$H(\omega_1, \omega_2) = \begin{cases} 1, & \omega_1^2 + \omega_2^2 \leq R^2 < \pi^2 \\ 0, & \text{otherwise} \end{cases} \qquad (1.87)$$

This frequency response, which is not separable, is shown in Figure 1.21. In this example,

$$h(n_1, n_2) = \frac{1}{4\pi^2} \iint_A \exp(j\omega_1 n_1 + j\omega_2 n_2)\, d\omega_1\, d\omega_2 \qquad (1.88)$$

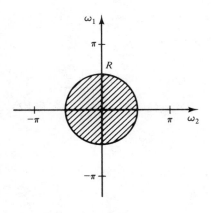

**Figure 1.21** Frequency response of the ideal circularly symmetric lowpass filter of Example 6.

The integral over the circular region $A$ is more easily

performed if $\omega_1$ and $\omega_2$ are replaced by polar coordinate variables. Therefore, we define

$$\omega \triangleq \sqrt{\omega_1^2 + \omega_2^2}, \qquad \phi \triangleq \tan^{-1}\frac{\omega_2}{\omega_1}, \qquad \theta \triangleq \tan^{-1}\frac{n_2}{n_1}$$

With these definitions, (1.88) becomes

$$\begin{aligned} h(n_1, n_2) &= \frac{1}{4\pi^2} \int_0^R \int_0^{2\pi} \omega \exp\left[j\omega\sqrt{n_1^2 + n_2^2}\cos(\theta - \phi)\right] d\phi \, d\omega \\ &= \frac{1}{2\pi} \int_0^R \omega J_0(\omega\sqrt{n_1^2 + n_2^2}) \, d\omega \qquad\qquad (1.89) \\ &= \frac{R}{2\pi} \frac{J_1(R\sqrt{n_1^2 + n_2^2})}{\sqrt{n_1^2 + n_2^2}} \end{aligned}$$

where $J_0(x)$ and $J_1(x)$ are the Bessel functions of the first kind of orders of 0 and 1, respectively. This impulse response is a sampled, circularly symmetric function. Along the $n_1$-axis, it has the form

$$h(n_1, 0) = \frac{R}{2\pi n_1} J_1(n_1 R) \qquad\qquad (1.90)$$

which is sketched in Figure 1.22.

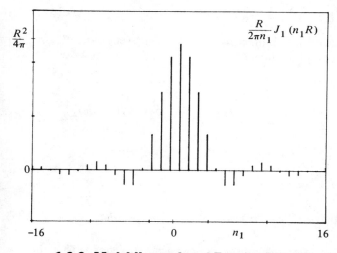

**Figure 1.22** Impulse response for the circularly symmetric lowpass filter of Example 6 evaluated along the $n_1$-axis.

### 1.3.3 Multidimensional Fourier Transform

In Section 1.2.1 we saw that an arbitrary 2-D sequence could be written as the sum of weighted and shifted impulses as in equation (1.25). An LSI system will respond to each impulse with its impulse response, appropriately weighted. Thus the output sequence can be interpreted as the superposition of the weighted and shifted impulse responses.

In this section we shall demonstrate that a 2-D sequence can, in most practical cases, be written as a weighted sum of complex sinusoids using the multidimensional

Fourier transform. Since we know the response of an LSI system to a sinusoidal input, we can write the output sequence as the superposition of the sinusoidal responses of the LSI system.

If we look carefully at the inverse frequency-response operator given by (1.84), we see that, in addition to providing a formula for $h(n_1, n_2)$, it also represents the sequence $h$ as a superposition of complex sinusoids. Let us use a similar representation for the input sequence $x$ and write

$$x(n_1, n_2) = \frac{1}{4\pi^2} \int_{-\pi}^{\pi} \int_{-\pi}^{\pi} X(\omega_1, \omega_2) \exp\left(j\omega_1 n_1 + j\omega_2 n_2\right) d\omega_1 \, d\omega_2 \qquad (1.91)$$

The complex function $X$, which is called the *2-D Fourier transform* of $x$, can be evaluated using

$$X(\omega_1, \omega_2) = \sum_{n_1 = -\infty}^{\infty} \sum_{n_2 = -\infty}^{\infty} x(n_1, n_2) \exp\left(-j\omega_1 n_1 - j\omega_2 n_2\right) \qquad (1.92)$$

With this definition, we see that the frequency response of an LSI system is the Fourier transform of the system's impulse response.

Now, suppose that we have a 2-D LSI system $L[\cdot]$ which possesses an impulse response $h(n_1, n_2)$ and a frequency response $H(\omega_1, \omega_2)$. We know that

$$L[\exp\left(j\omega_1 n_1 + j\omega_2 n_2\right)] = H(\omega_1, \omega_2) \exp\left(j\omega_1 n_1 + j\omega_2 n_2\right) \qquad (1.93)$$

Using the property of linearity in conjunction with the representation of $x(n_1, n_2)$ as the integral of weighted complex sinusoids, equation (1.91), we can write

$$
\begin{aligned}
y(n_1, n_2) &= L[x(n_1, n_2)] \\
&= L\left[\frac{1}{4\pi^2} \int_{-\pi}^{\pi} \int_{-\pi}^{\pi} X(\omega_1, \omega_2) \exp\left(j\omega_1 n_1 + j\omega_2 n_2\right) d\omega_1 \, d\omega_2\right] \\
&= \frac{1}{4\pi^2} \int_{-\pi}^{\pi} \int_{-\pi}^{\pi} X(\omega_1, \omega_2) L[\exp\left(j\omega_1 n_1 + j\omega_2 n_2\right)] \, d\omega_1 \, d\omega_2
\end{aligned}
\qquad (1.94)
$$

Finally, using equation (1.93), we get

$$y(n_1, n_2) = \frac{1}{4\pi^2} \int_{-\pi}^{\pi} \int_{-\pi}^{\pi} H(\omega_1, \omega_2) X(\omega_1, \omega_2) \exp\left(j\omega_1 n_1 + j\omega_2 n_2\right) d\omega_1 \, d\omega_2 \qquad (1.95)$$

We have tacitly assumed that $X(\omega_1, \omega_2)$ and $H(\omega_1, \omega_2)$ are well defined. This allowed us to interchange the order of the integration and the linear operator $L[\cdot]$ in equation (1.94).

Equation (1.95) gives us an alternative way of expressing the output of an LSI system. The relative weighting of the complex sinusoidal components comprising the input sequence has been altered by multiplication with the system's frequency response $H(\omega_1, \omega_2)$. Naturally, the output sequence computed by equation (1.95) is identical to the output sequence computed by the convolution sum equations (1.36) and (1.37) (see Problem 1.18).

The output sequence $y(n_1, n_2)$ may also be written in terms of its Fourier transform $Y(\omega_1, \omega_2)$ as

$$y(n_1, n_2) = \frac{1}{4\pi^2} \int_{-\pi}^{\pi} \int_{-\pi}^{\pi} Y(\omega_1, \omega_2) \exp{(j\omega_1 n_1 + j\omega_2 n_2)}\, d\omega_1\, d\omega_2 \qquad (1.96)$$

A comparison of equations (1.95) and (1.96) implies that

$$Y(\omega_1, \omega_2) = H(\omega_1, \omega_2) X(\omega_1, \omega_2) \qquad (1.97)$$

if $y = h ** x$. This result, often referred to as the *convolution theorem*, is extremely important; the Fourier transform of the convolution of two 2-D sequences is the product of their Fourier transforms.

The Fourier transform defined by (1.92) can be shown to exist whenever the sequence $x(n_1, n_2)$ is absolutely summable.

$$\sum_{n_1=-\infty}^{\infty} \sum_{n_2=-\infty}^{\infty} |x(n_1, n_2)| = S_1 < \infty \qquad (1.98)$$

If it exists, it can be shown to be continuous and analytic. This means that the frequency response of an LSI system exists only if the system is stable. It is occasionally useful to talk about a system, such as an ideal lowpass filter, whose frequency response is not continuous and whose impulse response does not satisfy (1.98). Although such an impulse response is not absolutely summable, it is square summable. Sequences that satisfy the weaker condition

$$\sum_{n_1} \sum_{n_2} |x(n_1, n_2)|^2 = S_2 < \infty \qquad (1.99)$$

rather than (1.98) may not possess continuous Fourier transforms, but their Fourier transforms are well defined except at points of discontinuity.

### 1.3.4 Other Properties of the 2-D Fourier Transform

We shall use the notation

$$x \longleftrightarrow X \qquad (1.100)$$

to indicate that $x(n_1, n_2)$ and $X(\omega_1, \omega_2)$ are a Fourier transform pair. Using this shorthand, we see that the convolution theorem becomes

$$y = h ** x \longleftrightarrow Y = HX \qquad (1.101)$$

The 2-D Fourier transform operator has a number of other useful properties which are straightforward extensions of properties of the 1-D transform. These are summarized below.

**Linearity.**    If

$$x_1 \longleftrightarrow X_1 \qquad \text{and} \qquad x_2 \longleftrightarrow X_2$$

then

$$ax_1 + bx_2 \longleftrightarrow aX_1 + bX_2 \qquad (1.102)$$

for any complex numbers $a$ and $b$.

**Spatial shift.**    If

$$x(n_1, n_2) \longleftrightarrow X(\omega_1, \omega_2)$$

then

$$x(n_1 - m_1, n_2 - m_2) \longleftrightarrow \exp(-j\omega_1 m_1 - j\omega_2 m_2) X(\omega_1, \omega_2) \qquad (1.103)$$

Shifting a sequence $x(n_1, n_2)$ by an amount $(m_1, m_2)$ corresponds to multiplying its Fourier transform $X(\omega_1, \omega_2)$ by the linear-phase term $\exp(-j\omega_1 m_1 - j\omega_2 m_2)$.

**Modulation.**

$$x(n_1, n_2) \exp(j\theta_1 n_1 + j\theta_2 n_2) \longleftrightarrow X(\omega_1 - \theta_1, \omega_2 - \theta_2) \qquad (1.104)$$

Multiplying a sequence by a complex sinusoidal sequence corresponds to shifting its Fourier transform.

**Multiplication.**

$$c(n_1, n_2) x(n_1, n_2) \longleftrightarrow \frac{1}{4\pi^2} \int_{-\pi}^{\pi} \int_{-\pi}^{\pi} X(\theta_1, \theta_2) C(\omega_1 - \theta_1, \omega_2 - \theta_2)\, d\theta_1\, d\theta_2$$

$$= \frac{1}{4\pi^2} \int_{-\pi}^{\pi} \int_{-\pi}^{\pi} X(\omega_1 - \theta_1, \omega_2 - \theta_2) C(\theta_1, \theta_2)\, d\theta_1, d\theta_2 \qquad (1.105)$$

The multiplication of two sequences results in the convolution of their Fourier transforms as indicated in (1.105). Note that the convolution integral has a special form; the integrand is doubly periodic and the integral extends over exactly one period of the integrand. The property of modulation (1.104) can be regarded as a special case of the multiplication of two sequences.

**Differentiation of the Fourier transform.**

$$-j n_1 x(n_1, n_2) \longleftrightarrow \frac{\partial X(\omega_1, \omega_2)}{\partial \omega_1} \qquad (1.106a)$$

$$-j n_2 x(n_1, n_2) \longleftrightarrow \frac{\partial X(\omega_1, \omega_2)}{\partial \omega_2} \qquad (1.106b)$$

$$-n_1 n_2 x(n_1, n_2) \longleftrightarrow \frac{\partial^2 X(\omega_1, \omega_2)}{\partial \omega_1\, \partial \omega_2} \qquad (1.106c)$$

**Transposition.**

$$x(n_2, n_1) \longleftrightarrow X(\omega_2, \omega_1) \qquad (1.107)$$

**Reflection.**

$$x(-n_1, n_2) \longleftrightarrow X(-\omega_1, \omega_2) \qquad (1.108a)$$

$$x(n_1, -n_2) \longleftrightarrow X(\omega_1, -\omega_2) \qquad (1.108b)$$

$$x(-n_1, -n_2) \longleftrightarrow X(-\omega_1, -\omega_2) \qquad (1.108c)$$

**Complex conjugation.**

$$x^*(n_1, n_2) \longleftrightarrow X^*(-\omega_1, -\omega_2) \tag{1.109}$$

**Real and imaginary parts.**

$$\text{Re}\,[x(n_1, n_2)] \longleftrightarrow \tfrac{1}{2}[X(\omega_1, \omega_2) + X^*(-\omega_1, -\omega_2)] \tag{1.110a}$$

$$j\,\text{Im}\,[x(n_1, n_2)] \longleftrightarrow \tfrac{1}{2}[X(\omega_1, \omega_2) - X^*(-\omega_1, -\omega_2)] \tag{1.110b}$$

$$\tfrac{1}{2}[x(n_1, n_2) + x^*(-n_1, -n_2)] \longleftrightarrow \text{Re}\,[X(\omega_1, \omega_2)] \tag{1.111a}$$

$$\tfrac{1}{2}[x(n_1, n_2) - x^*(-n_1, -n_2)] \longleftrightarrow j\,\text{Im}\,[X(\omega_1, \omega_2)] \tag{1.111b}$$

In the special case when $x(n_1, n_2)$ is a *real-valued* sequence, these relationships imply that

$$X(\omega_1, \omega_2) = X^*(-\omega_1, -\omega_2) \tag{1.112a}$$

$$\text{Re}\,[X(\omega_1, \omega_2)] = \text{Re}\,[X(-\omega_1, -\omega_2)] \tag{1.112b}$$

$$\text{Im}\,[X(\omega_1, \omega_2)] = -\text{Im}\,[X(-\omega_1, -\omega_2)] \tag{1.112c}$$

The real part of the Fourier transform possesses even symmetry with respect to the origin, and the imaginary part possesses odd symmetry with respect to the origin. When $x(n_1, n_2)$ is real-valued, the left sides of (1.111a) and (1.111b) become the even and odd parts of $x(n_1, n_2)$, respectively.

**Parseval's theorem.**    If

$$x(n_1, n_2) \longleftrightarrow X(\omega_1, \omega_2) \qquad \text{and} \qquad w(n_1, n_2) \longleftrightarrow W(\omega_1, \omega_2)$$

then

$$\sum_{n_1} \sum_{n_2} x(n_1, n_2) w^*(n_1, n_2) = \frac{1}{4\pi^2} \int_{-\pi}^{\pi} \int_{-\pi}^{\pi} X(\omega_1, \omega_2) W^*(\omega_1, \omega_2)\, d\omega_1\, d\omega_2 \tag{1.113}$$

This remarkable relationship can be interpreted and applied in a variety of ways. The left side of equation (1.113) defines an inner product between two 2-D sequences while the right side defines an inner product between two 2-D Fourier transforms. Parseval's theorem says that inner products are preserved by the Fourier transform operation.

Equation (1.113) reduces to the convolution theorem when $w(n_1, n_2)$ is chosen to be $h^*(m_1 - n_1, m_2 - n_2)$, as in Problem 1.19.

Another important special case occurs when $w(n_1, n_2) = x(n_1, n_2)$, so that equation (1.113) becomes

$$\sum_{n_1} \sum_{n_2} |x(n_1, n_2)|^2 = \frac{1}{4\pi^2} \int_{-\pi}^{\pi} \int_{-\pi}^{\pi} |X(\omega_1, \omega_2)|^2\, d\omega_1\, d\omega_2 \tag{1.114}$$

The left side of equation (1.114) can be interpreted as the total energy in the discrete signal $x(n_1, n_2)$. The function $|X(\omega_1, \omega_2)|^2$ can be interpreted as the energy-density spectrum, since its integral is equal to the total energy in the signal.

## 1.4 SAMPLING CONTINUOUS 2-D SIGNALS

Nearly all discrete sequences are formed in an attempt to represent some underlying continuous signal. Many discrete representations of continuous signals are possible—Fourier series expansions, Taylor series expansions, and expansions in terms of nonsinusoidal orthogonal functions, for example—but periodic sampling is by far the representation used most often, partly due to the simplicity of its implementation. In this section we look at the relationships between continuous signals and the discrete sequences which are obtained from them by periodic sampling. We do this twice—first for the specific case of rectangular periodic sampling, and then for a more general case where different geometries are chosen for the sampling locations.

### 1.4.1 Periodic Sampling with Rectangular Geometry

Of the several ways to generalize 1-D periodic sampling to the 2-D case, the most straightforward is periodic sampling in rectangular coordinates, which we will simply call *rectangular sampling*. If $x_a(t_1, t_2)$ is a 2-D continuous waveform, the discrete signal $x(n_1, n_2)$ obtained from it by rectangular sampling is given by

$$x(n_1, n_2) = x_a(n_1 T_1, n_2 T_2) \tag{1.115}$$

where $T_1$ and $T_2$ are positive real constants known as the horizontal and vertical *sampling intervals* or *periods*. The sample locations in the $(t_1, t_2)$-plane are shown in Figure 1.23. For a sequence formed in this fashion, we would like to determine two

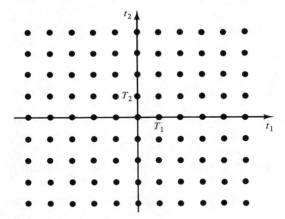

**Figure 1.23** Sampling locations in the $(t_1, t_2)$-plane for rectangular sampling.

things: Can the waveform $x_a(t_1, t_2)$ be recovered from $x(n_1, n_2)$? And how is the Fourier transform of $x$ related to the Fourier transform of $x_a$?

First, let us define the 2-D Fourier transform relations for continuous signals:

$$X_a(\Omega_1, \Omega_2) \triangleq \int_{-\infty}^{\infty} \int_{-\infty}^{\infty} x_a(t_1, t_2) \exp(-j\Omega_1 t_1 - j\Omega_2 t_2) \, dt_1 \, dt_2 \tag{1.116}$$

$$x_a(t_1, t_2) = \frac{1}{4\pi^2} \int_{-\infty}^{\infty} \int_{-\infty}^{\infty} X_a(\Omega_1, \Omega_2) \exp(j\Omega_1 t_1 + j\Omega_2 t_2) \, d\Omega_1 \, d\Omega_2 \tag{1.117}$$

Since $x(n_1, n_2) \triangleq x_a(n_1 T_1, n_2 T_2)$, we can use (1.117) to write

$$x(n_1, n_2) = \frac{1}{4\pi^2} \int_{-\infty}^{\infty} \int_{-\infty}^{\infty} X_a(\Omega_1, \Omega_2) \exp(j\Omega_1 n_1 T_1 + j\Omega_2 n_2 T_2) \, d\Omega_1 \, d\Omega_2 \qquad (1.118)$$

Next, we will manipulate this expression into the form of an inverse Fourier transform for discrete signals. We can begin by making the substitutions $\omega_1 = \Omega_1 T_1$ and $\omega_2 = \Omega_2 T_2$ to get the exponential terms into the correct form. This yields

$$x(n_1, n_2) = \frac{1}{4\pi^2} \int_{-\infty}^{\infty} \int_{-\infty}^{\infty} \frac{1}{T_1 T_2} X_a\left(\frac{\omega_1}{T_1}, \frac{\omega_2}{T_2}\right) \exp(j\omega_1 n_1 + j\omega_2 n_2) \, d\omega_1 \, d\omega_2 \qquad (1.119)$$

The double integral over the entire $(\omega_1, \omega_2)$-plane can be broken into an infinite series of integrals, each of which is over a square of area $4\pi^2$. Let $SQ(k_1, k_2)$ represent the square $-\pi + 2\pi k_1 \le \omega_1 < \pi + 2\pi k_1$; $-\pi + 2\pi k_2 \le \omega_2 < \pi + 2\pi k_2$. Then equation (1.119) can be written as

$$x(n_1, n_2) = \frac{1}{4\pi^2} \sum_{k_1} \sum_{k_2} \iint_{SQ(k_1, k_2)} \frac{1}{T_1 T_2} X_a\left(\frac{\omega_1}{T_1}, \frac{\omega_2}{T_2}\right) \exp(j\omega_1 n_1 + j\omega_2 n_2) \, d\omega_1 \, d\omega_2$$

Replacing $\omega_1$ by $\omega_1 - 2\pi k_1$ and $\omega_2$ by $\omega_2 - 2\pi k_2$ we can remove the dependence of the limits of integration on $k_1$ and $k_2$, giving

$$x(n_1, n_2) = \frac{1}{4\pi^2} \int_{-\pi}^{\pi} \int_{-\pi}^{\pi} \left[ \frac{1}{T_1 T_2} \sum_{k_1} \sum_{k_2} X_a\left(\frac{\omega_1 - 2\pi k_1}{T_1}, \frac{\omega_2 - 2\pi k_2}{T_2}\right) \right]$$
$$\cdot \exp(j\omega_1 n_1 + j\omega_2 n_2) \exp(-j2\pi k_1 n_1 - j2\pi k_2 n_2) \, d\omega_1 \, d\omega_2 \qquad (1.120)$$

The second exponential factor in equation (1.120) is seen to be equal to one for all values of the integer variables $n_1, k_1, n_2$ and $k_2$. Equation (1.120) now has the same form as an inverse Fourier transform, so we conclude that

$$X(\omega_1, \omega_2) = \frac{1}{T_1 T_2} \sum_{k_1} \sum_{k_2} X_a\left(\frac{\omega_1 - 2\pi k_1}{T_1}, \frac{\omega_2 - 2\pi k_2}{T_2}\right) \qquad (1.121)$$

or alternatively,

$$X(\Omega_1 T_1, \Omega_2 T_2) = \frac{1}{T_1 T_2} \sum_{k_1} \sum_{k_2} X_a\left(\Omega_1 - \frac{2\pi k_1}{T_1}, \Omega_2 - \frac{2\pi k_2}{T_2}\right) \qquad (1.122)$$

Equation (1.122) gives us the relation we seek between the Fourier transforms of the continuous and discrete signals. The right side of this expression can be interpreted as a periodic extension of $X_a(\Omega_1, \Omega_2)$, which yields the periodic function $X(\Omega_1 T_1, \Omega_2 T_2)$.

Equation (1.122) can be further simplified in the case where the continuous signal $x_a(t_1, t_2)$ is bandlimited. The Fourier transform $X_a(\Omega_1, \Omega_2)$ of a bandlimited signal is equal to zero outside some region of finite extent in the $(\Omega_1, \Omega_2)$-plane. For simplicity, let us assume that the sampling periods $T_1$ and $T_2$ are chosen small enough so that

$$X_a(\Omega_1, \Omega_2) = 0 \qquad \text{for } |\Omega_1| \ge \frac{\pi}{T_1}, \quad |\Omega_2| \ge \frac{\pi}{T_2} \qquad (1.123)$$

Then equation (1.122) becomes simply

$$X(\Omega_1 T_1, \Omega_2 T_2) = \frac{1}{T_1 T_2} X_a(\Omega_1, \Omega_2) \qquad (1.124)$$

for

$$|\Omega_1| \leq \frac{\pi}{T_1} \quad \text{and} \quad |\Omega_2| \leq \frac{\pi}{T_2}$$

The values of $X(\Omega_1 T_1, \Omega_2 T_2)$ outside this region are given by the periodicity of $X(\Omega_1 T_1, \Omega_2 T_2)$.

In Figure 1.24(a), we see a sketch of the Fourier transform of a continuous bandlimited signal. Forming the periodic extension of this transform gives us the periodic function pictured in Figure 1.24(b).

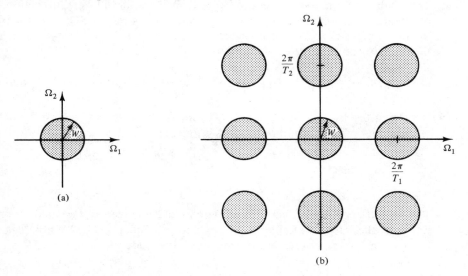

(a)

(b)

**Figure 1.24**  (a) Fourier transform of a continuous bandlimited signal. (b) Periodic extension of the transform.

As long as $X_a(\Omega_1, \Omega_2)$ satisfies equation (1.123), $X_a(\Omega_1, \Omega_2)$ can be recovered from $X(\Omega_1 T_1, \Omega_2 T_2)$ by inverting equation (1.124) to get

$$X_a(\Omega_1, \Omega_2) = \begin{cases} T_1 T_2 X(\Omega_1 T_1, \Omega_2 T_2), & |\Omega_1| < \dfrac{\pi}{T_1}, \ |\Omega_2| < \dfrac{\pi}{T_2} \\ 0, & \text{otherwise} \end{cases} \qquad (1.125)$$

Consequently, it is possible in this case to recover the continuous signal $x_a(t_1, t_2)$ from the discrete signal. To demonstrate this, we express $x_a(t_1, t_2)$ in terms of its Fourier transform.

$$\begin{aligned} x_a(t_1, t_2) &= \frac{1}{4\pi^2} \int_{-\infty}^{\infty} \int_{-\infty}^{\infty} X_a(\Omega_1, \Omega_2) \exp(j\Omega_1 t_1 + j\Omega_2 t_2) \, d\Omega_1 \, d\Omega_2 \\ &= \frac{1}{4\pi^2} \int_{-W_1}^{W_1} \int_{-W_2}^{W_2} T_1 T_2 X(\Omega_1 T_1, \Omega_2 T_2) \exp(j\Omega_1 t_1 + j\Omega_2 t_2) \, d\Omega_2 \, d\Omega_1 \end{aligned} \qquad (1.126)$$

For notational convenience, we have defined $W_1 \triangleq \pi/T_1$ and $W_2 \triangleq \pi/T_2$. Now, we proceed by expressing $X(\Omega_1 T_1, \Omega_2 T_2)$ in terms of $x(n_1, n_2)$.

$$x_a(t_1, t_2) = \frac{1}{4\pi^2} \int_{-W_1}^{W_1} \int_{-W_2}^{W_2} T_1 T_2 [\sum_{n_1} \sum_{n_2} x(n_1, n_2) \exp(-j\Omega_1 T_1 n_1 - j\Omega_2 T_2 n_2)]$$

$$\cdot \exp(j\Omega_1 t_1 + j\Omega_2 t_2) \, d\Omega_2 \, d\Omega_1$$

$$= \frac{T_1 T_2}{4\pi^2} \sum_{n_1} \sum_{n_2} x(n_1, n_2) \int_{-W_1}^{W_1} \int_{-W_2}^{W_2} \exp[j\Omega_1(t_1 - n_1 T_1) \qquad (1.127)$$

$$+ j\Omega_2(t_2 - n_2 T_2)] \cdot d\Omega_2 \, d\Omega_1$$

$$= \sum_{n_1} \sum_{n_2} x(n_1, n_2) \frac{\sin[W_1(t_1 - n_1 T_1)]}{W_1(t_1 - n_1 T_1)} \frac{\sin[W_2(t_2 - n_2 T_2)]}{W_2(t_2 - n_2 T_2)}$$

Equations (1.115), (1.125), and (1.127), taken together, form the basis of the 2-D *sampling theorem*. It states that a bandlimited continuous signal may be reconstructed from its sample values. The sampling periods $T_1$ and $T_2$ must be small enough, or equivalently the *sampling frequencies* $2W_1$ and $2W_2$ must be large enough, to ensure that condition (1.123) is true.

A continuous signal that is not bandlimited may still be sampled, of course, but in this case, equations (1.124) and (1.125) will not be true since contributions from other replicas of $X_a(\Omega_1, \Omega_2)$ in the periodic extension (1.122) will fold into the region $|\Omega_1 T_1| < \pi$, $|\Omega_2 T_2| < \pi$. As in one-dimensional signal processing, this condition is called *aliasing*, since high-frequency components of $X_a(\Omega_1, \Omega_2)$ will masquerade as low-frequency components in $X(\Omega_1 T_1, \Omega_2 T_2)$.

### 1.4.2 Periodic Sampling with Arbitrary Sampling Geometries

The concept of rectangular sampling can be generalized in a straightforward manner. If we define two linearly independent vectors $\mathbf{v}_1 = (v_{11}, v_{21})'$ and $\mathbf{v}_2 = (v_{12}, v_{22})'$, we can write the locations of a doubly periodic set of samples in the $(t_1, t_2)$-plane as

$$t_1 = v_{11} n_1 + v_{12} n_2 \qquad (1.128a)$$

$$t_2 = v_{21} n_1 + v_{22} n_2 \qquad (1.128b)$$

Using vector notation, we can express these relations as

$$\mathbf{t} = \mathbf{Vn}, \qquad (1.129)$$

where $\mathbf{t} = (t_1, t_2)'$, $\mathbf{n} = (n_1, n_2)'$, and $\mathbf{V}$ is a matrix made up of the sampling vectors $\mathbf{v}_1$ and $\mathbf{v}_2$.

$$\mathbf{V} = [\mathbf{v}_1 \mid \mathbf{v}_2] \qquad (1.130)$$

Because $\mathbf{v}_1$ and $\mathbf{v}_2$ were chosen to be linearly independent, the determinant of $\mathbf{V}$ is nonzero. We shall refer to $\mathbf{V}$ as the *sampling matrix*.

Sampling a continuous signal $x_a(\mathbf{t})$ produces the discrete signal

$$x(\mathbf{n}) \triangleq x_a(\mathbf{Vn}) \qquad (1.131)$$

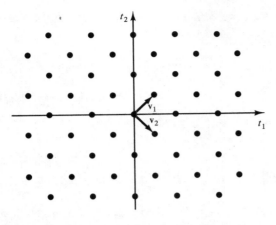

**Figure 1.25** Sample locations in the $(t_1, t_2)$-plane determined by the vectors $\mathbf{v}_1$ and $\mathbf{v}_2$ which comprise the sampling matrix $\mathbf{V}$.

The sampling locations are shown in Figure 1.25. Again, we can ask the questions: How are the Fourier transforms of $x(\mathbf{n})$ and $x_a(\mathbf{t})$ related? Under what circumstances can we reconstruct $x_a(\mathbf{t})$ from its samples $x(\mathbf{n})$? We proceed as before, by first defining the two-dimensional Fourier transform

$$X_a(\mathbf{\Omega}) \triangleq \int_{-\infty}^{\infty} x_a(\mathbf{t}) \exp{(-j\mathbf{\Omega}'\mathbf{t})} \, d\mathbf{t} \qquad (1.132a)$$

and observing that

$$x_a(\mathbf{t}) = \frac{1}{4\pi^2} \int_{-\infty}^{\infty} X_a(\mathbf{\Omega}) \exp{(j\mathbf{\Omega}'\mathbf{t})} \, d\mathbf{\Omega} \qquad (1.132b)$$

where the frequency vector $\mathbf{\Omega}$ is given by $(\Omega_1, \Omega_2)'$. [Note that the integrals in (1.132) are really double integrals because the differential quantities $d\mathbf{t}$ and $d\mathbf{\Omega}$ are vectors.] The Fourier transform of $x(\mathbf{n})$ can be written in vector notation as

$$X(\mathbf{\omega}) \triangleq \sum_{\mathbf{n}} x(\mathbf{n}) \exp{(-j\mathbf{\omega}'\mathbf{n})} \qquad (1.133a)$$

where $\mathbf{\omega} = (\omega_1, \omega_2)'$. Then

$$x(\mathbf{n}) = \frac{1}{4\pi^2} \int_{-\pi}^{\pi} X(\mathbf{\omega}) \exp{(j\mathbf{\omega}'\mathbf{n})} \, d\mathbf{\omega} \qquad (1.133b)$$

Since $x(\mathbf{n})$ is obtained from $x_a(\mathbf{t})$ by sampling, we can write

$$x(\mathbf{n}) = x_a(\mathbf{V}\mathbf{n}) = \frac{1}{4\pi^2} \int_{-\infty}^{\infty} X_a(\mathbf{\Omega}) \exp{(j\mathbf{\Omega}' \mathbf{V}\mathbf{n})} \, d\mathbf{\Omega}$$

Making the substitution $\mathbf{\omega} = \mathbf{V}'\mathbf{\Omega}$ yields

$$x(\mathbf{n}) = \frac{1}{4\pi^2} \int_{-\infty}^{\infty} \frac{1}{|\det \mathbf{V}|} X_a(\mathbf{V}'^{-1}\mathbf{\omega}) \exp{(j\mathbf{\omega}'\mathbf{n})} \, d\mathbf{\omega} \qquad (1.134)$$

As before, we perform the integration over the $\mathbf{\omega}$-plane as an infinite series of integrations over square areas. The result is analogous to equation (1.120).

$$x(\mathbf{n}) = \frac{1}{4\pi^2} \int_{-\pi}^{\pi} \left[ \frac{1}{|\det \mathbf{V}|} \sum_{\mathbf{k}} X_a(\mathbf{V}'^{-1}(\boldsymbol{\omega} - 2\pi\mathbf{k})) \cdot \exp(j\boldsymbol{\omega}'\mathbf{n}) \exp(-j2\pi\mathbf{k}'\mathbf{n}) \right] d\boldsymbol{\omega}$$

(1.135)

where $\mathbf{k}$ is an integer-valued vector. Again the second exponential factor is always equal to 1, so that a comparison of equations (1.135) and (1.133b) implies that

$$X(\boldsymbol{\omega}) = \frac{1}{|\det \mathbf{V}|} \sum_{\mathbf{k}} X_a(\mathbf{V}'^{-1}(\boldsymbol{\omega} - 2\pi\mathbf{k}))$$

(1.136)

or equivalently,

$$X(\mathbf{V}'\boldsymbol{\Omega}) = \frac{1}{|\det \mathbf{V}|} \sum_{\mathbf{k}} X_a(\boldsymbol{\Omega} - \mathbf{Uk})$$

(1.137)

where $\mathbf{U}$ is a matrix that satisfies

$$\mathbf{U}'\mathbf{V} = 2\pi\mathbf{I}$$

(1.138)

and $\mathbf{I}$ is the $2 \times 2$ identity matrix. Equation (1.137) provides the desired relation between the Fourier transforms of $x(\mathbf{n})$ and $x_a(\mathbf{t})$.

When rectangular sampling is used, the matrices $\mathbf{V}$ and $\mathbf{U}$ become

$$\mathbf{V} = \begin{bmatrix} T_1 & 0 \\ 0 & T_2 \end{bmatrix}; \quad \det \mathbf{V} = T_1 T_2$$

$$\mathbf{U} = \begin{bmatrix} \dfrac{2\pi}{T_1} & 0 \\ 0 & \dfrac{2\pi}{T_2} \end{bmatrix} = \begin{bmatrix} 2W_1 & 0 \\ 0 & 2W_2 \end{bmatrix}$$

and equation (1.137) reduces to equation (1.122).

$X(\mathbf{V}'\boldsymbol{\Omega})$ can again be interpreted as a periodic extension of $X_a(\boldsymbol{\Omega})$, but now the periodicity is described by the general matrix $\mathbf{U}$, which can be thought of as a set of two periodicity vectors $\mathbf{u}_1$ and $\mathbf{u}_2$.

$$\mathbf{U} = [\mathbf{u}_1 \mid \mathbf{u}_2]$$

(1.139)

Since $X(\boldsymbol{\omega})$ is periodic in both $\omega_1$ and $\omega_2$ with a period of $2\pi$, it follows that $X(\mathbf{V}'\boldsymbol{\Omega})$ is periodic in $\boldsymbol{\Omega}$ with the periodicity matrix $\mathbf{U}$.

$$X(\mathbf{V}'(\boldsymbol{\Omega} + \mathbf{Uk})) = X(\mathbf{V}'\boldsymbol{\Omega} + 2\pi\mathbf{k}) = X(\mathbf{V}'\boldsymbol{\Omega})$$

Consider the continuous signal $x_a(\mathbf{t})$ whose Fourier transform was sketched in Figure 1.24(a). If we sample $x_a(\mathbf{t})$ with the sampling matrix

$$\mathbf{V} = \begin{bmatrix} 1 & 1 \\ 1 & -1 \end{bmatrix}$$

(1.140)

which corresponds to the sample locations shown in Figure 1.25, $X_a(\boldsymbol{\Omega})$ will be periodically extended with a periodicity matrix $\mathbf{U}$ given by

$$\mathbf{U} = \begin{bmatrix} \pi & \pi \\ \pi & -\pi \end{bmatrix}$$

(1.141)

Thus $X(\mathbf{V}'\boldsymbol{\Omega})$ given by (1.137) has the form shown in Figure 1.26 when plotted as a function of $\boldsymbol{\Omega}$.

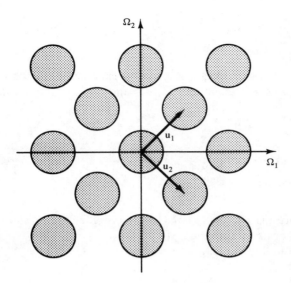

**Figure 1.26** Periodic function $X(\mathbf{V}'\boldsymbol{\Omega})$, plotted as a function of $\boldsymbol{\Omega}$ for the case where the sampling matrix $\mathbf{V}$ is given by equation (1.140).

At this point, we again consider the important case where the continuous signal $x_a(\mathbf{t})$ is bandlimited. The Fourier transform $X_a(\boldsymbol{\Omega})$ is identically zero outside a region of finite extent $B$, which we shall call the *baseband*. By varying the sampling matrix $\mathbf{V}$, we can adjust the periodicity matrix $\mathbf{U}$ so that there is no overlap among the periodically repeated versions of $X_a(\boldsymbol{\Omega})$ on the right side of equation (1.137).

By varying $\mathbf{U}$ in this manner, we ensure that there is no aliasing. Then equation (1.137) becomes simply

$$X(\mathbf{V}'\boldsymbol{\Omega}) = \frac{1}{|\det \mathbf{V}|} X_a(\boldsymbol{\Omega}) \tag{1.142}$$

for values of $\mathbf{V}'\boldsymbol{\Omega}$ lying in the square centered on the origin with sides of length $2\pi$. Consequently, $X_a(\boldsymbol{\Omega})$ can be recovered from $X(\mathbf{V}'\boldsymbol{\Omega})$ and therefore the continuous signal $x_a(\mathbf{t})$ can be recovered from the sequence $x(\mathbf{n})$. We can write

$$X_a(\boldsymbol{\Omega}) = \begin{cases} |\det \mathbf{V}| \cdot X(\mathbf{V}'\boldsymbol{\Omega}), & \boldsymbol{\Omega} \in B \\ 0, & \text{otherwise} \end{cases} \tag{1.143}$$

By taking the inverse Fourier transform of both sides and expressing $X(\mathbf{V}'\boldsymbol{\Omega})$ in terms of the sample values $x(\mathbf{n}_1, \mathbf{n}_2)$, we get an equation analogous to (1.127): namely,

$$x_a(\mathbf{t}) = \frac{|\det \mathbf{V}|}{4\pi^2} \sum_{\mathbf{n}} x(\mathbf{n}) \int_B \exp\left[ j\boldsymbol{\Omega}'(\mathbf{t} - \mathbf{V}\mathbf{n}) \right] d\boldsymbol{\Omega} \tag{1.144}$$

where the integral is over the baseband $B$ in the frequency plane. For notational ease, let us rewrite (1.144) as

$$x_a(\mathbf{t}) = \sum_{\mathbf{n}} x(\mathbf{n}) f(\mathbf{t} - \mathbf{V}\mathbf{n}) \tag{1.145}$$

where

$$f(\mathbf{t}) = \frac{|\det \mathbf{V}|}{4\pi^2} \int_B \exp\left( j\boldsymbol{\Omega}'\mathbf{t} \right) d\boldsymbol{\Omega} \tag{1.146}$$

The *interpolating function* $f(\mathbf{t})$ allows us to construct the values of $x_a(\mathbf{t})$ at points in between the sample locations given by $\mathbf{t} = \mathbf{Vn}$.

Let us summarize this more general derivation. We have a bandlimited, continuous signal $x_a(\mathbf{t})$. Its Fourier transform $X_a(\mathbf{\Omega})$ is zero outside region $B$ in the $\mathbf{\Omega}$ frequency plane. We want to represent $x_a(\mathbf{t})$ by a sequence of sample values $x(\mathbf{n})$. To do this, we must find an appropriate sampling matrix $\mathbf{V}$ that will allow us to recover $x_a(\mathbf{t})$ from $x(\mathbf{n})$ using equation (1.145).

From equation (1.137) we see that the Fourier transform of the discrete signal $x(\mathbf{n})$ is equal to a scaled and periodically repeated version of $X_a(\mathbf{\Omega})$. The periodicity matrix $\mathbf{U}$ represents the two linearly independent directions in which $X_a(\mathbf{\Omega})$ will be replicated. To achieve our goal, we must choose $\mathbf{U}$ so that there is no overlap among the replicated versions of $X_a(\mathbf{\Omega})$, thus avoiding aliasing. In this case, $X_a(\mathbf{\Omega})$ satisfies equation (1.143).

The choice of the periodicity matrix $\mathbf{U}$ determines the sampling matrix $\mathbf{V}$, since $\mathbf{U}$ and $\mathbf{V}$ are related by equation (1.138). The choice of $\mathbf{U}$ is not unique, in general; an adequate density of samples in the $\mathbf{t}$-plane will allow us to represent any bandlimited signal with several sampling geometries. However, it is often desirable to represent $x_a(\mathbf{t})$ with as few samples as possible. It can be shown that the density of samples per unit area is given by $1/|\det \mathbf{V}|$. Minimizing this quantity is equivalent to minimizing $|\det \mathbf{U}|$. Therefore, to provide an efficient sampling scheme for a bandlimited signal, we choose the periodicity matrix $\mathbf{U}$ which has the smallest value of $|\det \mathbf{U}|$ and which avoids aliasing for the particular shape of the signal's baseband $B$.

The derivation of the generalized sampling theorem is easily extended to $M$-dimensional signals. Since we have used vector notation, the only significant change in the equations would be the replacement of the constant $4\pi^2$ by the more general constant $(2\pi)^M$.

### Example 7

The preparation of marine seismic maps can result in data which is periodically sampled in an unusual fashion. In Figure 1.27, a boat towing a line of sensors moves with a speed $B$ knots in a direction perpendicular to an ocean current of speed $C$ knots. The sensors are uniformly spaced along the line with a spacing $D$ and all are

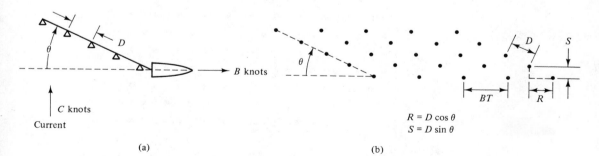

$$R = D \cos \theta$$
$$S = D \sin \theta$$

(a)                    (b)

**Figure 1.27**   (a) Scenario for Example 7. (b) Resulting sampling raster.

periodically sampled and digitally recorded. Denote the temporal sampling period by
$T$. How is the underlying spatial process sampled?

To a first approximation, the sensors remain in a straight line, but that line is
directed away from the direction of the boat movement, at an angle

$$\theta = \tan^{-1} \frac{C}{B}$$

The resulting sampling grid is shown in Figure 1.27(b). This grid corresponds to the
sampling matrix

$$\mathbf{V} = \begin{bmatrix} BT & -D\cos\theta \\ 0 & D\sin\theta \end{bmatrix} = \begin{bmatrix} BT & -\dfrac{BD}{\sqrt{B^2 + C^2}} \\ 0 & \dfrac{CD}{\sqrt{B^2 + C^2}} \end{bmatrix}$$

### 1.4.3 Comparison of Rectangular and Hexagonal Sampling

For any bandlimited waveform, there is an infinite number of possible choices for
the periodicity matrix $\mathbf{U}$ and the sampling matrix $\mathbf{V}$. Among these numerous possibilities, however, only two sampling strategies are at all common—rectangular
sampling and hexagonal sampling. We have already studied rectangular sampling in
Section 1.4.1. It corresponds to the case where the sampling matrix $\mathbf{V}$ is diagonal.

Hexagonal sampling is the term we give to sampling matrices of the form

$$\mathbf{V} = \begin{bmatrix} T_1 & T_1 \\ T_2 & -T_2 \end{bmatrix} \tag{1.147}$$

A set of sample locations for a hexagonal sampling matrix is shown in Figure 1.28.

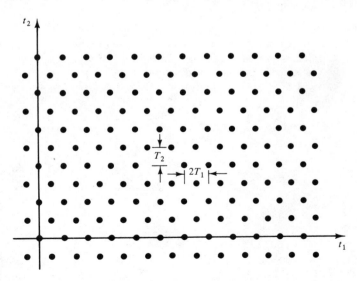

**Figure 1.28**   Hexagonal sampling raster.

The alternate rows of this raster are identical, and the odd-indexed rows are staggered one-half sample interval with respect to the even-indexed rows. The term "hexagonal" is used since each sample location will have six nearest neighbors when $T_2 = T_1\sqrt{3}$.

It is easy to show that the corresponding periodicity matrix $\mathbf{U}$ has the same form, namely

$$\mathbf{U} = \begin{bmatrix} u_1 & u_1 \\ u_2 & -u_2 \end{bmatrix} \tag{1.148}$$

where $u_1 \triangleq \pi/T_1$ and $u_2 \triangleq \pi/T_2$.

There are several regions which, when periodically extended according to equation (1.137), cover the $(\Omega_1, \Omega_2)$-plane with no overlap. Figure 1.29 shows four examples. We are particularly interested in the hexagonal region, which is shown in

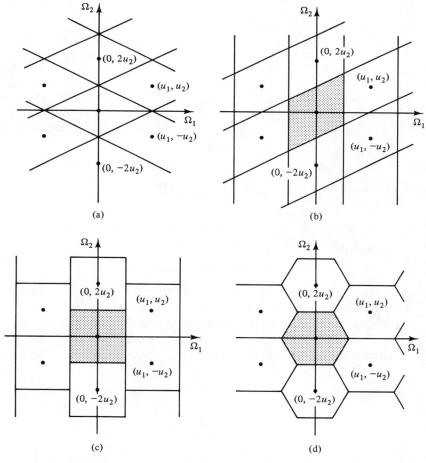

**Figure 1.29**  Four examples of regions which, when periodically extended in a hexagonal fashion, cover the frequency plane.

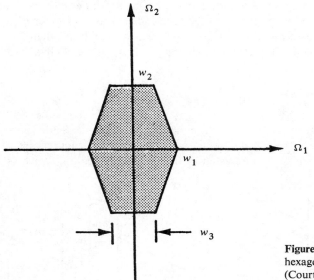

**Figure 1.30** Baseband of a hexagonally bandlimited waveform. (Courtesy of Russell M. Mersereau, *Proc. IEEE,* © 1979 IEEE.)

detail in Figure 1.30 with parameters $W_1$, $W_2$, and $W_3$ [6]. These parameters are related to the periodicity parameters $u_1$ and $u_2$ as follows:

$$u_1 = W_1 + \tfrac{1}{2}W_3 \tag{1.149a}$$

$$u_2 = W_2 \tag{1.149b}$$

Let us compare the relative efficiencies of rectangular and hexagonal sampling schemes when they are used to sample a 2-D continuous signal which is circularly bandlimited. The 2-D Fourier transform of such a signal has the property

$$X_a(\Omega_1, \Omega_2) = 0 \quad \text{for} \quad \Omega_1^2 + \Omega_2^2 \geq W^2 \tag{1.150}$$

The circular baseband, such as the one shown earlier in Figure 1.24(a), can be inscribed in a square with sides of length $2W$ or a hexagon with sides of length $2W/\sqrt{3}$. Consequently, this signal can be embedded in either a square or hexagonal baseband.

Periodic replication of the square baseband corresponds to sampling the continuous signal on a rectangular raster. In this case, the sampling matrix $\mathbf{V}$ is given by

$$\mathbf{V}_{\text{rect}} = \begin{bmatrix} \dfrac{\pi}{W} & 0 \\[2mm] 0 & \dfrac{\pi}{W} \end{bmatrix} \tag{1.151}$$

and

$$|\det \mathbf{V}_{\text{rect}}| = \frac{\pi^2}{W^2} \tag{1.152}$$

Alternatively, we could use the hexagonal sampling matrix

$$\mathbf{V}_{hex} = \begin{bmatrix} \dfrac{\pi}{W\sqrt{3}} & \dfrac{\pi}{W\sqrt{3}} \\ \dfrac{\pi}{W} & \dfrac{-\pi}{W} \end{bmatrix} \qquad (1.153)$$

for which

$$|\det \mathbf{V}_{hex}| = \frac{\pi^2}{W^2} \frac{2}{\sqrt{3}} \qquad (1.154)$$

Since the sampling density is proportional to $1/|\det \mathbf{V}|$, we see by taking the ratio of equations (1.152) and (1.154) that hexagonal sampling requires 13.4% fewer samples than rectangular sampling to represent the same circularly bandlimited continuous signal. In fact, it can be shown [5] that there is no more efficient sampling scheme for circularly bandlimited signals than hexagonal sampling.

The notions of hexagonal and rectangular sampling can be generalized to the $M$-dimensional case. Rectangular sampling corresponds to sampling on a hypercubic lattice. The sampling locations for the $M$-dimensional generalization of hexagonal sampling are at the centers of densely packed $M$-dimensional hyperspheres. In Table 1.1 the relative efficiency of hexagonal sampling to rectangular sampling is shown for various values of $M$ [5]. It should be noted that in the 1-D case, both rectangular and hexagonal sampling become 1-D periodic sampling. In Section 1.5 we consider a few properties of hexagonally sampled signals and systems.

**TABLE 1.1**  RATIO OF THE EFFICIENCY OF AN
$M$-DIMENSIONAL CUBIC LATTICE TO A
HYPERSPHERICAL LATTICE AS A FUNCTION
OF INCREASING $M$

| $M$ | Efficiency |
|:---:|:---:|
| 1 | 1.000 |
| 2 | 0.866 |
| 3 | 0.705 |
| 4 | 0.499 |
| 5 | 0.353 |
| 6 | 0.217 |
| 7 | 0.125 |
| 8 | 0.062 |

## *1.5 PROCESSING CONTINUOUS SIGNALS WITH DISCRETE SYSTEMS

In converting a continuous signal to a discrete sequence by periodic sampling, we allow ourselves the flexibility of processing signals with digital systems. Although we saw in the preceding section that there are different geometries for sampling multi-dimensional signals, most digital signal processing algorithms assume that sample values are taken on a rectangular sampling grid. This is somewhat unfortunate, for, as we have seen, an alternative sampling geometry can often reduce the sampling density and hence the amount of digital memory needed to store the representation of a signal. Furthermore, a lower sampling density can lead to a reduction in the amount of computation needed to implement a filtering operation.

In this section we examine the use of a discrete system to process continuous multidimensional signals in a very general and cursory manner. Our goal is to highlight some of the issues associated with multidimensional signal processing. The reader who is interested in learning more about the specific case of signals sampled and processed on a hexagonal grid is directed to [6].

### 1.5.1 Relationship between the System Input and Output Signals

Consider the system shown in Figure 1.31. A bandlimited signal $x_a(t)$ is sampled with a sampling matrix $V$ to generate the discrete signal $x(n)$. The discrete signal is processed by a multidimensional LSI system, such as those discussed in Section 1.2, which is

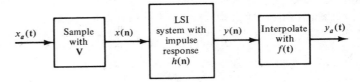

**Figure 1.31**  Discrete multidimensional system that can be used to process band-limited continuous signals.

characterized by the impulse response $h(n)$. The result is an output sequence $y(n)$ given by the convolution sum

$$y(n) = \sum_k x(k)h(n - k) = \sum_k h(k)x(n - k) \qquad (1.155)$$

This output sequence is used to generate a continuous signal $y_a(t)$ by applying the interpolation formula (1.145). In this manner, we can use a discrete LSI system to process a continuous bandlimited signal.

The sampling matrix $V$ is determined by the periodicity matrix $U$ through the relation $V' = 2\pi U^{-1}$. The periodicity matrix, in turn, is chosen so that $X_a(\Omega)$, which is zero outside some finite-area region $B$, can be recovered from its periodic extension. $U$ should also be chosen so that it has a minimum determinant. In the optimal case, the periodic extension of the region $B$ will cover the $\Omega$-plane without gaps or areas of overlap.

By combining equations (1.145) and (1.155), we can express the continuous output signal as

$$y_a(t) = \sum_n y(n)f(t - Vn)$$
$$= \sum_n \sum_k h(k)x(n - k)f(t - Vn) \qquad (1.156)$$
$$= \sum_k h(k)x_a(t - Vk)$$

Equation (1.156) represents the relationship between the continuous input signal $x_a(t)$ and the continuous output signal $y_a(t)$. Although it is tempting to try to define an impulse response for the entire system shown in Figure 1.31, an impulsive signal is not an admissible input since it is not bandlimited. Instead, let the input to the system be the interpolation function $f(t)$, which appeared in (1.146). The function $f(t)$ is limited to the baseband $B$ and, when sampled with sampling matrix $\mathbf{V}$, it gives the unit impulse sequence $\delta(\mathbf{n})$. If $f(t)$ is used to drive the system, the continuous output signal will be

$$y_a(t) = \sum_{\mathbf{k}} h(\mathbf{k}) f(t - \mathbf{Vk})$$

$$\triangleq h_a(t) \tag{1.157}$$

The response $h_a(t)$ can be thought of as the system's *bandlimited impulse response*. Using the definition of $h_a(t)$, we can derive an alternate version of the input–output relation (1.156).

$$y_a(t) = \sum_{\mathbf{k}} x(\mathbf{k}) h_a(t - \mathbf{Vk}) \tag{1.158}$$

The form of the convolution sum (1.155) is independent of the sampling matrix $\mathbf{V}$. If that sum is used as a realization for a filter, which can be done if the filter's impulse response $h(\mathbf{n})$ has finite support, then the same hardware or software could be used to evaluate the convolution of sequences which represent bandlimited signals sampled on rectangular, hexagonal, or other periodic girds. There may be an important difference in the amount of computation which needs to be performed, however. For example, if $x(\mathbf{n})$ represents samples of a circularly bandlimited signal which has been sampled hexagonally, then 13.4% fewer samples are required in both $x(\mathbf{n})$ and $y(\mathbf{n})$ compared to the rectangularly sampled case. This results in 25% fewer additions and multiplications to be performed in evaluating the convolution sum.

### 1.5.2 System Frequency Response

An obvious question to ask is: What is the output of the system in Figure 1.31 if the input signal is a complex sinusoid? Let us set

$$x_a(t) = \exp(j\mathbf{\Omega}'t), \qquad \mathbf{\Omega} \in B \tag{1.159}$$

Then

$$x(\mathbf{n}) = \exp(j\mathbf{\Omega}'\mathbf{Vn}) \tag{1.160}$$

and

$$y(\mathbf{n}) = \sum_{\mathbf{k}} h(\mathbf{k}) x(\mathbf{n} - \mathbf{k})$$

$$= \exp(j\mathbf{\Omega}'\mathbf{Vn}) \left[ \sum_{\mathbf{k}} h(\mathbf{k}) \exp(-j\mathbf{\Omega}'\mathbf{Vk}) \right] \qquad (1.161)$$

$$= \exp(j\mathbf{\Omega}\mathbf{Vn}) H(\mathbf{V}'\mathbf{\Omega})$$

Finally, interpolating $y(\mathbf{n})$, we get

$$y_a(t) = H(\mathbf{V}'\mathbf{\Omega}) \exp(j\mathbf{\Omega}'t), \qquad \mathbf{\Omega} \in B \tag{1.162}$$

The response of the system to a complex sinusoid is a complex sinusoid of the same frequency but weighted by the complex gain factor $H(\mathbf{V}'\mathbf{\Omega})$.

Since the function $H(\mathbf{V}'\mathbf{\Omega})$ is frequently plotted as a function of $\mathbf{\Omega}$, it is convenient to define

$$H_V(\mathbf{\Omega}) \triangleq H(\mathbf{V}'\mathbf{\Omega}) = \sum_{\mathbf{k}} h(\mathbf{k}) \exp(-j\mathbf{\Omega}'\mathbf{V}\mathbf{k}) \qquad (1.163)$$

Using this definition and (1.156), we can relate the Fourier transforms of the continuous input and output signals, giving us

$$Y_a(\mathbf{\Omega}) = H_V(\mathbf{\Omega}) X_a(\mathbf{\Omega}) \qquad \text{for } \mathbf{\Omega} \in B \qquad (1.164)$$

Thus $H_V(\mathbf{\Omega})$ can be interpreted as the frequency response of the system shown in Figure 1.31.

### 1.5.3 Alternative Definition of the Fourier Transform for Discrete Signals

When dealing with discrete signals and systems, we customarily use the Fourier transform definition

$$X(\boldsymbol{\omega}) \triangleq \sum_{\mathbf{n}} x(\mathbf{n}) \exp(-j\boldsymbol{\omega}'\mathbf{n})$$

which we studied in Section 1.3. When we are interested in the relationship between continuous signals and the discrete signals obtained by sampling them, it is sometimes more convenient to make the substitution $\boldsymbol{\omega} = \mathbf{V}'\mathbf{\Omega}$ to obtain the alternative Fourier transform

$$X_V(\mathbf{\Omega}) \triangleq \sum_{\mathbf{n}} x(\mathbf{n}) \exp(-j\mathbf{\Omega}'\mathbf{V}\mathbf{n}) = X(\mathbf{V}'\mathbf{\Omega}) \qquad (1.165)$$

The inverse transform can be derived as follows:

$$\begin{aligned}
x(\mathbf{n}) &= \frac{1}{4\pi^2} \int_{-\pi}^{\pi} X(\boldsymbol{\omega}) \exp(j\boldsymbol{\omega}'\mathbf{n}) \, d\boldsymbol{\omega} \\
&= \frac{|\det \mathbf{V}|}{4\pi^2} \int_{B} X(\mathbf{V}'\mathbf{\Omega}) \exp(j\mathbf{\Omega}'\mathbf{V}\mathbf{n}) \, d\mathbf{\Omega} \qquad (1.166) \\
&= \frac{|\det \mathbf{V}|}{4\pi^2} \int_{B} X_V(\mathbf{\Omega}) \exp(j\mathbf{\Omega}'\mathbf{V}\mathbf{n}) \, d\mathbf{\Omega}
\end{aligned}$$

The region $B$ over which the integral is evaluated results from mapping the square region $-\pi \leq \omega_1 \leq \pi, -\pi \leq \omega_2 \leq \pi$ into the $\mathbf{\Omega}$-plane by the substitution of variables $\boldsymbol{\omega} = \mathbf{V}'\mathbf{\Omega}$.

The transform $X_V(\mathbf{\Omega})$ is periodic with the periodicity matrix $\mathbf{U}$, where, as before,

$$\mathbf{U}'\mathbf{V} = 2\pi\mathbf{I}$$

This is easily seen by writing

$$\begin{aligned}
X_V(\mathbf{\Omega} + \mathbf{U}\mathbf{k}) &= \sum_{\mathbf{n}} x(\mathbf{n}) \exp(-j\mathbf{\Omega}'\mathbf{V}\mathbf{n} - j\mathbf{k}'\mathbf{U}'\mathbf{V}\mathbf{n}) \\
&= \sum_{\mathbf{n}} x(\mathbf{n}) \exp(-j\mathbf{\Omega}'\mathbf{V}\mathbf{n}) \exp(-j2\pi\mathbf{k}'\mathbf{n}) \qquad (1.167)
\end{aligned}$$

Since $\mathbf{k}$ and $\mathbf{n}$ are integer-valued vectors, the second exponential factor is always equal to 1, and we see that

$$X_V(\mathbf{\Omega} + \mathbf{Uk}) = X_V(\mathbf{\Omega}) \qquad\qquad (1.168)$$

Because of this periodicity, the integral in equation (1.166) may be evaluated over any region in the $\mathbf{\Omega}$-plane which contains exactly one period of $X_V(\mathbf{\Omega})$ without altering the result.

Because $X_V(\mathbf{\Omega})$ and $X(\mathbf{\omega})$ are so closely related, the properties satisfied by $X(\mathbf{\omega})$ are reflected in the properties satisfied by $X_V(\mathbf{\Omega})$. As examples, we list a few of the properties below. Proofs are left to the reader.

**Linearity.**    If

$$x(\mathbf{n}) \longleftrightarrow X_V(\mathbf{\Omega}) \qquad \text{and} \qquad w(\mathbf{n}) \longleftrightarrow W_V(\mathbf{\Omega})$$

then

$$ax(\mathbf{n}) + bw(\mathbf{n}) \longleftrightarrow aX_V(\mathbf{\Omega}) + bW_V(\mathbf{\Omega})$$

for any complex constants $a$ and $b$.

**Shift property.**

$$x(\mathbf{n} - \mathbf{m}) \longleftrightarrow X_V(\mathbf{\Omega}) \exp(-j\mathbf{\Omega}'\mathbf{Vm})$$

**Convolution theorem.**

$$x(\mathbf{n}) ** h(\mathbf{n}) \longleftrightarrow X_V(\mathbf{\Omega})H_V(\mathbf{\Omega})$$

**Parseval's theorem.**

$$\sum_{\mathbf{n}} x(\mathbf{n})w^*(\mathbf{n}) = \frac{|\det \mathbf{V}|}{4\pi^2} \int_B X_V(\mathbf{\Omega})\, W_V^*(\mathbf{\Omega})\; d\mathbf{\Omega}$$

## PROBLEMS

**1.1. (a)** Determine a periodicity matrix that describes the periodicity of the array shown in Figure P1.1.

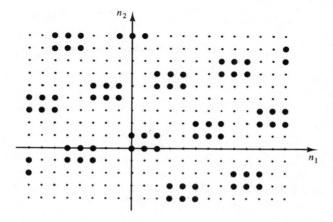

**Figure P1.1**

   **(b)** Determine a second periodicity matrix for this array.

   **(c)** Show that the absolute values of the determinants of these two arrays are equal.

**1.2.** For each of the systems described below, determine whether the system is (1) linear and (2) shift invariant.

   **(a)** $y(n_1, n_2) = x(n_1, n_2)x(n_1 - N, n_2)$

   **(b)** $y(n_1, n_2) = \sum\limits_{k_2=-\infty}^{\infty} x(n_1, k_2)$

   **(c)** $y(n_1, n_2) = \sum\limits_{k_2=-1}^{1} x(n_1, k_2)$

**1.3.** Consider the sequence $x$ defined by

$$x(n_1, n_2) = \begin{cases} 1, & n_1 \geq 0, \quad n_2 \geq n_1 \\ 0, & \text{otherwise} \end{cases}$$

Determine the convolution of $x$ with itself.

**1.4.** Through the use of the convolution sum, verify that the operation of convolution is

   **(a)** Commutative, that is, $x ** y = y ** x$.

   **(b)** Associative, that is, $(x ** y) ** z = x ** (y ** z)$.

   **(c)** Distributive, that is, $x ** (y + z) = x ** y + x ** z$.

**1.5.** Let $x$ and $h$ be the two sequences of finite extent depicted in Figure 1.14. If each sequence is 1 within its region of support and 0 outside it, evaluate the convolution of $x$ with $h$. The shape of the region of support for that convolution was determined in Section 1.2.4.

**1.6.** **(a)** Convolve the sequence $x(n_1, n_2) = a^{n_1}b^{n_2}u(n_1, n_2)$ with the sequence

$$y(n_1, n_2) = \sum\limits_{r=-\infty}^{\infty} \delta(n_1 - rN_1, n_2 - rN_2)$$

   **(b)** Convolve the sequence $x$ above with the sequence shown in Figure P1.6. (The dark circles represent samples of value 1, the dots represent samples of value zero.) [*Hint:* Use the result of part (a).]

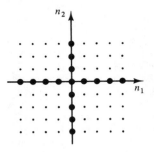

**Figure P1.6**

**1.7.** In this chapter we showed that an arbitrary 2-D array could be expressed as a linear combination of shifted impulse arrays. Using this representation we then showed that the output of a linear shift-invariant system could be expressed as the same linear combination of impulse *responses* of the system. A similar result can be obtained using step responses.

(a) Consider an arbitrary array $x(n_1, n_2)$. Show that $x$ can be expressed as an infinite linear combination of 2-D step functions.

(b) Suppose now that $x$ is the input to a linear shift-invariant system with step response $s(n_1, n_2)$. Derive an expression for the output of the system in terms of the sample values of $s$ and $x$.

(c) Is the operation of "step convolution" commutative? That is, would the answer to part (b) be the same if the arrays $s$ and $x$ were interchanged?

**1.8.** (a) All that is known about the sequences $x(n_1, n_2)$ and $y(n_1, n_2)$ is the fact that each has a region of support which is restricted to the first quadrant. That is,

$$x(n_1, n_2) = 0 \quad \text{if } n_1 < 0 \quad \text{or} \quad n_2 < 0$$

$$y(n_1, n_2) = 0 \quad \text{if } n_1 < 0 \quad \text{or} \quad n_2 < 0$$

Show that the region of support of their convolution is also limited to that quadrant.

(b) Repeat part (a) assuming this time that $x$ and $y$ are each confined to the third quadrant, $\{(n_1, n_2): \quad n_1 \leq 0 \text{ and } n_2 \leq 0\}$.

(c) Suppose now that $x$ and $y$ are each one-quadrant sequences but that each has support on a different quadrant. What statements, if any, can be made about the region of support of their convolution?

**1.9.** Consider two 2-D sequences that are separable:

$$\alpha(n_1, n_2) = a(n_1)b(n_2)$$

$$\beta(n_1, n_2) = c(n_1)d(n_2)$$

(a) Show that their convolution is a separable signal.

(b) Express that convolution in terms of $a$, $b$, $c$, and $d$.

**1.10.** (a) Two linear shift-invariant systems are connected in parallel as shown in Figure P1.10(a). Show that the overall system is linear and shift invariant.

(b) Two linear shift-invariant systems are connected in cascade as shown in Figure P1.10(b). Show that the overall system is linear and shift invariant.

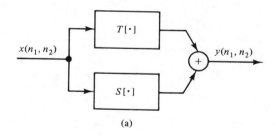

(a)

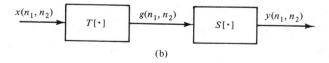

(b)

**Figure P1.10**

**1.11.** Prove the following statement:

A separable LSI system with impulse response

$$h(n_1, n_2) = f(n_1)g(n_2)$$

is stable if and only if $f(n_1)$ and $g(n_2)$ are absolutely summable sequences.

**1.12.** Show that the two-dimensional frequency response is periodic. That is, show that

$$H(\omega_1 + 2\pi k_1, \omega_2 + 2\pi k_2) = H(\omega_1, \omega_2)$$

for any ordered pair of integers $(k_1, k_2)$.

**1.13.** (a) Show that the frequency response of a separable system is a separable function; that is, if

$$h(n_1, n_2) = f(n_1)g(n_2)$$

then

$$H(\omega_1, \omega_2) = F(\omega_1)G(\omega_2)$$

where

$$F(\omega) = \sum_n f(n) \exp(-j\omega n)$$

and

$$G(\omega) = \sum_n g(n) \exp(-j\omega n)$$

(b) Show that this property holds for $M$-dimensional systems.
(c) Determine the Fourier transform of the impulse response

$$h(n_1, n_2, n_3) = f(n_1)g(n_2, n_3)$$

in terms of $F(\omega_1)$ and $G(\omega_2, \omega_3)$.

**1.14.** Find the impulse response of a filter with frequency response (one period)

$$H(\omega_1, \omega_2, \omega_3) = \begin{cases} 1, & |\omega_1| < W, \quad |\omega_2| < W, \quad |\omega_3| < W \\ 0, & \text{otherwise} \end{cases}$$

**1.15.** Find the Fourier transforms of the following sequences.
(a) $x(n_1, n_2) = a^{2n_1 + n_2}u(n_1, n_2)$, $\quad |a| < 1$
(b) $x(n_1, n_2) = a^{n_1}b^{n_2}\delta(4n_1 - n_2)u(n_1)$, $\quad |a| < 1$, $\quad |b| < 1$
(c) $x(n_1, n_2, n_3) = a^{n_1}v(n_2, n_3)u(n_1)$, $\quad |a| < 1$

where

$$v(n_2, n_3) = \begin{cases} 1, & 0 \le n_2 \le N - 1, \quad 0 \le n_3 \le M - 1 \\ 0, & \text{otherwise} \end{cases}$$

**1.16.** If $X(\omega_1, \omega_2)$ is the Fourier transform of $x(n_1, n_2)$, find the Fourier transform of the sequence

$$x(an_1 + bn_2, cn_1 + dn_2), \quad ad - bc = 1$$

if $a$, $b$, $c$, and $d$ are integers.

**1.17.** Consider 2-D signals with regions of support as shown in Figure P1.17. If each sequence is 1 on the samples marked with heavy dots and zero on the samples with the light dots, for which sequences is

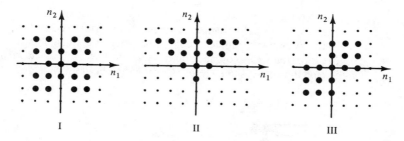

Figure P1.17

(a) $X(\omega_1, \omega_2)$ real?

(b) $X(\omega_1, \omega_2) = X(\omega_2, \omega_1)$?

(c) $X(\omega_1, \omega_2) = X(-\omega_1, \omega_2)$?

**1.18.** Let

$$y(n_1, n_2) = \sum_{k_1} \sum_{k_2} x(k_1, k_2) h(n_1 - k_1, n_2 - k_2)$$

By using the Fourier transforms of $x(n_1, n_2)$ and $h(n_1, n_2)$, show that

$$y(n_1, n_2) = \frac{1}{4\pi^2} \int_{-\pi}^{\pi} \int_{-\pi}^{\pi} H(\omega_1, \omega_2) X(\omega_1, \omega_2) \exp(j\omega_1 n_1 + j\omega_2 n_2)\, d\omega_1\, d\omega_2$$

**1.19.** (a) If $h \longleftrightarrow H$ is a Fourier transform pair, what is the Fourier transform of $h^*(m_1 - n_1, m_2 - n_2)$ where $m_1$ and $m_2$ are integer constants?

(b) Prove Parseval's theorem:

$$\sum_{n_1} \sum_{n_2} x(n_1, n_2) w^*(n_1, n_2) = \frac{1}{4\pi^2} \int_{-\pi}^{\pi} \int_{-\pi}^{\pi} X(\omega_1, \omega_2) W^*(\omega_1, \omega_2)\, d\omega_1\, d\omega_2 \quad \text{(P1.19)}$$

(c) When $w(n_1, n_2) = h^*(m_1 - n_1, m_2 - n_2)$, what does the right side of equation (P1.19) become?

**1.20.** If $h(n_1, n_2, n_3)$ is the impulse response corresponding to the ideal spherical lowpass filter for which one period of the frequency response is

$$H(\omega_1, \omega_2, \omega_3) = \begin{cases} 1, & \omega_1^2 + \omega_2^2 + \omega_3^2 < W^2 \\ 0, & \text{otherwise} \end{cases}$$

determine the value of $S$ if

$$S = \sum_{n_1} \sum_{n_2} \sum_{n_3} h^2(n_1, n_2, n_3)$$

**1.21.** (a) Consider a sequence $x(n_1, n_2)$ of the form

$$x(n_1, n_2) = \sum_{p=0}^{N-1} \delta(n_1 - pm_1, n_2 - pm_2)$$

Sketch this sequence. Your sketch should be labeled in terms of the parameters $m_1$ and $m_2$.

(b) Determine and sketch the Fourier transform of this sequence.

(c) Now consider a 2-D LSI digital filter with the frequency response shown in Figure P1.21(a) (the shaded region has value 1, the white region has value 0). The input to this system is a sampled photograph which is completely white (sample values = 0) except for a series of black lines (sample values = 1) as shown in Figure P1.21(b). Sketch the output signal. [*Hint:* Consider the Fourier transform of Figure P1.21(b).]

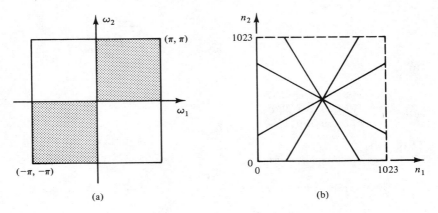

(a)                                        (b)

**Figure P1.21**

**1.22.** Consider an analog signal that is bandlimited. Its Fourier transform is nonzero over the shaded region in Figure P1.22.

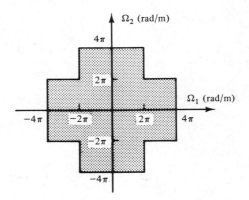

**Figure P1.22**

(a) If this signal is sampled rectangularly, find the minimum sampling density (in samples per square meter) which will permit an exact reconstruction of the analog waveform.

(b) Repeat part (a) if the signal is sampled hexagonally.

(c) The minimum sampling density possible with any sampling strategy is 12 samples per square meter. Sketch the sampling raster that corresponds to this optimal case.

**1.23.** Analog bandlimited waveforms have Fourier transforms with the regions of support indicated in Figure P1.23. For each determine the minimum sampling density (in samples per square meter) which will permit an exact reconstruction of the analog waveform. In each case sketch the optimal sampling raster.

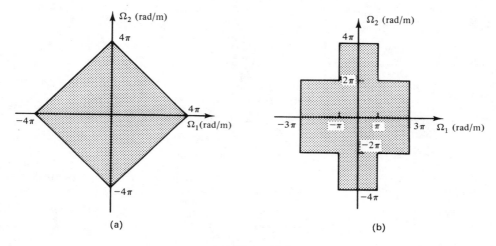

(a)                                                    (b)

**Figure P1.23**

**1.24.** Determine the convolution of the hexagonally sampled signal shown in Figure P1.24 with itself.

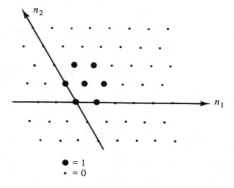

**Figure P1.24**

**1.25.** Find the impulse response of the ideal hexagonal lowpass filter defined by

$$H(\omega_1, \omega_2) = \begin{cases} 1, & \omega_1^2 + \omega_2^2 \leq W^2 \\ 0, & \omega_1^2 + \omega_2^2 > W^2, \\ \text{periodic}, & \text{otherwise} \end{cases} \quad (\omega_1, \omega_2) \in R_H$$

where $R_H$ is the region of the Fourier plane sketched in Figure P1.25.

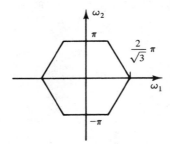

**Figure P1.25**

**1.26.** A bilinear system represents a generalization of the class of linear systems (Figure P1.26). Although they operate with 1-D inputs and outputs, they are closely related

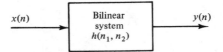

**Figure P1.26**

to 2-D systems. The output of a bilinear system can be expressed as

$$y(n) = \sum_{m_1=-\infty}^{\infty} \sum_{m_2=-\infty}^{\infty} x(m_1)x(m_2)h(n - m_1, n - m_2)$$

If a bilinear system is characterized by the 2-D Fourier transform of $h(n_1, n_2)$, denoted by $H(\omega_1, \omega_2)$, find the response of the system to the input

$$x(n) = a \exp{(j\omega_1 n)} + b \exp{(j\omega_2 n)}$$

## REFERENCES

1. Alan V. Oppenheim and Ronald W. Schafer, *Digital Signal Processing* (Englewood Cliffs, N.J.: Prentice-Hall, Inc., 1975).

2. L. R. Rabiner and B. Gold, *Theory and Applications of Digital Signal Processing* (Englewood Cliffs, N.J.: Prentice-Hall, Inc., 1975).

3. Alan V. Oppenheim and Alan S. Willsky with Ian T. Young; *Signals and Systems* (Englewood Cliffs, N. J.: Prentice-Hall, Inc., 1983).

4. Brian T. O'Connor and Thomas S. Huang, "Stability of General Two-Dimensional Recursive Filters," *IEEE Trans. Acoustics, Speech, and Signal Processing*, ASSP-26, no. 6 (Dec. 1978), 550–60.

5. D. P. Petersen and D. Middleton, "Sampling and Reconstruction of Wave-Number Limited Functions in *N*-Dimensional Euclidean Spaces," *Information and Control*, 5 (1962), 279–323.

6. Russell M. Mersereau, "The Processing of Hexagonally Sampled Two-Dimensional Signals," *Proc. IEEE*, 67, no. 6 (June 1979), 930–49.

# 2

# DISCRETE FOURIER ANALYSIS
# OF MULTIDIMENSIONAL SIGNALS

A fundamental advance in the field of digital signal processing occurred in the mid-1960s with the discovery of an efficient algorithm for computing the sampled spectrum of a signal [1,2]. The discovery of the fast Fourier transform (FFT) algorithm made it possible for the Fourier analysis of signals to be performed using digital hardware and digital computers instead of analog filter banks and spectrum analyzers. The computational savings made possible by the FFT are even more important in the multidimensional case, where the task is more complex and the volume of data is greater. In this chapter we examine the multidimensional *discrete Fourier transform* (DFT) and look at its relationship to the Fourier transform. We also develop several FFT algorithms for evaluating the discrete Fourier transform of a finite-extent sequence.

The multidimensional DFT is really two things: it is an exact Fourier representation for finite-extent sequences, and it is also a Fourier series expansion for multidimensional periodic sequences. This dual nature of the transform is important; it accounts for many of the properties that the DFT possesses. In addition, the latent periodicity of this transform can be exploited to develop computationally efficient algorithms for its evaluation. We begin our study of the DFT with a discussion of multidimensional periodic sequences and their Fourier series representations.

We saw in Chapter 1 that there are a number of ways in which a multidimensional sequence can be periodic. We shall see that this implies the existence of many possible DFT algorithms. For simplicity, we begin by restricting our attention to rectangularly periodic signals and the rectangular DFT. Later we consider the more general case.

## 2.1 DISCRETE FOURIER SERIES REPRESENTATION OF RECTANGULARLY PERIODIC SEQUENCES

A 2-D sequence $\tilde{x}(n_1, n_2)$ is *rectangularly periodic* if

$$\tilde{x}(n_1, n_2) = \tilde{x}(n_1 + N_1, n_2)$$
$$= \tilde{x}(n_1, n_2 + N_2) \tag{2.1}$$

for all $(n_1, n_2)$. The numbers $N_1$ and $N_2$ are positive integers. If they are the smallest possible positive integers for which equation (2.1) holds, they are called the *horizontal* and *vertical periods* of $\tilde{x}$. Any periodic array with horizontal and vertical periods $N_1$ and $N_2$ is completely specified by $N_1 N_2$ independent samples. This is readily seen by observing that every sample in the periodic array $\tilde{x}(n_1, n_2)$ is equal to one of the samples in the region $0 \leq n_1 \leq N_1 - 1$, $0 \leq n_2 \leq N_2 - 1$. We shall use the term *fundamental period* and the symbol $R_{N_1 N_2}$ to denote this rectangular region. Thus formally we define

$$R_{N_1 N_2} \triangleq \{(n_1, n_2): \quad 0 \leq n_1 \leq N_1 - 1, 0 \leq n_2 \leq N_2 - 1\} \tag{2.2}$$

A periodic sequence and its fundamental period are shown in Figure 2.1.

Any periodic array $\tilde{x}(n_1, n_2)$ with horizontal and vertical periods $N_1$ and $N_2$ can be expressed as a finite sum of harmonically related complex sinusoids. This relationship, the *2-D discrete Fourier series*, is given by

$$\tilde{x}(n_1, n_2) = \frac{1}{N_1 N_2} \sum_{k_1=0}^{N_1-1} \sum_{k_2=0}^{N_2-1} \tilde{X}(k_1, k_2) \exp \left( j \frac{2\pi}{N_1} n_1 k_1 + j \frac{2\pi}{N_2} n_2 k_2 \right) \tag{2.3}$$

The complex sinusoid

$$\exp \left( j \frac{2\pi}{N_1} n_1 k_1 + j \frac{2\pi}{N_2} n_2 k_2 \right)$$

is rectangularly periodic with horizontal period $N_1$ and vertical period $N_2$ for all integer values of the parameters $k_1$ and $k_2$. The numbers $\tilde{X}(k_1, k_2)$ are called the *Fourier series coefficients*. They can be computed from $\tilde{x}$ by means of the relation

$$\tilde{X}(k_1, k_2) = \sum_{n_1=0}^{N_1-1} \sum_{n_2=0}^{N_2-1} \tilde{x}(n_1, n_2) \exp \left( -j \frac{2\pi}{N_1} n_1 k_1 - j \frac{2\pi}{N_2} n_2 k_2 \right) \tag{2.4}$$

The demonstration that equations (2.3) and (2.4) form a mathematical identity is left as an exercise of the reader (see Problem 2.1).

The Fourier series representation for a multidimensional periodic sequence resembles in form the expression for the Fourier transform of a sequence that we encountered in Chapter 1, but there are several important differences. First, it must be remembered that a periodic sequence (except for the zero sequence) does not

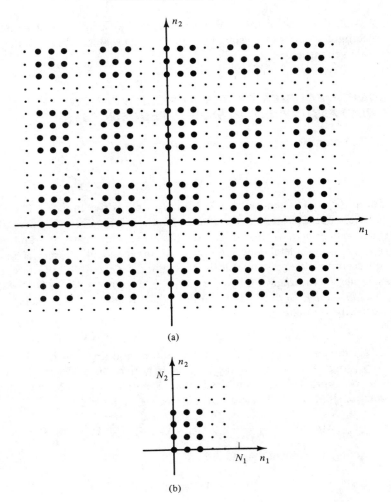

**Figure 2.1** (a) Periodic two-dimensional sequence. (b) Fundamental period of that sequence.

even possess a Fourier transform in the formal sense since it is not absolutely summable. In contrast to the sum that defines the Fourier transform, the limits of summation in (2.4) are finite and the frequency variables $k_1$ and $k_2$ are integers. Furthermore, only $N_1 N_2$ values of $\tilde{X}(k_1, k_2)$ are needed to specify $\tilde{x}(n_1, n_2)$. The discrete Fourier series is thus a computable transform. Using the definition in equation (2.4), it can be computed with $N_1^2 N_2^2$ complex multiplications and complex additions.

From equation (2.4) we see that $\tilde{X}(k_1, k_2)$ is itself a periodic sequence with horizontal period $N_1$ and vertical period $N_2$. The operation of computing Fourier series coefficients can be interpreted as a transformation from one periodic sequence to another with the same periodicity.

**Example 1**

Consider the determination of the discrete Fourier series coefficients for the sequence with periods $N_1 = 5$, $N_2 = 4$ which is described by

$$\tilde{x}(n_1, n_2) = \delta(n_1, n_2), \qquad 0 \le n_1 \le 4, \quad 0 \le n_2 \le 3 \qquad (2.5)$$

This periodic signal is shown in Figure 2.2.

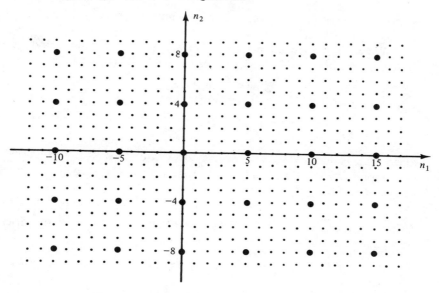

**Figure 2.2**   Periodic signal used in the example.

To compute the Fourier series coefficients we can substitute (2.5) into (2.4).

$$\tilde{X}(k_1, k_2) = \sum_{n_1=0}^{4} \sum_{n_2=0}^{3} \delta(n_1, n_2) \exp\left(-j\frac{2\pi}{5} n_1 k_1 - j\frac{2\pi}{4} n_2 k_2\right)$$
$$= 1 \qquad \text{for all } (k_1, k_2) \qquad (2.6)$$

Thus we can write $\tilde{x}(n_1, n_2)$ as the 2-D Fourier series

$$\tilde{x}(n_1, n_2) = \frac{1}{20} \sum_{k_1=0}^{4} \sum_{k_2=0}^{3} \exp\left(j\frac{2\pi}{5} n_1 k_1 + j\frac{2\pi}{4} n_2 k_2\right) \qquad (2.7)$$

## 2.2  MULTIDIMENSIONAL DISCRETE FOURIER TRANSFORM

### 2.2.1  Definitions

A periodic sequence can be easily generated from a finite-extent sequence. For example, let $x(n_1, n_2)$ be a finite-extent sequence with the region of support $R_{N_1 N_2}$ defined in equation (2.2). We can then define $\tilde{x}(n_1, n_2)$ as the sequence

$$\tilde{x}(n_1, n_2) = \sum_{r_1=-\infty}^{\infty} \sum_{r_2=-\infty}^{\infty} x(n_1 - r_1 N_1, n_2 - r_2 N_2) \qquad (2.8)$$

This signal is rectangularly periodic with horizontal period $N_1$ and vertical period $N_2$ and, in addition, it has the property that in the region $R_{N_1 N_2}$, $\tilde{x}(n_1, n_2)$ and $x(n_1, n_2)$ are equal. The array $\tilde{x}(n_1, n_2)$ is called a *periodic extension* of $x(n_1, n_2)$. Since

$$x(n_1, n_2) = \begin{cases} \tilde{x}(n_1, n_2), & (n_1, n_2) \in R_{N_1 N_2} \\ 0, & \text{otherwise} \end{cases} \tag{2.9}$$

we see that any finite-extent sequence is completely specified by its periodic extension and its region of support.

We can also view the periodic sequence $\tilde{X}(k_1, k_2)$ as the periodic extension of a finite-extent sequence of Fourier series coefficients $X(k_1, k_2)$. Thus

$$\tilde{X}(k_1, k_2) = \sum_{r_1} \sum_{r_2} X(k_1 - r_1 N_1, k_2 - r_2 N_2) \tag{2.10}$$

and

$$X(k_1, k_2) = \begin{cases} \tilde{X}(k_1, k_2), & 0 \le k_1 \le N_1 - 1, \ 0 \le k_2 \le N_2 - 1 \\ 0, & \text{otherwise} \end{cases} \tag{2.11}$$

Since $\tilde{x}(n_1, n_2)$ and $\tilde{X}(k_1, k_2)$ are related by equations (2.3) and (2.4), it is possible to compute $x(n_1, n_2)$ and $X(k_1, k_2)$ from each other through a reversible set of operations

$$x(n_1, n_2) \longleftrightarrow \tilde{x}(n_1, n_2) \longleftrightarrow \tilde{X}(k_1, k_2) \longleftrightarrow X(k_1, k_2) \tag{2.12}$$

Using equations (2.3), (2.4), (2.9), and (2.11), we can write the discrete Fourier transform relations.

$$X(k_1, k_2) = \sum_{n_1=0}^{N_1-1} \sum_{n_2=0}^{N_2-1} x(n_1, n_2) \exp\left(-j\frac{2\pi}{N_1} n_1 k_1 - j\frac{2\pi}{N_2} n_2 k_2\right) \tag{2.13}$$

$$\text{for } 0 \le k_1 \le N_1 - 1, \quad 0 \le k_2 \le N_2 - 1$$

$$x(n_1, n_2) = \frac{1}{N_1 N_2} \sum_{k_1=0}^{N_1-1} \sum_{k_2=0}^{N_2-1} X(k_1, k_2) \exp\left(j\frac{2\pi}{N_1} n_1 k_1 + j\frac{2\pi}{N_2} n_2 k_2\right) \tag{2.14}$$

$$\text{for } 0 \le n_1 \le N_1 - 1, \quad 0 \le n_2 \le N_2 - 1$$

Effectively, all that has been done is to remove the tildas from the discrete Fourier series relations. The fact that the discrete Fourier series and the discrete Fourier transform (DFT) are really the same means that many of the properties of the DFT make sense only if we think of it as a Fourier series. For example, if equation (2.14) is used to evaluate samples of $x$ outside the region $R_{N_1 N_2}$ we get, not samples of $x$, but samples of $\tilde{x}$.

The Fourier transform of a finite-extent sequence supported on $R_{N_1 N_2}$ is given by

$$X(\omega_1, \omega_2) = \sum_{n_1=0}^{N_1-1} \sum_{n_2=0}^{N_2-1} x(n_1, n_2) \exp\left(-j\omega_1 n_1 - j\omega_2 n_2\right) \tag{2.15}$$

Comparing (2.15) and (2.13) we see that the DFT consists of samples of the Fourier transform.

$$X(k_1, k_2) = X(\omega_1, \omega_2) \Big|_{\omega_1 = 2\pi k_1/N_1, \, \omega_2 = 2\pi k_2/N_2} \tag{2.16}$$

Equation (2.16) highlights an abuse of notation since we have used $X$ to represent both the Fourier transform and the discrete Fourier transform. It should be clear in future discussions which of these two transforms is being considered so that confusion should not result.

From our discussion of the sampling theorem in Chapter 1, we know that bandlimited signals can be represented exactly by sample values from the spatial domain. Here we see that signals which are spatially limited (i.e., finite-extent sequences) can be represented exactly by samples of their Fourier transforms. Going one step further, we know that if a nonbandlimited signal is sampled, its spectrum will be aliased. Similarly, if the Fourier transform of a non-spatially limited signal is sampled to yield $\tilde{X}(k_1, k_2)$, and then equation (2.3) is used to compute $\tilde{x}(n_1, n_2)$, the result will be the spatially aliased signal given by

$$\tilde{x}(n_1, n_2) = \sum_{r_1=-\infty}^{\infty} \sum_{r_2=-\infty}^{\infty} x(n_1 - r_1 N_1, n_2 - r_2 N_2) \tag{2.17}$$

The sequence $x(n_1, n_2)$ can be recovered from $\tilde{x}(n_1, n_2)$ only if it is spatially limited to a region no larger than $R_{N_1,N_2}$. The DFT thus represents a reenactment of the sampling theorem where a Fourier transform rather than a continuous waveform is sampled.

We can extend the definition of the DFT to multidimensional sequences. Assume that we have an $M$-dimensional sequence with support over the region $R_N$ defined by

$$R_N = \{\mathbf{n}: \ 0 \le n_i \le N_i - 1, i = 1, 2, \dots, M\} \tag{2.18}$$

Let the matrix $\mathbf{N}$ be a diagonal matrix whose $i$th diagonal element is $N_i$:

$$\mathbf{N} = \begin{bmatrix} N_1 & & & 0 \\ & N_2 & & \\ & & \ddots & \\ 0 & & & N_M \end{bmatrix} \tag{2.19}$$

With this notation, the $M$-dimensional DFT can be written as

$$X(\mathbf{k}) = \sum_{\mathbf{n} \in R_N} x(\mathbf{n}) \exp\left[-j\mathbf{k}'(2\pi\mathbf{N}^{-1})\mathbf{n}\right] \tag{2.20}$$

$$x(\mathbf{n}) = \frac{1}{|\det \mathbf{N}|} \sum_{\mathbf{k} \in R_N} X(\mathbf{k}) \exp\left[j\mathbf{k}'(2\pi\mathbf{N}^{-1})\mathbf{n}\right] \tag{2.21}$$

(As before, $\mathbf{k}'$ is the transpose of the vector variable $\mathbf{k}$.)

**Example 2**

Let us compute the three-dimensional DFT of the $N_1 \times N_2 \times N_3$-point array defined by

$$x(n_1, n_2, n_3) = \begin{cases} 1, & 0 \le n_1 \le N_1 - 1, \ n_2 = 0, \ n_3 = 1 \\ 0, & \text{otherwise} \end{cases} \tag{2.22}$$

The DFT of this sequence can be computed by recognizing that over the region $R_N$ the sequence is equivalent to

$$x(n_1, n_2, n_3) = \delta(n_2, n_3 - 1), \quad (n_1, n_2, n_3) \in R_N$$

Applying equation (2.20) and using the shorthand notation $W_N$ for the complex exponential $\exp(-j2\pi/N)$, we get

$$X(k_1, k_2, k_3) = \sum_{n_1=0}^{N_1-1} \sum_{n_2=0}^{N_2-1} \sum_{n_3=0}^{N_3-1} \delta(n_2, n_3 - 1) W_{N_1}^{n_1 k_1} W_{N_2}^{n_2 k_2} W_{N_3}^{n_3 k_3}$$

$$= W_{N_3}^{k_3} \sum_{n_1=0}^{N_1-1} \sum_{n_2=0}^{N_2-1} \delta(n_2) W_{N_1}^{n_1 k_1} W_{N_2}^{n_2 k_2}$$

$$= W_{N_3}^{k_3} \sum_{n_1=0}^{N_1-1} W_{N_1}^{n_1 k_1} \qquad\qquad (2.23)$$

$$X(k_1, k_2, k_3) = \begin{cases} N_1 W_{N_3}^{k_3}, & k_1 = 0, \quad (k_1, k_2, k_3) \in R_\mathbf{N} \\ 0, & \text{otherwise} \end{cases}$$

**Example 3**

Consider the inverse 2-D DFT of the $N_1 \times N_2$-point sequence given by

$$X(k_1, k_2) = \begin{cases} 1, & 0 \leq k_1 \leq M_1 - 1, \quad 0 \leq k_2 \leq M_2 - 1 \\ 0, & \text{otherwise} \end{cases} \qquad (2.24)$$

where $N_1 \geq M_1$ and $N_2 \geq M_2$. The sequence $x(n_1, n_2)$ is given by

$$x(n_1, n_2) = \frac{1}{N_1 N_2} \sum_{k_1=0}^{M_1-1} \sum_{k_2=0}^{M_2-1} W_{N_1}^{-n_1 k_1} W_{N_2}^{-n_2 k_2}$$

$$= \frac{1}{N_1} \sum_{k_1=0}^{M_1-1} W_{N_1}^{-n_1 k_1} \frac{1}{N_2} \sum_{k_2=0}^{M_2-1} W_{N_2}^{-n_2 k_2}$$

$$= \frac{1}{N_1} \frac{1 - W_{N_1}^{-n_1 M_1}}{1 - W_{N_1}^{-n_1}} \frac{1}{N_2} \frac{1 - W_{N_2}^{-n_2 M_2}}{1 - W_{N_2}^{-n_2}} \qquad (2.25)$$

$$= \exp\left[ -j 2\pi \left( \frac{n_1(M_1 - 1)}{2N_1} + \frac{n_2(M_2 - 1)}{2N_2} \right) \right]$$

$$\cdot \frac{\sin(\pi n_1 M_1 / N_1) \sin(\pi n_2 M_2 / N_2)}{N_1 N_2 \sin(\pi n_1 / N_1) \sin(\pi n_2 / N_2)}$$

This sequence is complex because of the complex exponential factor in equation (2.25). We shall ignore this factor for the moment and plot the remaining factor as a function of $(n_1, n_2)$ in Figure 2.3. It can be shown that Figure 2.3 represents a spatially aliased version of the impulse response of an ideal lowpass filter [equation (1.86)].

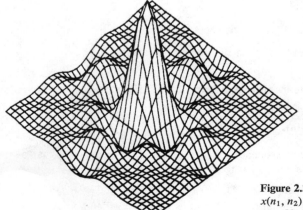

**Figure 2.3** Amplitude of the signal $x(n_1, n_2)$ given in Example 3 for the case $N_1 = N_2 = 32$, $M_1 = M_2 = 8$.

### 2.2.2 Properties of the Discrete Fourier Transform

We shall briefly discuss several properties of the multidimensional DFT using the 2-D case for our examples.

   **Linearity.**    From the defining relation for the DFT it should be clear that if two sequences both have support on $R_{N_1 N_2}$, the DFT of their sum must be the sum of their DFTs. More generally, if $x_1(n_1, n_2)$ and $x_2(n_1, n_2)$ are arbitrary finite-extent sequences and if $a$ and $b$ are arbitrary complex constants, then

$$ax_1(n_1, n_2) + bx_2(n_1, n_2) \longleftrightarrow aX_1(k_1, k_2) + bX_2(k_1, k_2) \qquad (2.26)$$

This property is also true in reverse; if two DFTs are added together, the inverse DFT of the sum is the sum of the individual inverse DFTs. The only caveat associated with the use of this property is that all the DFTs must be the same size. Furthermore, the size must be sufficient to enclose the whole region of support of the sequence $ax_1(n_1, n_2) + bx_2(n_1, n_2)$. This does not mean that $x_1$ and $x_2$ must have the same region of support, since the region of support of either sequence can be enlarged by appending samples of zero value. For example, let $x_1$ be defined over a region of support $R_{M_1 M_2}$ and let $x_2$ have the region of support $R_{N_1 N_2}$. Let $P_1 = \max(M_1, N_1)$, $P_2 = \max(M_2, N_2)$ and define two augmented arrays

$$x_1'(n_1, n_2) = \begin{cases} x_1(n_1, n_2), & (n_1, n_2) \in R_{M_1 M_2} \\ 0, & (n_1, n_2) \in R_{P_1 P_2}, (n_1, n_2) \notin R_{M_1 M_2} \end{cases} \qquad (2.27a)$$

$$x_2'(n_1, n_2) = \begin{cases} x_2(n_1, n_2), & (n_1, n_2) \in R_{N_1 N_2} \\ 0, & (n_1, n_2) \in R_{P_1 P_2}, (n_1, n_2) \notin R_{N_1 N_2} \end{cases} \qquad (2.27b)$$

The sequences $x_1'$ and $x_2'$ have support over $R_{P_1 P_2}$ and

$$ax_1' + bx_2' = ax_1 + bx_2 \qquad (2.28)$$

The size of the region $R_{P_1 P_2}$ now determines the parameters of the DFT.

   **Circular shifts.**    We saw in Chapter 1 that if a signal is linearly shifted, its Fourier transform is multiplied by a complex exponential. The DFT has an analogous property; if a finite-extent sequence is *circularly shifted*, the DFT is multiplied by a complex exponential.

   Consider a periodic sequence $\tilde{x}(n_1, n_2)$ with horizontal and vertical periods $N_1$ and $N_2$ and discrete Fourier series coefficients $\tilde{X}(k_1, k_2)$. Let $\tilde{y}$ be a shifted version of $\tilde{x}$.

$$\tilde{y}(n_1, n_2) = \tilde{x}(n_1 - m_1, n_2 - m_2) \qquad (2.29)$$

The discrete Fourier series coefficients of $\tilde{x}$ and $\tilde{y}$ are related by

$$\tilde{Y}(k_1, k_2) = W_{N_1}^{m_1 k_1} W_{N_2}^{m_2 k_2} \tilde{X}(k_1, k_2) \qquad (2.30)$$

Since $\tilde{y}$ and $\tilde{x}$ are related by equation (2.29), the finite-extent sequence $y$, defined as

$$y(n_1, n_2) \triangleq \begin{cases} \tilde{y}(n_1, n_2), & (n_1, n_2) \in R_{N_1 N_2} \\ 0, & \text{otherwise} \end{cases} \qquad (2.31)$$

is a circularly shifted version of $x(n_1, n_2)$. That is,

$$y(n_1, n_2) = x(\,((n_1 - m_1))_{N_1}, ((n_2 - m_2))_{N_2}), (n_1, n_2) \in R_{N_1 N_2} \quad (2.32)$$

where the notation $((n))_N$ means that the integer variable $n$ is evaluated modulo $N$. [If $p = ((n))_N$, then $0 \le p \le N - 1$ and there exists an integer $I$ such that $n = p + IN$.] The term " circular shift" is carried over from 1-D digital signal processing terminology; it conveys the notion that sample values which are shifted off the left (or top) edge of the region of support reappear on the right (or bottom) edge.

The DFT $Y(k_1, k_2)$ is defined as

$$\begin{aligned} Y(k_1, k_2) &= \tilde{Y}(k_1, k_2) \qquad \text{for } (k_1, k_2) \in R_{N_1 N_2} \\ &= W_{N_1}^{m_1 k_1} W_{N_2}^{m_2 k_2} \tilde{X}(k_1, k_2) \\ &= W_{N_1}^{m_1 k_1} W_{N_2}^{m_2 k_2} X(k_1, k_2) \end{aligned} \quad (2.33)$$

Combining equations (2.32) and (2.33) gives us the DFT pair

$$x(\,((n_1 - m_1))_{N_1}, ((n_2 - m_2))_{N_2}) \longleftrightarrow W_{N_1}^{m_1 k_1} W_{N_2}^{m_2 k_2} X(k_1, k_2) \quad (2.34)$$

**Symmetry properties of real $x(n_1, n_2)$.** In discussing the DFT to this point, we have not assumed that our finite-extent array contains only real sample values. Indeed, all of our results are valid whether $x$ is real or complex. As with 1-D discrete Fourier transforms, however, if $x(n_1, n_2)$ is known to consist of a real sequence of data, the DFT will satisfy certain symmetry relations. If $x(n_1, n_2)$ is real, we see that

$$X^*(k_1, k_2) = \sum_{n_1=0}^{N_1-1} \sum_{n_2=0}^{N_2-1} x(n_1, n_2) \{W_{N_1}^{n_1 k_1} W_{N_2}^{n_2 k_2}\}^* \quad (2.35)$$

where the asterisk denotes the operation of complex conjugation. However, we know that

$$[W_N^{nk}]^* = W_N^{-nk} = W_N^{n(N-k)}$$

so

$$\begin{aligned} X^*(k_1, k_2) &= \sum_{n_1=0}^{N_1-1} \sum_{n_2=0}^{N_2-1} x(n_1, n_2) W_{N_1}^{n_1(N_1-k_1)} W_{N_2}^{n_2(N_2-k_2)} \\ &= X(((N_1 - k_1))_{N_1}, ((N_2 - k_2))_{N_2}) \end{aligned} \quad (2.36)$$

Thus if a signal is purely real, its DFT is Hermitian symmetric in this special sense. Equation (2.36) implies that

$$\text{Re}\,[X(k_1, k_2)] = \text{Re}\,[X(((N_1 - k_1))_{N_1}, ((N_2 - k_2))_{N_2})] \quad (2.37a)$$

$$\text{Im}\,[X(k_1, k_2)] = -\text{Im}\,[X(((N_1 - k_1))_{N_1}, ((N_2 - k_2))_{N_2})] \quad (2.37b)$$

In a similar vein, we can define the Hermitian symmetric and antisymmetric components of a complex finite-extent sequence as

$$x_s(n_1, n_2) \triangleq \tfrac{1}{2}[x(n_1, n_2) + x^*(((N_1 - n_1))_{N_1}, ((N_2 - n_2))_{N_2})] \quad (2.38a)$$

$$x_a(n_1, n_2) \triangleq \tfrac{1}{2}[x(n_1, n_2) - x^*(((N_1 - n_1))_{N_1}, ((N_2 - n_2))_{N_2})] \quad (2.38b)$$

Since

$$x^*( ((N_1 - n_1))_{N_1}, ((N_2 - n_2))_{N_2}) \longleftrightarrow X^*(k_1, k_2) \tag{2.39}$$

(see Problem 2.7), it follows that

$$x_s(n_1, n_2) \longleftrightarrow \tfrac{1}{2}[X(k_1, k_2) + X^*(k_1, k_2)] = \text{Re}\,[X(k_1, k_2)] \tag{2.40a}$$
$$x_a(n_1, n_2) \longleftrightarrow \tfrac{1}{2}[X(k_1, k_2) - X^*(k_1, k_2)] = j\,\text{Im}\,[X(k_1, k_2)] \tag{2.40b}$$

Thus the Hermitian symmetric portion of a signal is transformed into the real part of the DFT and the Hermitian antisymmetric portion of a signal is transformed into $j$ times the imaginary part of the DFT.

**Reflection.**    These properties are basically the same as the analogous properties of the Fourier transform in Chapter 1, once the periodicity of the DFT is taken into account. It can be easily shown that if

$$x(n_1, n_2) \longleftrightarrow X(k_1, k_2)$$

then

$$x(n_2, n_1) \longleftrightarrow X(k_2, k_1) \tag{2.41a}$$
$$x( ((N_1 - n_1))_{N_1}, n_2) \longleftrightarrow X( ((N_1 - k_1))_{N_1}, k_2) \tag{2.41b}$$
$$x(n_1, ((N_2 - n_2))_{N_2}) \longleftrightarrow X(k_1, ((N_2 - k_2))_{N_2}) \tag{2.41c}$$
$$x( ((N_1 - n_1))_{N_1}, ((N_2 - n_2))_{N_2}) \longleftrightarrow X( ((N_1 - k_1))_{N_1}, ((N_2 - k_2))_{N_2}) \tag{2.41d}$$

**Parseval's theorem.**

$$\sum_{n_1=0}^{N_1-1} \sum_{n_2=0}^{N_2-1} x(n_1, n_2)y^*(n_1, n_2) = \frac{1}{N_1 N_2} \sum_{k_1=0}^{N_1-1} \sum_{k_2=0}^{N_2-1} X(k_1, k_2)Y^*(k_1, k_2) \tag{2.42}$$

**Duality.**    If $X(k_1, k_2)$ is the DFT of $x(n_1, n_2)$, what is the DFT of $X(n_1, n_2)$? Because of the similarity of the forward and inverse DFT expressions, we would expect that the result would be closely related to $x$. We can multiply both sides of (2.14) by $N_1 N_2$ and take complex conjugates to get

$$N_1 N_2 x^*(n_1, n_2) = \sum_{k_1=0}^{N_1-1} \sum_{k_2=0}^{N_2-1} X^*(k_1, k_2)W_{N_1}^{n_1 k_1}W_{N_2}^{n_2 k_2} \tag{2.43}$$

Equation (2.43) now has the same form as equation (2.13). Thus if

$$x(n_1, n_2) \longleftrightarrow X(k_1, k_2)$$

then

$$X^*(n_1, n_2) \longleftrightarrow N_1 N_2 x^*(k_1, k_2) \tag{2.44}$$

This property is known as the duality property.

**Modulation.**    The modulation property states that if a sequence is multiplied by a complex exponential, its DFT is circularly shifted. By evaluating the dual of equation (2.34), we see that

$$W_{N_1}^{-n_1 l_1} W_{N_2}^{-n_2 l_2} x(n_1, n_2) \longleftrightarrow X( ((k_1 - l_1))_{N_1}, ((k_2 - l_2))_{N_2}) \tag{2.45}$$

### 2.2.3 Circular Convolution

We saw in Chapter 1 that the Fourier transform of the convolution of two sequences was the product of their Fourier transforms. In this section we would like to derive an equivalent statement for DFTs. In particular, what sequence is the inverse DFT of the product of two DFTs?

Suppose that we have two finite-extent sequences, $x(n_1, n_2)$ and $h(n_1, n_2)$, with support on $R_{N_1N_2}$. Each sequence has a DFT,

$$x(n_1, n_2) \longleftrightarrow X(k_1, k_2); \qquad h(n_1, n_2) \longleftrightarrow H(k_1, k_2)$$

Let $Y(k_1, k_2)$ be the $N_1 \times N_2$ DFT formed by

$$Y(k_1, k_2) = H(k_1, k_2) X(k_1, k_2) \tag{2.46}$$

and let us seek to determine $y(n_1, n_2)$.

We shall begin by considering the periodically extended sequences $\tilde{x}, \tilde{h}, \tilde{y}, \tilde{X}, \tilde{H}$, and $\tilde{Y}$. Since

$$\tilde{Y}(k_1, k_2) = \tilde{H}(k_1, k_2)\tilde{X}(k_1, k_2) \tag{2.47}$$

we can apply the inverse discrete Fourier series (2.3) and write

$$\tilde{y}(n_1, n_2) = \frac{1}{N_1 N_2} \sum_{k_1=0}^{N_1-1} \sum_{k_2=0}^{N_2-1} \tilde{H}(k_1, k_2)\tilde{X}(k_1, k_2) W_{N_1}^{-n_1 k_1} W_{N_2}^{-n_2 k_2} \tag{2.48}$$

By expressing $\tilde{X}(k_1, k_2)$ as

$$\tilde{X}(k_1, k_2) = \sum_{m_1=0}^{N_1-1} \sum_{m_2=0}^{N_2-1} \tilde{x}(m_1, m_2) W_{N_1}^{m_1 k_1} W_{N_2}^{m_2 k_2} \tag{2.49}$$

substituting it into equation (2.48), and rearranging terms, we get

$$\tilde{y}(n_1, n_2) = \sum_{m_1=0}^{N_1-1} \sum_{m_2=0}^{N_2-1} \tilde{x}(m_1, m_2) \frac{1}{N_1 N_2} \sum_{k_1=0}^{N_1-1} \sum_{k_2=0}^{N_2-1} \tilde{H}(k_1, k_2) W_{N_1}^{-(n_1-m_1)k_1} W_{N_2}^{-(n_2-m_2)k_2}$$

$$= \sum_{m_1=0}^{N_1-1} \sum_{m_2=0}^{N_2-1} \tilde{x}(m_1, m_2)\tilde{h}(n_1 - m_1, n_2 - m_2) \tag{2.50}$$

Except for the limits of summation, this relation has the form of an ordinary (linear) 2-D convolution sum. Instead of summing over all the samples of the product here we sum only those samples of the product sequence lying in the region $R_{N_1N_2}$. The sequence $\tilde{y}(n_1, n_2)$ is rectangularly periodic with periods $N_1$ and $N_2$.

Since

$$y(n_1, n_2) = \begin{cases} \tilde{y}(n_1, n_2), & (n_1, n_2) \in R_{N_1N_2} \\ 0, & \text{otherwise} \end{cases} \tag{2.51}$$

we can write

$$y(n_1, n_2) = \sum_{m_1=0}^{N_1-1} \sum_{m_2=0}^{N_2-1} x(m_1, m_2)\tilde{h}(n_1 - m_1, n_2 - m_2) \qquad \text{for } (n_1, n_2) \in R_{N_1N_2}$$

$$= \sum_{m_1=0}^{N_1-1} \sum_{m_2=0}^{N_2-1} x(m_1, m_2) h(((n_1 - m_1))_{N_1}, ((n_2 - m_2))_{N_2}) \tag{2.52}$$

We will say that $y$ is the *circular convolution* of $h$ and $x$. The term "circular convolution," also carried over from 1-D signal processing terminology, indicates that $h$ is circularly shifted past $x$, in contrast to ordinary linear convolution, where $h$ is simply shifted past $x$ linearly. The circular convolution can also be written in the alternative form

$$y(n_1, n_2) = \sum_{m_1=0}^{N_1-1} \sum_{m_2=0}^{N_2-1} h(m_1, m_2) x(\,((n_1 - m_1))_{N_1}, ((n_2 - m_2))_{N_2})  \qquad (2.53)$$

As with ordinary convolution, circular convolution is commutative, associative, and distributive over addition. Using the symbol $\circledast$ to denote the operation of 2-D circular convolution, we can write

$$y = h \circledast x = x \circledast h \longleftrightarrow H(k_1, k_2) X(k_1, k_2) \qquad (2.54)$$

Notice that the result of a circular convolution depends not only on the two finite-extent sequences being circularly convolved but also on the periods $N_1$ and $N_2$ which define the size of the DFT.

The DFT has the advantage over the Fourier transform of being computable with $N_1^2 N_2^2$ complex multiplications and additions, as we saw earlier. Unfortunately, the product of two DFTs corresponds to the circular convolution of two sequences, and in modeling and simulating LSI systems, it is the linear convolution of two sequences that interests us. As in one dimension, however, there is a simple remedy that allows us to use a circular convolution to compute a linear convolution.

Assume that $x(n_1, n_2)$ has a region of support $R_{P_1 P_2}$ and that $h(n_1, n_2)$ has a region of support $R_{Q_1 Q_2}$. Let $w(n_1, n_2)$ denote the result of the linear convolution

$$w(n_1, n_2) = \sum_{m_1} \sum_{m_2} h(m_1, m_2) x(n_1 - m_1, n_2 - m_2) \qquad (2.55)$$

The sequence $w(n_1, n_2)$ has the region of support

$$\begin{aligned} 0 \leq n_1 \leq P_1 + Q_1 - 1 \\ 0 \leq n_2 \leq P_2 + Q_2 - 1 \end{aligned} \qquad (2.56)$$

Now, using a DFT of size $N_1 \geq \max(P_1, Q_1)$ and $N_2 \geq \max(P_2, Q_2)$, we can compute the circular convolution $y = h \circledast x$, which can be written as

$$y(n_1, n_2) = \sum_{m_1=0}^{Q_1-1} \sum_{m_2=0}^{Q_2-1} h(m_1, m_2) \tilde{x}(n_1 - m_1, n_2 - m_2) \qquad \text{for } (n_1, n_2) \in R_{N_1 N_2} \qquad (2.57)$$

The periodic extension $\tilde{x}(n_1 - m_1, n_2 - m_2)$ can be written in terms of the finite-extent sequence $x(n_1, n_2)$ as

$$\tilde{x}(n_1 - m_1, n_2 - m_2) = \sum_{r_1} \sum_{r_2} x(n_1 - m_1 - r_1 N_1, n_2 - m_2 - r_2 N_2) \qquad (2.58)$$

Using this expression in equation (2.57) and rearranging terms, we get

$$y(n_1, n_2) = \sum_{r_1} \sum_{r_2} \left[ \sum_{m_1=0}^{Q_1-1} \sum_{m_2=0}^{Q_2-1} h(m_1, m_2) x(n_1 - m_1 - r_1 N_1, n_2 - m_2 - r_2 N_2) \right]$$
$$\text{for } (n_1, n_2) \in R_{N_1 N_2} \qquad (2.59)$$

Because of the limited extent of $h(n_1, n_2)$, the term in brackets is a linear convolution which may be written in terms of $w(n_1, n_2)$ to yield

$$y(n_1, n_2) = \sum_{r_1} \sum_{r_2} w(n_1 - r_1 N_1, n_2 - r_2 N_2) \qquad \text{for } (n_1, n_2) \in R_{N_1 N_2} \qquad (2.60)$$

The sequence $y(n_1, n_2)$ is equal to a spatially aliased version of the sequence $w(n_1, n_2)$ in the region $R_{N_1 N_2}$. However, since $w(n_1, n_2)$ is a finite-extent sequence, if we pick $N_1$ and $N_2$ large enough, specifically

$$\begin{aligned} N_1 &\geq P_1 + Q_1 - 1 \\ N_2 &\geq P_2 + Q_2 - 1 \end{aligned} \qquad (2.61)$$

then the replicas of $w(n_1, n_2)$ in equation (2.60) will not overlap, and equation (2.60) becomes

$$y(n_1, n_2) = w(n_1, n_2) \qquad \text{for } (n_1, n_2) \in R_{N_1 N_2} \qquad (2.62)$$

Consequently, the result of the circular convolution is equal to the result of the linear convolution as desired.

This result is summarized in the following procedure for computing a linear convolution using DFTs.

1. Choose $N_1$ and $N_2$ to satisfy equation (2.61).
2. Augment $h(n_1, n_2)$ with sufficient samples of zero value to fill the region $R_{N_1 N_2}$.
3. Augment $x(n_1, n_2)$ in a similar manner.
4. Compute the $(N_1 \times N_2)$-point DFTs of both $h(n_1, n_2)$ and $x(n_1, n_2)$.
5. Form the product $H(k_1, k_2) X(k_1, k_2)$.
6. Compute the $(N_1 \times N_2)$-point inverse DFT of $H(k_1, k_2) X(k_1, k_2)$. The result of this operation is the desired linear convolution.

In this procedure, the reader should recognize that steps 4–6 implement the circular convolution $y = h \circledast x$, which is equal to the linear convolution $w = h * * x$, thanks to the judicious choice of $N_1$ and $N_2$.

### Example 4

As a numerical example, let us consider the particularly simple case where $x(n_1, n_2)$ and $h(n_1, n_2)$ are the $2 \times 2$ arrays shown below:

$$x(n_1, n_2) = \begin{array}{|cc|} \hline 1 & 0 \\ 2 & 1 \\ \hline \end{array} \qquad h(n_1, n_2) = \begin{array}{|cc|} \hline 1 & 0 \\ 1 & 1 \\ \hline \end{array} \qquad (2.63)$$

We have explicitly represented the sequences as arrays of sample values, using $n_1$ as the column index and $n_2$ as the row index. The sample at $(0, 0)$ is in the lower left-hand corner.

The $2 \times 2$ DFT can be written as

$$X(0, 0) = x(0, 0) + x(1, 0) + x(0, 1) + x(1, 1)$$
$$X(1, 0) = x(0, 0) - x(1, 0) + x(0, 1) - x(1, 1)$$
$$X(0, 1) = x(0, 0) + x(1, 0) - x(0, 1) - x(1, 1)$$
$$X(1, 1) = x(0, 0) - x(1, 0) - x(0, 1) + x(1, 1)$$

(2.64)

Substituting into (2.64), we get

$$X(k_1, k_2) = \begin{vmatrix} 2 & 0 \\ 4 & 2 \end{vmatrix}$$

(2.65)

$$H(k_1, k_2) = \begin{vmatrix} 1 & -1 \\ 3 & 1 \end{vmatrix}$$

(2.66)

$$H(k_1, k_2) X(k_1, k_2) = \begin{vmatrix} 2 & 0 \\ 12 & 2 \end{vmatrix}$$

(2.67)

Taking the inverse $2 \times 2$ DFT of (2.67) we arrive at the $2 \times 2$ circular convolution

$$y(n_1, n_2) = \begin{vmatrix} 3 & 2 \\ 4 & 3 \end{vmatrix}$$

(2.68)

Now let us compute the linear convolution of the same two sequences by using $(4 \times 4)$-point DFTs. The linear convolution of two $2 \times 2$ arrays will be a $3 \times 3$ array. Since the DFT is larger in each dimension than the anticipated size of the linear convolution, the $(4 \times 4)$-point circular convolution and the linear convolution will be identical. Augmenting both $x$ and $h$ to fill the $4 \times 4$ region of support, we get

$$x(n_1, n_2) = \begin{vmatrix} 0 & 0 & 0 & 0 \\ 0 & 0 & 0 & 0 \\ 1 & 0 & 0 & 0 \\ 2 & 1 & 0 & 0 \end{vmatrix}$$

$$h(n_1, n_2) = \begin{vmatrix} 0 & 0 & 0 & 0 \\ 0 & 0 & 0 & 0 \\ 1 & 0 & 0 & 0 \\ 1 & 1 & 0 & 0 \end{vmatrix}$$

Evaluating the $(4 \times 4)$-point DFTs, we find

$$X(k_1, k_2) = \begin{vmatrix} 3+j & 2 & 1+j & 2+2j \\ 2 & 1-j & 0 & 1+j \\ 3-j & 2-2j & 1-j & 2 \\ 4 & 3-j & 2 & 3+j \end{vmatrix}$$

$$H(k_1, k_2) = \begin{vmatrix} 2+j & 1 & j & 1+2j \\ 1 & -j & -1 & j \\ 2-j & 1-2j & -j & 1 \\ 3 & 2-j & 1 & 2+j \end{vmatrix} \qquad (2.69)$$

$$H(k_1, k_2)X(k_1, k_2) = \begin{vmatrix} 5+5j & 2 & -1+j & -2+6j \\ 2 & -1-j & 0 & -1+j \\ 5-5j & -2-6j & -1-j & 2 \\ 12 & 5-5j & 2 & 5+5j \end{vmatrix}$$

Taking the $4 \times 4$ inverse DFT of $H(k_1, k_2) X(k_1, k_2)$, we get

$$w(n_1, n_2) = \begin{vmatrix} 0 & 0 & 0 & 0 \\ 1 & 0 & 0 & 0 \\ 3 & 2 & 0 & 0 \\ 2 & 3 & 1 & 0 \end{vmatrix} \qquad (2.70)$$

If this array is spatially aliased to produce a $2 \times 2$ array, the result is identical to (2.68).

If you verified the numbers in this example you undoubtedly observed that these calculations are very tedious and that considerably more effort was expended than would be needed to compute the linear convolution directly. In Section 2.3 we explore fast, efficient algorithms for the evaluation of the $M$-dimensional DFT which will allow us to compute the result of a convolution with fewer arithmetic operations than we would need to evaluate the convolution sum directly.

## 2.3 CALCULATION OF THE DISCRETE FOURIER TRANSFORM

The DFT can be put to several important uses because it can be calculated in a finite number of arithmetic operations. For example, since the DFT samples are equal to samples of the Fourier transform, the DFT can be used to analyze the frequency content of multidimensional signals. As we saw in the preceding section, the DFT can also be used to evaluate linear convolutions, if care is taken to avoid spatial aliasing. Consequently, the DFT can be useful both for simulating and implementing discrete linear shift-invariant systems. We shall explore this use of the DFT further in Chapter 3. In this section we examine three algorithms for calculating multi-dimensional DFTs which vary considerably in their computational complexity.

### 2.3.1 Direct Calculation

The direct calculation of the 2-D DFT is simply the evaluation of the double sum

$$X(k_1, k_2) = \sum_{n_1=0}^{N_1-1} \sum_{n_2=0}^{N_2-1} x(n_1, n_2) W_{N_1}^{n_1 k_1} W_{N_2}^{n_2 k_2}$$

$$\text{for } 0 \leq k_1 \leq N_1 - 1 \quad \text{and} \quad 0 \leq k_2 \leq N_2 - 1$$

(2.71)

where, as before,

$$W_N \triangleq \exp\left(-j\frac{2\pi}{N}\right)$$

If we assume that the complex exponentials in equation (2.71) have been precomputed and stored in a table, then the direct evaluation of one sample of $X(k_1, k_2)$ requires $N_1 N_2$ complex multiplications and a like number of complex additions. Since the entire DFT involves $N_1 N_2$ output samples, the total number of complex multiplications and complex additions needed to evaluate the DFT by direct calculation is $N_1^2 N_2^2$. In the $M$-dimensional case the total computation is $N_1^2 N_2^2 \cdots N_M^2$ complex multiplications and additions.

This approach is somewhat naive, however, for we know that an $N$-point 1-D DFT can be computed with far fewer than $N^2$ multiplications by using a fast Fourier transform (FFT) algorithm. As we shall see, fast algorithms can be developed for calculating higher-dimensional DFTs as well.

### 2.3.2 Row–Column Decompositions

The DFT sum in (2.71) can be rewritten as

$$X(k_1, k_2) = \sum_{n_1=0}^{N_1-1} \left[ \sum_{n_2=0}^{N_2-1} x(n_1, n_2) W_{N_2}^{n_2 k_2} \right] W_{N_1}^{n_1 k_1}$$

(2.72)

The quantity in brackets is a 2-D sequence which we shall denote $G(n_1, k_2)$. Equation (2.72) can then be expressed as the pair of relations

$$G(n_1, k_2) = \sum_{n_2=0}^{N_2-1} x(n_1, n_2) W_{N_2}^{n_2 k_2}$$

(2.73a)

$$X(k_1, k_2) = \sum_{n_1=0}^{N_1-1} G(n_1, k_2) W_{N_1}^{n_1 k_2}$$

(2.73b)

Each column of $G$ is the 1-D DFT of the corresponding column of $x$. Each row of $X$ is the 1-D DFT of the corresponding row of $G$. Thus we can compute a 2-D DFT by decomposing it into row and column DFTs; we first compute the DFT of each column of $x$, put the results into an intermediate array, and then compute the DFT of each row of the intermediate array. Alternatively, we could do the row DFTs first and the column DFTs second.

An $M$-dimensional DFT can be evaluated in the same fashion. First the 1-D DFT is computed with respect to one of the variables, say $n_M$, for each value of the remaining variables. This requires a total of $N_1 N_2 \cdots N_{M-1}$ one-dimensional DFTs. Next, one-dimensional DFTs are computed with respect to the variable $n_{M-1}$ for all values of the $(M - 1)$-tuple $(n_1, \ldots, n_{M-2}, k_M)$. We continue in this fashion until 1-D DFTs have been evaluated with respect to all the spatial variables.

How much computation has been saved by this procedure? We have already seen that

$$C_{\text{direct}} = N_1^2 N_2^2 \cdots N_M^2 \qquad (2.74)$$

complex multiplications and additions are required in a direct calculation. If a direct calculation is used to compute the 1-D DFTs in a row–column decomposition, then the evaluation of a multidimensional DFT requires

$$C_{\text{r/c direct}} = N_1 \cdot N_2 \cdots N_M (N_1 + N_2 + \cdots + N_M) \qquad (2.75)$$

complex multiplications and additions. If each of the $N_i$ is a power of 2, so that 1-D FFTs can be used, the number of complex multiplications is further reduced to

$$C_{\text{r/c FFT}} = N_1 N_2 \cdots N_M \frac{\log_2 N_1 N_2 \cdots N_M}{2} \qquad (2.76)$$

The number of complex additions needed is twice this number. Other fast algorithms for evaluating the 1-D DFT can also be used with the row–column decomposition to compute a multidimensional DFT.

To get a feeling for the numerical savings involved, consider the evaluation of a $1024 \times 1024$ 2-D DFT by each of these approaches. For this problem we find that

$$C_{\text{direct}} = 2^{40} \approx 10^{12} \text{ complex multiplications}$$

$$C_{\text{r/c direct}} = 2^{31} \approx 2 \times 10^9 \text{ complex multiplications}$$

$$C_{\text{r/c FFT}} = 10 \times 2^{20} \approx 10^7 \text{ complex multiplications}$$

For this problem the row–column decomposition reduces the number of calculations by a factor of 500. Using the 1-D FFT reduces the computation by a factor of $10^5$. (There are approximately $10^5$ seconds in a day.)

### 2.3.3 Vector-Radix Fast Fourier Transform [3,4]

The 1-D fast Fourier transform algorithm achieves its computational efficiency through a "divide and conquer" strategy. If the DFT length is, for example, a power of 2, the DFT can be expressed as a combination of two half-length DFTs, each of which can be expressed in turn as a combination of two quarter-length DFTs, and so on. The 2-D vector-radix FFT algorithm is philosophically identical. A 2-D DFT is broken down into successively smaller 2-D DFTs until, ultimately, only trivial 2-D DFTs need to be evaluated.

We can derive the decimation-in-time version of the algorithm by expressing an $(N_1 \times N_2)$-point DFT in terms of four $N_1/2 \times N_2/2$ DFTs (if $N_1$ and $N_2$ are divisible by 2). For simplicity let us assume that $N_1 = N_2 = N$. The DFT summation of (2.71) can be decomposed into four summations: one over those samples of $x$ for which both $n_1$ and $n_2$ are even, one for which $n_1$ is even and $n_2$ is odd, one for which $n_1$ is odd and $n_2$ is even, and one for which both $n_1$ and $n_2$ are odd. This gives us

$$X(k_1, k_2) = S_{00}(k_1, k_2) + S_{01}(k_1, k_2)W_N^{k_2}$$
$$+ S_{10}(k_1, k_2)W_N^{k_1} + S_{11}(k_1, k_2)W_N^{k_1+k_2} \tag{2.77a}$$

where

$$S_{00}(k_1, k_2) \triangleq \sum_{m_1=0}^{N/2-1} \sum_{m_2=0}^{N/2-1} x(2m_1, 2m_2)W_N^{2m_1k_1+2m_2k_2} \tag{2.77b}$$

$$S_{01}(k_1, k_2) \triangleq \sum_{m_1=0}^{N/2-1} \sum_{m_2=0}^{N/2-1} x(2m_1, 2m_2 + 1)W_N^{2m_1k_1+2m_2k_2} \tag{2.77c}$$

$$S_{10}(k_1, k_2) \triangleq \sum_{m_1=0}^{N/2-1} \sum_{m_2=0}^{N/2-1} x(2m_1 + 1, 2m_2)W_N^{2m_1k_1+2m_2k_2} \tag{2.77d}$$

$$S_{11}(k_1, k_2) \triangleq \sum_{m_1=0}^{N/2-1} \sum_{m_2=0}^{N/2-1} x(2m_1 + 1, 2m_2 + 1)W_N^{2m_1k_1+2m_2k_2} \tag{2.77e}$$

The arrays $S_{00}$, $S_{01}$, $S_{10}$, and $S_{11}$ are each periodic in $(k_1, k_2)$ with horizontal and vertical periods $N/2$. Using this fact and the fact that $W_N^{N/2} = -1$, we can derive the following identities from equation (2.77a).

$$X(k_1, k_2) = S_{00}(k_1, k_2) + W_N^{k_2}S_{01}(k_1, k_2) + W_N^{k_1}S_{10}(k_1, k_2)$$
$$+ W_N^{k_1+k_2}S_{11}(k_1, k_2) \tag{2.78a}$$

$$X\left(k_1 + \frac{N}{2}, k_2\right) = S_{00}(k_1, k_2) + W_N^{k_2}S_{01}(k_1, k_2) - W_N^{k_1}S_{10}(k_1, k_2)$$
$$- W_N^{k_1+k_2}S_{11}(k_1, k_2) \tag{2.78b}$$

$$X\left(k_1, k_2 + \frac{N}{2}\right) = S_{00}(k_1, k_2) - W_N^{k_2}S_{01}(k_1, k_2) + W_N^{k_1}S_{10}(k_1, k_2)$$
$$- W_N^{k_1+k_2}S_{11}(k_1, k_2) \tag{2.78c}$$

$$X\left(k_1 + \frac{N}{2}, k_2 + \frac{N}{2}\right) = S_{00}(k_1, k_2) - W_N^{k_2}S_{01}(k_1, k_2) - W_N^{k_1}(k_1, k_2)$$
$$+ W_N^{k_1+k_2}S_{11}(k_1, k_2) \tag{2.78d}$$

These equations tell us how to compute the four DFT points $X(k_1, k_2)$, $X(k_1 + N/2, k_2)$, $X(k_1, k_2 + N/2)$, and $X(k_1 + N/2, k_2 + N/2)$ for a particular value of $(k_1, k_2)$ from the four points $S_{00}(k_1, k_2)$, $S_{01}(k_1, k_2)$, $S_{10}(k_1, k_2)$, and $S_{11}(k_1, k_2)$. We can also see that the relevant samples of $S_{00}(k_1, k_2)$ can be obtained by evaluating an $(N/2 \times N/2)$-point DFT (and similarly for the other $S_{ij}$ arrays). Thus equation (2.77) expresses the samples of the $N \times N$ DFT, $X(k_1, k_2)$, in terms of four $N/2 \times N/2$ DFTs. By analogy with the corresponding equations from the 1-D case, the computation represented by (2.78) is called a *butterfly*, or more properly a

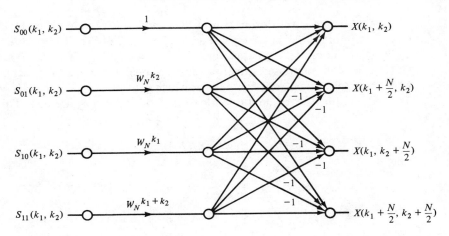

**Figure 2.4** Isolated radix-(2 × 2) butterfly. To calculate the outputs from the inputs requires three complex multiplications and eight complex additions.

*radix-(2 × 2) butterfly.* An isolated radix-(2 × 2) butterfly is shown in Figure 2.4. Each butterfly requires that three complex multiplications and eight complex additions be performed (see Problem 2.11) and to compute all the samples of $X$ from $S_{00}$, $S_{01}$, $S_{10}$, and $S_{11}$ requires the calculation of $N^2/4$ butterflies. Figure 2.5 shows a flowchart which summarizes these calculations.

If $N/2$ is itself a multiple of 2, this same decomposition can be used to compute the $N/2 \times N/2$ transforms $S_{00}$, $S_{01}$, $S_{10}$, and $S_{11}$. This decimation process can be continued until only $2 \times 2$ arrays need to be transformed; these can be performed using equation (2.64), which requires no multiplication at all. A complete vector-radix transform for a $4 \times 4$ array is shown in Figure 2.6.

This decimation procedure can be performed $\log_2 N$ times if $N$ is a power of 2. Each stage of decimations consists of $N^2/4$ butterflies, and each butterfly involves three complex multiplications and eight complex additions. Thus the number of complex multiplications that need to be performed during the computation of an $(N \times N)$-point radix-(2 × 2) FFT is

$$C_{\text{vr}(2 \times 2)} = \frac{3N^2}{4} \log_2 N \tag{2.79}$$

which is 25% fewer than are needed for the row–column decomposition used in conjunction with the 1-D FFT. The vector radix algorithm also requires $2N^2 \log_2 N$ complex additions, the same as the row–column algorithm.

The flowchart in Figure 2.6 shows the computations that must be performed in the calculation of the transform. We can go one step further, however, and assume that it also represents the order in which the data are to be stored. In this regard we are hampered by the necessity of drawing the flowchart as if the input and output arrays are 1-D, when, in fact, they are 2-D. The output array in the flowchart is a normally ordered 2-D array.

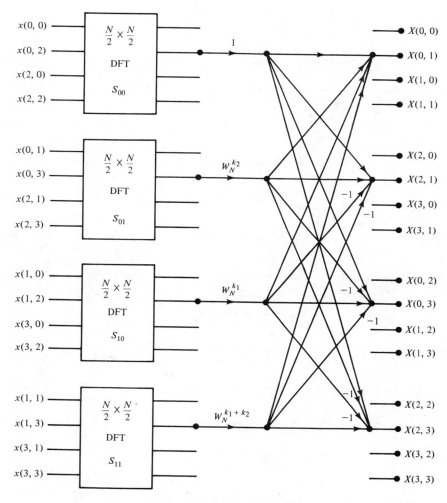

**Figure 2.5**  First stage of decimation of a radix-$(2 \times 2)$ FFT. Only one of four butterflies is shown to minimize confusion.

Because of the successive separations into even- and odd-indexed groups, the data in a 1-D DFT are accessed in bit-reversed order. For example, with $N = 16$ the index $n = 12$ would be classified in successive decimations as even, even, odd, and odd because the remainders of 12 when it is successively divided by two are 0, 0, 1, and 1. Of course, 0011 is just the reverse of the 4-bit binary representation of the number 12 (1100). In the 2-D case, because of the successive separation of $(k_1, k_2)$ into even–even, odd–even, and so on, parts the input data will be accessed in bit-reversed fashion on both indices $k_1$ and $k_2$. This order is shown in Figure 2.7.

The radix-$(2 \times 2)$ FFT represents an in-place computation. The outputs from

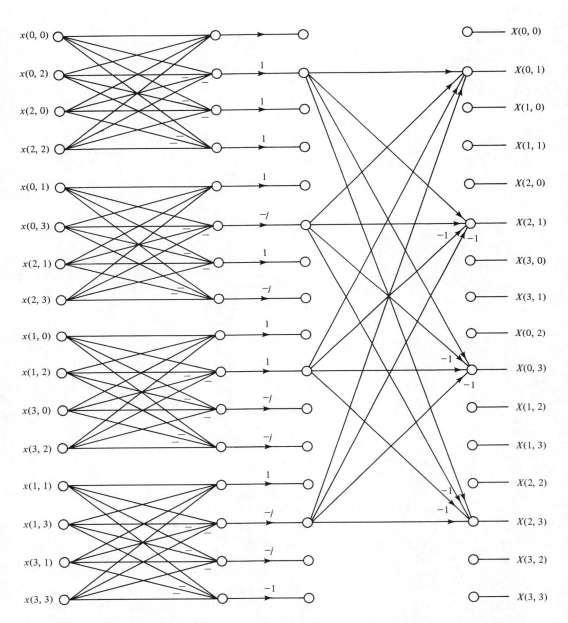

**Figure 2.6**  Complete (4 × 4)-point radix-(2 × 2) DFT. Only one of the four butterflies in the second stage is shown.

| $n_2$ | | | | | | | |
|---|---|---|---|---|---|---|---|
| (0, 7) | (4, 7) | (2, 7) | (6, 7) | (1, 7) | (5, 7) | (3, 7) | (7, 7) |
| (0, 3) | (4, 3) | (2, 3) | (6, 3) | (1, 3) | (5, 3) | (3, 3) | (7, 3) |
| (0, 5) | (4, 5) | (2, 5) | (6, 5) | (1, 5) | (5, 5) | (3, 5) | (7, 5) |
| (0, 1) | (4, 1) | (2, 1) | (6, 1) | (1, 1) | (5, 1) | (3, 1) | (7, 1) |
| (0, 6) | (4, 6) | (2, 6) | (6, 6) | (1, 6) | (5, 6) | (3, 6) | (7, 6) |
| (0, 2) | (4, 2) | (2, 2) | (6, 2) | (1, 2) | (5, 2) | (3, 2) | (7, 2) |
| (0, 4) | (4, 4) | (2, 4) | (6, 4) | (1, 4) | (5, 4) | (3, 4) | (7, 4) |
| (0, 0) | (4, 0) | (2, 0) | (6, 0) | (1, 0) | (5, 0) | (3, 0) | (7, 0) $\longrightarrow$ $n_1$ |

**Figure 2.7**   2-D bit-reversed arrangement of samples for $N = 8$.

any butterfly can be stored in place of the inputs for that butterfly since those inputs are never required again. The output of the FFT will then be stored in place of the input. The operation of bit reversal is also an in-place operation.

The vector-radix decomposition can be generalized in a number of ways. If $N_1 = R_1^i$, $N_2 = R_2^i$ for some integer $i$, a decomposition can be performed with respect to the vector radix $(R_1 \times R_2)$. In this case the basic computational module (or butterfly) is an $(R_1 \times R_2)$-point DFT. This concept can also be extended to the $M$-dimensional case in which the butterfly is an $(R_1 \times R_2 \times \cdots \times R_M)$-point DFT. Mixed-radix decompositions are also possible if not all of the $N_i$ are equal to a common power of their respective radices. In this case different butterflies are used at different stages of the decomposition. Rather than derive explicit FFT algorithms for each of these generalizations, we shall simply comment that they, in turn, are simply special cases of the generalized FFT algorithms considered in Section 2.4.

### 2.3.4 Computational Considerations in DFT Calculations

There are a variety of methods available for measuring the complexity of an $M$-dimensional FFT algorithm. We could, for example, count the number of multiplications and additions, measure the storage, or measure the total computer time needed to compute a particular DFT. The latter is probably the most informative but it is both machine and programmer dependent. Instead of looking for a single, definitive measure of complexity, we shall simply compare algorithms with respect to several different measures of complexity. For purposes of analysis, let us assume that we wish to compute the $M$-dimensional DFT of an $(N \times N \times \cdots \times N)$-point array where $N$ is a power of 2.

A row–column transform requires the evaluation of $MN^{M-1}$ $N$-point 1-D DFTs. In applications where an array processor will be used, this is the critical measure. If a radix-2 1-D FFT is used to evaluate the row–column transforms, we will have a total of

$$C_{\text{r/c FFT}} = \frac{M}{2} N^M \log_2 N \qquad \text{complex multiplications.} \qquad (2.80)$$

and

$$A_{\text{r/c FFT}} = MN^M \log_2 N \qquad \text{complex additions} \qquad (2.81)$$

For a radix-$(2 \times 2 \times \cdots \times 2)$ transform, each butterfly requires $2^M - 1$ multiplications and in each stage there are $(N/2)^M$ butterflies, resulting in a total of

$$C_{\text{vr}(2 \times 2)} = \frac{2^M - 1}{2^M} N^M \log_2 N \qquad (2.82)$$

complex multiplications and

$$A_{\text{vr}(2 \times 2)} = MN^M \log_2 N \qquad (2.83)$$

complex additions. Thus the vector-radix algorithm requires the same number of additions but fewer multiplications. The savings as a function of $M$ are tabulated in Table 2.1.

**TABLE 2.1**  COMPARISON OF THE NUMBER OF COMPLEX MULTIPLICATIONS REQUIRED FOR TWO $M$-DIMENSIONAL FFT ALGORITHMS[a]

| $M$ | $C_{\text{vr}(2 \times 2)}/C_{\text{r/c FFT}}$ |
|---|---|
| 2 | 0.75 |
| 3 | 0.58 |
| 4 | 0.47 |
| 5 | 0.39 |

[a]A hypercubic region of support for the array under consideration is assumed.

The amount of computation needed to evaluate an $M$-dimensional DFT may also be reduced somewhat by using a larger radix. For example, if $N$ is a power of 4, we may use a radix-4 row–column decomposition or a vector radix-$(4 \times 4)$ algorithm to evaluate an $(N \times N)$-point DFT. The butterflies for these algorithms will still involve only the trivial multiplications by $+1$, $+j$, $-1$, and $-j$ and since the number of butterflies is reduced relative to the radix-2 algorithms, the total computation is also reduced (see Problem 2.13).

With both row–column and vector-radix FFTs, the number of complex words of storage is equal to the size of the data array—$N^M$ complex words. For many problems of interest this is greater than the capacity of the primary memory of a minicomputer. Consequently, the signal must reside on a secondary storage device, such as a disk, and be accessed one section at a time. The resulting input–output (I/O) difficulties can severely affect the efficiency of the FFT algorithm.

Consider as an example the problem of computing the 2-D row–column FFT of a $(1024 \times 1024)$-point image on a minicomputer with 16K $(K = 1024)$ complex words of memory available for data storage. We will assume that the entire signal is stored on a disk, and that, although disk "blocks" may be accessed at random, at least one entire block must be read or written during a data transfer operation between the disk and memory. For the purposes of our example, a block will be 1024 complex words long. In reality, blocks may be much shorter, but virtually all disks have the characteristic that some minimum number of words must be transfered. (Even if this were not the case, imposing such a restriction would serve to greatly reduce I/O overhead, usually one of the most time-consuming parts of an executing program.)

If the signal is stored row by row, that is, the signal samples for the $(i + 1)$st row immediately follow those for the $i$th row, we may read in rows one at a time from the disk, perform the row FFTs, and write them back out to the disk. In our particular case, since we have 16K words of memory we can fit 16 rows of data into memory simultaneously. Data transfers to and from the disk can be done in sections which are 16 blocks long. The portion of the program that computes the row FFTs would consist of repeating the following loop 64 times: read in 16 rows from the disk, do 16 1-D FFTs, and write the results back on the disk.

Now, however, there is a serious problem. The column transforms must now be computed, but since the data are stored by rows, accessing the $n$th column means accessing only the $n$th word of each 1K block, which is forbidden. Fortunately, there exists an efficient algorithm due to Eklundh [5] for the transposition of a matrix that resides on secondary storage. After performing the row transforms, the intermediate data array can be transposed, resulting in a data array which is now stored column-wise, and on which the column DFTs can be efficiently computed. The array of results is now stored columnwise; if this is unsatisfactory, a second transposition can be performed.

Eklundh's transposition algorithm is based on a "divide and conquer" strategy. Let the 2-D sequence to be transposed be denoted by $\mathbf{A}$. If $\mathbf{A}$ is $2^{\alpha} \times 2^{\alpha}$ points in extent, it can be partitioned into four smaller sequences each of which is $2^{\alpha-1} \times 2^{\alpha-1}$ points in extent.

$$\mathbf{A} = \begin{array}{|cc|} \hline \mathbf{A}_{12} & \mathbf{A}_{22} \\ \mathbf{A}_{11} & \mathbf{A}_{21} \\ \hline \end{array} \qquad (2.84)$$

($\mathbf{A}$ represents a 2-D sequence so its origin is taken as the lower left-hand corner.) The transpose of $\mathbf{A}$ can be written in terms of the partitions as

$$\mathbf{A}' = \begin{array}{|cc|} \hline \mathbf{A}'_{21} & \mathbf{A}'_{22} \\ \mathbf{A}'_{11} & \mathbf{A}'_{12} \\ \hline \end{array} \qquad (2.85)$$

Each of the partitions is itself transposed and the upper left and lower right partitions are interchanged. Each of the smaller transpositions can now be performed using a repetition of the same algorithm. We see that the complete algorithm will require a total of $\alpha$ such decompositions. The stages of an $8 \times 8$ transposition are shown in Figure 2.8.

The major requirement of this algorithm is that we have room in memory for at least two complete rows of data. In this case we will need to read and write the entire array $\alpha$ times. The row FFTs can be combined with the first stage of the transposition algorithm and the column FFTs can be combined with the last stage, so that $\alpha = \log_2 N$ read/write cycles of the data array are required for the entire 2-D FFT with the results stored columnwise. The number of passes through the data can be reduced, of course, if there is room in memory to store more than two rows at a time. For example, let the number $C$ be the largest power of 2 such that $C$ rows of data may be stored simultaneously in the primary memory of the computer. Then

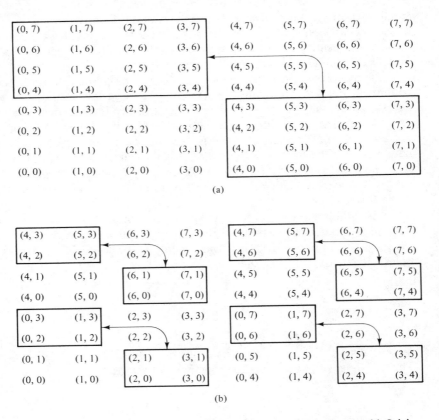

**Figure 2.8** Eklundh's transposition algorithm [5] for an $8 \times 8$ array. (a) Original sequence and exchanges for the first stage. (b) Results of the first stage and the exchanges for the second. (c) Results of the second stage and the exchanges for the third. (d) Results of the third stage.

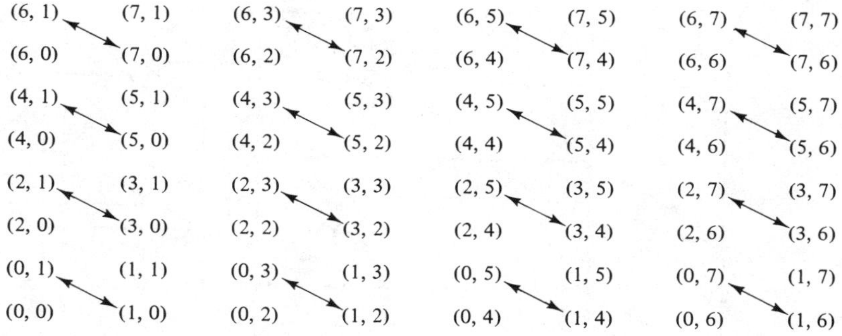

(6, 1)    (7, 1)    (6, 3)    (7, 3)    (6, 5)    (7, 5)    (6, 7)    (7, 7)

(6, 0)    (7, 0)    (6, 2)    (7, 2)    (6, 4)    (7, 4)    (6, 6)    (7, 6)

(4, 1)    (5, 1)    (4, 3)    (5, 3)    (4, 5)    (5, 5)    (4, 7)    (5, 7)

(4, 0)    (5, 0)    (4, 2)    (5, 2)    (4, 4)    (5, 4)    (4, 6)    (5, 6)

(2, 1)    (3, 1)    (2, 3)    (3, 3)    (2, 5)    (3, 5)    (2, 7)    (3, 7)

(2, 0)    (3, 0)    (2, 2)    (3, 2)    (2, 4)    (3, 4)    (2, 6)    (3, 6)

(0, 1)    (1, 1)    (0, 3)    (1, 3)    (0, 5)    (1, 5)    (0, 7)    (1, 7)

(0, 0)    (1, 0)    (0, 2)    (1, 2)    (0, 4)    (1, 4)    (0, 6)    (1, 6)

(c)

| (7, 0) | (7, 1) | (7, 2) | (7, 3) | (7, 4) | (7, 5) | (7, 6) | (7, 7) |
| (6, 0) | (6, 1) | (6, 2) | (6, 3) | (6, 4) | (6, 5) | (6, 6) | (6, 7) |
| (5, 0) | (5, 1) | (5, 2) | (5, 3) | (5, 4) | (5, 5) | (5, 6) | (5, 7) |
| (4, 0) | (4, 1) | (4, 2) | (4, 3) | (4, 4) | (4, 5) | (4, 6) | (4, 7) |
| (3, 0) | (3, 1) | (3, 2) | (3, 3) | (3, 4) | (3, 5) | (3, 6) | (3, 7) |
| (2, 0) | (2, 1) | (2, 2) | (2, 3) | (2, 4) | (2, 5) | (2, 6) | (2, 7) |
| (1, 0) | (1, 1) | (1, 2) | (1, 3) | (1, 4) | (1, 5) | (1, 6) | (1, 7) |
| (0, 0) | (0, 1) | (0, 2) | (0, 3) | (0, 4) | (0, 5) | (0, 6) | (0, 7) |

(d)

**Figure 2.8**  (*Continued*)

the number of passes required to transpose an $N \times N$ array can be reduced to the smallest integer greater than $\log_2 N / \log_2 C$.

The case we have studied above involved the transposition of an $N \times N$ array where $N$ is a power of 2. In general, an algorithm could be similarly derived for $N$ a power of an arbitrary radix $R$. As with the FFT, it is also possible to conceive of a composite transposition algorithm based on a prime factorization of $N$ when it is not the power of an integer.

The problem of array transposition becomes even more severe for $M$-dimensional row–column decompositions. Consider the final stage in such an algorithm where $N(M-1)$-dimensional DFTs have been computed. The data must be arranged so that the final $N^{M-1}$ 1-D DFTs can be computed from data stored in contiguous locations in secondary storage. This rearrangement requires $N^{M-2}$ 2-D transpositions, as illustrated in Figure 2.9 for the 3-D case. If $T_N(M)$ represents the number of 2-D transpositions needed to perform an $M$-dimensional DFT, it must satisfy the following recursion:

$$T_N(M) = NT_N(M-1) + N^{M-2} \tag{2.86}$$

with the initial condition

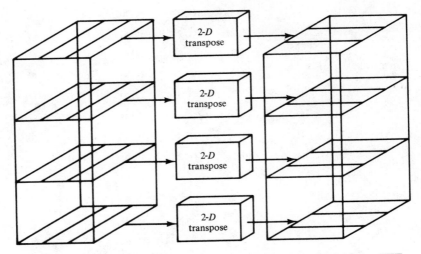

**Figure 2.9**  $N$ 2-D transpositions are needed to prepare for the final row FFTs in a 3-D row-column FFT ($N = 4$).

$$T_N(2) = 1$$

Solving this recursive equation, we see that the total number of 2-D transpositions needed for an $M$-dimensional row–column FFT is

$$T_N(M) = (M - 1)N^{M-2} \tag{2.87}$$

Because each 2-D transposition accesses only $N^2$ of the data points, the number of read/write cycles of the *entire* data array ($N^M$ points) required for an $M$-dimensional row–column decomposition is

$$(M - 1)\frac{\log_2 N}{\log_2 C} \tag{2.88}$$

where, as before, $C$ is the number of rows which can be stored simultaneously in memory.

The multidimensional vector-radix FFT requires one read/write cycle of the entire data array to accomplish the bit-reversed reordering of the data, followed by $\log_2 N$ additional read/write cycles to perform the $\log_2 N$ stages of decimation in the vector-radix algorithm, giving a total of $1 + \log_2 N$ read/write cycles. However, the $M$-dimensional vector-radix FFT also requires that $2^{M-1}$ rows of data be stored simultaneously in the primary memory of the computer in order to compute the butterflies efficiently. If this quantity of storage is available to the $M$-dimensional row–column FFT, we can set $C = 2^{M-1}$ in expression (2.88), implying that only $\log_2 N$ read/write cycles are required for the row–column FFT, one less than the number required for the vector-radix FFT. Depending on the values of the transform parameters and the relative speeds of computation and I/O on the particular machine in question, the computational advantage of the vector-radix decomposition may or may not be offset by the need to perform an additional read/write cycle.

## *2.4 DISCRETE FOURIER TRANSFORMS FOR GENERAL PERIODICALLY SAMPLED SIGNALS

In Section 1.5 we showed that several signal processing algorithms could be extended to apply to signals that have been sampled with a general periodic sampling geometry. In this section we apply the same reasoning that led to the DFT and FFT algorithms to the general case of periodically sampled signals. We will show that such signals possess discrete Fourier series and discrete Fourier transform representations. We will also relate these DFTs to sampled Fourier transforms and we will discover a general family of fast Fourier transform algorithms which include the row–column decomposition and the vector-radix decomposition as special cases.

### 2.4.1 DFT Relations for General Periodically Sampled Signals

Let us consider a periodic sequence $\tilde{x}(\mathbf{n})$ with periodicity matrix $\mathbf{N}$. For such a sequence

$$\tilde{x}(\mathbf{n}) = \tilde{x}(\mathbf{n} + \mathbf{Nr}) \tag{2.89}$$

for any integer vector $\mathbf{r}$. Let $I_\mathbf{N}$ denote a region in the $(n_1, n_2)$-plane which contains exactly one period of this sequence. We shall call this region the *fundamental period* of the array. It contains $|\det \mathbf{N}|$ samples of $\tilde{x}$. ($I_\mathbf{N}$ is a generalization of $R_{N_1 N_2}$ used earlier.)

By analogy with the rectangular case, let us hypothesize that $\tilde{x}(\mathbf{n})$ can be uniquely represented as a finite sum of harmonically related complex sinusoids.

$$\tilde{x}(\mathbf{n}) = \sum_{\mathbf{k} \in J_\mathbf{N}} a(\mathbf{k}) \exp(j\mathbf{k}'\mathbf{R}'\mathbf{n}) \tag{2.90}$$

where $\mathbf{k}$ is an integer vector. $J_\mathbf{N}$ denotes a finite-extent region in the $\mathbf{k}$-domain. Since the sequence $\tilde{x}$ is periodic, we see that

$$
\begin{aligned}
\tilde{x}(\mathbf{n}) = \tilde{x}(\mathbf{n} + \mathbf{Nr}) &= \sum_{\mathbf{k} \in J_\mathbf{N}} a(\mathbf{k}) \exp[j\mathbf{k}'\mathbf{R}'(\mathbf{n} + \mathbf{Nr})] \\
&= \sum_{\mathbf{k} \in J_\mathbf{N}} a(\mathbf{k}) \exp(j\mathbf{k}'\mathbf{R}'\mathbf{Nr}) \exp(j\mathbf{k}'\mathbf{R}'\mathbf{n})
\end{aligned}
\tag{2.91}
$$

Since the right sides of (2.90) and (2.91) must be equal for all values of $\mathbf{n}$ and $\mathbf{r}$, we must have

$$\exp(j\mathbf{k}'\mathbf{R}'\mathbf{Nr}) = 1 \tag{2.92}$$

for all integer vectors $\mathbf{r}$ and $\mathbf{k}$.† For nontrivial $\mathbf{R}'$ and $\mathbf{N}$, this implies that

$$\mathbf{R}'\mathbf{N} = 2\pi\mathbf{I}$$

or

$$\mathbf{R}' = 2\pi\mathbf{N}^{-1} \tag{2.93}$$

---

†The validity of this argument rests on the orthogonality of the harmonically related complex exponentials over one period.

If we substitute for $\mathbf{R}'$ and let $a(\mathbf{k}) = (1/|\det \mathbf{N}|)\tilde{X}(\mathbf{k})$, we arrive at the following expression:

$$x(\mathbf{n}) = \frac{1}{|\det \mathbf{N}|} \sum_{\mathbf{k} \in J_\mathbf{N}} \tilde{X}(\mathbf{k}) \exp\left[j\mathbf{k}'(2\pi\mathbf{N}^{-1})\mathbf{n}\right] \tag{2.94}$$

Since the complex exponentials in this sum are periodic in both $\mathbf{n}$ (periodicity matrix $\mathbf{N}$) and $\mathbf{k}$ (periodicity matrix $\mathbf{N}'$), we see that at most $|\det \mathbf{N}|$ samples of $\tilde{X}(\mathbf{k})$ can be independent. Thus the region $J_\mathbf{N}$, like $I_\mathbf{N}$, will contain only $|\det \mathbf{N}|$ samples. If $\tilde{X}(\mathbf{k})$ is defined as

$$\tilde{X}(\mathbf{k}) \triangleq \sum_{\mathbf{n} \in I_\mathbf{N}} \tilde{x}(\mathbf{n}) \exp\left[-j\mathbf{k}'(2\pi\mathbf{N}^{-1})\mathbf{n}\right] \tag{2.95}$$

we can establish the existence of a Fourier series relation for any periodic sequence. It is straightforward to verify that (2.94) and (2.95) constitute an identity. It is also straightforward to establish the uniqueness of (2.95) due to the orthogonality of the complex exponentials $\exp\left[-j\mathbf{k}'(2\pi\mathbf{N}^{-1})\mathbf{n}\right]$ over the region $I_\mathbf{N}$. We also note that $\tilde{X}(\mathbf{k})$ is periodic with periodicity matrix $\mathbf{N}'$:

$$\tilde{X}(\mathbf{k}) = \tilde{X}(\mathbf{k} + \mathbf{N}'\mathbf{r}) \tag{2.96}$$

If $x(\mathbf{n})$ is a finite-extent sequence with support confined to $I_\mathbf{N}$, we can use the foregoing Fourier series relations to define a discrete Fourier transform (DFT)

$$X(\mathbf{k}) = \sum_{\mathbf{n} \in I_\mathbf{N}} x(\mathbf{n}) \exp\left[-j\mathbf{k}'(2\pi\mathbf{N}^{-1})\mathbf{n}\right] \tag{2.97}$$

$$x(\mathbf{n}) = \frac{1}{|\det \mathbf{N}|} \sum_{\mathbf{k} \in J_\mathbf{N}} X(\mathbf{k}) \exp\left[j\mathbf{k}'(2\pi\mathbf{N}^{-1})\mathbf{n}\right] \tag{2.98}$$

These relations are similar to equations (2.20) and (2.21). The only difference is that the matrix $\mathbf{N}$ is not restricted to be diagonal.

Recall that $\mathbf{N}$ is the spatial domain periodicity matrix; it relates the finite-extent sequence to its periodic extension. The periodic extension of $x$ is not unique; any finite-extent sequence may have several periodic extensions from which it can be recovered. Consider, for example, a rectangularly sampled signal with the $(N_1 \times N_2)$-point region of support shown in Figure 2.10(a). It can be periodically extended in a rectangular fashion [Figure 2.10(b)] by means of the periodicity matrix

$$\mathbf{N} = \begin{bmatrix} N_1 & 0 \\ 0 & N_2 \end{bmatrix} \tag{2.99}$$

or it can be extended hexagonally [Figure 2.10(c)] using the matrix

$$\mathbf{N} = \begin{bmatrix} N_1 & N_1 \\ \dfrac{N_2}{2} & -\dfrac{N_2}{2} \end{bmatrix} \tag{2.100}$$

under the assumption that $N_2$ is divisible by 2 [6]. The reader should be able to find other extensions as well.

Each choice of $\mathbf{N}$ leads to a different periodic extension and thus each leads to a different DFT. How are all of these similar? They are all transforms of the same

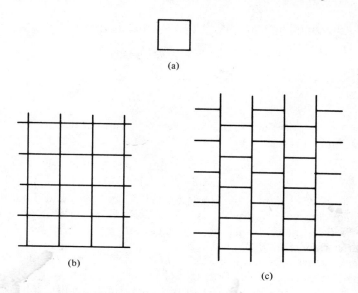

**Figure 2.10** Rectangularly sampled finite-extent sequence and two periodic extensions of that sequence.

sequence, and consequently they all correspond to samples of the Fourier transform of that sequence. How are they different? They differ in the manner in which the Fourier transform is sampled. Comparing (2.97) with (1.133a), which defines the Fourier transform $X(\omega)$, we see that

$$X(\mathbf{k}) = X(\omega)\Big|_{\omega = (2\pi\mathbf{N}^{-1})'\mathbf{k} = \mathbf{R}\mathbf{k}} \qquad (2.101)$$

The matrix $\mathbf{R} \triangleq (2\pi\mathbf{N}^{-1})'$ is the Fourier-domain sampling matrix.

In our derivation of the sampling theorem in Chapter 1 we defined two matrices $\mathbf{U}$ and $\mathbf{V}$ which were related by

$$\mathbf{U}'\mathbf{V} = 2\pi\mathbf{I}$$

$\mathbf{V}$ was the sampling matrix which indicated where the samples of a bandlimited analog signal should be taken. $\mathbf{U}$ represented how the Fourier transform of the original signal was periodically extended to give the Fourier transform of the sampled signal. One interpretation of the DFT is that it represents a sampled Fourier transform. The matrix that defines the frequency samples, $\mathbf{R}$, is thus analogous to $\mathbf{V}$, except that the spatial and frequency domains are reversed. Similarly, $\mathbf{N}$, which satisfies

$$\mathbf{N}'\mathbf{R} = 2\pi\mathbf{I}$$

is analogous to $\mathbf{U}$. It tells how the sequence in the other domain, in this case the spatial domain, must be periodically extended (or aliased).

If we begin with a continuous, bandlimited signal $x_a(t)$ that has a continuous Fourier transform $X_a(\mathbf{\Omega})$, the discrete signal $x(\mathbf{n})$ can be generated by sampling $x_a(t)$

using the sampling matrix $\mathbf{V}$. As we saw in Chapter 1, if

$$x(\mathbf{n}) = x_a(\mathbf{Vn}) \tag{2.102}$$

then

$$X(\boldsymbol{\omega}) = \frac{1}{|\det \mathbf{V}|} X_a(\mathbf{V}'^{-1}\boldsymbol{\omega}) \tag{2.103}$$

[The assumption that $x_a(t)$ is bandlimited removes the problem of aliasing.] Using equation (2.101), we get

$$X(\mathbf{k}) = \frac{1}{|\det \mathbf{V}|} X_a(2\pi(\mathbf{VN})'^{-1}\mathbf{k}) \tag{2.104}$$

The DFT $X(\mathbf{k})$ thus corresponds to scaled samples of the Fourier transform of the original continuous signal at the frequencies $\boldsymbol{\Omega} = 2\pi(\mathbf{VN})'^{-1}\mathbf{k}$. We can interpret the matrix $\mathbf{S} \triangleq 2\pi(\mathbf{VN})'^{-1}$ as the sampling matrix that determines how the continuous Fourier transform $X_a(\boldsymbol{\Omega})$ is sampled.

In general, $\mathbf{N}$ and $\mathbf{R}$ are $M \times M$ matrices where $M$ is the dimensionality of the signals being considered. The matrix $\mathbf{N}$ must be invertible, and the elements of $\mathbf{N}$ (but not of $\mathbf{R}$) must be integers. With this understanding all the formulas in this section are equally valid for signals of any positive dimensionality.

### 2.4.2 Fast Fourier Transform Algorithms for General Periodically Sampled Signals [7]

In this section we consider efficient algorithms for the evaluation of a DFT of the form

$$X(\mathbf{k}) = \sum_{\mathbf{n} \in I_\mathbf{N}} x(\mathbf{n}) \exp\left[-j\mathbf{k}'(2\pi\mathbf{N}^{-1})\mathbf{n}\right] \tag{2.105}$$

Fast Fourier transform algorithms exist whenever $\mathbf{N}$ is a composite matrix, that is, whenever $\mathbf{N}$ can be factored into a nontrivial product of integer matrices. This is consistent with the existence condition for a 1-D FFT, which requires that the length of the 1-D DFT be a composite integer. As in the 1-D case, we will see that the more factors that can be found for $\mathbf{N}$, the greater the computational savings.

In the remainder of this section all the matrices that we will use will be understood to have only integer elements. Any matrix $\mathbf{E}$ for which $|\det \mathbf{E}| = 1$ will be called a *unimodular matrix*. It should be noted that $\mathbf{E}^{-1}$ is also a unimodular matrix. Unimodular matrices are the only integer matrices whose inverses are also integer matrices.

If $|\det \mathbf{N}|$ is a prime number, we will say that $\mathbf{N}$ is a *prime matrix*. If $\mathbf{N}$ is neither a prime nor unimodular, we will say that it is *composite*. [It should be remembered that $|\det \mathbf{N}|$ is always an integer and that it is equal to the number of samples in $I_\mathbf{N}$, the region of support of $x(\mathbf{n})$.]

If $\mathbf{N}$ is composite, it can be decomposed into the product of two matrices

$$\mathbf{N} = \mathbf{PQ} \tag{2.106}$$

where neither $\mathbf{P}$ nor $\mathbf{Q}$ is unimodular. It should be noted that such a factorization is not unique since

$$\mathbf{N} = [\mathbf{PE}][\mathbf{E}^{-1}\mathbf{Q}] \qquad (2.107)$$

provides another factorization of $\mathbf{N}$ for any unimodular matrix $\mathbf{E}$.

We will say that two integer vectors $\mathbf{m}$ and $\mathbf{n}$ are *congruent* ($\mathbf{m} \equiv \mathbf{n}$) with respect to the matrix modulus $\mathbf{N}$ if

$$\mathbf{m} = \mathbf{n} + \mathbf{Nr} \qquad (2.108)$$

for some integer vector $\mathbf{r}$. We will use the notation

$$\mathbf{m} = ((\mathbf{n}))_{\mathbf{N}}$$

to mean, first, that $\mathbf{m} \equiv \mathbf{n}$ and second, that $\mathbf{m} \in I_{\mathbf{N}}$. Every vector $\mathbf{n}$ in the periodic extension of $I_{\mathbf{N}}$ is congruent to a vector in $I_{\mathbf{N}}$.

Any vector $\mathbf{n}$ in the region $I_{\mathbf{N}}$ can be uniquely expressed as

$$\mathbf{n} = ((\mathbf{Pq} + \mathbf{p}))_{\mathbf{N}} \qquad (2.109)$$

where

$$\mathbf{p} \in I_{\mathbf{P}}$$

$$\mathbf{q} \in I_{\mathbf{Q}}$$

The set $I_{\mathbf{P}}$ contains $|\det \mathbf{P}|$ integer vectors and the set $I_{\mathbf{Q}}$ contains $|\det \mathbf{Q}|$ integer vectors. Furthermore, any pair of vectors—one from $I_{\mathbf{P}}$ and one from $I_{\mathbf{Q}}$—determines a unique element from $I_{\mathbf{N}}$. The vector $\mathbf{q}$ can be interpreted as the "quotient" when $\mathbf{n}$ is "divided" by $\mathbf{P}$, and $\mathbf{p}$ can be interpreted as the "remainder."

In a similar fashion we can define

$$\mathbf{k}' = ((\mathbf{i}' + \mathbf{m}'\mathbf{Q}))_{\mathbf{N}} \qquad (2.110)$$

where

$$\mathbf{m} \in J_{\mathbf{P}}$$

$$\mathbf{i} \in J_{\mathbf{Q}}$$

With these definitions the DFT summation of (2.105) can be rewritten as

$$X(\mathbf{Q}'\mathbf{m} + \mathbf{i}) = \sum_{\mathbf{p}} \sum_{\mathbf{q}} x(((\mathbf{Pq} + \mathbf{p}))_{\mathbf{N}}) \exp\left[-j(\mathbf{i}' + \mathbf{m}'\mathbf{Q})(2\pi\mathbf{N}^{-1})(\mathbf{Pq} + \mathbf{p})\right] \qquad (2.111)$$

By expanding the exponential, this sum can be decomposed into two parts:

$$C(\mathbf{p}, \mathbf{i}) = \sum_{\mathbf{q} \in I_{\mathbf{Q}}} x(((\mathbf{Pq} + \mathbf{p}))_{\mathbf{N}}) \exp\left[-j\mathbf{i}'(2\pi\mathbf{Q}^{-1})\mathbf{q}\right] \qquad (2.112a)$$

$$X(\mathbf{Q}'\mathbf{m} + \mathbf{i}) = \sum_{\mathbf{p} \in I_{\mathbf{P}}} C(\mathbf{p}, \mathbf{i}) \exp\left[-j\mathbf{i}'(2\pi\mathbf{N}^{-1})\mathbf{p}\right] \exp\left[-j\mathbf{m}'(2\pi\mathbf{P}^{-1})\mathbf{p}\right] \qquad (2.112b)$$

These relations represent the first level of decomposition of a decimation-in-time Cooley–Tukey FFT algorithm. It is helpful to consider the two equations separately. The sequence $C(\mathbf{p}, \mathbf{i})$ is periodic in $\mathbf{i}$ with periodicity matrix $\mathbf{Q}'$. Thus the summation in (2.112a) represents a 2-D DFT of the array $x(((\mathbf{Pq} + \mathbf{p}))_{\mathbf{N}})$ taken with respect to the periodicity matrix $\mathbf{Q}$. The region of support for this sequence is $\mathbf{I_Q}$, which must be chosen to be one period of $x(((\mathbf{Pq} + \mathbf{p}))_{\mathbf{N}})$, interpreted as a function of $\mathbf{q}$. A different matrix-$\mathbf{Q}$ DFT must be evaluated for each value of the vector $\mathbf{p}$. This means that $|\det \mathbf{P}|$ such transforms need to be evaluated in all.

The summation in (2.112b) shows how the results of these matrix-$\mathbf{Q}$ DFTs should be combined to produce the matrix-$\mathbf{N}$ DFT. The numbers $C(\mathbf{p}, \mathbf{i})$ are first multiplied by the factors $\exp\left[-j\mathbf{i}'(2\pi\mathbf{N}^{-1})\mathbf{p}\right]$ (which are sometimes called *twiddle factors*), and the products are combined in a series of matrix-$\mathbf{P}$ DFTs or *butterflies*. The number of twiddle-factor multiplications is $|\det \mathbf{N}|$ and the number of matrix-$\mathbf{P}$ butterflies is $|\det \mathbf{Q}|$. If either $\mathbf{P}$ or $\mathbf{Q}$ is composite, either set of smaller DFTs can be decomposed further.

The vectors

$$\mathbf{n} = ((\mathbf{Pq}))_{\mathbf{N}}, \qquad \mathbf{q} \in I_{\mathbf{Q}} \tag{2.113}$$

form that subset of the region of support which is created by sampling $I_{\mathbf{N}}$ with the sampling matrix $\mathbf{P}$. For a fixed value of $\mathbf{p}$, the samples

$$\mathbf{n} = ((\mathbf{Pq} + \mathbf{p}))_{\mathbf{N}}, \qquad \mathbf{q} \in I_{\mathbf{Q}} \tag{2.114}$$

form a coset with respect to this subset. Since each coset has the same size as the subset, there are $|\det \mathbf{N}|/|\det \mathbf{Q}| = |\det \mathbf{P}|$ cosets in all. The members of any coset are congruent to one another with respect to the modulus $\mathbf{P}$. The region $I_{\mathbf{P}}$ should be chosen to consist of one member from each coset—$|\det \mathbf{P}|$ elements in all.

The regions $J_{\mathbf{P}}$ and $J_{\mathbf{Q}}$ can be chosen similarly. Since the discrete Fourier transform is periodic in $\mathbf{k}$ with periodicity matrix $\mathbf{N}'$ and

$$\mathbf{N}' = \mathbf{Q}'\mathbf{P}'$$
$$\mathbf{k} = ((\mathbf{Q}'\mathbf{m} + \mathbf{i}))_{\mathbf{N}'} \tag{2.115}$$

we see that in the frequency domain $\mathbf{Q}'$ plays a role that is analogous to $\mathbf{P}$ in the spatial domain, and $\mathbf{P}'$ plays a role that is analogous to $\mathbf{Q}$. Thus we can identify samples of the form

$$\mathbf{k} = ((\mathbf{Q}'\mathbf{m}))_{\mathbf{N}'}, \qquad \mathbf{m} \in J_{\mathbf{P}} \tag{2.116}$$

as a subset of samples from the frequency domain. The set $J_{\mathbf{P}}$ should be chosen to consist of a set of $|\det \mathbf{P}|$ vectors which will generate that subset. For fixed values of $\mathbf{i}$, vectors of the form

$$\mathbf{k} = ((\mathbf{Q}'\mathbf{m} + \mathbf{i}))_{\mathbf{N}'} \tag{2.117}$$

can be sorted into $|\det \mathbf{Q}|$ cosets. $J_{\mathbf{Q}}$ must be chosen to consist of one member of each coset.

At this point some of these issues can perhaps be clarified through the consideration of a simple example. Consider the evaluation of a $4 \times 4$ rectangular DFT with the periodicity matrix

$$\mathbf{N} = \begin{bmatrix} 4 & 0 \\ 0 & 4 \end{bmatrix} \tag{2.118}$$

using the factorization

$$\mathbf{P} = \begin{bmatrix} 2 & 2 \\ 1 & -1 \end{bmatrix} \qquad \mathbf{Q} = \begin{bmatrix} 1 & 2 \\ 1 & -2 \end{bmatrix} \tag{2.119}$$

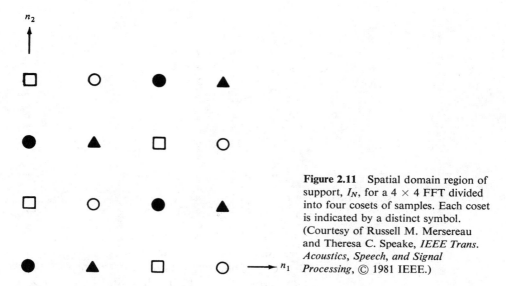

**Figure 2.11**  Spatial domain region of support, $I_N$, for a $4 \times 4$ FFT divided into four cosets of samples. Each coset is indicated by a distinct symbol. (Courtesy of Russell M. Mersereau and Theresa C. Speake, *IEEE Trans. Acoustics, Speech, and Signal Processing*, © 1981 IEEE.)

In Figure 2.11 we show the region $I_N$ divided into four cosets by the sampling matrix **P**; a different symbol is used to indicate the members of each coset. Notice that all four cosets have the same geometry when periodically extended. To define $I_Q$, we need a set of four vectors that will satisfy (2.113); that is, we need a set of four vectors, which through the use of (2.109) will span any one of the cosets. One set of vectors that will accomplish this is

$$I_Q = \{(0, 0)', (1, 0)', (2, 0)', (3, 0)'\} \qquad (2.120)$$

This set is not unique. The set $I_P$ must be chosen to consist of one member of each coset. Thus one possibility for $I_P$ is

$$I_P = \{(0, 0)', (1, 0)', (2, 0)', (3, 0)'\} \qquad (2.121)$$

In Figure 2.12 we show the region of support for the DFT, $J_N$, divided into four cosets by the frequency-domain sampling matrix $\mathbf{Q}'$. Through an examination of this figure we see that possible choices for $J_P$ and $J_Q$ are

$$J_P = \{(0, 0)', (1, 0)', (2, 0)', (3, 0)'\} \qquad (2.122)$$

$$J_Q = \{(0, 0)', (0, 1)', (0, 2)', (0, 3)'\} \qquad (2.123)$$

Once these four sets have been chosen, a partial flowchart for the algorithm can be drawn, which is done in Figure 2.13. Each matrix-**Q** DFT operates on one of the cosets of the input, which were shown in Figure 2.11, to produce an intermediate array $C(m_1, m_2)$. [This array is the same as $C(\mathbf{p}, \mathbf{i})$ in (2.112a), but is expressed in terms of two real indices instead of two vector ones.] That array is multiplied by the twiddle factors (which for this FFT are all 1) and the results are fed to the matrix-**P** DFTs. Each of the latter DFTs produces one of the output cosets shown in Figure 2.12.

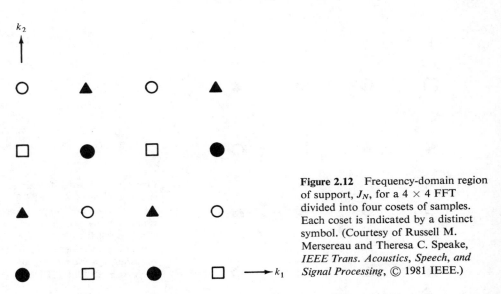

**Figure 2.12** Frequency-domain region of support, $J_N$, for a $4 \times 4$ FFT divided into four cosets of samples. Each coset is indicated by a distinct symbol. (Courtesy of Russell M. Mersereau and Theresa C. Speake, *IEEE Trans. Acoustics, Speech, and Signal Processing,* © 1981 IEEE.)

The four outputs of the matrix-**Q** DFTs can be computed directly from the four inputs. If the inputs are denoted by $w$, $x$, $y$, and $z$ and the outputs by $A$, $B$, $C$, and $D$, the direct evaluation of these DFTs corresponds to setting

$$A = w + x + y + z \tag{2.124a}$$

$$B = w - jx - y + jz \tag{2.124b}$$

$$C = w - x + y - z \tag{2.124c}$$

$$D = w + jx - y - jz \tag{2.124d}$$

Alternatively, since **Q** is composite, we could use the factorization

$$\mathbf{Q} = \begin{bmatrix} 1 & 2 \\ 1 & -2 \end{bmatrix} = \begin{bmatrix} 1 & 1 \\ 1 & -1 \end{bmatrix} \begin{bmatrix} 1 & 0 \\ 0 & 2 \end{bmatrix} \tag{2.125}$$

The flowchart of a four-point DFT based on this factorization is shown in Figure 2.14.

For this particular example the matrix-**P** DFTs are very similiar. If the inputs to these transforms are denoted as $w$, $x$, $y$, and $z$ and the outputs are denoted $A$, $B$, $C$, and $D$, the direct evaluation results in the same equations as for the matrix-**Q** DFTs given in (2.124). In this example the inputs and outputs may be arranged in such a way that the flowchart in Figure 2.14 describes both the matrix-**Q** DFTs and the matrix-**P** DFTs.

If $C_\mathbf{N}$ denotes the computational complexity of a matrix-**N** FFT algorithm measured in terms of the number of complex multiplications required, then

$$C_\mathbf{N} \le |\det \mathbf{P}| \, C_\mathbf{Q} + |\det \mathbf{Q}| \, C_\mathbf{P} + |\det \mathbf{N}| \tag{2.126}$$

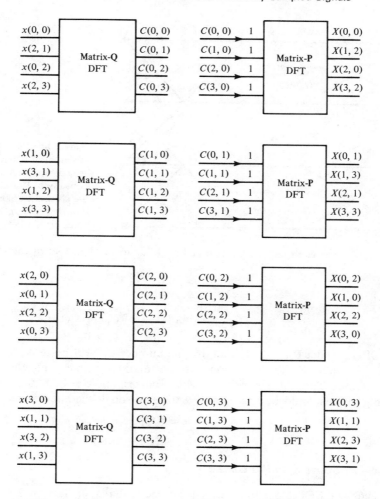

**Figure 2.13** Flowchart of the $4 \times 4$ FFT for the example defined by equation (2.119). (Courtesy of Russell M. Mersereau and Theresa C. Speake, *IEEE Trans. Acoustics, Speech, and Signal Processing*, © 1981 IEEE.)

The first term represents the number of complex multiplications in $|\det \mathbf{P}|$ matrix-$\mathbf{Q}$ DFTs; the second term represents the contribution from the $|\det \mathbf{Q}|$ matrix-$\mathbf{P}$ DFTs; and the final term represents the contribution from the multiplications by the twiddle factors. This expression is presented as an inequality because in some instances the number of complex multiplications required may be less. This occurs when some of the coefficients in the algorithm reduce to $1, -1, j,$ or $-j$. In the example given above there were no multiplications due to the twiddle factors, for example. (In fact, for that transform there were no multiplications at all—only complex additions and subtractions.)

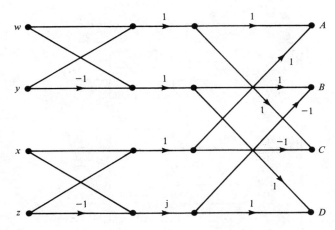

**Figure 2.14** Flowchart for the evaluation of a matrix-**Q** DFT using the decomposition in equation (2.125). (Courtesy of Russell M. Mersereau and Theresa C. Speake, *IEEE Trans. Acoustics, Speech, and Signal Processing,* © 1981 IEEE.)

This result can be generalized if **N** has more than two prime factors. If

$$\mathbf{N} = \prod_{i=1}^{v} \mathbf{P}_i \tag{2.127}$$

then

$$C_\mathbf{N} \le v \,|\det \mathbf{N}| + \sum_{i=1}^{v} C_{\mathbf{P}_i} \prod_{\substack{j=1 \\ j \ne i}}^{v} |\det \mathbf{P}_j| \tag{2.128}$$

Often, if $|\det \mathbf{P}_i| = 2, 4, 8,$ or 16, the number $C_{\mathbf{P}_i}$ will be zero.

These DFTs correspond to decimation-in-time algorithms. A similar but different class of algorithms, which correspond to 1-D decimation-in-frequency algorithms could be analogously derived through the alternative substitutions

$$\mathbf{n} = ((\mathbf{Qp} + \mathbf{q}))_\mathbf{N} \tag{2.129a}$$

$$\mathbf{k} = ((\mathbf{P'i} + \mathbf{m}))_{\mathbf{N'}} \tag{2.129b}$$

### 2.4.3 Some Special Cases

The two FFT algorithms which we encountered in Section 2.3—the row–column decomposition and the vector-radix algorithm—can both be derived as special cases of the general DFT algorithm. The DFT of a rectangularly sampled array, which we defined earlier in this chapter, corresponds to a diagonal periodicity matrix **N** of the form

$$\mathbf{N} = \begin{bmatrix} N_1 & 0 \\ 0 & N_2 \end{bmatrix} \tag{2.130}$$

The row–column algorithms correspond to the factorizations

$$\mathbf{N} = \mathbf{P}_1 \mathbf{Q}_1 = \begin{bmatrix} N_1 & 0 \\ 0 & 1 \end{bmatrix} \begin{bmatrix} 1 & 0 \\ 0 & N_2 \end{bmatrix} \tag{2.131}$$

$$\mathbf{N} = \mathbf{P}_2 \mathbf{Q}_2 = \begin{bmatrix} 1 & 0 \\ 0 & N_2 \end{bmatrix} \begin{bmatrix} N_1 & 0 \\ 0 & 1 \end{bmatrix} \tag{2.132}$$

With the former factorization the column transforms are performed before the row transforms; with the latter, vice versa. With the first factorization, any integer vector $(n_1, n_2)'$ in the region

$$I_N = \{(n_1, n_2)': \quad 0 \leq n_1 < N_1 \quad \text{and} \quad 0 \leq n_2 < N_2\}$$

can be written in the form

$$\begin{bmatrix} n_1 \\ n_2 \end{bmatrix} = \begin{bmatrix} N_1 & 0 \\ 0 & 1 \end{bmatrix} \begin{bmatrix} 0 \\ n_2 \end{bmatrix} + \begin{bmatrix} n_1 \\ 0 \end{bmatrix} \tag{2.133}$$

Thus the set $I_Q$ contains the $N_2$ vectors

$$I_Q = \{(0, n_2)': \quad 0 \leq n_2 < N_2\}$$

Similarly, we can identify the set $I_P$ as

$$I_P = \{(n_1, 0)': \quad 0 \leq n_1 < N_1\}$$

If $N_1$ and $N_2$ are each powers of 2, we can further decompose **P** and **Q** to get

$$N = \begin{bmatrix} 2 & 0 \\ 0 & 1 \end{bmatrix} \begin{bmatrix} 2 & 0 \\ 0 & 1 \end{bmatrix} \cdots \begin{bmatrix} 2 & 0 \\ 0 & 1 \end{bmatrix} \begin{bmatrix} 1 & 0 \\ 0 & 2 \end{bmatrix} \begin{bmatrix} 1 & 0 \\ 0 & 2 \end{bmatrix} \cdots \begin{bmatrix} 1 & 0 \\ 0 & 2 \end{bmatrix} \tag{2.134}$$

This decomposition corresponds to the use of a 1-D radix-2 FFT to evaluate the row and column transforms.

If $N_1$ and $N_2$ are divisible by 2, it is also possible to perform the factorization $N = PQ$ as follows:

$$\begin{bmatrix} N_1 & 0 \\ 0 & N_2 \end{bmatrix} = \begin{bmatrix} 2 & 0 \\ 0 & 2 \end{bmatrix} \begin{bmatrix} \dfrac{N_1}{2} & 0 \\ 0 & \dfrac{N_2}{2} \end{bmatrix} \tag{2.135}$$

Then any integer vector in the set $I_N$ can be represented by

$$\begin{bmatrix} n_1 \\ n_2 \end{bmatrix} = \begin{bmatrix} 2 & 0 \\ 0 & 2 \end{bmatrix} \mathbf{q} + \mathbf{p} \tag{2.136}$$

In this case the set $I_P = J_P$ contains the four elements $(0, 0)'$, $(0, 1)'$, $(1, 0)'$, and $(1, 1)'$ and the set $I_Q = J_Q$ consists of the $N_1 N_2/4$ vectors of the form $(n_1, n_2)'$ with $0 \leq n_1 < N_1/2$ and $0 \leq n_2 < N_2/2$.

This factorization of **N** corresponds to the first stage of decimation in the vector-radix algorithm derived in Section 2.3.3. If $N_1 = N_2 = N$ and $N$ is a power of 2, the complete factorization of **N** for the radix-$(2 \times 2)$ FFT is

$$N = \begin{bmatrix} 2 & 0 \\ 0 & 2 \end{bmatrix} \begin{bmatrix} 2 & 0 \\ 0 & 2 \end{bmatrix} \cdots \begin{bmatrix} 2 & 0 \\ 0 & 2 \end{bmatrix} \tag{2.137}$$

This approach can be readily applied to the cases of other radices, mixed radices, or higher-dimensional transforms.

We saw in Chapter 1 that, after rectangularly sampled signals, the next most important class of sequences is the class of hexagonally sampled signals. A DFT that will relate a hexagonally sampled signal to hexagonal samples of its Fourier transform [6] is given by

$$X(k_1, k_2) = \sum_{n_1=0}^{3N-1} \sum_{n_2=0}^{N-1} x(n_1, n_2) \exp\left[-j\frac{2\pi}{3N}\{(2n_1 - n_2)k_1 + 3n_2k_2\}\right] \quad (2.138)$$

This corresponds to the periodicity matrix

$$\mathbf{N} = \begin{bmatrix} 2N & N \\ N & 2N \end{bmatrix} \quad (2.139)$$

If the 2-D discrete signal $x(n_1, n_2)$ was obtained by sampling the bandlimited function $x_a(t_1, t_2)$ using the hexagonal sampling matrix

$$\mathbf{V}_{\text{hex}} = \begin{bmatrix} T_1 & T_1 \\ T_2 & -T_2 \end{bmatrix}$$

then the matrix $\mathbf{S}$ which relates $X(k_1, k_2)$ and $X_a(\Omega_1, \Omega_2)$ is given by

$$\mathbf{S} = 2\pi(\mathbf{VN})'^{-1} = \begin{bmatrix} \dfrac{\pi}{3NT_1} & \dfrac{\pi}{3NT_1} \\ \dfrac{\pi}{NT_2} & \dfrac{-\pi}{NT_2} \end{bmatrix} \triangleq \begin{bmatrix} S_1 & S_1 \\ S_2 & -S_2 \end{bmatrix} \quad (2.140)$$

$\mathbf{S}$ has the form of a hexagonal sampling matrix with $S_1 = \pi/3NT_1$ and $S_2 = \pi/NT_2$.

The periodicity matrix $\mathbf{N}$ can be decomposed as

$$\mathbf{N} = \begin{bmatrix} 2N & N \\ N & 2N \end{bmatrix} = \begin{bmatrix} 2 & 0 \\ 0 & 2 \end{bmatrix} \begin{bmatrix} N & \dfrac{N}{2} \\ \dfrac{N}{2} & N \end{bmatrix} \quad (2.141)$$

This decomposition leads to a vector-radix type of algorithm for a hexagonal DFT. The complete flowchart is shown in Figure 2.15(a). The regions $I_P$ and $I_Q$ for the first stage of this algorithm are

$$I_P = \{(0, 0)', (0, 1)', (1, 0)', (1, 1)'\} \quad (2.142)$$

$$I_Q = \left\{(q_1, q_2): \ 0 \le q_1 \le \frac{3N}{2} - 1, \ 0 \le q_2 \le \frac{N}{2} - 1\right\} \quad (2.143)$$

If $N$ is a power of 2, we can get the more complete decomposition

$$\mathbf{N} = \begin{bmatrix} 2 & 0 \\ 0 & 2 \end{bmatrix} \begin{bmatrix} 2 & 0 \\ 0 & 2 \end{bmatrix} \cdots \begin{bmatrix} 2 & 0 \\ 0 & 2 \end{bmatrix} \begin{bmatrix} 2 & 1 \\ 1 & 2 \end{bmatrix} \quad (2.144)$$

The butterflies for all the stages except the first contain four inputs and four outputs and they are all alike. The first stage contains three-input, three-output butterflies, one of which is shown in Figure 2.15(b).

The matrix $\mathbf{N}$ can also be factored in the form

$$\mathbf{N} = \begin{bmatrix} 2 & 1 \\ 1 & 2 \end{bmatrix} \begin{bmatrix} N & 0 \\ 0 & 1 \end{bmatrix} \begin{bmatrix} 1 & 0 \\ 0 & N \end{bmatrix} \quad (2.145)$$

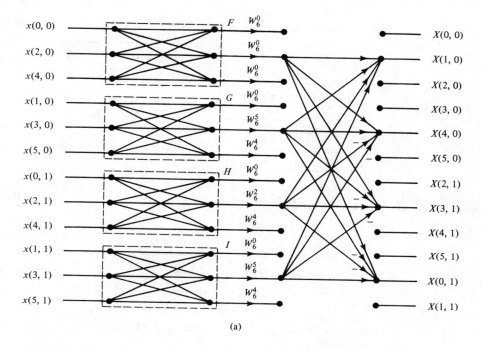

(a)

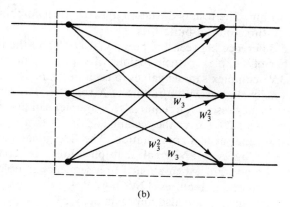

(b)

**Figure 2.15**    (a) Complete flowchart for the $N = 2$ hexagonal FFT, when **N** is factored as in equation (2.141). One of the 3-point butterflies from the first stage is shown with its multiplications in (b). (Courtesy of Russell M. Mersereau and Theresa C. Speake, *IEEE Trans. Acoustics, Speech, and Signal Processing*, © 1981 IEEE.)

This gives a row–column implementation which is shown in Figure 2.16. Three $N \times N$ DFTs, which are identical to rectangular DFTs, are performed on the data once the data have been sorted into three groups and reindexed. These DFTs can thus be evaluated using a rectangular row–column algorithm or a rectangular vector-

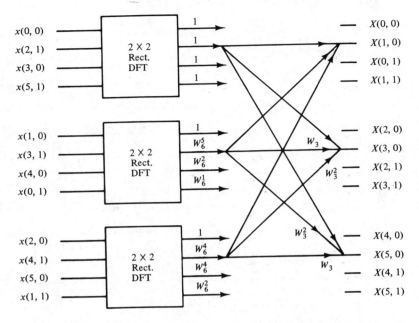

**Figure 2.16**  Flowchart for the $N = 2$ hexagonal FFT, when N is factored as in equation (2.145).

radix algorithm. The results of these DFTs are then combined using a single stage of three-input, three-output butterflies.

One difference between the hexagonal DFT and the rectangular DFT concerns the number of samples in their regions of support. The HDFT is a transformation between $3N^2$ complex sample values in each domain. These regions can be chosen to be hexagonally shaped with a radius of $N$ samples. The rectangular DFT with comparable frequency resolution requires $4N^2$ complex sample values. Thus one advantage of the hexagonal DFT over the rectangular one is that it requires 25% less storage.

It also requires less computation. The hexagonal vector-radix FFT requires a total of $9N^2 \log_2 N + 8N^2$ real multiplications. In comparison, the rectangular vector-radix algorithm, when applied to a sequence sampled to produce comparable frequency resolution, requires $12N^2 \log_2 N + 12N^2$ real multiplications. Thus the computational savings are also approximately 25%.

## *2.5 INTERRELATIONSHIP BETWEEN M-DIMENSIONAL AND ONE-DIMENSIONAL DFTs

While the 1-D DFT is a special case of the $M$-dimensional DFT, the $M$-dimensional DFT can be considered to be a special case of the 1-D DFT. As paradoxical as that statement sounds, it is not that difficult to understand, for the DFT is simply a transformation from one set of numbers to another. Whether those numbers are arranged in a row and addressed by a single index or whether they are arranged in an array with multiple indices is up to us; the choice between one representation or the other is generally made for our own convenience. In this section we want to explore the fuzzy area where 1-D and *M*-dimensional representations mix. We will first consider an *M*-dimensional DFT which is really a 1-D DFT and then we will interpret a 1-D DFT as an *M*-dimensional DFT.

### 2.5.1 *Slice DFT [8]*

The Fourier transform of a finite-extent array with a rectangular region of support is given by

$$X(\omega_1, \omega_2) = \sum_{n_1=0}^{N_1-1} \sum_{n_2=0}^{N_2-1} x(n_1, n_2) \exp\left(-j\omega_1 n_1 - j\omega_2 n_2\right) \qquad (2.146)$$

It is a 2-D trigonometric polynomial of degree $N_1 - 1$ in the variable $\omega_1$ and degree $N_2 - 1$ in the variable $\omega_2$. A 2-D DFT consists of a set of $N_1 N_2$ independent samples of that polynomial; by varying the periodicity matrix **N**, we vary the locations of these samples.

A particularly interesting set of Fourier domain samples arises through the use of the periodicity matrix

$$\mathbf{N} = \begin{bmatrix} N_1 & -1 \\ 0 & N_2 \end{bmatrix} \qquad (2.147)$$

which samples the Fourier transform at the points

$$\omega_1 = \frac{2\pi k}{N_1}; \qquad \omega_2 = \frac{2\pi k}{N_1 N_2}, \qquad k = 0, 1, \ldots, N_1 N_2 - 1 \qquad (2.148)$$

shown in Figure 2.17. These samples are seen to lie on a single line in the Fourier plane, which makes an angle $\theta = \tan^{-1}(1/N_2)$ with the $\omega_1$-axis and which cuts across several periods of $X(\omega_1, \omega_2)$. However, when the underlying periodicity of the Fourier transform is taken into account, these samples can also be considered to lie on a set of $N_2$ parallel line segments. Thus these samples simultaneously represent samples of the 2-D Fourier transform and also samples of a 1-D function.

Expressing these DFT samples, which we shall call the *slice DFT*, in terms of the single index $k$, we get

$$X_p(k) \triangleq \sum_{n_1=0}^{N_1-1} \sum_{n_2=0}^{N_2-1} x(n_1, n_2) \exp\left[-j\frac{2\pi k}{N_1 N_2}(N_2 n_1 + n_2)\right]$$
$$k = 0, 1, \ldots, N_1 N_2 - 1 \qquad (2.149)$$

If a new variable, $n$, is defined by

$$n \triangleq N_2 n_1 + n_2 \qquad (2.150)$$

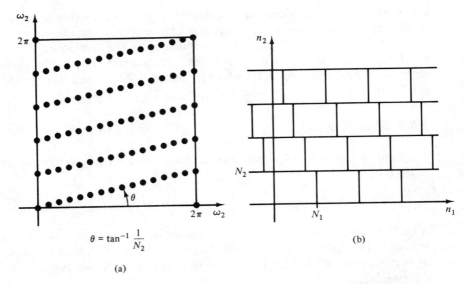

$$\theta = \tan^{-1} \frac{1}{N_2}$$

(a)

**Figure 2.17** (a) Samples of the 2-D Fourier transform corresponding to the slice DFT. (b) Periodic extension of $x(n_1, n_2)$ used to give the slice DFT.

then, as $n_1$ assumes values over the range 0 to $N_1 - 1$ and $n_2$ assumes values over the range 0 to $N_2 - 1$, $n$ will range from 0 to $N_1 N_2 - 1$. Furthermore, any value of $n$ over this range is associated with a unique ordered pair $(n_1, n_2)$. We can thus unambiguously define the sequence

$$x_p(N_2 n_1 + n_2) \triangleq x(n_1, n_2) \qquad (2.151)$$

and we recognize that $x_p(n)$ and $X_p(k)$ are a 1-D DFT pair.

$$x_p(n) \longleftrightarrow X_p(k) \qquad (2.152)$$

Since the array $x(n_1, n_2)$ has finite extent, it can be uniquely recovered from $x_p(n)$. In fact, the 1-D sequence is seen to be the concatenation of the $N_1$ columns of the 2-D array $x(n_1, n_2)$. Thus if the columns of a 2-D array are concatenated to form a 1-D sequence and a 1-D DFT of the concatenation is computed, the resulting 1-D DFT values can be interpreted as samples of the 2-D Fourier transform on the set of parallel line segments.

It is straightforward to extend this DFT to higher-dimensional arrays. Consider an $M$-dimensional signal which is nonzero only for $0 \leq n_i \leq N_i - 1$, $i = 1, 2, \ldots$, $M$. We can form a 1-D signal by letting

$$x_p(n) = x(n_1, n_2, \ldots, n_M) \qquad (2.153)$$

where

$$n = (\cdots ((n_1 N_2 + n_2) N_3 + n_3) N_4 + \cdots + n_M) \qquad (2.154)$$

Then, as in the 2-D case,

$$X_p(k) = \sum_{n_1=0}^{N_1-1} \cdots \sum_{n_M=0}^{N_M-1} x(n_1, n_2, \ldots, n_M) \exp\left(-j\frac{2\pi k n_1}{N_1}\right) \cdots \exp\left(-j\frac{2\pi k n_M}{\prod_i N_i}\right) \quad (2.155)$$

which is the evaluation of the $M$-dimensional continuous Fourier transform at the points

$$\omega_1 = \frac{2\pi k}{N_1}, \quad \omega_2 = \frac{2\pi k}{N_1 N_2}, \quad \ldots, \quad \omega_M = \frac{2\pi k}{N_1 N_2 \cdots N_M}, \quad 0 \le k < \prod_{i=1}^{M} N_i \quad (2.156)$$

### 2.5.2 Good's Prime Factor Algorithm for Decomposing a 1-D DFT [9, 10]

The discussion above has demonstrated one way in which a modified $M$-dimensional DFT is equivalent to a 1-D DFT. Another way to demonstrate that the DFT is, in some sense, independent of dimension is to consider Good's method [9] of decomposing 1-D DFTs. Consider the DFT of an $N$-point signal $x_1(n)$:

$$X_1(k) = \sum_{n=0}^{N-1} x_1(n) W_N^{nk} \quad (2.157)$$

where $N$ can be expressed as a product of integer factors which are relatively prime:

$$N = N_1 N_2 \cdots N_M$$

From the Chinese Remainder Theorem of number theory, any integer in the range $[0, N-1]$ can be uniquely represented by the $m$-tuple $(n_1, n_2, \ldots, n_M)$, where

$$n_i = ((n))_{N_i}, \; i = 1, \ldots, M \quad (2.158)$$

(Recall that double sets of parentheses indicate that $n$ is evaluated modulo $N_i$.) The integer $n$ can be represented by means of the remainders obtained when it is divided by each of the factors of $N$. The frequency index $k$ can be similarly represented as an $M$-tuple $(k_1, k_2, \ldots, k_M)$, where $k_i = ((k))_{N_i}$. By representing the integer variables $n$ and $k$ this way, we have mapped them from a 1-D space into an $M$-dimensional space.

These $M$-tuple representations are invertible. One way to derive the inversion formula is to postulate the existence of a set of synthesis numbers $\mu_i$ such that

$$((\mu_i))_{N_j} = \begin{cases} 1 & \text{if } i = j \\ 0 & \text{if } i \neq j \end{cases} \quad (2.159)$$

If numbers with this special property exist, it is fairly straightforward to show that the variable $n$ can be computed from its $M$-tuple representation $(n_1, n_2, \ldots, n_M)$ by using the formula

$$n = \left(\left(\sum_{i=1}^{M} n_i \mu_i\right)\right)_N \quad (2.160)$$

Results from number theory tell us that the numbers $\mu_i$ do exist and that they can be expressed as

$$\mu_i = \left(\frac{N}{N_i}\right)^{\phi(N_i)} \tag{2.161}$$

where the function $\phi(N_i)$, Euler's totient function, is equal to the number of integers in the set $1, 2, \ldots, N_i - 1$ which have no factors in common with $N_i$. For example, $\phi(2) = 1$, $\phi(3) = 2$, $\phi(5) = 4$, and $\phi(7) = 6$. In general, $\phi(p) = p - 1$ if $p$ is prime. For $N = 210 = 2 \times 3 \times 5 \times 7$ we would have

$$\mu_1 = \left(\frac{210}{2}\right)^1 = 105$$

$$\mu_2 = \left(\frac{210}{3}\right)^2 = 4900$$

$$\mu_3 = \left(\frac{210}{5}\right)^4 = 3,111,696$$

$$\mu_4 = \left(\frac{210}{7}\right)^6 = 729,000,000$$

It is evident that the $\mu_i$ can become quite large. In practice, however, these numbers can be replaced by their remainders when divided by $N$, giving the reduced values

$$\mu_1 = ((105))_{210} = 105$$
$$\mu_2 = ((4900))_{210} = 70$$
$$\mu_3 = ((3,111,696))_{210} = 126$$
$$\mu_4 = ((729,000,000))_{210} = 120$$

Carrying on with our example we see that the number $n = 111$ can be represented by $(1, 0, 1, 6)$ and reconstructed by

$$n = ((1 \times 105 + 0 \times 70 + 1 \times 126 + 6 \times 120))_{210}$$
$$= ((951))_{210} = 111$$

Lest it appear that we have gotten far afield, let us now apply these ideas to our original problem of converting a 1-D DFT into an $M$-dimensional DFT. We begin with

$$X_1(k) = \sum_{n=0}^{N-1} x_1(n) W_N^{nk} \tag{2.162}$$

Using the $M$-tuple representations for $n$ and $k$ to define $M$-dimensional arrays for input and output, we get

$$X_M(k_1, k_2, \ldots, k_M) = \sum_{n_1=0}^{N_1-1} \sum_{n_2=0}^{N_2-1} \cdots \sum_{n_M=0}^{N_M-1} x_M(n_1, n_2, \ldots, n_M) W_N^{(\sum_{i=1}^{M} \mu_i n_i)(\sum_{j=1}^{M} \mu_j k_j)} \tag{2.163}$$

However, since

$$W_N^m = W_N^{((m))_N}$$

we can evaluate the exponent in equation (2.163) by writing

$$\left(\left(\sum_{i=1}^{M} \mu_i n_i \sum_{j=1}^{M} \mu_j k_j\right)\right)_N = \left(\left(\sum_{i=1}^{M} \sum_{j=1}^{M} ((\mu_i \mu_j))_N n_i k_j\right)\right)_N$$
$$= \left(\left(\sum_{i=1}^{M} \mu_i^2 n_i k_i\right)\right)_N \qquad (2.164)$$

Because of the definition of $\mu_i$, the product $\mu_i \mu_j$ is an integer multiple of $N$ when $i \neq j$. The term $\mu_i^2$ can be written as

$$\mu_i^2 = \left(\frac{N}{N_i}\right)^{2\phi(N_i)} \triangleq \left(\frac{N}{N_i}\right) R_i \qquad (2.165)$$

where we have defined $R_i$ as

$$R_i \triangleq \left(\frac{N}{N_i}\right)^{2\phi(N_i)-1} \qquad (2.166)$$

Now (2.163) becomes

$$X_M(k_1, k_2, \ldots, k_M) = \sum_{n_1=0}^{N_1-1} \sum_{n_2=0}^{N_2-1} \cdots \sum_{n_M=0}^{N_M-1} x_M(n_1, n_2, \ldots, n_M) W_{N_1}^{R_1 n_1 k_1} W_{N_2}^{R_2 n_2 k_2}$$
$$\cdots W_{N_M}^{R_M n_M k_M}$$

By defining

$$m_i \triangleq ((R_i n_i))_{N_i} \qquad (2.167)$$

this becomes

$$X_M(k_1, k_2, \ldots, k_M) = \sum_{m_1=0}^{N_1-1} \sum_{m_2=0}^{N_2-1} \cdots \sum_{m_M=0}^{N_M-1} \hat{x}_M(m_1, m_2, \ldots, m_M) W_{N_1}^{m_1 k_1} W_{N_2}^{m_2 k_2} \cdots W_{N_M}^{m_M k_M}$$

$$(2.168)$$

Since $R_i$ and $N_i$ are mutually prime for each $i$, as $n_i$ takes on all values in the range zero to $N_i - 1$, $m_i$ will also take on all values in this range and the sequence $\hat{x}_M$ is simply a reindexing of $x_M$.

For our example, the $\{R_i\}$ are

$$R_1 = \left(\frac{210}{2}\right)^1 = 105; \quad ((R_1))_{N_1} = 1$$

$$R_2 = \left(\frac{210}{3}\right)^3 = 343{,}000; \quad ((R_2))_{N_2} = 1$$

$$R_3 = \left(\frac{210}{5}\right)^7 = 230{,}539{,}333{,}248; \quad ((R_3))_{N_3} = 3$$

$$R_4 = \left(\frac{210}{7}\right)^{11} = 17{,}714{,}700{,}000{,}000{,}000; \quad ((R_4))_{N_4} = 4$$

In summary, the steps that must be performed to convert a 210-point 1-D DFT into a $2 \times 3 \times 5 \times 7$-point 4-D DFT are the following:

1. Compute the $R_i$'s. ($R_1 = R_2 = 1$, $R_3 = 3$, $R_4 = 4$.)
2. Compute the $\mu_i$'s. ($\mu_1 = 105$, $\mu_2 = 70$, $\mu_3 = 126$, $\mu_4 = 120$.)
3. Use the $R_i$'s when computing the indices $m_i$ from the 1-D index $n$.

$$(m_1, m_2, m_3, m_4) = (\ ((R_1 n))_2, ((R_2 n))_3, ((R_3 n))_5, ((R_4 n))_7)$$

For example,

$$n = 1 \quad \longrightarrow (m_1, m_2, m_3, m_4) = (1, 1, 3, 4)$$
$$n = 17 \quad \longrightarrow (m_1, m_2, m_3, m_4) = (1, 2, 1, 5)$$
$$n = 198 \longrightarrow (m_1, m_2, m_3, m_4) = (0, 0, 4, 1)$$

This maps $x_1(n)$ into $\hat{x}_4(m_1, m_2, m_3, m_4)$.

4. Take the 4-D DFT of $\hat{x}_4(m_1, m_2, m_3, m_4)$ to get $X_4(k_1, k_2, k_3, k_4)$.
5. Map $X_4(k_1, k_2, k_3, k_4)$ into $X_1(k)$ by the formula

$$k = ((\mu_1 k_1 + \mu_2 k_2 + \mu_3 k_3 + \mu_4 k_4))_N$$
$$= ((105 k_1 + 70 k_2 + 126 k_3 + 120 k_4))_{210}$$

Thus

$$(k_1, k_2, k_3, k_4) = (1, 1, 1, 1) \longrightarrow k = 1$$
$$(k_1, k_2, k_3, k_4) = (0, 2, 0, 5) \longrightarrow k = 110$$

## PROBLEMS

**2.1.** Verify that the Fourier series coefficients $\tilde{X}(k_1, k_2)$ defined by (2.4) when substituted into the Fourier synthesis formula (2.3) produce the sequence $\tilde{x}(n_1, n_2)$.

**2.2.** If $\tilde{x}(n_1, n_2)$ is a rectangularly periodic sequence with periodicity matrix $\begin{bmatrix} N_1 & 0 \\ 0 & N_2 \end{bmatrix}$ and discrete Fourier series coefficients $\tilde{X}(k_1, k_2)$, find the Fourier series coefficients of the following sequences.
   (a) $\tilde{x}(n_1 - m_1, n_2 - m_2)$, $m_1, m_2$ = constants
   (b) $\tilde{x}(n_2, n_1)$   (Assume that $N_1 = N_2 = N$.)
   (c) $\tilde{x}^*(n_1, n_2)$
   (d) $\tilde{x}(-n_1, -n_2)$

**2.3.** Suppose that $\tilde{x}_2(n_1, n_2)$ is a rectangularly periodic sequence as in Problem 2.2. The sequence $\tilde{x}_2(n, n)$ is then a 1-D periodic sequence.
   (a) What is the period of the sequence $\tilde{x}_1(n) = \tilde{x}_2(n, n)$? How does your answer depend on any prime factors that are common to $N_1$ and $N_2$?
   (b) If $N_1$ and $N_2$ are relatively prime and the 2-D sequence has Fourier series coefficients $\tilde{X}_2(k_1, k_2)$, find the Fourier series coefficients of $\tilde{X}_1(k)$.

**2.4.** Find the Fourier series coefficients of the sequences shown in Figure P2.4.

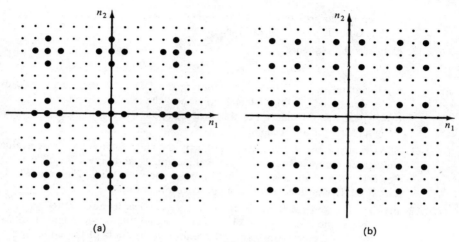

(a)                                                                          (b)

**Figure P2.4**

**2.5.** Compute the discrete Fourier transforms of the following sequences.

(a) $x(n_1, n_2) = \alpha^{n_1 - n_2}, \ 0 \leq n_1, n_2 < N$

(b) $x(n_1, n_2, n_3) = \alpha^{n_1 - n_2}, \ 0 \leq n_1, n_2, n_3 < N$

(c) $x(n_1, n_2) = \sum_{k=0}^{N-1} y(n_1, k), \ 0 \leq n_1, n_2 < N$

[Assume that $Y(k_1, k_2)$ is known.]

**2.6.** The Fourier transform of the signal $x(n_1, n_2) = \alpha^{n_1} \beta^{n_2} u(n_1, n_2)$ is sampled at points

$$(\omega_1, \omega_2) = \left( \frac{\pi}{2} k_1, \frac{\pi}{4} k_2 \right) \qquad \text{for } 0 \leq k_1 \leq 3, 0 \leq k_2 \leq 7$$

The inverse DFT of these samples is then taken. Determine the resulting spatial domain signal.

**2.7.** Show that the following properties of the DFT are true.

(a) $x^*(n_1, n_2) \longleftrightarrow X^*( \, ((N_1 - k_1))_{N_1}, ((N_2 - k_2))_{N_2})$

(b) $x^*( \, ((N_1 - n_1))_{N_1}, ((N_2 - n_2))_{N_2}) \longleftrightarrow X^*(k_1, k_2)$

(c) $x( \, ((N_1 - n_1))_{N_1}, ((N_2 - n_2))_{N_2}) \longleftrightarrow X( \, ((N_1 - k_1))_{N_1}, ((N_2 - k_2))_{N_2})$

**2.8.** (a) Compute the circular convolution of the two arrays

$$x_1(n_1, n_2) = \delta(n_1), \qquad 0 \leq n_1 < N_1, \quad 0 \leq n_2 < N_2$$

$$x_2(n_1, n_2) = \delta(n_2), \qquad 0 \leq n_1 < N_1, \quad 0 \leq n_2 < N_2$$

(b) Compute the linear convolution of these arrays.

(c) Repeat parts (a) and (b) but replace $x_2(n_1, n_2)$ by

$$x_3(n_1, n_2) = \delta(n_1 - n_2)$$

and assume that $N_1 = N_2 = N$.

**2.9. (a)** Two 2-D sequences, each of which is $3 \times 4$ points in extent, are circularly convolved using $(6 \times 6)$-point 2-D DFTs. Which samples of the $(6 \times 6)$-point output array are identical to samples of the linear convolution of the two input arrays, and which are different?

**(b)** Repeat the problem of part (a), where now one of the input arrays is $3 \times 4$ and the other is $4 \times 3$ points in extent.

**2.10.** Consider the sequence

$$x(n_1, n_2) = a^{n_1}, \qquad 0 \le n_1 < N_1, \quad 0 \le n_2 < N_2$$

**(a)** Evaluate the DFT by first computing the 1-D DFT along each row of $x$. Then evaluate the 1-D DFT of each column of the resulting array.

**(b)** Repeat this procedure but this time compute the column DFTs first. Verify that your answers are the same.

**2.11.** Show that the $2 \times 2$ vector-radix butterfly shown in Figure 2.4 can be evaluated using only three complex multiples and eight complex additions. Sketch a more detailed flowchart that indicates these additions explicitly. (*Hint:* The butterfly is really a $2 \times 2$ DFT. Consider a row–column decomposition.)

**2.12. (a)** Sketch a general butterfly for a $2 \times 2 \times 2$ vector-radix butterfly.

**(b)** What is the minimum number of complex additions and complex multiplications that must be performed to evaluate all eight outputs of the butterfly?

**2.13. (a)** How many complex multiplications and complex additions are required to evaluate an $M$-dimensional row–column DFT if radix-4 1-D FFTs are used? (Input size $=$ $N \times N \times \cdots \times N$, where $N$ is a power of 4.)

**(b)** How many complex multiplications and complex additions are required to evaluate a radix-$(4 \times 4 \times \cdots \times 4)$ $M$-dimensional vector-radix FFT? (Assume an $N \times N \times \cdots \times N$ input, where $N$ is a power of 4.)

**\*2.14.** Write a radix-$(2 \times 2)$ 2-D FFT in FORTRAN or some other high-level language. Assume that the data array is resident in memory and that the transform is to be stored in the same location in memory as the original data.

**2.15. (a)** We wish to perform a 2-D $(N \times N)$-point row–column FFT where $N = R^2$. Assuming that at least $R$ rows of data can reside in memory, generalize Eklundh's transposition procedure.

**(b)** Verify that your procedure works on the array

| | | | | | | | | |
|---|---|---|---|---|---|---|---|---|
| $x(0, 8)$ | $x(1, 8)$ | $x(2, 8)$ | $x(3, 8)$ | $x(4, 8)$ | $x(5, 8)$ | $x(6, 8)$ | $x(7, 8)$ | $x(8, 8)$ |
| $x(0, 7)$ | $x(1, 7)$ | $x(2, 7)$ | $x(3, 7)$ | $x(4, 7)$ | $x(5, 7)$ | $x(6, 7)$ | $x(7, 7)$ | $x(8, 7)$ |
| $x(0, 6)$ | $x(1, 6)$ | $x(2, 6)$ | $x(3, 6)$ | $x(4, 6)$ | $x(5, 6)$ | $x(6, 6)$ | $x(7, 6)$ | $x(8, 6)$ |
| $x(0, 5)$ | $x(1, 5)$ | $x(2, 5)$ | $x(3, 5)$ | $x(4, 5)$ | $x(5, 5)$ | $x(6, 5)$ | $x(7, 5)$ | $x(8, 5)$ |
| $x(0, 4)$ | $x(1, 4)$ | $x(2, 4)$ | $x(3, 4)$ | $x(4, 4)$ | $x(5, 4)$ | $x(6, 4)$ | $x(7, 4)$ | $x(8, 4)$ |
| $x(0, 3)$ | $x(1, 3)$ | $x(2, 3)$ | $x(3, 3)$ | $x(4, 3)$ | $x(5, 3)$ | $x(6, 3)$ | $x(7, 3)$ | $x(8, 3)$ |
| $x(0, 2)$ | $x(1, 2)$ | $x(2, 2)$ | $x(3, 2)$ | $x(4, 2)$ | $x(5, 2)$ | $x(6, 2)$ | $x(7, 2)$ | $x(8, 2)$ |
| $x(0, 1)$ | $x(1, 1)$ | $x(2, 1)$ | $x(3, 1)$ | $x(4, 1)$ | $x(5, 1)$ | $x(6, 1)$ | $x(7, 1)$ | $x(8, 1)$ |
| $x(0, 0)$ | $x(1, 0)$ | $x(2, 0)$ | $x(3, 0)$ | $x(4, 0)$ | $x(5, 0)$ | $x(6, 0)$ | $x(7, 0)$ | $x(8, 0)$ |

**2.16.** Determine a periodicity matrix that describes the periodicity of the array shown in Figure P2.16.

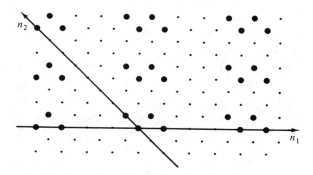

**Figure P2.16**

**2.17.** A signal $x(n_1, n_2)$ with finite support can be exactly recovered from its DFT taken with respect to the periodicity matrix

$$\mathbf{N} = \begin{bmatrix} 2N & N \\ N & 2N \end{bmatrix}$$

(a) Give an upper limit to the number of nonzero samples of $x(n_1, n_2)$.
(b) Determine the DFT of $x(n_1, n_2)$ if

$$x(n_1, n_2) = \delta(n_1), \quad 0 \leq n_1 \leq 3N - 1, \quad 0 \leq n_2 \leq N - 1$$

**2.18.** If a signal is separable in the sense that $x(n_1, n_2) = x_1(n_1)x_2(n_2)$, what, if anything, can be said about the separability of its DFT taken with respect to an arbitrary periodicity matrix

$$\mathbf{N} = \begin{bmatrix} N_{00} & N_{01} \\ N_{10} & N_{11} \end{bmatrix}?$$

**2.19.** The sequence $x(n_1, n_2)$ has finite support which is confined to the region $0 \leq n_1 \leq 2$, $0 \leq n_2 \leq 1$. It has a DFT $X(k_1, k_2)$ which is defined by the periodicity matrix

$$\mathbf{N} = \begin{bmatrix} 3 & 0 \\ 0 & 2 \end{bmatrix}$$

The DFT is confined to the region $0 \leq k_1 \leq 2$, $0 \leq k_2 \leq 1$. The DFT is evaluated with a two-stage FFT using the factorization $\mathbf{N} = \mathbf{PQ}$, where

$$\mathbf{P} = \begin{bmatrix} 3 & -3 \\ -1 & 2 \end{bmatrix} \quad \mathbf{Q} = \begin{bmatrix} 2 & 2 \\ 1 & 2 \end{bmatrix}$$

Sketch a flowchart for this FFT algorithm.

**2.20.** Suppose that we intend to compute the DFT $X(\mathbf{k})$ of an array $x(\mathbf{n})$ with respect to the periodicity matrix $\mathbf{N}$. If $\mathbf{N}$ can be expressed as the product

$$\mathbf{N} = \mathbf{EDF}$$

where

$$\mathbf{D} \text{ is a diagonal matrix}$$

$$|\det \mathbf{E}| = 1$$

$$|\det \mathbf{F}| = 1$$

show that the DFT can be computed in the following three steps.

1. Rearrange the elements of the input array according to

$$\hat{\mathbf{n}} = \mathbf{E}^{-1}\mathbf{n}$$

2. Compute a row–column DFT of the array $x(\hat{\mathbf{n}})$. Call the result $X(\hat{\mathbf{k}})$.
3. Rearrange the elements of the array $X(\hat{\mathbf{k}})$ to form $X(\mathbf{k})$ according to

$$\mathbf{k} = \mathbf{F}'\hat{\mathbf{k}}$$

(*Note:* Such a decomposition exists for any integer matrix **N**. This result thus establishes a row–column algorithm for any periodicity matrix.)

**2.21.** Consider the following algorithm for doing a 2-D linear convolution. Suppose that the two sequences $x(n_1, n_2)$ and $h(n_1, n_2)$ are to be convolved. Assume that $x(n_1, n_2)$ is $N_X \times N_X$ points and that $h(n_1, n_2)$ is $N_H \times N_H$ points.

1. Define

$$\hat{x}(n_1, n_2) = \begin{cases} x(n_1, n_2), & 0 \le n_1, n_2 < N_X \\ 0, & \text{otherwise}, 0 \le n_1, n_2 < N_X + N_H \end{cases}$$

2. Define

$$\hat{h}(n_1, n_2) = \begin{cases} h(n_1, n_2), & 0 \le n_1, n_2 < N_H \\ 0, & \text{otherwise}, 0 \le n_1, n_2 < N_X + N_H \end{cases}$$

3. Form one-dimensional sequences from $\hat{x}$ and $\hat{h}$ by arranging their columns end to end; that is, define

$$p_X(q) = p_X(n_1 N + n_2) \triangleq \hat{x}(n_1, n_2)$$

$$p_H(q) = p_H(n_1 N + n_2) \triangleq \hat{h}(n_1, n_2)$$

where $N \triangleq N_X + N_H$.

4. Perform a 1-D convolution of $p_X(q)$ and $p_H(q)$.

$$p_X(q) * p_H(q) = p_Y(q)$$

5. Define $y(n_1, n_2)$ by the relation

$$y(n_1, n_2) \triangleq \begin{cases} p_Y(n_1 N + n_2), & 0 \le n_1, n_2 \le N - 1 \\ 0, & \text{otherwise} \end{cases}$$

6. The result, it is claimed, is the 2-D convolution of $x$ and $h$.

**(a)** Using this algorithm, convolve the two sequences

$$x = \begin{bmatrix} b & d \\ a & c \end{bmatrix} \qquad h = \begin{bmatrix} 2 & 4 \\ 1 & 3 \end{bmatrix}$$

and compare your answer with the 2-D convolution obtained by more standard means.

**(b)** Relate the Fourier transform of $p_X(q)$ to the 2-D Fourier transform of $x(n_1, n_2)$.

(c) Show that this technique always works if $N$ is chosen to be sufficiently large. (The argument is easier in the frequency domain.) How large must $N$ be?

(d) Explain how this computation could be performed using a 1-D DFT with the foregoing algorithm.

**2.22.** We would like to compute the 1-D DFT of the 20-point sequence $x(n)$ using the Good prime factor algorithm. This is done by mapping that sequence into a 2-D (5 × 4) array $x_2(n_1, n_2)$, computing the DFT $X_2(k_1, k_2)$ of that array, and mapping it into the 1-D DFT $X(k)$.

(a) Determine the 2-D array $x_2(n_1, n_2)$ in terms of the samples of $x(n)$. Use $N_1 = 5$, $N_2 = 4$.

(b) Determine the 1-D array $X(k)$ in terms of the samples of $X_2(k_1, k_2)$.

# REFERENCES

1. James W. Cooley and John W. Tukey, "An Algorithm for the Machine Calculation of Complex Fourier Series," *Mathematics of Computation*, 19, no. 90 (1965), 297–301.

2. James W. Cooley, Peter A. W. Lewis, and Peter D. Welch, "Historical Notes on the Fast Fourier Transform," *IEEE Trans. Audio Electroacoustics*, AU-15, no. 2 (June 1967), 76–79.

3. David B. Harris, James H. McClellan, David S. K. Chan, and Hans W. Schuessler, "Vector Radix Fast Fourier Transform," *Proc. IEEE Int. Conf. Acoustics, Speech, and Signal Processing* (May 1977), 548–51.

4. Glenn K. Rivard, "Direct Fast Fourier Transform of Bivariate Functions," *IEEE Trans. Acoustics, Speech, and Signal Processing*, ASSP-25, no. 3 (June 1977), 250–52.

5. J. O. Eklundh, "A Fast Computer Method for Matrix Transposing," *IEEE Trans. Computers*, C-21 (July 1972), 801–3.

6. Russell M. Mersereau, "The Processing of Hexagonally Sampled Two-Dimensional Signals," *Proc. IEEE*, 61, no. 6 (June 1979), 930–49.

7. Russell M. Mersereau and Theresa C. Speake, "A Unified Treatment of Cooley-Tukey Algorithms for the Evaluation of the Multidimensional DFT," *IEEE Trans. Acoustics, Speech, and Signal Processing*, ASSP-29, no. 5 (Oct. 1981), 1011–18.

8. Russell M. Mersereau and Dan E. Dudgeon, "The Representation of Two-Dimensional Sequences as One-Dimensional Sequences," *IEEE Trans. Acoustics, Speech, and Signal Processing*, ASSP-22, no. 5 (Oct. 1974), 320–25.

9. I. J. Good, "The Interaction Algorithm and Practical Fourier Analysis," *J. Royal Statistical Society B*, 20 (1960), 361–72.

10. James H. McClellan and Charles M. Rader, *Number Theory in Digital Signal Processing* (Englewood Cliffs, N.J.: Prentice-Hall, Inc., 1979).

# 3

# DESIGN AND IMPLEMENTATION
# OF TWO-DIMENSIONAL FIR FILTERS

There is an important difference between the 1-D and 2-D digital filter design problems. In the 1-D case the filter design and filter implementation issues are distinct and decoupled. The filter can first be designed and then, through the appropriate manipulations of the transfer function, the coefficients required by a particular network structure can be determined. In the 2-D case the situation is quite different, because multidimensional polynomials cannot be factored in general. This means that an arbitrary transfer function can generally not be manipulated into a form required by a particular implementation. If our implementation can realize only factorable transfer functions, our design algorithm must be tailored to design only filters of this class. This has the effect of complicating the design problem and also limiting the number of practical implementations.

## 3.1 FIR FILTERS

An *FIR* (finite-extent impulse response), or *nonrecursive*, filter is one whose impulse response possesses only a finite number of nonzero samples. For such a filter, the impulse response is always absolutely summable and thus FIR filters are always stable. FIR filters also possess the advantage of being very well understood in both the 1-D and multidimensional cases.

An *IIR* (infinite-extent impulse response), or *recursive*, filter is one whose input and output satisfy a multidimensional difference equation of finite order. These filters may or may not be stable, but in many cases they may be less complex to realize than equivalent FIR filters. The design of a 2-D recursive filter is quite different from the design of a 1-D filter, however. This is due in part to the increased difficulty of assuring stability. Difference equations and IIR filters form the subjects of Chapters 4 and 5.

One of the biggest advantages that FIR filters enjoy over IIR filters is the fact that implementable FIR filters can be designed to have purely real frequency responses. Such filters are called *zero-phase*.† In the frequency domain the zero-phase condition can be expressed as

$$H(\omega_1, \omega_2) = H^*(\omega_1, \omega_2) \tag{3.1}$$

By computing the inverse Fourier transform of both sides of (3.1) we get the spatial-domain symmetry requirement for the impulse response of a zero-phase filter:

$$h(n_1, n_2) = h^*(-n_1, -n_2) \tag{3.2}$$

Clearly, an FIR filter can be constrained to satisfy this relation if its region of support is centered about the origin.

A zero-phase response for digital filters is important to many applications of multidimensional digital signal processing. In image processing, for example, non-zero-phase responses tend to destroy lines and edges. To see why this might be so, recall from our discussions of Fourier transforms that any signal can be represented as a superposition of complex sinusoids. A linear shift-invariant filter with a non-trivial frequency response will selectively amplify or attenuate some of these sinusoidal components and delay some components with respect to others. The amount of delay at any frequency depends on the value of the phase response at that frequency. A nonlinear phase response thus tends to disperse those sinusoidal components of a signal that are precisely aligned, such as those that occur at bright spots, lines, and edges.

A zero-phase filter provides other rewards as well. Since its frequency response is purely real, the filter design problem is simplified. In addition, the symmetry constraint on the impulse response of the filter can be exploited in the implementation of the filter to reduce the number of multiplications needed for its realization.

## 3.2 IMPLEMENTATION OF FIR FILTERS

### 3.2.1 Direct Convolution

We know from Chapter 1 that the output of any LSI filter can be determined from its input by means of the convolution sum

$$y(n_1, n_2) = \sum_{k_1} \sum_{k_2} h(k_1, k_2) x(n_1 - k_1, n_2 - k_2) \tag{3.3}$$

---

†Strictly speaking, a purely real frequency response can have a negative amplitude at some frequencies, corresponding to a phase of $\pi$, not zero. In spite of this, the term "zero-phase" has come to include all purely real frequency responses.

For an FIR filter the impulse response has only a finite number of nonzero samples and the limits of summation in (3.3) are finite. In this case the convolution sum represents an algorithm that allows us to compute the successive output samples of the filter. For example, if we assume that the filter has support over the region $\{(n_1, n_2): \ 0 \leq n_1 < N_1, 0 \leq n_2 < N_2\}$, the output samples can be computed using

$$y(n_1, n_2) = \sum_{k_1=0}^{N_1-1} \sum_{k_2=0}^{N_2-1} h(k_1, k_2)x(n_1 - k_1, n_2 - k_2) \qquad (3.4)$$

If all the input samples are available, the output samples can be computed in any order or they can be computed simultaneously. If only selected samples of the output signal are desired, only those samples need to be computed. For each output sample that we desire, however, we must evaluate $N_1 N_2$ multiplications and $N_1 N_2 - 1$ additions.

The computation of $y(n_1, n_2)$ depends on input samples from the $N_1 - 1$ "previous" columns of the input and the $N_2 - 1$ "previous" rows. If the input samples arrive row by row we need sufficient storage to store $N_2$ rows of the input sequence. If the input is available column by column instead, we need to store $N_1$ columns of the input.

A zero-phase filter with a real impulse response satisfies $h(n_1, n_2) = h(-n_1, -n_2)$, which means that each sample can be paired with another of identical value. In this case we can use the arithmetic distributive law to interchange some of the multiplications and additions in (3.3) to reduce the number of multiplications necessary to implement the filter, but the number of required multiplications is still proportional to the filter order. Specifically, if the region of support for the filter is assumed to be rectangular and centered at the origin, we have

$$y(n_1, n_2) = \sum_{k_1=-N_1}^{N_1} \sum_{k_2=-N_2}^{N_2} h(k_1, k_2)x(n_1 - k_1, n_2 - k_2) \qquad (3.5)$$

$$= \sum_{k_1=-N_1}^{N_1} \sum_{k_2=1}^{N_2} h(k_1, k_2)[x(n_1 - k_1, n_2 - k_2) + x(n_1 + k_1, n_2 + k_2)]$$

$$+ \sum_{k_1=1}^{N_1} h(k_1, 0)[x(n_1 - k_1, n_2) + x(n_1 + k_1, n_2)] \qquad (3.6)$$

$$+ h(0, 0)x(n_1, n_2)$$

Using equation (3.6) to implement an FIR filter requires roughly one-half the number of multiplications of an implementation based on (3.5), although both implementations require the same number of additions and the same amount of storage. Should the impulse response of an FIR filter possess other symmetries, they can be exploited in a similar fashion to reduce further the number of required multiplications.

### 3.2.2 Discrete Fourier Transform Implementations of FIR Filters

Any FIR filter can also be implemented by means of the discrete Fourier transform. This can be particularly appealing for high-order filters because the various fast Fourier transform algorithms permit the efficient evaluation of the DFT.

Let $w(n_1, n_2)$ be the linear convolution of a finite-extent sequence $x(n_1, n_2)$ with the impulse response $h(n_1, n_2)$ of an FIR filter.

$$w(n_1, n_2) = x(n_1, n_2) ** h(n_1, n_2) \tag{3.7}$$

Computing Fourier transforms of both sides of this expression, we get

$$W(\omega_1, \omega_2) = X(\omega_1, \omega_2)H(\omega_1, \omega_2) \tag{3.8}$$

We saw in Chapter 2 that there were many possible definitions of the 2-D discrete Fourier transform, and that all of these correspond to sets of samples of the 2-D Fourier transform; these DFTs can be used to perform convolutions as long as their assumed region of support contains the support for $w(n_1, n_2)$. For the purposes of this discussion, let us assume that $W(\omega_1, \omega_2)$ is sampled on an $N_1 \times N_2$ rectangular lattice of samples, and let

$$W(k_1, k_2) = W(\omega_1, \omega_2)\Big|_{\omega_1 = 2\pi k_1/N_1, \omega_2 = 2\pi k_2/N_2} \tag{3.9}$$

with the obvious abuse in notation. Then

$$W(k_1, k_2) = X(k_1, k_2)H(k_1, k_2) \tag{3.10}$$

Let us define $y(n_1, n_2)$ to be the sequence resulting from the inverse DFT of the product $H(k_1, k_2)X(k_1, k_2)$, as we did in Section 2.2.3. Then $y(n_1, n_2)$ will be the circular convolution of $h(n_1, n_2)$ and $x(n_1, n_2)$ as defined in equations (2.52) and (2.53). If $N_1$ and $N_2$ are chosen to be sufficiently large, then $y(n_1, n_2) = w(n_1, n_2)$, as desired. To compute $(N_1 \times N_2)$-point DFTs of $x$ and $h$ requires that both sequences have their regions of support extended with samples of value zero.

Discrete Fourier transform implementations of FIR filters are efficient with respect to computation but prodigal with respect to storage. They require sufficient storage to contain all $N_1 N_2$ points of the signal $x(n_1, n_2)$. In addition, we must store the filter response coefficients $H(k_1, k_2)$, which doubles the required storage. With direct convolution the number of rows of the input that needs to be stored depends on the order of the filter. With the DFT implementation the whole input must be stored regardless of the filter order.

To calculate the number of multiplications that must be performed to evaluate all the output samples, assume that $H(k_1, k_2)$ has been precomputed and stored. Then two DFTs (one forward, one inverse) must be performed to compute the desired samples of the output array. If a row–column algorithm is used which exploits the fact that $x(n_1, n_2)$ and $y(n_1, n_2)$ are real, the total number of real multiplications needed to compute $y(n_1, n_2)$ is

$$2N_1 N_2 \log_2 N_1 N_2 + 2N_1 N_2 \tag{3.11}$$

when $N_1$ and $N_2$ are powers of 2. If the region of support for the linear convolution $w(n_1, n_2)$ is an $M_1 \times M_2$ rectangle, then

$$\frac{2N_1 N_2 \log_2 N_1 N_2 + 2N_1 N_2}{M_1 M_2}$$

real multiplications per output sample are needed. When the extent of the filter impulse response is small with respect to that of the input signal, this number is effectively independent of the filter order.

### 3.2.3 Block Convolution

We saw in the preceding section that the arithmetic complexity of the DFT implementation of an FIR filter is effectively independent of the order of the filter, while the complexity of a direct convolutional implementation is proportional to the filter order. When the filter order is low, we would expect the convolution implementation to be more efficient, but as the filter order increases, we would expect that the DFT implementation would eventually become more efficient. For very high order filters the savings can amount to orders of magnitude.

The problem with the DFT implementation is that it requires a great deal of storage. Block convolution methods offer a compromise. With these approaches the convolutions are performed on sections or blocks of data using DFT methods. Limiting the size of these blocks limits the amount of storage required and using transform methods maintains the efficiency of the procedure.

The simplest block convolution method to understand is called the *overlap-add* technique. We begin by partitioning a 2-D array, $x(n_1, n_2)$, into $(N_1 \times N_2)$-point sections, where the section indexed by the pair $(k_1, k_2)$ is defined as

$$x_{k_1 k_2}(n_1, n_2) \triangleq \begin{cases} x(n_1, n_2), & k_1 N_1 \leq n_1 < (k_1 + 1)N_1, \\ & k_2 N_2 \leq n_2 < (k_2 + 1)N_2 \\ 0 & \text{otherwise} \end{cases} \qquad (3.12)$$

The region of support for one of these sections is shown in Figure 3.1(a). The regions of support for the different sections do not overlap, and collectively they cover the entire region of support of the array $x(n_1, n_2)$; thus

$$x(n_1, n_2) = \sum_{k_1} \sum_{k_2} x_{k_1 k_2}(n_1, n_2) \qquad (3.13)$$

Because the operation of discrete convolution distributes with respect to addition,

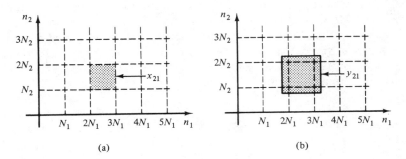

(a)                                                                    (b)

**Figure 3.1**   Overlap-add method of block convolution. (a) Section of the input array, $x_{21}(n_1, n_2)$. (b) Region of support of the convolution of that section with $h$.

we can write the following:

$$y(n_1, n_2) = x(n_1, n_2) ** h(n_1, n_2)$$
$$= \{\sum_{k_1} \sum_{k_2} x_{k_1 k_2}(n_1, n_2)\} ** h(n_1, n_2) \tag{3.14}$$
$$= \sum_{k_1} \sum_{k_2} \{x_{k_1 k_2}(n_1, n_2) ** h(n_1, n_2)\}$$
$$= \sum_{k_1} \sum_{k_2} y_{k_1 k_2}(n_1, n_2)$$

The block output $y_{k_1 k_2}(n_1, n_2)$ is the convolution of $h(n_1, n_2)$ with block $(k_1, k_2)$ of $x(n_1, n_2)$. The results of these block convolutions must be added together to produce the complete filter output $y(n_1, n_2)$. Because the support of $y_{k_1 k_2}(n_1, n_2)$ is greater than the support of $x_{k_1 k_2}(n_1, n_2)$, the output blocks will of necessity overlap, but the degree of that overlap is limited. In Figure 3.1(b) we show the region of support of one of these output blocks.

The convolutions of the $x_{k_1 k_2}(n_1, n_2)$ with $h(n_1, n_2)$ can be evaluated by means of discrete Fourier transforms, provided that the size of the transform is large enough to support $y_{k_1 k_2}(n_1, n_2)$. By controlling the block size we can limit the size of the DFTs, which reduces the required storage. Generally this is done, however, with a concomitant loss in efficiency.

The *overlap-save* method is an alternative block convolution technique. If we look again at Figure 3.1 we see that when the block size is considerably larger than the support of $h$, the samples of $y$ in the center of each block are not overlapped by samples from neighboring blocks. Similarly, when a sequence $x$ is circularly convolved with another, $h$, which has a much smaller region of support, only a subset of the samples of that circular convolution will show the effects of spatial aliasing. The remaining samples of the circular convolution will be identical to the samples of the linear convolution. The locations of these points are shown in Figure 3.2. Thus if an $(N_1 \times N_2)$-point section of $x(n_1, n_2)$ is circularly convolved with an $(M_1 \times M_2)$-point impulse response using an $(N_1 \times N_2)$-point DFT, the resulting circular con-

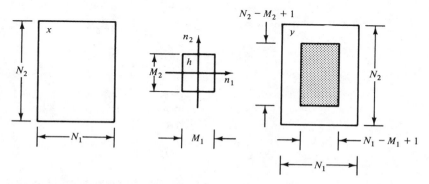

**Figure 3.2**  Overlap-save method. The shaded region gives those samples of $y$ for which both the $N_1 \times N_2$ circular convolution and the linear convolution of $x$ with $h$ are identical.

volution will contain a cluster of $(N_1 - M_1 + 1) \times (N_2 - M_2 + 1)$ samples which are identical to samples of the linear convolution, $y$. The whole output array can be constructed from these "good" samples by carefully choosing the regions of support for the input sections. If the input sections are allowed to overlap, the "good" samples of the various blocks can be made to abut. The overlap-save method thus involves overlapping input sections, whereas the overlap-add method involves overlapping output sections.

With both the overlap-add and overlap-save procedures, the choice of block size seriously affects the efficiency of the resulting implementation. First, in an obvious way, it affects the amount of storage needed, but it also affects the amount of computation. From Figure 3.2 we see that the fraction of usable samples from the circular convolution increases as the block size increases relative to the size of the impulse response. Although it is difficult to make any general statements about how large the input blocks should be because the results are quite machine dependent, experiments by Twogood *et al.* [1] suggest that filters whose regions of support fall in the range between $30 \times 30$ samples and $80 \times 80$ samples should be implemented with a block size of $256 \times 256$. This size is too large for many minicomputer installations. Thus the available memory will generally limit the speed of the algorithm. When this is the case the input blocks should be made as large as possible.

## 3.3 DESIGN OF FIR FILTERS USING WINDOWS

### 3.3.1 Description of the Method

The window method for the design of multidimensional FIR filters is philosophically identical to its 1-D counterpart. It is a spatial domain method, which is to say that we try to approximate an ideal impulse response rather than an ideal frequency response. Let $i(n_1, n_2)$ and $I(\omega_1, \omega_2)$ denote the impulse response and frequency response for the ideal filter and let $h(n_1, n_2)$ and $H(\omega_1, \omega_2)$ denote the impulse response and frequency response of the filter designed by the algorithm. The support of $h(n_1, n_2)$ is confined to some finite-extent region $R$. With the window method the coefficients $h(n_1, n_2)$ are given by

$$h(n_1, n_2) = i(n_1, n_2)w(n_1, n_2) \tag{3.15}$$

The sequence $w(n_1, n_2)$ is called a *window function* or *window array*. By confining the support of $w$ to $R$ we confine the support of $h$ to $R$ also. Since $h$ is formed as the product of $i$ and $w$, the frequency response $H(\omega_1, \omega_2)$ is related to $I(\omega_1, \omega_2)$ by a frequency-domain convolution. Specifically,

$$H(\omega_1, \omega_2) = \frac{1}{(2\pi)^2} \int_{-\pi}^{\pi} \int_{-\pi}^{\pi} I(\Omega_1, \Omega_2)W(\omega_1 - \Omega_1, \omega_2 - \Omega_2) \, d\Omega_1 \, d\Omega_2 \tag{3.16}$$

where $W(\omega_1, \omega_2)$ is the Fourier transform of the window array. The frequency response $H(\omega_1, \omega_2)$ is a smoothed version of the ideal frequency response, where the smoothing function is the Fourier transform of the window array.

For many filter design problems, the desired filter behavior is specified through $I(\omega_1, \omega_2)$ rather than $i(n_1, n_2)$. In these cases we must either compute $i(n_1, n_2)$ analytically or approximate it by sampling $I(\omega_1, \omega_2)$ and then performing an inverse DFT. Since the support of $i(n_1, n_2)$ is generally of infinite extent, the result of this procedure is a spatially aliased approximation to $i(n_1, n_2)$. To minimize the aliasing error, the extent of the support of the inverse DFT should be several times larger than the extent of $R$.

### 3.3.2 Choosing the Window Function

The choice of the window function is governed by three requirements. First, it must have the region of support $R$. Second, for $H(\omega_1, \omega_2)$ to approximate $I(\omega_1, \omega_2)$ closely, $W(\omega_1, \omega_2)$ should approximate a 2-D impulse function. Finally, if $h(n_1, n_2)$ is to be zero-phase, the window should satisfy the zero-phase relation

$$w(n_1, n_2) = w^*(-n_1, -n_2) \tag{3.17}$$

Because these requirements are similar to the requirements for a 1-D window function, 1-D windows are often used as a basis for generating 2-D windows. There are two means by which this is commonly done. The first method forms a 2-D window with a square or rectangular region of support by taking the outer product of two 1-D windows.

$$w_R(n_1, n_2) = w_1(n_1)w_2(n_2) \tag{3.18}$$

The second, which was introduced by Huang [2], forms a 2-D window by sampling a circularly rotated, 1-D, continuous window function.

$$w_C(n_1, n_2) = w(\sqrt{n_1^2 + n_2^2}) \tag{3.19}$$

The resulting 2-D windows have nearly circular regions of support.

The Fourier transform of $w_R(n_1, n_2)$ is the outer product of the Fourier transforms of $w_1(n_1)$ and $w_2(n_2)$. Thus

$$W_R(\omega_1, \omega_2) = W_1(\omega_1)W_2(\omega_2) \tag{3.20}$$

The Fourier transform of $w_C(n_1, n_2)$ resembles, but differs in detail from, a circularly rotated version of the 1-D Fourier transform of $w(t)$.

If $w(t)$, $w_1(n_1)$, and $w_2(n_2)$ are good 1-D windows (i.e., satisfy the three criteria we have given), then $w_C$ and $w_R$ will be good 2-D windows. Any of a number of 1-D windows can be used. Among the most popular, however, are the rectangular window

$$w(t) = \begin{cases} 1, & |t| < \tau \\ 0, & \text{otherwise} \end{cases} \tag{3.21}$$

the Hanning window

$$w(t) = \begin{cases} \frac{1}{2}\left(1 + \cos\frac{\pi t}{\tau}\right), & |t| < \tau \\ 0, & \text{otherwise} \end{cases} \tag{3.22}$$

and the Kaiser, or $I_0$-sinh, window [3]

$$w(t) = \begin{cases} \dfrac{I_0(\alpha\sqrt{1 - (t/\tau)^2})}{I_0(\alpha)}, & |t| < \tau \\ 0, & \text{otherwise} \end{cases} \qquad (3.23)$$

where $I_0(x)$ is the modified Bessel function of the first kind of order zero. All these windows have been presented as continuous windows with support on the interval $[-\tau, \tau]$.

Figure 3.3 shows the Fourier transform of a typical 1-D window which has been normalized to have unit area in the frequency domain. It has a large concentra-

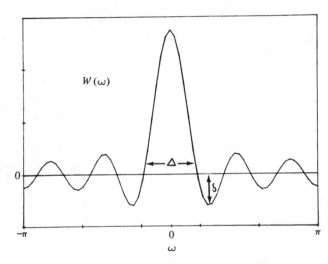

**Figure 3.3** Fourier transform of a typical one-dimensional window function.

tion of low-frequency energy and it exhibits oscillations at high frequencies. In this figure $\Delta$ denotes the width of the low-frequency main lobe and $\delta$ represents the height of the highest of the side lobes. The smaller these quantities become, the better the window and the closer $H(\omega_1, \omega_2)$ will approximate $I(\omega_1, \omega_2)$. For a window with a fixed region of support, there is a trade-off between these two quantities; the only way to decrease both of them is to increase the number of samples in $R$. The parameter $\alpha$ in the expression for the Kaiser window, equation (3.23), is used to adjust the trade-off between $\Delta$ and $\delta$. It is the presence of this parameter that makes the Kaiser window particularly versatile.

### 3.3.3  Design Example

As a simple example, let us consider the design of an $(11 \times 11)$-point FIR filter to approximate the ideal frequency response

$$I(\omega_1, \omega_2) = \begin{cases} 1, & \omega_1^2 + \omega_2^2 \leq (0.4\pi)^2 \\ 0, & \text{otherwise}, \quad -\pi \leq \omega_1, \omega_2 \leq \pi \end{cases} \qquad (3.24)$$

Since our ideal response is purely real, the design should be zero-phase. This means that the origin should be a point of symmetry in the region $R$. We will thus choose $R$ to be the set of samples

$$R = \{(n_1, n_2): \quad -5 \leq n_1, n_2 \leq 5\} \tag{3.25}$$

We will design the filter using both 2-D window formulations based upon a 1-D Kaiser window prototype [3]. With the rotated formulation the actual region of support for the design will be a circular subregion of $R$.

The ideal impulse response, $i(n_1, n_2)$, can be found by computing the inverse Fourier transform of $I(\omega_1, \omega_2)$. We did this in Chapter 1 and obtained the result

$$i(n_1, n_2) = \frac{0.2J_1(0.4\pi\sqrt{n_1^2 + n_2^2})}{\sqrt{n_1^2 + n_2^2}} \tag{3.26}$$

where $J_1(x)$ is the Bessel function of the first kind of order 1. Evaluating the two windows, we get

$$w_R(n_1, n_2) = \begin{cases} \dfrac{I_0[\alpha\sqrt{1 - (n_1/5)^2}]\, I_0[\alpha\sqrt{1 - (n_2/5)^2}]}{I_0^2[\alpha]}, & |n_1| \leq 5, \quad |n_2| \leq 5 \\ 0, & \text{otherwise} \end{cases} \tag{3.27}$$

$$w_C(n_1, n_2) = \begin{cases} \dfrac{I_0[\alpha\sqrt{1 - ((n_1^2 + n_2^2)/25)]}}{I_0[\alpha]}, & n_1^2 + n_2^2 \leq 25 \\ 0, & \text{otherwise} \end{cases} \tag{3.28}$$

We then get our two filters by multiplying either (3.27) or (3.28) by (3.26).

The frequency responses for the two filters are shown in Figure 3.4 for the case $\alpha = 0.0$ [4]. In each case each frequency response is presented both as a perspective plot and as a contour plot. The former provides a quick feeling for the quality of the filter, but the latter is more helpful for measuring the circularity of the passband and stopband. The filter designed using $w_R(n_1, n_2)$ has a peak passband error of 0.2914 and a peak stopband error of 0.1341. For the filter designed using $w_C(n_1, n_2)$, the peak passband error is 0.1468 and the peak stopband error is 0.1105.

The frequency responses of the filters we have designed differ from the ideal response in two ways. The response is not flat in either the passband or the stopband and the filter cutoff is not perfectly sharp. The former is due to the side lobes in the Fourier transform of the window; the latter is due to the finite width of the main lobe of that Fourier transform. Varying the window parameter $\alpha$ in the Kaiser window allows us to trade off the sharpness of the cutoff for the smoothness of the frequency response in the passband and stopband.

A circularly symmetric lowpass filter has the form shown in Figure 3.5. Ideally, the response has value 1 in the passband and value 0 in the stopband, but in practice, the actual values obtained will deviate from these nominal values. Let $\delta_p$ denote the maximum passband error and let $\delta_s$ similarly denote the maximum stopband error. The passbands and stopbands themselves are defined by the passband and stopband cutoff frequencies $\omega_p$ and $\omega_s$. As $\delta_p$, $\delta_s$, and $\omega_s - \omega_p$ become smaller, the quality of

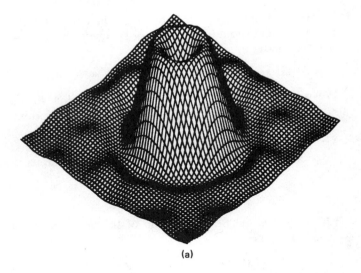

(a)

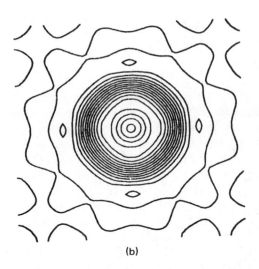

(b)

**Figure 3.4** Frequency responses of two 11 × 11 FIR lowpass filters designed using windows. (a) Perspective plot of a design based on an outer product window. (b) Contour plot. (c) Perspective plot of a design based on a rotated window. (d) Contour plot. The contours are shown for values of $H$ between $-0.1$ and $1.1$ in increments of 0.1. (Courtesy of Russell M. Mersereau, from *Two-Dimensional Digital Signal Processing I*, Thomas S. Huang, ed., *Topics in Applied Physics*, Vol. 42, © 1981 Springer-Verlag.)

the filter improves. One of the prime motivations for limiting this discussion to circular lowpass filters is that the measure of filter quality is contained in these three numbers. We will use the filter order, $N$, as a measure of complexity.

Following the approach of Kaiser [3] for the 1-D case, a formula for the filter order $N$ (impulse response area $= N \times N$) in terms of its specifications $\delta_p$, $\delta_s$, and $\Delta\omega = \omega_s - \omega_p$ can be experimentally developed [5]. The order $N$ can be approximated by

$$N_R \simeq \frac{-20 \log_{10}\sqrt{\delta_p\delta_s} - 8}{2.10\Delta\omega} \tag{3.29}$$

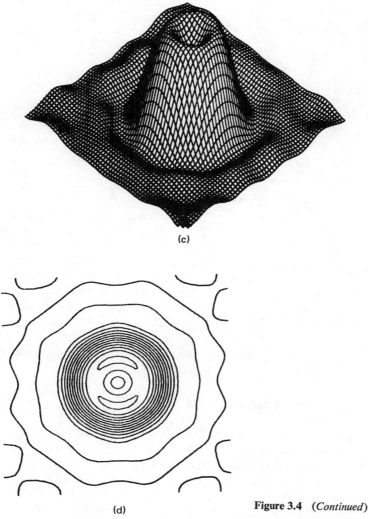

(c)

(d)        **Figure 3.4**   (*Continued*)

for the outer product window and by

$$N_C \simeq \frac{-20 \log_{10}\sqrt{\delta_p \delta_s} - 7}{2.18 \Delta \omega} \tag{3.30}$$

for the rotated window. In terms of the required filter order, the two windows are essentially the same.

The window parameter $\alpha$ can be determined from the desired attenuation of the filter, where the attenuation ATT is defined as

$$\text{ATT} \triangleq -20 \log_{10}\sqrt{\delta_p \delta_s} \tag{3.31}$$

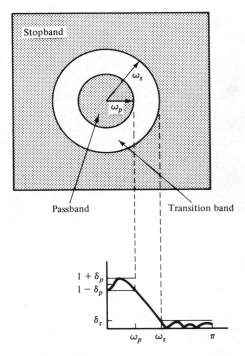

**Figure 3.5**  Specifications for a lowpass filter.

For the outer product windows, $\alpha$ has been experimentally determined to be

$$\alpha \simeq \begin{cases} 0.42(\text{ATT} - 19.3)^{0.4} + 0.089(\text{ATT} - 19.3), \\ \qquad\qquad\qquad\qquad\qquad 20 < \text{ATT} < 60 \\ 0, \qquad\qquad\qquad\qquad\qquad \text{ATT} < 20 \end{cases}$$

$$(3.32)$$

For the rotated window, $\alpha$ is given approximately by

$$\alpha \simeq \begin{cases} 0.56(\text{ATT} - 20.2)^{0.4} + 0.083(\text{ATT} - 20.2), \\ \qquad\qquad\qquad\qquad\qquad 20 < \text{ATT} < 60 \\ 0, \qquad\qquad\qquad\qquad\qquad \text{ATT} < 20 \end{cases}$$

$$(3.33)$$

For circular lowpass filters, both window formulations lead to filters of essentially the same order. This represents an advantage for the rotated window since its circular region of support has fewer nonzero samples than the square region of support of the outer product window. The reduced number of coefficients can be exploited for computational savings when the filter is implemented. It should be stated, however, that for filters which are not circularly symmetric, this conclusion probably does not apply.

### 3.3.4 Image Processing Example

Windows can be used for signal processing applications other than the design of digital filters. For example, a procedure developed by McClellan [6] for image enhancement uses windowing to reduce the effects of data-dependent processing artifacts. McClellan's procedure is a variation on a technique known as alpha-rooting [7], which involves taking the $\alpha$th power ($\alpha < 1$) of the magnitude of the Fourier transform of an image while retaining the original phase. This operation reduces the dynamic range of the magnitude of the Fourier transform, boosting low-energy regions relative to high-energy ones. For most images the net effect is a high spatial frequency boost which enhances the detail in the image. There is one difficulty with the alpha-rooting process, however; the output image suffers from artifacts which are correlated with the structure of the image being enhanced. This can be seen in the example shown in Figure 3.6 for which $\alpha = 0.5$.

The alpha-rooting operation can be considered to be a filtering operation with the frequency response of the filter depending on the input image. Thus alpha-rooting produces the same result that would be obtained if $x(n_1, n_2)$ were filtered by a 2-D filter with the frequency response

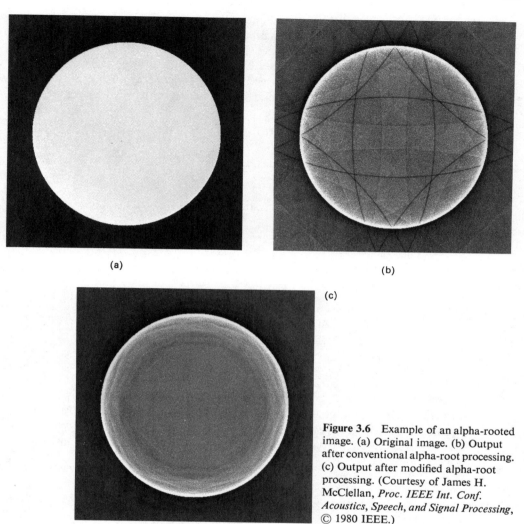

(a)

(b)

(c)

**Figure 3.6**  Example of an alpha-rooted image. (a) Original image. (b) Output after conventional alpha-root processing. (c) Output after modified alpha-root processing. (Courtesy of James H. McClellan, *Proc. IEEE Int. Conf. Acoustics, Speech, and Signal Processing,* © 1980 IEEE.)

$$H(\omega_1, \omega_2) = |X(\omega_1, \omega_2)|^{\alpha-1} \tag{3.34}$$

Although the alpha-rooting procedure is implemented in the frequency domain, the final result could be obtained by convolving $x(n_1, n_2)$ with the inverse Fourier transform of (3.34). This convolution is circular, however, since alpha-rooting is generally implemented in the frequency domain using sampled transforms. In fact, the annoying artifacts that are in evidence in Figure 3.6(b) can be attributed to the use of circular rather than linear 2-D convolution.

Since $\alpha < 1$, any zeros in the spectrum $X(\omega_1, \omega_2)$ become peaks in $|X(\omega_1, \omega_2)|^{\alpha-1}$.

Because of these peaks, the inverse Fourier transform of (3.34) does not decay quickly and $h(n_1, n_2)$ does not have finite support. Thus the circular convolution of $x$ with $h$ is a heavily aliased linear convolution, causing artifacts in the final output.

McClellan's procedure [6] uses a window to guarantee the finite support of $h(n_1, n_2)$. This variation adds extra computation since it requires explicit evaluation of $h(n_1, n_2)$. The steps of the modified procedure are depicted in Figure 3.7. The

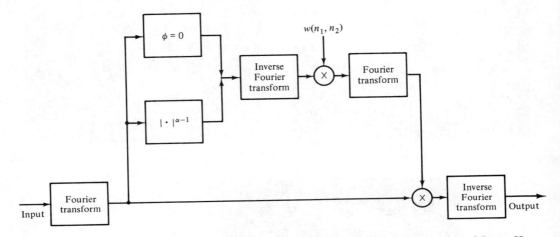

**Figure 3.7**  Block diagram of the modified alpha-root processor. (Courtesy of James H. McClellan, *Proc. IEEE Int. Conf. Acoustics, Speech, and Signal Processing*, © 1980 IEEE.)

frequency response is formed from $x(n_1, n_2)$ using a DFT. Then it is inverse-transformed to yield the sequence $h(n_1, n_2)$. This sequence is multiplied by a window to produce a finite-extent sequence $\hat{h}(n_1, n_2)$, which is then convolved with $x(n_1, n_2)$ to give the final result. A typical result using a Gaussian window is shown in Figure 3.6(c).

## *3.4 OPTIMAL FIR FILTER DESIGN

Filters with finite support have frequency responses which can only approximate most ideal frequency specifications. Typically there will be an error between the frequency response that we get and the one we would like, which is given by

$$E(\omega_1, \omega_2) = H(\omega_1, \omega_2) - I(\omega_1, \omega_2) \qquad (3.35)$$

One approach to filter design is to choose the coefficients of the filter to minimize some function of this error, such as its $L_2$-norm (mean-squared value)†

---

†To be completely consistent with the other error norms, we should define the $L_2$-norm as a root-mean-squared value. In practice, however, mean-squared error is typically used because its partial derivatives have a simple form.

$$E_2 = \frac{1}{4\pi^2} \int_{-\pi}^{\pi} \int_{-\pi}^{\pi} |E(\omega_1, \omega_2)|^2 \, d\omega_1 \, d\omega_2 \tag{3.36}$$

its $L_p$ norm

$$E_p = \sqrt[p]{\frac{1}{4\pi^2} \int_{-\pi}^{\pi} \int_{-\pi}^{\pi} |E(\omega_1, \omega_2)|^p \, d\omega_1 \, d\omega_2} \tag{3.37}$$

or its Chebyshev ($L_\infty$) norm

$$E_\infty = \max_{(\omega_1, \omega_2)} |E(\omega_1, \omega_2| \tag{3.38}$$

Since filters designed using different error criteria can be quite different, we look at several examples in this section. We will restrict our attention to zero-phase filters [for which $H(\omega_1, \omega_2)$ is real].

The frequency response of an FIR filter with support on $R$ is given by

$$H(\omega_1, \omega_2) = \sum \sum_{(n_1, n_2) \in R} h(n_1, n_2) \exp(-j\omega_1 n_1 - j\omega_2 n_2) \tag{3.39}$$

Substituting into (3.35), we obtain

$$E(\omega_1, \omega_2) = \left[ \sum \sum_{(n_1, n_2) \in R} h(n_1, n_2) \exp(-j\omega_1 n_1 - j\omega_2 n_2) \right] - I(\omega_1, \omega_2) \tag{3.40}$$

We see that the error is a linear function of the unknown filter coefficients. This fact makes the general FIR filter design problem easier than some of the specialized design problems to be discussed in Section 3.5 or some of the IIR design techniques to be considered in Chapter 5.

For a real zero-phase filter, $h(n_1, n_2)$ and $h(-n_1, -n_2)$ are equal, allowing equation (3.39) to be rewritten as

$$H(\omega_1, \omega_2) = h(0, 0) + \sum \sum_{(n_1, n_2) \in R'} 2h(n_1, n_2) \cos(\omega_1 n_1 + \omega_2 n_2) \tag{3.41}$$

where $R'$ contains approximately half as many samples as $R$. For the purposes of performing a linear approximation, it is convenient to simplify (3.41) further and write it in the form

$$H(\omega_1, \omega_2) = \sum_{i=1}^{F} a(i)\phi_i(\omega_1, \omega_2) \tag{3.42}$$

where $i$ is an index that denotes some ordering of the samples $(n_1, n_2)$ in $R'$ and $F$ is the number of independent samples in the impulse response, or degrees of freedom in the approximation. The coefficients $a(i)$ are simply the impulse response values to be determined,

$$a(i) = h(n_1, n_2) \tag{3.43}$$

and the functions $\{\phi_i(\omega_1, \omega_2)\}$, often called the *basis functions* of the approximation, are given by

$$\phi_i(\omega_1, \omega_2) = \begin{cases} 2\cos(\omega_1 n_1 + \omega_2 n_2), & (n_1, n_2) \neq (0, 0) \\ 1, & (n_1, n_2) = (0, 0) \end{cases} \tag{3.44}$$

This formulation allows us to handle linear constraints among the impulse response coefficients. For example, if we wish to impose the constraint $a(i) = a(j)$, we can simply replace $\phi_i(\omega_1, \omega_2)$ by $\phi_i(\omega_1, \omega_2) + \phi_j(\omega_1, \omega_2)$ in (3.42) and omit the term containing $a(j)$. This has the effect of reducing the number of degrees of freedom by 1. We can similarly treat constraints of the form $a(i) = k$ by replacing $I(\omega_1, \omega_2)$ by $I(\omega_1, \omega_2) - k\phi_i(\omega_1, \omega_2)$ and omitting the $a(i)$ term from the sum defining $H(\omega_1, \omega_2)$. This also has the effect of reducing the number of degrees of freedom by 1. At this point the reader should be able to verify that constraints of the form $a(i) + a(j) = k$ can also be accommodated. Again the number of degrees of freedom will be reduced by 1.

As an example of how the number of degrees of freedom can be reduced, consider a $[(2N + 1) \times (2N + 1)]$-point FIR filter which is to have eightfold (octal) symmetry. This can be assured by imposing the conditions

$$h(n_1, n_2) = h(\pm n_1, \pm n_2) = h(\pm n_2, \pm n_1) \tag{3.45}$$

For such a filter the frequency response can be written as in (3.42) with

$$i = \frac{n_1(n_1 + 1)}{2} + n_2 + 1, \qquad 0 \le n_1 \le N, \quad 0 \le n_2 \le n_1$$

$$\phi_i(\omega_1, \omega_2) = \cos \omega_1 n_1 \cos \omega_2 n_2 + \cos \omega_1 n_2 \cos \omega_2 n_1 \tag{3.46}$$

$$F = \frac{(N + 1)(N + 2)}{2}$$

### 3.4.1 Least-Squares Designs

In this section we look at algorithms for choosing $h(n_1, n_2)$ to minimize the mean-squared error, $E_2$ in (3.36), and some related error functions. For the most part these algorithms are quite simple, requiring little more than the solution of a number of linear equations.

The filter coefficients that minimize $E_2$ can be obtained using a design algorithm we have already seen—the window method with a window that is flat over $R$. To see this we can begin with the definition of $E_2$.

$$E_2 = \frac{1}{4\pi^2} \int_{-\pi}^{\pi} \int_{-\pi}^{\pi} |E(\omega_1, \omega_2)|^2 \, d\omega_1 \, d\omega_2$$

Through the use of Parseval's relation, $E_2$ can be expressed in terms of spatial domain quantities as

$$\begin{aligned}
E_2 &= \sum_{n_1} \sum_{n_2} [h(n_1, n_2) - i(n_1, n_2)]^2 \\
&= \sum_{(n_1, n_2) \in R} [h(n_1, n_2) - i(n_1, n_2)]^2 + \sum_{(n_1, n_2) \notin R} [h(n_1, n_2) - i(n_1, n_2)]^2 \\
&= \sum_{(n_1, n_2) \in R} [h(n_1, n_2) - i(n_1, n_2)]^2 + \sum_{(n_1, n_2) \notin R} i^2(n_1, n_2)
\end{aligned} \tag{3.47}$$

In writing the last expression we recognize that $h(n_1, n_2)$ is zero for any $(n_1, n_2)$ not in $R$. Since both of the summations in (3.47) are positive and only the first can be altered by selecting the filter coefficients, $h(n_1, n_2)$, it follows that $E_2$ is minimized by setting

$$h(n_1, n_2) = \begin{cases} i(n_1, n_2), & (n_1, n_2) \in R \\ 0, & (n_1, n_2) \notin R \end{cases} \qquad (3.48)$$

which is the filter that results from the window method with a constant window over $R$.

In the slightly more general case where linear constraints may be present and the filter frequency response is given by (3.42), we get

$$E_2 = \frac{1}{4\pi^2} \int_{-\pi}^{\pi} \int_{-\pi}^{\pi} \left\{ \sum_{i=1}^{F} a(i)\phi_i(\omega_1, \omega_2) - I(\omega_1, \omega_2) \right\}^2 d\omega_1 \, d\omega_2$$

We can minimize $E_2$ by taking its derivative with respect to each of the $a(k)$, setting these derivatives to zero, and solving the resulting equations. Since the partial derivatives $\{\partial E_2/\partial a(k)\}$ are all linear functions of the unknown coefficients, this requires at most the solution of $F$ linear equations. These equations can be written as

$$\sum_{i=1}^{F} a(i)\phi_{ik} = I_k, \qquad k = 1, 2, \ldots, F \qquad (3.49)$$

where

$$\phi_{ik} \triangleq \frac{1}{4\pi^2} \int_{-\pi}^{\pi} \int_{-\pi}^{\pi} \phi_i(\omega_1, \omega_2)\phi_k(\omega_1, \omega_2) \, d\omega_1 \, d\omega_2 \qquad (3.50a)$$

$$I_k \triangleq \frac{1}{4\pi^2} \int_{-\pi}^{\pi} \int_{-\pi}^{\pi} I(\omega_1, \omega_2)\phi_k(\omega_1, \omega_2) \, d\omega_1 \, d\omega_2 \qquad (3.50b)$$

In the frequently occurring special case that the $\{\phi_i(\omega_1, \omega_2)\}$ are orthogonal ($\phi_{ik} = 0$, $i \neq k$), the solution of (3.49) is simply $a(i) = I_i/\phi_{ii}$. The number of simultaneous linear equations to be solved places an effective upper limit on the number of degrees of freedom that can be accommodated.

The error measure $E_2$ weights errors equally at all frequencies. From our experience with the window design method, we know that filters which are designed to minimize $E_2$ may not always be satisfactory; they suffer from large passband and stopband ripple. The integrals in (3.50b) may also be difficult to evaluate if $I(\omega_1, \omega_2)$ is not simple. These difficulties can be partially alleviated by replacing the error $E_2$ by $E_2'$, where

$$E_2' \triangleq \sum_m W_m[H(\omega_{1m}, \omega_{2m}) - I(\omega_{1m}, \omega_{2m})]^2 \qquad (3.51)$$

The set of frequencies $\{(\omega_{1m}, \omega_{2m})\}$, called the *constraint frequencies*, correspond to a finite number of discrete locations in the 2-D frequency plane and the (positive) numbers $W_m$ denote weighting values. With this error measure, in regions of the frequency plane where we demand little error we can increase the density of the constraint frequencies and/or increase their weights. As a practical matter, the number of constraint frequencies should be several times the number of degrees of freedom.

Finding the coefficients $\{a(k)\}$ that minimize $E_2'$ again involves the solution of $F$ linear equation in $F$ unknowns given by

$$\sum_{i=1}^{F} a(i)\phi_{ik} = I_k, \qquad k = 1, 2, \ldots, F \qquad (3.52)$$

with

$$\phi_{ik} = \sum_m W_m \phi_i(\omega_{1m}, \omega_{2m}) \phi_k(\omega_{1m}, \omega_{2m}) \tag{3.53a}$$

$$I_k = \sum_m W_m I(\omega_{1m}, \omega_{2m}) \phi_k(\omega_{1m}, \omega_{2m}) \tag{3.53b}$$

Although the design of filters to minimize an $L_p$ error norm ($p \neq 2$) can be approached similarly, the equations that result are not linear and their solution is not simple. The more common approach is to use an iterative algorithm, such as the method of steepest descent, to get successively better approximations to the desired filter coefficients. With the method of steepest descent, the coefficients $\{a^{(k)}(i)\}$ at the $k$th iteration are determined from

$$a^{(k)}(i) = a^{(k-1)}(i) - \alpha_k \frac{\partial E_p^{(k-1)}}{\partial a^{(k-1)}(i)}, \qquad i = 1, 2, \cdots, F \tag{3.54}$$

The parameter $\alpha_k$ is known as the step size. Different strategies exist for its selection.

### 3.4.2 Design of Zero-Phase Equiripple FIR Filters

We have observed that there is a trade-off between the width of the transition bands in a filter and the magnitude of the error in the passbands and stopbands. Of all filters with given passbands and stopbands, the filters with the smallest passband and stopband ripple are equiripple designs. These filters are chosen to minimize the Chebyshev (or $L_\infty$) error criterion

$$E_\infty = \max_{(\omega_1, \omega_2) \in K} |H(\omega_1, \omega_2) - I(\omega_1, \omega_2)| \tag{3.55}$$

Here $K$ is a compact region of the 2-D frequency plane. This region is normally chosen to consist of the union of the desired passbands and stopbands of the filter (including their boundaries). As before, $H(\omega_1, \omega_2)$ is assumed to be a zero-phase FIR filter with a frequency response of the form

$$H(\omega_1, \omega_2) = \sum_{(n_1, n_2) \in R} h(n_1, n_2) \exp[-j\omega_1 n_1 - j\omega_2 n_2] \tag{3.56}$$

Because of symmetry or other constraints, all the coefficients of the filter may not be independent. We prefer to isolate the independent parameters of the design and write the frequency response as

$$H(\omega_1, \omega_2) = \sum_{i=1}^{F} a(i) \phi_i(\omega_1, \omega_2) \tag{3.57}$$

where the basis functions $\{\phi_i(\omega_1, \omega_2)\}$ are real.

The filters that minimize the Chebyshev error criterion are called *minimax* because they minimize the maximum value of the error. They are also called *equiripple* designs because the final error waveform, $E = H - I$, contains many peaks or ripples which have the same magnitude. Figure 3.8 shows an example of an equiripple filter with an $(11 \times 11)$-point impulse response [11]. For this filter the passband region is a disk of radius $0.4\pi$ and the stopband is the exterior of a disk of radius $0.6\pi$. The peak error magnitude is 0.0569 and the nominal passband amplitude is 1.

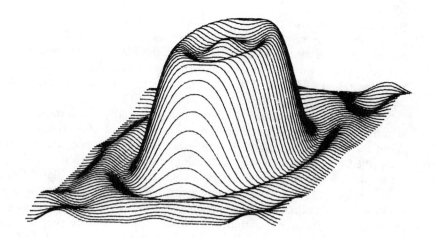

**Figure 3.8**   An $(11 \times 11)$-point zero-phase equiripple FIR filter. (Courtesy of David B. Harris and Russell M. Mersereau, *IEEE Trans. Acoustics, Speech, and Signal Processing,* © 1977 IEEE.)

The optimization problem can be solved using iterative algorithms which are guaranteed to converge in a finite number of steps. These algorithms are 2-D analogs of the 1-D Parks–McClellan algorithm [8], which, in turn, is an adaptation of the second algorithm of Remez [9]. In the 1-D case the equiripple solution can be proven to be unique, there are no numerical difficulties to limit the filter order, and the algorithm converges rapidly. For higher-dimensional cases, due primarily to the absence of a factorization theorem, the algorithms are slower to converge, more difficult to understand, and somewhat limited in their capabilities [10, 11].

An *extremal point* is a point $\boldsymbol{\omega} = (\omega_1, \omega_2)'$ in the domain of approximation $K$ on which the error function, $|H - I|$, attains its maximum value. The *extremal point set* is the set of all extremal points. It generally consists of $F + 1$ independent members. An extremal point set is a *critical point set* if the errors at the points of the set satisfy certain sign conditions. The free parameters, $\{a(i)\}$, of the filter which minimize $E_\infty$ are completely determined by the critical point set, and the optimal value of $E_\infty$ is the magnitude of the error at each member of that set.

If the domain of approximation $K$ consists of a finite number of frequency samples, the approximation can be performed by searching for a critical point set. This search is aided by two results which are generally stated as theorems [11]. First, if $H(\boldsymbol{\omega})$ is a best approximation to $I(\boldsymbol{\omega})$ on $K$ and $C$ is a critical point set associated with the error $H - I$, then $H$ is a best approximation to $I$ on $C$. Second, if $C$ contains $F + 1$ points, then $H$ is an approximation of *maximum* error among all best approximations to $I$ on subsets of $K$ containing $F + 1$ points.

These two results form the basis for the ascent algorithm, which is equivalent to the dual form of the simplex algorithm of linear programming. The ascent algorithm repeatedly finds best approximations on sets of $F + 1$ points. Successive sets of $F + 1$

points are chosen so that on each the Chebyshev error will be greater than on the preceding set. According to the second result above, when this norm attains its maximum value on a set of $F + 1$ points, the best Chebyshev approximation on $K$ has been found and the algorithm terminates. The details of how the ascent algorithm works and the proof that it will, in fact, converge are given in a number of references on approximation theory, such as [9,12].

The ascent algorithm requires a large number of iterations, each of which requires that the error function be evaluated over all of $K$. It is extremely time consuming. The algorithms of Kamp and Thiran [10] and Hersey and Mersereau (described in [11]), which are specifically tailored for FIR filter design, can converge up to an order of magnitude faster. The Kamp and Thiran algorithm changes some of the inner workings of the ascent algorithm by modifying the way in which error information is utilized. The Hersey and Mersereau algorithm defines an intermediate sparse grid of frequency samples, $G$, which contains more than $F + 1$ samples but considerably fewer than $K$. The ascent algorithm can rather quickly find a best approximation on $G$, which is periodically updated. The details are described in [11]. We will say nothing more here about the inner workings of these algorithms.

Filters designed to minimize an $L_p$-norm approach equiripple designs in the limit as $p \rightarrow \infty$. Lodge and Fahmy [13] have used this approach to design very good approximations to equiripple filters. Their algorithm, which uses the method of parallel tangents to perform the error minimization, is considerably more efficient than the true equiripple design algorithms described above and it can be used to design filters for which $F$ is large.

## 3.5 DESIGN OF FIR FILTERS FOR SPECIAL IMPLEMENTATIONS

We commented in the introduction to this chapter that the design and implementation of multidimensional filters are intimately coupled. Whereas any FIR filter can be implemented using either the convolution sum or the DFT once its impulse response is known, other implementations, which can be quite efficient, typically restrict the class of implementable filters and require specialized design algorithms. We look at a few of these specialized designs and their implementations in this section.

The simplest of these specialized implementations is the cascade implementation. For someone with extensive experience with 1-D filters, it might seem unusual to classify the cascade structure as a specialized implementation, but recall that multidimensional polynomials usually cannot be factored. Thus multidimensional cascades are very special indeed!

### 3.5.1 Cascaded FIR Filters

Consider a 1-D causal FIR filter, $h(n)$, of length $N$ with frequency response

$$H(\omega) = \sum_{n=0}^{N-1} h(n) \exp(-j\omega n) \qquad (3.58)$$

Since $H(\omega)$ is a polynomial in $\exp(-j\omega)$ of degree $N - 1$, it can be factored. Denote one factorization of $H$ by

$$H(\omega) = F(\omega)G(\omega) \tag{3.59}$$

where $F(\omega)$ and $G(\omega)$ are themselves polynomials in $\exp(-j\omega)$ of degree less than $N - 1$. We can thus say that the original filter is equivalent to two FIR filters in cascade —one with frequency response $F(\omega)$ and one with frequency response $G(\omega)$. If $f(n)$ is $M$ samples long, then $g(n)$ is $N - M + 1$ samples long.

To compare the direct with the cascade implementation, we recall that the direct implementation requires one multiplication per output sample for each sample of the impulse response—$N$ multiplications per output sample in all. Using the cascade implementation requires $M + (N - M + 1) = N + 1$ multiplications per output sample. [This number can be reduced by one since either $f(n)$ or $g(n)$ can be chosen to have a leading coefficient of 1.] In the 1-D case the cascade and direct implementations require the same number of numerical operations.

The situation is different in the multidimensional case. Consider a factorable 2-D FIR frequency response which can be written as

$$H(\omega_1, \omega_2) = F(\omega_1, \omega_2)G(\omega_1, \omega_2) \tag{3.60}$$

If $f(n_1, n_2)$ is an $(M \times M)$-point array and $g(n_1, n_2)$ is an $[(N - M + 1) \times (N - M + 1)]$-point array, then $h(n_1, n_2)$ will have an $(N \times N)$-point region of support. To implement $h(n_1, n_2)$ as a direct convolution requires $N^2$ multiplications per output sample, whereas to implement the filter in the cascade implied by (3.60) requires only $M^2 + (N - M + 1)^2$ multiplications. The savings can be as much as 50%. In those cases when further factorizations are possible or when we are dealing with higher-dimensional signals, the savings can be even greater. A frequency response of the form of (3.60) has only $M^2 + (N - M + 1)^2 - 1$ degrees of freedom. Since an arbitrary $N \times N$ frequency response has $N^2$ degrees of freedom, it is not surprising that arbitrary frequency responses are not factorable. Factorable multidimensional polynomials actually represent a degenerate special case.

To design a filter that is to be realized as a cascade, the frequency response must first be expressed in factored form:

$$H(\omega_1, \omega_2) = \prod_i \left[ \sum_{n_1} \sum_{n_2} h_i(n_1, n_2) \exp(-j\omega_1 n_1 - j\omega_2 n_2) \right] \tag{3.61}$$

The coefficients of the filter $\{h_i(n_1, n_2)\}$ can then be chosen to minimize an error functional such as a weighted $L_p$ error norm,

$$E_p = \sqrt[p]{\sum_k W_k |H(\omega_{1k}, \omega_{2k}) - I(\omega_{1k}, \omega_{2k})|^p} \tag{3.62}$$

which is similar to some of the error norms we encountered in the preceding section. Since the error is a nonlinear function of the unknown coefficients, an iterative minimization of $E_p$ is normally performed. The coefficients on the $j$th iteration can be determined from the coefficients of the $(j - 1)$st using

$$h_i^{(j)}(n_1, n_2) = h_i^{(j-1)}(n_1, n_2) - \alpha_j \frac{\partial E_p^{(j-1)}}{\partial h_i^{(j-1)}(n_1, n_2)} \tag{3.63}$$

The initial choice of filter coefficients is somewhat arbitrary, but should be as accurate as possible.

An alternative approach to the design of cascade filters is an iterative one where at each iteration every stage except one is fixed and the remaining stage is optimized. At the next iteration these new coefficients are fixed and a different stage is optimized. This procedure, though simple, is not guaranteed to converge, and it may converge to a local minimum.

### 3.5.2 Parallel FIR Filters

If two FIR filters with impulse responses $h_1(n_1, n_2)$ and $h_2(n_1, n_2)$ are connected in parallel, they are equivalent to a single filter with the impulse response

$$h(n_1, n_2) = h_1(n_1, n_2) + h_2(n_1, n_2) \qquad (3.64)$$

Going the other way, we see that any filter with impulse response $h(n_1, n_2)$ can be decomposed into the parallel connection of an $h_1$ and an $h_2$. In general, we would expect the region of support for $h$ to be the union of the regions of support of $h_1$ and $h_2$, although it could be smaller than this if cancellations occur. Unfortunately, this decomposition rarely offers a computational savings and may, in fact, involve extra computation.

Computational savings are possible, however, when restrictions are placed on $h_1$ and $h_2$, for example, when $h_1$ and $h_2$ are restricted to be separable filters. Separable filters are very efficient to implement, but they can only approximate separable ideal responses accurately. However, a parallel connection of two separable filters gives a nonseparable filter. This allows the possibility of approximating nonseparable transfer functions with easily implementable filters. This idea was originally proposed by Treitel and Shanks [14], who called these filters *multistage separable filters.*

We recall from Chapter 1 that a separable filter is one whose impulse response factors into a sequence in the horizontal index and one in the vertical index. Thus

$$h(n_1, n_2) = r(n_1)c(n_2) \qquad (3.65)$$

is the impulse response of a separable system. The frequency response of this filter is the outer product of the two 1-D frequency responses.

$$H(\omega_1, \omega_2) = R(\omega_1)C(\omega_2) \qquad (3.66)$$

This filter can be implemented by convolving each row of the input with $r(n_1)$ and then convolving each column of the resulting array with $c(n_2)$. If the impulse response $h(n_1, n_2)$ contains $N_1 N_2$ samples, corresponding to an $N_1$-point sequence $r(n_1)$ and an $N_2$-point sequence $c(n_2)$, the evaluation of each sample of the output will require $N_1$ multiplications and additions from the row convolutions and $N_2$ multiplications and additions from the column convolutions. The total number of operations, $N_1 + N_2$, compares favorably with the $N_1 N_2$ that would be required if $h$ were not separable.

As a generalization, consider a filter whose impulse response is a parallel combination of separable filters, as indicated in Figure 3.9. The impulse response of

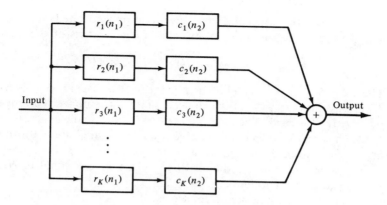

**Figure 3.9**  Multistage separable FIR filter with $K$ separable stages.

such a filter is

$$h(n_1, n_2) = \sum_{k=1}^{K} r_k(n_1)c_k(n_2) \tag{3.67}$$

and its frequency response is

$$H(\omega_1, \omega_2) = \sum_{k=1}^{K} R_k(\omega_1)C_k(\omega_2) \tag{3.68}$$

where $K$ represents the number of separable filters so combined. If each of the $r_k(n_1)$ is $N_1$ points long and each of the $c_k(n_2)$ is $N_2$ points long, the evaluation of a single sample of the output array is seen to require the execution of $K(N_1 + N_2)$ multiplications and additions. These filters offer a savings in implementation if

$$K(N_1 + N_2) < N_1 N_2 \tag{3.69}$$

all other factors being equal. Separable filters correspond to the special case $K = 1$.

The design problem consists of finding values for $r_k(n_1)$, $c_k(n_2)$, $k = 1, 2, \ldots, K$ so that the response of the filter will approximate the ideal response for a moderate value of $K$. (If $K$ becomes comparable to $N_1$ or $N_2$, the implementation is not efficient.) Treitel and Shanks tried a spatial domain approximation; that is, they tried to approximate an $N_1 \times N_2$ ideal impulse response $i(n_1, n_2)$ with $h(n_1, n_2)$ of the form of (3.67).

The approximation involves performing an eigenvector expansion on $i(n_1, n_2)$. If we assume, without loss of generality, that $N_2 \geq N_1$ and that the columns of $i(n_1, n_2)$ are linearly independent, we can write

$$i(n_1, n_2) = \sum_{k=1}^{N_1} \sqrt{\lambda_k} r_k(n_1)c_k(n_2) \tag{3.70}$$

where the $\{\lambda_k\}$ are the (real and positive) eigenvalues of the matrix

$$S(n_1, n_2) = \sum_{k=0}^{N_2-1} i(n_1, k)i(n_2, k) \tag{3.71}$$

The $r_k(n_1)$ are the normalized row eigenvectors associated with the corresponding $\lambda_k$ and the $c_k(n_2)$ are the normalized column eigenvectors of the matrix

$$Q(n_1, n_2) = \sum_{k=0}^{N_1-1} i(k, n_1)i(k, n_2) \tag{3.72}$$

The array $S$ is $N_1 \times N_1$ and it has $N_1$ eigenvalues. The array $Q$ is $N_2 \times N_2$ and it has $N_2$ eigenvalues, only $N_1$ of which are nonzero. The nonzero eigenvalues are equal to the eigenvalues of $S$. The row eigenvector from $S$ and the column eigenvector from $Q$ which correspond to the same eigenvalue are paired in (3.70). In implementing the filter, $\sqrt{\lambda_k}$ can be combined with either $r_k(n_1)$ or $c_k(n_2)$ to reduce slightly the number of computations. This decomposition is exact if $K$ is equal to $N_1$, but in this case there is no computational savings.

Assume that the eigenvalues are listed in the order of decreasing size $\lambda_1, \lambda_2, \ldots, \lambda_{N_1}$, so that $\lambda_1$ is the largest and $\lambda_i \geq \lambda_{i+1}$. Any repeated eigenvalues should be listed separately. If $E_0$ is the sum of the squares of the elements of $i(n_1, n_2)$ and $E_K$ is the sum of the squares of the elements in the error array measured as the difference between the ideal and a $K$-stage approximation, the normalized error is given by

$$\frac{E_K}{E_0} = 1 - \frac{\lambda_1 + \lambda_2 + \ldots + \lambda_K}{\lambda_1 + \lambda_2 + \ldots + \lambda_{N_1}} \tag{3.73}$$

Several possible extensions to this work suggest themselves. One could consider expansions where not all of the $r_k(n_1)$ and $c_k(n_2)$ have the same length or where frequency-domain error criteria are used to design the free parameters of the filter. The method can also be extended to the design of multistage separable IIR filters. To the authors' knowledge, none of this work has been done.

**Example 1**

As a very simple example, consider the ideal 2-D impulse response given by

$$i(0, 0) = 0.2; \quad i(1, 0) = 0.2; \quad i(0, 1) = 2.3; \quad i(1, 1) = 1.2 \tag{3.74}$$

which can be written in matrix form as

$$i(n_1, n_2) = \begin{bmatrix} 0.2 & 0.2 \\ 2.3 & 1.2 \end{bmatrix} \tag{3.75}$$

We would like to find a separable approximation to this ideal. The corresponding $S$ and $Q$ matrices are given by

$$S = \begin{bmatrix} 5.33 & 2.80 \\ 2.80 & 1.48 \end{bmatrix}; \quad Q = \begin{bmatrix} 0.08 & 0.70 \\ 0.70 & 6.73 \end{bmatrix} \tag{3.76}$$

The eigenvalues $\lambda_1$, $\lambda_2$ must satisfy either of the two equations

$$\det [\lambda_k I - S] = 0 \quad \text{or} \quad \det [\lambda_k I - Q] = 0 \tag{3.77}$$

Thus $\lambda_1 = 6.803$ and $\lambda_2 = 0.007$. The row and column eigenvectors satisfy the relations

$$r_1 S = \lambda_1 r_1 \tag{3.78a}$$

$$r_2 S = \lambda_2 r_2 \tag{3.78b}$$

$$\mathbf{Qc}_1 = \lambda_1 \mathbf{c}_1 \tag{3.78c}$$

$$\mathbf{Qc}_2 = \lambda_2 \mathbf{c}_2 \tag{3.78d}$$

Solving these equations under the constraint that the eigenvectors are normalized to have unit length, gives

$$\mathbf{r}_1 = (0.885, 0.466) \tag{3.79a}$$

$$\mathbf{r}_2 = (0.466, -0.885) \tag{3.79b}$$

$$\mathbf{c}_1 = (0.104, 0.994)' \tag{3.79c}$$

$$\mathbf{c}_2 = (0.994, -0.104)' \tag{3.79d}$$

Approximating $i(n_1, n_2)$ by $\sqrt{\lambda_1} r_1(n_1) c_1(n_2)$ then gives the separable approximation

$$h(n_1, n_2) = \begin{bmatrix} 0.240 & 0.126 \\ 2.294 & 1.208 \end{bmatrix} \tag{3.80}$$

which has a normalized error of

$$\frac{E_1}{E_0} = 0.001 \tag{3.81}$$

Adding $\sqrt{\lambda_2} r_2(n_1) c_2(n_2)$ will realize the ideal exactly.

For a $2 \times 2$ filter there is no computational savings with a separable approximation. Larger examples, however, are not particularly amenable to hand calculations.

### 3.5.3 Design of FIR Filters Using Transformations

The idea of converting a 1-D zero-phase FIR filter into a multidimensional one through a substitution of variables is an attractive one for a number of reasons. The designs are easy to perform; 1-D design methods are well understood; and by using optimal 1-D filters it might be possible to design optimal multidimensional filters. For these reasons, the transformation method has become a popular method for designing multidimensional FIR filters. As a bonus, it was later discovered that these filters also possess an efficient implementation. For filters of moderate order, this implementation can be considerably more efficient than either direct convolution or the use of the DFT.

The design procedure was developed by McClellan [15] and bears his name, but the existence of an efficient implementation was not discovered until some years later by Mecklenbräuker and Mersereau [16, 17]. The improved implementation described in the next section is due to McClellan and Chan [18]. Because the substitution of variables which is made with this procedure is not the obvious one, we begin our development with a discussion of 1-D zero-phase FIR filters.

The impulse response, $h(n)$, of a 1-D zero-phase filter is Hermitian symmetric:

$$h(n) = h^*(-n) \tag{3.82}$$

Since this implies that the support of such a filter must be centered at the origin and must contain an odd number of samples, consider the support of $h(n)$ to be the interval $-N \leq n \leq N$. Let us also assume that the samples $h(n)$ are all real and that their Fourier transform is $H(\omega)$.

Then we can write

$$H(\omega) = h(0) + \sum_{n=1}^{N} h(n)[\exp(-j\omega n) + \exp(j\omega n)] \qquad (3.83)$$

$$= \sum_{n=0}^{N} a(n) \cos(\omega n)$$

In writing this last expression, we have defined

$$a(n) \triangleq \begin{cases} h(0), & n = 0 \\ 2h(n), & n > 0 \end{cases} \qquad (3.84)$$

The function $\cos \omega n$ can be expressed as a polynomial of degree $n$ in the variable $\cos \omega$. The resulting polynomial is the $n$th Chebyshev polynomial, $T_n[\cdot]$. The first few Chebyshev polynomials and their inverse relationships are given in Tables 3.1 and 3.2. Thus we can write

$$\cos \omega n = T_n[\cos \omega] \qquad (3.85)$$

Substituting this relation into equation (3.83) permits us to write $H(\omega)$ in the form

$$H(\omega) = \sum_{n=0}^{N} a(n) T_n[\cos \omega] \qquad (3.86)$$

With the frequency response in this form, we can now make a transformation of variables. For example, if we make the substitution

**TABLE 3.1**  FIRST SEVEN
CHEBYSHEV POLYNOMIALS

$T_0[x] = 1$
$T_1[x] = x$
$T_2[x] = 2x^2 - 1$
$T_3[x] = 4x^3 - 3x$
$T_4[x] = 8x^4 - 8x^2 + 1$
$T_5[x] = 16x^5 - 20x^3 + 5x$
$T_6[x] = 32x^6 - 48x^4 + 18x^2 - 1$

$\vdots$

$T_n[x] = 2x T_{n-1}[x] - T_{n-2}[x]$

**TABLE 3.2**  INVERSES OF
THE FIRST FIVE CHEBYSHEV
POLYNOMIALS

$1 = T_0[x]$
$x = T_1[x]$
$x^2 = \frac{1}{2}(T_0[x] + T_2[x])$
$x^3 = \frac{1}{4}(3T_1[x] + T_3[x])$
$x^4 = \frac{1}{8}(3T_0[x] + 4T_2[x] + T_4[x])$

$$F(\omega_1, \omega_2) \longrightarrow \cos \omega \qquad (3.87)$$

we obtain the 2-D frequency response

$$H(\omega_1, \omega_2) = \sum_{n=0}^{N} a(n) T_n[F(\omega_1, \omega_2)] \qquad (3.88)$$

The function $F(\omega_1, \omega_2)$ is called a *transformation function*. An $M$-dimensional frequency response results if an $M$-dimensional transformation function is substituted for $\cos \omega$.

What are good choices for the transformation function $F(\omega_1, \omega_2)$? First, $F(\omega_1, \omega_2)$ must itself be the frequency response of a 2-D FIR filter to ensure that $H(\omega_1, \omega_2)$ is, in fact, the frequency response of a 2-D FIR filter. Second, the nature of $H(\omega_1, \omega_2)$ must be predictable from a knowledge of $H(\omega)$ and $F(\omega_1, \omega_2)$. Finally, a procedure must exist for choosing a particular $H(\omega)$ and $F(\omega_1, \omega_2)$ to satisfy a set of frequency specifications on $H(\omega_1, \omega_2)$. Clearly, not all of this can be accomplished without imposing some constraints on the class of allowable transformation functions.

The simplest choice for $F(\omega_1, \omega_2)$ is the frequency response of a zero-phase $3 \times 3$ filter. In this case we can write

$$
\begin{aligned}
&F(\omega_1, \omega_2) \\
&= A + B \cos \omega_1 + C \cos \omega_2 + D \cos (\omega_1 - \omega_2) + E \cos (\omega_1 + \omega_2)
\end{aligned}
\qquad (3.89)
$$

where $A$, $B$, $C$, $D$, and $E$ are free parameters. Since $T_n[x]$ is a polynomial of degree $n$ in $x$, it follows that $H(\omega_1, \omega_2)$ is a polynomial of degree $N$ in $F$. Thus a filter with frequency response $H(\omega_1, \omega_2)$ can be realized as cascade and parallel combinations of subnetworks which have the frequency response, $F(\omega_1, \omega_2)$. As an example, Figure 3.10 shows a realization of a filter with the overall response

$$
\begin{aligned}
H(\omega_1, \omega_2) &= T_2[F(\omega_1, \omega_2)] + 4T_1[F(\omega_1, \omega_2)] + 2T_0[F(\omega_1, \omega_2)] \\
&= 2F^2(\omega_1, \omega_2) + 4F(\omega_1, \omega_2) + 1
\end{aligned}
\qquad (3.90)
$$

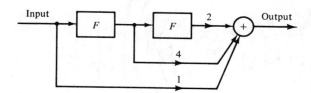

**Figure 3.10**    Realization of a filter whose frequency response is a quadratic function of $F$. (Courtesy of Russell M. Mersereau, from *Two-Dimensional Digital Signal Processing I*, Thomas S. Huang, ed., *Topics in Applied Physics*, Vol. 42, © 1981 Springer-Verlag.)

In the spatial domain, multiplication by $F(\omega_1, \omega_2)$ becomes convolution with $f(n_1, n_2)$, the inverse Fourier transform of $F(\omega_1, \omega_2)$. If the support of $f$ contains $(2P + 1) \times (2P + 1)$ samples and is centered at the origin, the $N$-fold convolution of $f(n_1, n_2)$ with itself will have a $[(2NP + 1) \times (2NP + 1)]$-sample region of support, as will the filter $h(n_1, n_2)$. We therefore see that if $f(n_1, n_2)$ is the impulse response of an FIR filter, $h(n_1, n_2)$ will correspond to an FIR filter. In addition, if $f(n_1, n_2)$ is zero-phase so that $F(\omega_1, \omega_2)$ is real, then $H(\omega_1, \omega_2)$ will also be real since the Chebyshev polynomials have real coefficients. Thus if $f(n_1, n_2)$ is zero-phase, $h(n_1, n_2)$ will be zero-phase.

Consider a locus of points in the $(\omega_1, \omega_2)$-plane such that

$$F(\omega_1, \omega_2) = \text{constant}$$

By analogy with lines of constant electromagnetic potential in field theory, these loci will be called *isopotentials*. A contour plot is nothing but a display of several super-imposed isopotentials. Any isopotential contours of $F(\omega_1, \omega_2)$ are also isopotentials of $H(\omega_1, \omega_2)$. Thus apart from the labeling of the contours, a contour plot of $H(\omega_1, \omega_2)$ looks identical to a contour plot of $F(\omega_1, \omega_2)$. Going one step further, we see that for the first-order transformation defined in (3.89), the shapes of the isopotentials of $H(\omega_1, \omega_2)$ depend only on the five parameters $A$, $B$, $C$, $D$, and $E$, no matter how large we make the order of the 1-D prototype. On the other hand, the value of $H(\omega_1, \omega_2)$ on a particular isopotential depends on both the value of $F(\omega_1, \omega_2)$ and the parameters of the prototype filter $\{a(n)\}$. If $F(\omega_1, \omega_2)$ satisfies the relation $|F(\omega_1, \omega_2)| \leq 1$, $H(\omega_1, \omega_2)$ can only assume values that are assumed by the prototype response, $H(\omega)$. With the transformation procedure, the design is decoupled into two parts. First, the isopotential shape is fixed by means of the transformation function; then the isopotential value is fixed by means of the prototype frequency response.

### Example 2

As an example, consider the first-order transformation with $A = -\frac{1}{2}$, $B = C = \frac{1}{2}$, $D = E = \frac{1}{4}$. This is the original transformation function proposed by McClellan [14]. Selected isopotentials for this transformation are shown in Figure 3.11. The

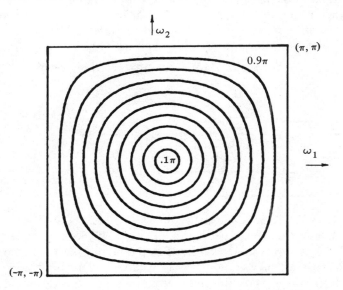

**Figure 3.11** Contours of constant value for the transformation of first order with $A = -\frac{1}{2}$, $B = C = \frac{1}{2}$, $D = E = \frac{1}{4}$. The contours are shown for values of $\omega$ in increments of $0.1\pi$. (Courtesy of Russell M. Mersereau, from *Two-Dimensional Digital Signal Processing I*, Thomas S. Huang, ed., *Topics in Applied Physics*, Vol. 42, © 1981 Springer-Verlag.)

isopotentials close to the center are nearly circular and those toward the outside become progressively more nearly square. For this example

$$F(\omega_1, \omega_2) = \tfrac{1}{2}(-1 + \cos \omega_1 + \cos \omega_2 + \cos \omega_1 \cos \omega_2) \qquad (3.91)$$

Letting $\omega_2 = 0$ gives us

$$F(\omega_1, 0) = \cos \omega_1$$

Thus

$$H(\omega_1, 0) = H(\omega_1) \qquad (3.92)$$

For this transformation function, the prototype frequency response becomes a cross-sectional slice of the 2-D frequency response. A 1-D lowpass thus produces a 2-D lowpass; a 1-D bandpass produces a 2-D "ring" bandpass, as shown in Figure 3.12.

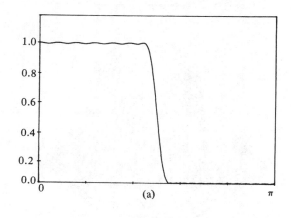

(a)

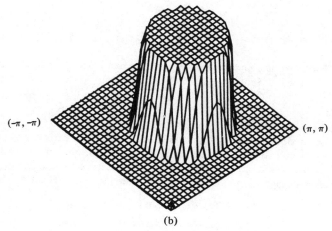

(b)

**Figure 3.12**  Parts (b) and (d) show the 2-D filters that result from the transformation function of Figure 3.11 used with the lowpass and bandpass prototypes shown in parts (a) and (c). (Courtesy of Russell M. Mersereau, from *Two-Dimensional Digital Signal Processing I*, Thomas S. Huang, ed., *Topics in Applied Physics*, Vol. 42, © 1981 Springer-Verlag.)

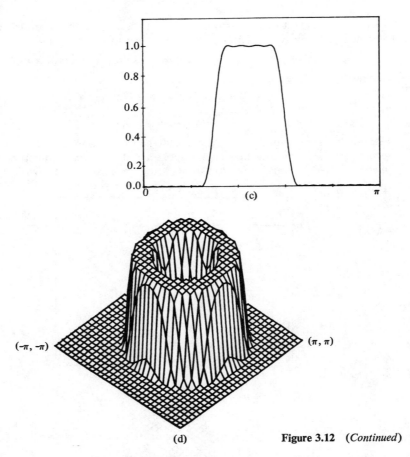

Figure 3.12  (*Continued*)

Furthermore, the height of the passband and stopband ripples in the prototype become the ripple heights in the final design.

**Example 3**

As a second example, consider the design of a fan filter whose frequency response approximates unity in quadrants I and III and approximates zero in quadrants II and IV (see Problem 1.21). This ideal response is shown in Figure 3.13(a). To approximate such a characteristic using a transformation we need an $F(\omega_1, \omega_2)$ which has the two axes as isopotential contours. One transformation function that has this property is

$$F(\omega_1, \omega_2) = \sin \omega_1 \sin \omega_2$$
$$= \tfrac{1}{2}[\cos (\omega_1 - \omega_2) - \cos (\omega_1 + \omega_2)] \tag{3.93}$$

This is a first-order transformation with $A = B = C = 0$, $D = \tfrac{1}{2}$, $E = -\tfrac{1}{2}$. This transformation function is positive in quadrants I and III and negative in quadrants II and IV. Since

$$\cos \omega = F(\omega_1, \omega_2) = \sin \omega_1 \sin \omega_2$$

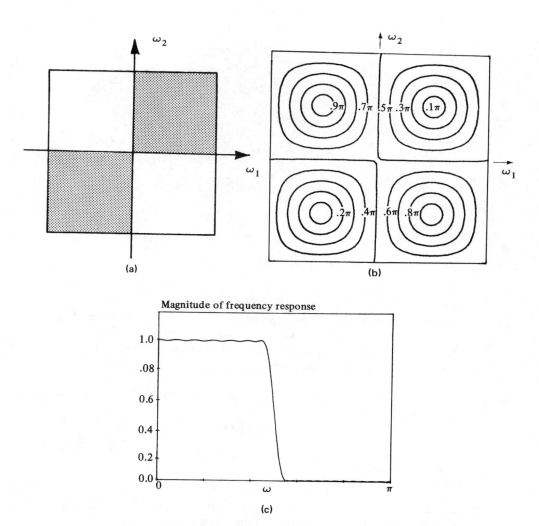

**Figure 3.13** (a) Ideal frequency response to be approximated. (b) Contours of the function $F(\omega_1, \omega_2)$ appropriate to this problem. (c) 1-D prototype filter. (d) Final digital frequency response of a $53 \times 53$ FIR filter designed using this transformation. (Courtesy of Russell M. Mersereau, *IEEE Trans. Acoustics, Speech, and Signal Processing,* © 1980 IEEE.)

is the substitution to be made, and $\cos \omega$ is positive for $|\omega| < \pi/2$ and negative for $\pi/2 < |\omega| < \pi$, values of the prototype response for $|\omega| < \pi/2$ will be mapped into quadrants I and III and values of the prototype response for $\pi/2 < |\omega| < \pi$ will be mapped into quadrants II and IV. The prototype filter must thus be a lowpass filter with a cutoff frequency of $\pi/2$ radians, as shown in Figure 3.13(c) for $N = 26$. The 2-D filter that results, shown in Figure 3.13(d), has a $(53 \times 53)$-point impulse response [19].

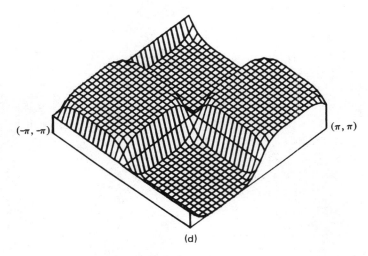

(d)

**Figure 3.13**  (*Continued*)

A variety of algorithms have been developed for determining the transformation function which is best for a particular design. These are surveyed in [16]. Since the transformation functions are themselves the frequency responses of zero-phase FIR filters, filter design methods such as windowing can be used. The ideal transformation functions should approximate 1 where the ultimate passbands are to be and should approximate $-1$ where the ultimate stopbands are to be, if the 1-D prototype filter is a lowpass.

Since $F(\omega_1, \omega_2)$ replaces $\cos \omega$ in the 1-D prototype frequency response, it should satisfy the relation

$$-1 \leq F(\omega_1, \omega_2) \leq 1 \tag{3.94}$$

for $-\pi \leq \omega_1, \omega_2 \leq \pi$. This fact need not complicate the transformation function selection process, however, since if

$$F(\omega_1, \omega_2) = C_1 F'(\omega_1, \omega_2) + C_2 \tag{3.95}$$

$F$ and $F'$ will have the same isopotentials. If $F'(\omega_1, \omega_2)$ is designed without regard to (3.94), we can determine its maximum and minimum values and then use the transformation function

$$F(\omega_1, \omega_2) = \frac{2}{F'_{max} - F'_{min}} F'(\omega_1, \omega_2) - \frac{F'_{max} + F'_{min}}{F'_{max} - F'_{min}} \tag{3.96}$$

which has the same isopotentials as $F'(\omega_1, \omega_2)$ and which satisfies (3.94) for the design of our filter.

### 3.5.4 Implementing Filters Designed Using Transformations

The FIR filters that result from the transformation design method, like any other FIR filters, can be implemented directly using the convolution sum, or they can be implemented using the DFT. In addition, for these filters there is a third implementation which exploits the structure inherent in their design. For filters of moderate order, this third implementation can be the most efficient and in itself may be sufficient motivation to justify the use of transformation designs in some applications.

The key to this special implementation is equation (3.88), which is reproduced below.

$$H(\omega_1, \omega_2) = \sum_{n=0}^{N} a(n) T_n[F(\omega_1, \omega_2)]$$

We can form a digital network to realize this frequency response by using the recurrence formula for the Chebyshev polynomials.

$$T_0[x] = 1 \tag{3.97a}$$

$$T_1[x] = x \tag{3.97b}$$

$$T_n[x] = 2x T_{n-1}[x] - T_{n-2}[x] \tag{3.97c}$$

If $x$ is replaced by $F(\omega_1, \omega_2)$, the recursion becomes

$$T_n[F(\omega_1, \omega_2)] = 2F(\omega_1, \omega_2) T_{n-1}[F(\omega_1, \omega_2)] - T_{n-2}[F(\omega_1, \omega_2)] \tag{3.98}$$

Using this relation we can produce a signal which has the Fourier transform $T_n[F(\omega_1, \omega_2)]$ if we have signals whose Fourier transforms are $T_{n-1}[F(\omega_1, \omega_2)]$ and $T_{n-2}[F(\omega_1, \omega_2)]$. Such a network is shown in Figure 3.14. Since each of these signals,

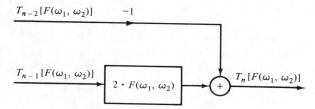

**Figure 3.14**  Network realization of the Chebyshev recursion.

in turn, can be generated from two lower-order signals we can build a ladder network with $N$ outputs such that the frequency response between the input and the $n$th output is $T_n[F(\omega_1, \omega_2)]$. By weighting these outputs according to (3.88) we have a realization for the filter $H(\omega_1, \omega_2)$. This realization is depicted in the network of Figure 3.15.

A subnetwork with frequency response $F$ represents a low-order, zero-phase FIR filter. For the first-order transformation given by (3.89), this filter has the $(3 \times 3)$-point impulse response

$$f(n_1, n_2) = \begin{cases} A, & n_1 = n_2 = 0 \\[2mm] \dfrac{B}{2}, & n_1 = \pm 1, \quad n_2 = 0 \\[2mm] \dfrac{C}{2}, & n_1 = 0, \quad n_2 = \pm 1 \\[2mm] \dfrac{D}{2}, & n_1 = \pm 1, \quad n_2 = n_1 \\[2mm] \dfrac{E}{2}, & n_1 = \pm 1, \quad n_2 = -n_1 \end{cases} \tag{3.99}$$

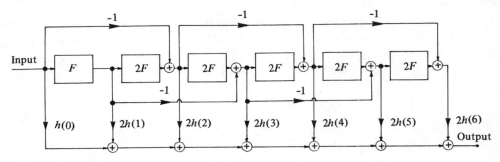

**Figure 3.15** Realization of an $M$-dimensional filter designed by means of a transformation. The coefficients $h(n)$ are the coefficients of the 1-D prototype filter. The filters $F$ define the transformation function. (Courtesy James H. McClellan and David S. K. Chan, *IEEE Trans. Circuits and Systems* © 1977 IEEE.)

Such a filter can be realized directly using five multiplications and eight additions per output sample. Since there are $N$ filters in the network, the total required computation is $6N + 1$ multiplications and $10N - 2$ additions per output sample.

One important feature of this implementation is its modularity. To change the 1-D prototype filter, we merely change the weighting coefficients $\{h(n)\}$. To change the transformation function, we change the coefficients inside each of the subfilters. To implement a filter designed by means of an $M$-dimensional transformation, we simply replace $f(n_1, n_2)$ by the impulse response corresponding to an $M$-dimensional transformation function. McClellan and Chan [18] have shown that these filters possess good arithmetic quantization properties when the filter must be realized using finite precision arithmetic.

The determination of how much storage is required is complicated by the fact that the overall filter and the subfilters within it are zero-phase. Since it is generally easier to realize first quadrant filters, let us consider using operators of the form

$$g(n_1, n_2) = f(n_1 - P, n_2 - P)$$

to realize the impulse response

$$\hat{h}(n_1, n_2) = h(n_1 - NP, n_2 - NP) \qquad (3.100)$$

where $P$ is the order of the transformation. In Figure 3.16 we have shown such a realization. The branches labeled $z_1^{-P}z_2^{-P}$ correspond to delays of $P$ samples in the $n_1$ variable and $P$ samples in the $n_2$ variable. Each of the transformation filters requires slightly more than $2P$ rows of storage. In addition, each branch labeled $z_1^{-P}z_2^{-P}$ requires approximately $P$ rows of storage and each branch labeled $z_1^{-2P}z_2^{-2P}$ requires approximately $2P$ rows of storage. The overall storage requirement is thus between $5NP$ and $6NP$ rows, roughly twice the storage requirement of a convolutional realization.

In the $M$-dimensional case, with a first-order transformation, $F$ becomes an $M$-dimensional filter of spatial extent $3 \times 3 \times \cdots \times 3$ and the remainder of the structure remains unchanged. An output sample from such a filter thus requires $[(3^M + 3)/2]N + 1$ multiplications and $[3^M + 1]N$ additions per output sample. Also, $5N$ hyperrows of dimension $(M - 1)$ of the input array must be stored. This

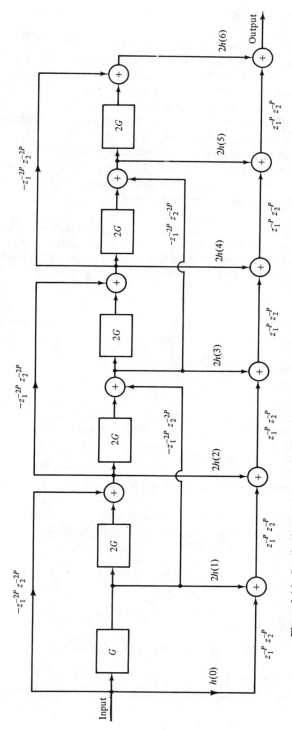

**Figure 3.16** Realization of a linear-phase transformation filter using a causal, linear-phase transformation function.

implies that $5NL^{M-1}$ words must be stored for a hypercubic array $L$ points on a side. The storage is approximately double that of direct convolution, but the number of multiplications per output grows only linearly with $N$. As with the direct convolution implementation, each input sample needs to be retrieved and deposited into secondary storage only once. This implementation offers distinct advantages over direct convolution for the two-dimensional case and the savings become significantly greater as the dimensionality of the problem increases.

In Table 3.3 we have assumed that we wish to filter a $(1024 \times 1024)$-point input array with a $(2N + 1) \times (2N + 1)$-point FIR filter designed using a transformation. The table gives the number of multiplications required per output point for the convolutional, FFT, and special implementations [4].

**TABLE 3.3** NUMBER OF MULTIPLICATIONS PER OUTPUT POINT REQUIRED TO FILTER A (1024 × 1024)-POINT INPUT ARRAY WITH A [(2N + 1) × (2N + 1)]-POINT ZERO-PHASE FILTER USING DIRECT CONVOLUTION, THE FFT, AND THE SPECIAL IMPLEMENTATION FOR TRANSFORMATION DESIGNS

| $N$ | Direct Convolution | FFT | Special Implementation |
|---|---|---|---|
| 5 | 36 | 46 | 31 |
| 10 | 121 | 46 | 61 |
| 20 | 441 | 46 | 121 |
| 40 | 1681 | 46 | 241 |

(Courtesy of Russell M. Mersereau from *Two-Dimensional Digital Signal Processing I*, Thomas S. Huang, ed., *Topics in Applied Physics*, Vol. 42, © 1981 Springer-Verlag.)

### 3.5.5 Filters with Small Generating Kernels

Faugeras and Abramatic [20] have proposed a structure for realizing zero-phase FIR filters using a tapped cascade implementation such as that shown in Figure 3.17. The frequency response of their filters is given by

$$H(\omega_1, \omega_2) = \sum_{k=0}^{K} \lambda_k F^{(k)}(\omega_1, \omega_2) \qquad (3.101)$$

where

$$F^{(0)}(\omega_1, \omega_2) \triangleq 1$$

and

$$F^{(k)}(\omega_1, \omega_2) \triangleq \prod_{i=1}^{k} F_i(\omega_1, \omega_2) \qquad (3.102)$$

Each of the filters $F_i(\omega_1, \omega_2)$ has a linear phase with a $[(2P + 1) \times (2P + 1)]$-point impulse response, where $P$ is assumed to be small, hence the name "small generating kernel" (SGK) filters.

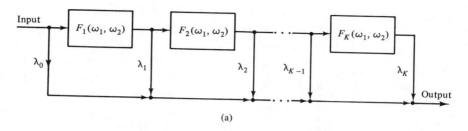

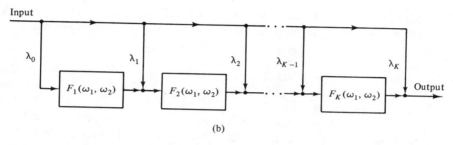

(a)

(b)

**Figure 3.17**  Implementation of SGK filters [20]. (a) Normal implementation. (b) Transposed implementation.

Cascade filters and McClellan transformation filters both correspond to special cases of SGK filters. The cascade realization corresponds to the special case where $\lambda_k = 0$ for $k < K$. McClellan transformations correspond to the special case where all of the $F_i(\omega_1, \omega_2)$ are identical and each is equal to the transformation function. [In this special case the structure of Figure 3.17 has poor sensitivity to arithmetic round-off noise. For this reason these structures were not considered in the preceding section. This is a much less serious problem when all the $\{F_i(\omega_1, \omega_2)\}$ are different.]

If $P = 1$, this structure requires $6K + 1$ multiplications per output sample (the same as the McClellan transformation) and slightly more than $3N$ rows of storage (less than the McClellan transformation). Furthermore, because the structure is modular and because the SGK structure uses small convolutional kernels, this structure is an attractive candidate for hardware realization.

The difficulty with these filters is their design. The frequency response is a highly nonlinear function of the filter parameters which requires iterative solution techniques. Several design algorithms have been proposed [20], but we will not examine them here.

## *3.6 FIR FILTERS FOR HEXAGONALLY SAMPLED SIGNALS

Conceptually, FIR filters for hexagonally sampled signals are no different than FIR filters for rectangularly sampled ones; both compute samples of the output sequence as linear combinations of input signals. In their finer details, however, there are some differences—in efficiency and in the forms for the resulting formulas and expressions.

The frequency response of a hexagonal filter can be expressed as

$$H(\omega_1, \omega_2) = \sum_{n_1} \sum_{n_2} h(n_1, n_2) \exp\left[-j\left(\frac{2n_1 - n_2}{\sqrt{3}}\omega_1 + n_2\omega_2\right)\right] \qquad (3.103)$$

The range of summation extends over the entire region of support of the impulse response of the filter. If $H(\omega_1, \omega_2)$ is purely real, the filter is zero-phase. In this case

$$h(n_1, n_2) = h^*(-n_1, -n_2) \qquad (3.104)$$

### 3.6.1 Implementation of Hexagonal FIR Filters

The two implementations for general FIR filters, direct convolution and the use of the DFT, extend readily to the hexagonal case. In fact, the convolution sum

$$y(n_1, n_2) = \sum_{k_1} \sum_{k_2} h(k_1, k_2)x(n_1 - k_1, n_2 - k_2) \qquad (3.105)$$

is identical in form to its rectangular counterpart. If the impulse response, $h(n_1, n_2)$, is symmetric, then, as in the rectangular case, this fact can be exploited in a direct convolution implementation to reduce the number of required multiplications.

One advantage of using hexagonal systems occurs when a circularly symmetric response is desired. A rectangular system can have eightfold symmetry in its impulse or frequency response—about both frequency axes and both diagonals. A hexagonal system, on the other hand, can have 12-fold symmetry in its impulse and frequency responses—symmetry about all six vertices of a hexagon and across the bisectors of all six sides. Thus, if all other factors are equal, the symmetric hexagonal FIR filter requires only about two-thirds the multiplications of the rectangular one. However, all other factors are not equal. The hexagonally sampled input and output waveforms also have fewer samples. As a result, the savings in the hexagonal case can be as high as 58%. In Table 3.4 we see the normalized mean-squared error for rectangular and hexagonal FIR lowpass filters designed using a flat window [21]. The error goes down with increased radius of the filter and the error for both the rectangular and hexagonal designs are seen to be roughly comparable. However, the number of coefficients in the filter, which is proportional to the number of additions, is about 25% less for the hexagonal case. Exploiting the symmetry in the impulse response (the ideal in this case is a circularly symmetric lowpass filter), we arrive at the number of distinct coefficients, which is proportional to the number of multiplications required. The number of distinct coefficients for the hexagonal case is asymptotically one-half the number required in the rectangular case.

The DFT can also be used to implement hexagonal FIR filters. This can be done using either two large transforms or using block convolution techniques. The interested reader should be able to work out the details.

### 3.6.2 Design of Hexagonal FIR Filters

Most design algorithms for 2-D FIR filter design which have been developed for the design of rectangularly sampled filters can be adapted to the design of hexagonal ones. The window method, for example, extends trivially. If $i(n_1, n_2)$ represents the ideal

**TABLE 3.4** NUMBER OF COEFFICIENTS AND NUMBER OF DISTINCT COEFFICIENTS FOR RECTANGULAR AND HEXAGONAL FIR FILTERS WITH VARIOUS RADII FOR THEIR REGIONS OF SUPPORT[a]

| Radius $N$ | Filter type | Number of coefficients | Number of distinct coefficients | Normalized error |
|---|---|---|---|---|
| 0 | Rect. | 1 | 1 | 0.804 |
| 0 | Hex. | 1 | 1 | 0.773 |
| 1 | Rect. | 9 | 3 | 0.200 |
| 1 | Hex. | 7 | 2 | 0.214 |
| 2 | Rect. | 25 | 6 | 0.158 |
| 2 | Hex. | 19 | 4 | 0.166 |
| 3 | Rect. | 49 | 10 | 0.098 |
| 3 | Hex. | 37 | 6 | 0.102 |
| 4 | Rect. | 81 | 15 | 0.084 |
| 4 | Hex. | 61 | 9 | 0.090 |

[a] The ideal filter is a circularly symmetric lowpass.
(Courtesy of Russell M. Mersereau, *Proc. IEEE*, © 1979 IEEE.)

hexagonal response, we can obtain a hexagonal design by letting

$$h(n_1, n_2) = i(n_1, n_2)w(n_1, n_2) \tag{3.106}$$

where the window sequence $w(n_1, n_2)$ has the same region of support as that desired for $h$. Two choices for $w$ which can be obtained from a 1-D window, $v(n)$ are

$$w(n_1, n_2) = v(n_1)v(n_2)v(n_1 - n_2) \tag{3.107}$$

which corresponds to an outer product formulation, and

$$w(n_1, n_2) = v\left(2\sqrt{\frac{n_1^2 + n_2^2 - n_1 n_2}{3}}\right) \tag{3.108}$$

which corresponds to a rotated formulation. (In the latter case a continuous 1-D window is used.) The former windows have a hexagonal region of support and the latter ones have a round region of support. For the hexagonal case it is not known at this time which is the better formulation. In Figure 3.18 we show an example of a filter designed with a hexagonal outer product window.

Equiripple hexagonal FIR filters can also be straightforwardly designed using the same principles as those used for the equiripple rectangular FIR filters, since the approximating frequency response depends linearly upon the unknown filter parameters. Thus the approximation is still of the form of (3.42), but the basis functions $\phi_i(\omega_1, \omega_2)$ are obviously different from those in the rectangular case. Their exact form will depend on which, if any, symmetries are to be imposed on the impulse response.

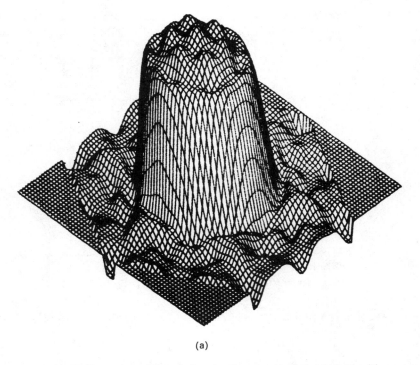

(a)

**Figure 3.18** FIR lowpass filter designed using a constant window with a hexagonally shaped region of support. The hexagonally sampled filter contains 363 samples and possesses 12-fold symmetry. (a) Perspective plot. (b) Contour plot. (Courtesy of Russell M. Mersereau, *Proc. IEEE*, © 1979 IEEE.)

The transformation approach can also be adapted to the design of hexagonal FIR filters. The method is very similar to that described in Section 3.5.3 except that the frequency response of a low-order zero-phase hexagonal filter is used in place of a rectangular one for the transformation function. In the first-order case this assumes the form

$$
\begin{aligned}
\cos \omega &= F_H(\omega_1, \omega_2) \\
&= A + B \cos \frac{2\omega_1}{\sqrt{3}} + C \cos \left( \frac{\omega_1}{\sqrt{3}} + \omega_2 \right) + D \cos \left( \frac{\omega_1}{\sqrt{3}} - \omega_2 \right)
\end{aligned}
\tag{3.109}
$$

Nearly circular contours can be obtained for small values of $(\omega_1, \omega_2)$ with the choice $A = -\frac{1}{3}$, $B = C = D = \frac{4}{9}$. In Figure 3.19, we show the isopotentials obtained for these values. In Figure 3.20 we show an $N = 26$ hexagonal lowpass filter designed from a 1-D lowpass prototype. The filter is designed to pass those frequency components for which $\omega_1^2 + \omega_2^2 \leq \pi^2/4$ and to reject those components outside this band. These filters also possess efficient implementations [21].

(b)

**Figure 3.18**    (*Continued*)

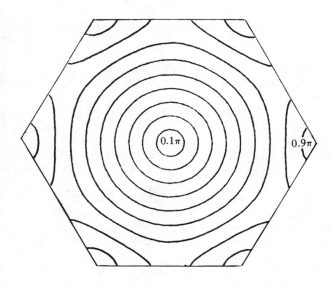

**Figure 3.19**    Contours of constant value for the transformation function of (3.109) with $A = -\frac{1}{3}$, $B = C = D = \frac{4}{9}$. (Courtesy of Russell M. Mersereau, *Proc. IEEE*, © 1979 IEEE.)

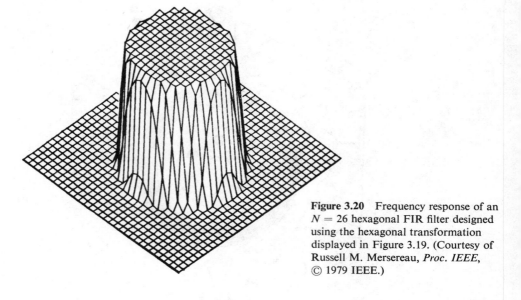

**Figure 3.20** Frequency response of an $N = 26$ hexagonal FIR filter designed using the hexagonal transformation displayed in Figure 3.19. (Courtesy of Russell M. Mersereau, *Proc. IEEE*, © 1979 IEEE.)

## PROBLEMS

**3.1.** If the impulse response $h(n_1, n_2)$ of a 2-D FIR filter satisfies the symmetry condition

$$h(n_1, n_2) = h(-n_1, -n_2)$$

then $H(\omega_1, \omega_2)$ is purely real and the filter is zero-phase. This fact can be exploited in realizing the filter.

**(a)** Assume that a FIR filter satisfies the alternative symmetry condition

$$h(n_1, n_2) = -h(-n_1, -n_2)$$

In this case, what can be inferred about its frequency response?

**(b)** Show how the direct implementation of this filter should be modified to exploit this symmetry.

**(c)** Repeat parts (a) and (b) for FIR filters that satisfy the symmetry conditions
  (i) $h(n_1, n_2) = h(-n_1, n_2)$
  (ii) $h(n_1, n_2) = -h(n_1, -n_2)$

**3.2.** We have an array of data whose region of support is $340 \times 340$ points. We wish to filter that array with a zero-phase FIR filter which is $(2N + 1) \times (2N + 1)$ points in extent.

**(a)** If $N = 50$, how many complex multiplications and how much storage will be required to implement the filter using the DFT implemented with a row–column algorithm? (Assume that the FFT size must be a power of 2.)

**(b)** How much computation and storage will be required if the filter is implemented directly, exploiting the zero-phase condition?

**(c)** If memory were in infinite supply and the only issue concerned the number of multiplications, for what minimum value of $N$ is the FFT approach to be preferred?

**3.3.** We would like to filter a 340 × 340 array of real data with a (23 × 23)-point zero-phase FIR filter. The primary memory of the computer available for storing samples of the input array will hold up to 4096 complex words. How would you implement the filter? Justify your answer.

**3.4.** We wish to filter a (256 × 256)-point image with a (16 × 16)-point real FIR filter. If this filter is to be implemented using the overlap-save method of block convolution, we might consider using

    1. (32 × 32)-point DFTs
    2. (64 × 64)-point DFTs
    3. (128 × 128)-point DFTs
    4. (256 × 256)-point DFTs
    5. (512 × 512)-point DFTs

    **(a)** For each case, determine how many DFTs need to be evaluated. (Assume two DFTs per convolution.)
    **(b)** For each of the five possibilities, how many complex multiplications are required? For the sake of this calculation, assume that an $(N \times N)$-point DFT requires $N^2 \log_2 N$ complex multiplications.
    **(c)** For this example, what is the best block size to use? How does the number of multiplications using this block size compare with the number required for a direct implementation of the convolution?

**3.5.** A 2-D window function is formed by taking the outer product of a 1-D window with itself.

$$w_2(n_1, n_2) = w_1(n_1)w_1(n_2)$$

The Fourier transform of the 1-D window has a main lobe width $\Delta$ and a highest side lobe height of $\delta$. What are the main lobe width and side lobe height of the 2-D window?

**3.6.** We wish to design a 2-D FIR filter using a rotated Kaiser window whose frequency response satisfies (approximately)

$$H(\omega_1, \omega_2) = \begin{cases} 1.0 \pm 0.1, & \omega_1^2 + \omega_2^2 \le (0.4\pi)^2 \\ 0.0 \pm 0.1, & \omega_1^2 + \omega_2^2 \ge (0.6\pi)^2 \end{cases}$$

Using the design formulas presented in this chapter:
**(a)** Estimate the order of the filter.
**(b)** Determine the value of the Kaiser window parameter $\alpha$.
**(c)** Determine an expression for the impulse response of the filter.
You may leave your answer in terms of the appropriate Bessel functions.

**3.7.** Let $H(\omega_1, \omega_2)$ denote the frequency response of a $[(2N + 1) \times (2N + 1)]$-point zero-phase filter. This filter is used in the configuration shown in Figure P3.7 to realize another zero-phase filter with the frequency response $G(\omega_1, \omega_2)$.

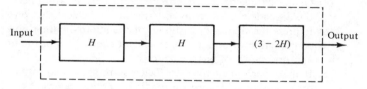

**Figure P3.7**

(a) What is the size of the impulse response $g(n_1, n_2)$?

(b) We would like the filter $G(\omega_1, \omega_2)$ to satisfy the following frequency-response conditions:

$$0.99 \leq G(\omega_1, \omega_2) \leq 1.01, \qquad \omega_1^2 + \omega_2^2 < (0.4\pi)^2$$

$$-0.01 \leq G(\omega_1, \omega_2) \leq 0.01, \qquad \omega_1^2 + \omega_2^2 > (0.6\pi)^2$$

If $H$ is designed using the window method with a rotated one-dimensional Kaiser window, what order filter is needed for $H$ in order that $G$ will satisfy the foregoing specifications? [*Hint:* To work this problem, you will need to determine how passband and stopband errors in $H(\omega_1, \omega_2)$ affect $G(\omega_1, \omega_2)$.]

(c) Suppose instead that we build a single filter, $G'$, using a rotated Kaiser window which uses the same number of multiplications for its implementation as the foregoing cascade implementation. How small would the passband and stopband errors be if the width of the transition band were the same as for $G$?

**3.8.** We would like to perform a $(5 \times 5)$-point least-squares design of a zero-phase fan filter whose frequency response, $I(\omega_1, \omega_2)$, is sketched in Figure P3.8. The frequency

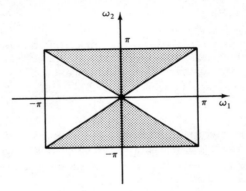

**Figure P3.8**

response is 1 in the shaded areas and is zero in the unshaded areas. To reduce the number of degrees of freedom in the approximation, we require that the approximating function satisfy:

1. $H(\omega_1, \omega_2)$ is purely real.
2. $H(\omega_1, \omega_2) = H(-\omega_1, \omega_2)$.
3. $H(\omega_1, \omega_2) = H(\omega_1, -\omega_2)$.

The filter $h(n_1, n_2)$ should have support over the region $-2 \leq n_1, n_2 \leq 2$.

(a) Of the 25 coefficients in the filter, how many are linearly independent? Express the dependent coefficients in terms of the independent ones.

(b) Express the frequency response of the approximating filter $H(\omega_1, \omega_2)$ as a linear combination of only the independent coefficients of the impulse response with an appropriate set of basis functions.

(c) If the independent filter coefficients are chosen to minimize the integral-squared error, derive an expression for their optimal values. You do not need to evaluate the integrals.

**3.9.** The frequency response of a family of FIR filters can be expressed in the form

$$H(\omega_1, \omega_2) = \sum_{i=1}^{F} a(i)\phi_i(\omega_1, \omega_2)$$

We wish to find the member of that family which will optimally approximate the ideal response $I(\omega_1, \omega_2)$, subject to the restriction that

$$a(l) + a(m) = K$$

Show that this constrained approximation problem can be solved by finding the response

$$H'(\omega_1, \omega_2) = \sum_{i=1}^{F'} a'(i)\phi_i'(\omega_1, \omega_2)$$

which optimally approximates $I'(\omega_1, \omega_2)$ with no constraints. Find $a'(i)$, $\phi_i'(\omega_1, \omega_2)$, $F'$, and $I'$ in terms of the unprimed quantities.

**3.10.** Consider the problem of approximating an ideal lowpass filter with the frequency response

$$I(\omega_1, \omega_2) = \begin{cases} 1, & |\omega_1| < a < \pi, \quad |\omega_2| < b < \pi \\ 0, & \text{otherwise} \end{cases}$$

by a filter whose impulse response is of the form

$$h(n_1, n_2) = \begin{cases} A, & n_1 = n_2 = 0 \\ B, & n_1 = \pm 1, \quad n_2 = 0 \\ C, & n_1 = 0, \quad n_2 = \pm 1 \\ 0, & \text{otherwise} \end{cases}$$

What values of $A$, $B$, and $C$ will minimize the error

$$E = \int_{-\pi}^{\pi} \int_{-\pi}^{\pi} |I(\omega_1, \omega_2) - H(\omega_1, \omega_2)|^2 \, d\omega_1 \, d\omega_2$$

**3.11.** Iterative algorithms for filter design such as the one embodied in equation (3.54) often require that the partial derivatives of the error with respect to the unknown parameters be computed.

**(a)** Let us define the error

$$E_p \triangleq \sum_m W_m[H(\omega_{1m}, \omega_{2m}) - I(\omega_{1m}, \omega_{2m})]^{2p}$$

Compute the gradient of $E_p$ with respect to the independent coefficients $\{a(i)\}$ if

$$H(\omega_1, \omega_2) = \sum_{i=1}^{F} a(i)\phi_i(\omega_1, \omega_2)$$

**(b)** Repeat this calculation for the coefficients $h_i(n_1, n_2)$ in a cascade realization of the filter where

$$H(\omega_1, \omega_2) = \prod_i \left[ \sum_{n_1} \sum_{n_2} h_i(n_1, n_2) \exp(-j\omega_1 n_1 - j\omega_2 n_2) \right]$$

**3.12.** Consider the $2 \times 2$ array given by

$$x(0, 0) = 0$$
$$x(1, 0) = 2$$
$$x(0, 1) = 9$$
$$x(1, 1) = 1$$

(a) Form a separable approximation to this array using the technique described in Section 3.5.2.

(b) What is the normalized mean-squared error of this approximation?

**3.13.** Consider the multistage separable filter shown in Figure 3.9. Assume that each of the $r_i(n_1)$ and $c_i(n_2)$ is $N$ points long.

(a) How many words of storage are needed to implement the filter if the input sequence has extent $L \times L$? Assume that the input array can only be accessed row-wise.

(b) How much storage is necessary if the filter is simply implemented directly as a 2-D FIR filter?

**3.14.** Although procedures have been developed for the design of transformation functions, ad hoc methods often work well since the transformation typically involves very few free parameters. Ad hoc methods may take the form of specifying the mapping function for a few key frequencies. As an example, consider a first-order transformation of the form

$$F(\omega_1, \omega_2) = A + B\cos\omega_1 + C\cos\omega_2 + D\cos\omega_1\cos\omega_2$$

to design a fan filter that approximates the ideal response shown in Figure P3.8.

(a) Find a reasonable set of values for $A$, $B$, $C$, and $D$. Justify your answer.

(b) Sketch the response of a 1-D prototype filter to be used with this transformation.

**3.15.** We would like to design a 3-D spherically symmetric, zero-phase lowpass filter using the method of transformations with a first-order transformation function of the form

$$F(\omega_1, \omega_2, \omega_3) = A + B\cos\omega_1 + C\cos\omega_2 + D\cos\omega_3 + E\cos\omega_1\cos\omega_2$$
$$+ F\cos\omega_1\cos\omega_3 + G\cos\omega_2\cos\omega_3 + H\cos\omega_1\cos\omega_2\cos\omega_3$$

This transformation will convert a 1-D zero-phase prototype frequency response $G(\omega)$ into the 3-D response $H(\omega_1, \omega_2, \omega_3)$ under the substitution

$$F(\omega_1, \omega_2, \omega_3) \longrightarrow \cos\omega$$

Choose a set of transformation parameters $A, B, \ldots, H$ so that the transformation will have the following properties:

1. $\omega = \pi$ will map to $(\omega_1, \omega_2, \omega_3) = (\pi, \pi, \pi)$.

2. $F(\omega_1, \omega_2 \, \omega_3) = F(\omega_2, \omega_3, \omega_1) = F(\omega_3, \omega_1, \omega_2) = F(\omega_1, \omega_3, \omega_2)$
   $= F(\omega_2, \omega_1, \omega_3) = F(\omega_3, \omega_2, \omega_1)$.

3. $F(\omega, 0, 0) = \cos\omega$. (This will guarantee that the response on each axis is the same as the prototype response.)

4. $F(\omega_1, \omega_2, \pi) = -1$.

**3.16.** A second-order transformation uses the frequency response of a $(5 \times 5)$-point zero-phase filter as its transformation function. Assume that such a transformation function is used with a $(2N + 1)$-point zero-phase 1-D filter to realize a 2-D filter.

(a) How many multiplications need to be performed to compute one sample of the output sequence?

(b) How many multiplications need to be performed per output sample if this same frequency response is implemented using direct convolution?

(c) How much storage is required to implement this filter?

**3.17.** Show that any filter designed by a McClellan transformation can be realized as a cascade.

**3.18.** Abramatic and Faugeras [20] have proposed the implementation for 2-D FIR filters shown in Figure P3.18. If $f$, $g$, and $h$ are all 2-D zero-phase FIR filters with impulse

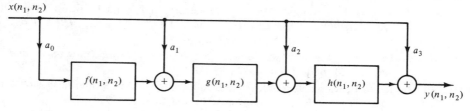

**Figure P3.18**

responses that are $3 \times 3$ points in extent (and centered at the origin):

**(a)** How large is the equivalent impulse response of the overall system?

**(b)** How many multiplications per output sample are required?

**(c)** How many multiplications per output sample would be required if the system is implemented using a single convolution with the equivalent impulse response?

**3.19.** We wish to design a hexagonally sampled zero-phase equiripple FIR filter. We want this filter to be as close to circularly symmetric as possible. This means that the impulse response should satisfy the symmetry conditions:

$$h(n_1, n_2) = h(-n_1, -n_2) = h(n_1 - n_2, n_1) = h(n_2 - n_1, -n_1)$$
$$= h(n_2, n_1) = h(-n_1 + n_2, n_2) = h(-n_2, n_1 - n_2)$$
$$= h(n_1, n_1 - n_2) = h(-n_2, -n_1) = h(n_2, n_2 - n_1)$$
$$= h(n_1 - n_2, -n_2) = h(-n_1, n_2 - n_1)$$

This approximation problem can be formulated as a linear approximation problem of the form

$$H(\omega_1, \omega_2) = \sum_{i=1}^{F} a(i)\phi_i(\omega_1, \omega_2)$$

Determine the coefficients $a(i)$ in terms of the impulse response coefficients of the filter and find the set of $\{\phi_i(\omega_1, \omega_2)\}$ that result in the minimum number of design parameters. Relate the number of degrees of freedom, $F$, to the size of the filter. [The rectangularly sampled equivalents of these results appear in equation (3.46).]

**3.20.** The transformation method for zero-phase FIR filter design can be applied to hexagonally sampled filters as well as to rectangularly sampled ones. This is done by making the substitution

$$F_H(\omega_1, \omega_2) \longrightarrow \cos \omega$$

into the 1-D zero-phase prototype frequency response. In the first-order case, this transformation function has the form

$$F_H(\omega_1, \omega_2) = A + B \cos \frac{2\omega_1}{\sqrt{3}} + C \cos \left(\frac{\omega_1}{\sqrt{3}} + \omega_2\right) + D \cos \left(\frac{\omega_1}{\sqrt{3}} - \omega_2\right)$$

**(a)** Determine the impulse response of a zero-phase (hexagonal) FIR filter that has the frequency response $F_H(\omega_1, \omega_2)$.

**(b)** If this filter is used in the transformation implementation shown in Figure 3.15, determine how many multiplications need to be performed for each sample of the output signal. Assume that the 1-D prototype filter is $2N + 1$ points long.

**(c)** How many samples are contained in the impulse response of the overall filter?

## REFERENCES

1. Richard E. Twogood, Michael P. Ekstrom, and Sanjit K. Mitra, "Optimal Sectioning Procedure for the Implementation of 2-D Digital Filters," *IEEE Trans. Circuits and Systems*, CAS-25, no. 5 (May 1978), 260–69.

2. Thomas S. Huang, "Two-Dimensional Windows," *IEEE Trans. Audio and Electroacoustics*, AU-20, no. 1 (Mar. 1972), 88–90.

3. James F. Kaiser, "Nonrecursive Digital Filter Design Using the $I_0$-sinh Window Function," *Proc. IEEE Int. Symp. Circuits and Systems* (1974), 20–23.

4. Russell M. Mersereau, "Two-Dimensional Nonrecursive Filter Design," in *Two-Dimensional Digital Signal Processing I*, ed. Thomas S. Huang, Topics in Applied Physics Series, Vol. 42 (New York: Springer-Verlag, 1981), 11–40.

5. Theresa C. Speake and Russell M. Mersereau, "A Note on the Use of Windows for Two-Dimensional FIR Filter Design," *IEEE Trans. Acoustics, Speech, and Signal Processing*, ASSP-29, no. 1 (Feb. 1981), 125–27.

6. James H. McClellan, "Artifacts in Alpha-Rooting of Images," *Proc. IEEE Int. Conf. Acoustics, Speech, and Signal Processing* (Apr. 1980), 449–52.

7. William K. Pratt, *Digital Image Processing* (New York: John Wiley & Sons, Inc., 1978). (See Figure 12.5-1.)

8. Thomas W. Parks and James H. McClellan, "Chebyshev Approximation for Nonrecursive Digital Filters with Linear Phase," *IEEE Trans. Circuit Theory*, CT-19, no. 2 (Mar. 1972), 189–94.

9. E. W. Cheney, *Introduction to Approximation Theory* (New York: McGraw-Hill Book Company, 1966).

10. Y. Kamp and J. P. Thiran, "Chebyshev Approximation for Two-Dimensional Nonrecursive Digital Filters," *IEEE Trans. Circuits and Systems*, CAS-22, no. 3 (Mar. 1975), 208–18.

11. David B. Harris and Russell M. Mersereau, "A Comparison of Algorithms for Minimax Design of Two-Dimensional Linear Phase FIR Digital Filters," *IEEE Trans. Acoustics, Speech, and Signal Processing*, ASSP-25, no. 6 (Dec. 1977), 492–500.

12. J. R. Rice, *The Approximation of Functions*, Vol. 2, *Nonlinear and Multivariate Theory* (Reading, Mass.: Addison-Wesley Publishing Company, Inc., 1969).

13. John H. Lodge and Moustafa M. Fahmy, "An Efficient $l_p$ Optimization Technique for the Design of Two-Dimensional Linear-Phase FIR Digital Filters," *IEEE Trans. Acoustics, Speech, and Signal Processing*, ASSP-28, no. 3 (June 1980), 308–13.

14. Sven Treitel and John L. Shanks, "The Design of Multistage Separable Planar Filters," *IEEE Trans. Geoscience Electronics*, GE-9, no. 1 (Jan. 1971), 10–27.

15. James H. McClellan, "The Design of Two-Dimensional Digital Filters by Transformations," *Proc. 7th Annual Princeton Conf. Information Sciences and Systems* (1973), 247–51.

16. Russell M. Mersereau, Wolfgang F. G. Mecklenbräuker, and Thomas F. Quatieri, Jr., "McClellan Transformation for 2-D Digital Filtering: I-Design," *IEEE Trans. Circuits and Systems*, CAS-23, no. 7 (July 1976), 405–14.

17. Wolfgang F. G. Mecklenbräuker and Russell M. Mersereau, "McClellan Transformations for 2-D Digital Filtering: II-Implementation," *IEEE Trans. Circuits and Systems*, CAS-23, no. 7 (July 1976), 414–22.

18. James H. McClellan and David S. K. Chan, "A 2-D FIR Filter Structure Derived from the Chebyshev Recursion," *IEEE Trans. Circuits and Systems*, CAS-24, no. 7 (July 1977), 372–78.

19. Russell M. Mersereau, "The Design of Arbitrary 2-D Zero-Phase FIR Filters Using Transformations," *IEEE Trans. Circuits and Systems*, CAS-27, no. 2 (Feb. 1980), 142–44.

20. O. D. Faugeras and J. F. Abramatic, "2-D FIR Filter Design from Independent 'Small' Generating Kernels Using a Mean Square and Tchebyshev Error Criterion," *Proc. IEEE Int. Conf. Acoustics, Speech, and Signal Processing* (Apr. 1979), 1–4.

21. Russell M. Mersereau, "The Processing of Hexagonally Sampled Two-Dimensional Signals," *Proc. IEEE*, 67, no. 6 (June 1979), 930–49.

# 4

# MULTIDIMENSIONAL RECURSIVE SYSTEMS

Linear shift-invariant discrete systems are generally implemented using difference equations. Although multidimensional difference equations represent a generalization of 1-D difference equations, we shall see that they are considerably more complex and that they are, in fact, quite different. A number of important issues associated with multidimensional difference equations, such as the direction of recursion and the ordering relation, are really not issues in the 1-D case. Other issues, such as stability, although present in the 1-D case, are far more difficult to understand for higher-dimensional systems. Despite these difficulties, however, multidimensional difference equations can be studied and, because they are fundamental, they are important to an understanding of multidimensional systems.

In this chapter we define and discuss multidimensional recursive systems and consider many of the issues associated with multidimensional difference equations. We also define the multidimensional $z$-transform and discuss the issue of stability for multidimensional systems. Chapter 5 is also concerned with aspects of recursive systems. There, attention is focused on the design and implementation of multidimensional recursive digital filters.

## 4.1 FINITE-ORDER DIFFERENCE EQUATIONS

A *difference equation* is an implicit relationship between the input, $x$, and the output, $y$, of a linear shift-invariant system of the form

$$\sum_{k_1} \sum_{k_2} b(k_1, k_2) y(n_1 - k_1, n_2 - k_2) = \sum_{r_1} \sum_{r_2} a(r_1, r_2) x(n_1 - r_1, n_2 - r_2) \quad (4.1)$$

If the output samples of an LSI system can be computed from the input samples with a finite amount of computation, the computation can be expressed in the form (4.1) with finite limits of summation. This latter condition implies that the coefficient arrays $a$ and $b$ have finite extent. In this case, we say that the difference equation has *finite order*.

Furthermore, if $b(0, 0) \neq 0$, we can normalize the coefficient arrays $a$ and $b$ by dividing both sides of (4.1) by $b(0, 0)$. This defines new coefficient arrays, allowing us to assume that $b(0, 0) = 1$ without loss of generality. This normalization will simplify some of the equations that follow.

The *order* of a difference equation is a measure of the size of the region of support for the array $b(k_1, k_2)$. The higher the order, the greater the degree of complexity. Unfortunately, there is no precise definition of order since the array $b$ can have any shape. The order can be quantitatively defined only after the shape of the $b$ array is specified. For example, in the special case where $b(k_1, k_2)$ has a rectangular shape so that (4.1) can be written as

$$\sum_{k_1=0}^{N_1} \sum_{k_2=0}^{N_2} b(k_1, k_2) y(n_1 - k_1, n_2 - k_2) = \sum_{r_1} \sum_{r_2} a(r_1, r_2) x(n_1 - r_1, n_2 - r_2) \quad (4.2)$$

we will say that the order is $N_1 \times N_2$.

The class of difference equations of order zero-by-zero is an important special case. For these systems, the array $b(k_1, k_2)$ consists only of a single sample at the origin and we can write

$$y(n_1, n_2) = \sum_{r_1} \sum_{r_2} a(r_1, r_2) x(n_1 - r_1, n_2 - r_2) \quad (4.3)$$

Comparing the difference equation above with the convolution sum, we see that the output array $y(n_1, n_2)$ is the convolution of the input array with the array of coefficients, $a(r_1, r_2)$, and that $a(r_1, r_2)$ can be identified as the impulse response of the filter. Since $a(r_1, r_2)$ possesses only a finite number of nonzero values, we see that difference equations of order zero-by-zero correspond to FIR filters, such as those discussed in Chapter 3. Finite-order difference equations whose order is not zero-by-zero correspond to IIR (infinite-extent impulse response) filters.

### 4.1.1 Realizing LSI Systems Using Difference Equations

Difference equations are important not only for defining certain LSI systems, but also because they can serve as computational algorithms for realizing those systems. This can be seen by rewriting (4.1) as

$$y(n_1, n_2) = \sum_{r_1} \sum_{r_2} a(r_1, r_2) x(n_1 - r_1, n_2 - r_2)$$

$$- \sum_{\substack{k_1 \\ (k_1, k_2) \neq (0,0)}} \sum_{k_2} b(k_1, k_2) y(n_1 - k_1, n_2 - k_2) \qquad (4.4)$$

In this form, the difference equation represents an algorithm for computing the sample of $y$ at $(n_1, n_2)$ under the assumptions that the required samples of the input are available and that those samples of $y$ which appear on the right-hand side of (4.4) have either been previously computed or have been specified as initial conditions.

We will say that systems for which the output samples can be computed in this manner are *recursively computable*. Not all systems specified by finite-order difference equations possess this property. Whether or not a system is recursively computable depends on the region of support of the coefficient array $b(k_1, k_2)$, the locations of the samples of the output array which are prespecified as initial conditions, and the order in which the output samples are to be computed. (In some cases an input sequence of infinite extent can prevent the output sequence from being computed recursively.) Only one of these pieces of information—the region of support of the array $b$—is actually provided by the difference equation. The others must be given by the system designer to specify the system completely. We will have much more to say about these issues in the sections that follow.

### 4.1.2 Recursive Computability

Equation (4.4), which provides us with a means for computing a sample of $y(n_1, n_2)$, can be rewritten as

$$y(n_1, n_2) = \sum_{r_1} \sum_{r_2} a(n_1 - r_1, n_2 - r_2) x(r_1, r_2)$$

$$- \sum_{\substack{k_1 \\ (k_1, k_2) \neq (n_1, n_2)}} \sum_{k_2} b(n_1 - k_1, n_2 - k_2) y(k_1, k_2) \qquad (4.5)$$

This equation is interpreted graphically in Figure 4.1. An *input mask*, or window of finite area whose shape is determined by the array $a(r_1, r_2)$, is positioned over the input array at a position that depends on $(n_1, n_2)$. Only a finite number of input samples are covered by the mask. The samples of $x$ that are covered are multiplied by the appropriate coefficients $a(r_1, r_2)$ and the resulting products are summed. Similarly and synchronously, the *output mask*, determined by the array $b(k_1, k_2)$, is swept over the output array. All of the output samples which are covered by this mask, except the one at $(n_1, n_2)$, are weighted by the coefficients $b(k_1, k_2)$, summed and subtracted from the sum derived from the input mask to produce the value $y(n_1, n_2)$. This number is stored in the output array, the two masks are moved to new locations, and the process is repeated. Pictorially, the location on the output mask which covers the point $(n_1, n_2)$ is indicated by a circle or hole. This serves two purposes; it reminds us that this sample is not involved in the computations and it shows us where the computed output value should be placed.

In order to compute the value of a particular output sample, the output mask

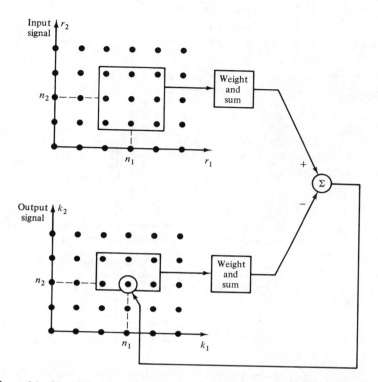

**Figure 4.1**   Use of input and output masks to compute successive samples of the output of a recursive filter.

must cover only sample values that are known (excluding, of course, the sample located under the hole.) This means that certain samples must be computed before others. If there is no possible ordering that allows the outputs to be computed sequentially given a set of initial conditions, the system is not recursively computable. If there are several possible orderings, we have a choice; some of these orderings may be better than others. The number of possible orderings depends on both the shape of the output mask and the location of its hole.

The simplest example of a 2-D output mask that allows the output to be computed recursively is the first-quadrant or "causal" filter. In this case, the coefficient array $b(k_1, k_2)$ is nonzero only in the finite region $\{0 \leq k_1 \leq N_1, 0 \leq k_2 \leq N_2\}$. The shape of the output mask is given in Figure 4.2. If the values for the output array are given as boundary conditions in the L-shaped region indicated by the open circles in that figure, the output mask can be swept upward column by column to generate the remainder of the output array without ever overlapping an unknown output sample. This is not, however, the only possible ordering that we can associate with this mask. We could also compute the output samples row by row, left to right, beginning with the lower rows and working up. In addition, the output array could be swept along diagonal lines. More generally, this mask can be swept over any family

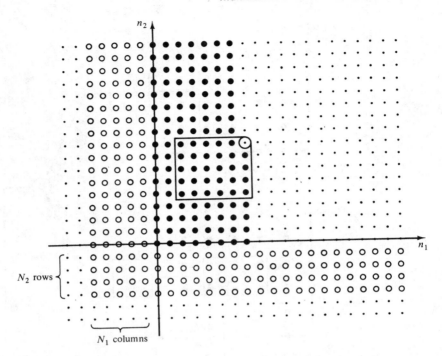

**Figure 4.2** Output mask corresponding to a recursively computable first-quadrant filter. The closed circles denote output samples which have been computed, the open circles represent initial conditions, and the dots represent samples that remain to be computed. The output mask is $(N_1 + 1) \times (N_2 + 1)$ points in extent.

of parallel lines provided only that the slope of those lines is negative as illustrated in Figure 4.3.

It is important to realize that the output signals computed for each of these sweep directions are identical. The same arithmetic operations are performed to compute corresponding output samples for each sweep direction; only the order in which the output values are computed varies with sweep direction.

In Figure 4.4 we present two masks that do not possess recursively computable solutions. As with the mask in Figure 4.2, the shape of the mask is rectangular but the position of the hole does not lie on a corner. For a mask with its hole in the interior of the mask, initial conditions or previously computed output samples must completely surround the point presently being evaluated. The recursion is thus not able to sustain itself. A similar phenomenon occurs when the mask has its hole somewhere along an edge. Here the recursion can proceed along one row or column but not beyond.

The one-quadrant mask with a hole on a corner, however, is not the only form for an output mask that corresponds to a recursively computable system. Consider the mask shown in Figure 4.5 for a nonsymmetric half-plane filter [1, 2]. This filter

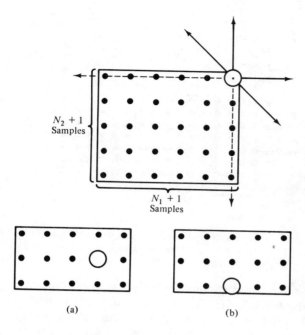

**Figure 4.3**  Allowable recursion directions for a first-quadrant output mask (shown shaded). The solid boundaries are acceptable directions, the dashed boundaries are not.

**Figure 4.4**  Two examples of output masks that are not recursively computable. (a) Mask with its hole in the center. (b) Mask with its hole on an edge (noncorner).

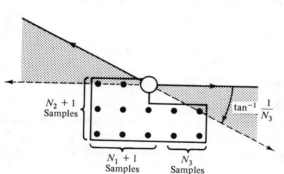

**Figure 4.5**  Allowable recursion directions for a nonsymmetric half-plane output mask (shown shaded). As in Figure 4.3, the solid boundary lines correspond to acceptable directions, the dashed ones do not.

has an impulse response that extends over a sector of the $(n_1, n_2)$-plane, where the sector angle is less than 180°. The exact sector angle depends on the dimensions of the mask. The natural ordering for computing output samples with this mask is to sweep each row from left to right, beginning at the row just above the initial conditions, and continuing upward row by row. Other sweep directions, however, are possible as indicated in Figure 4.5. The initial conditions required by this particular nonsymmetric half-plane (NSHP) filter are given in Figure 4.6. Notice that the shape of the band of initial conditions depends on the filter order. NSHP filters can be generalized further by either a reflection or a rotation of the output mask, or by a combination of the two operations.

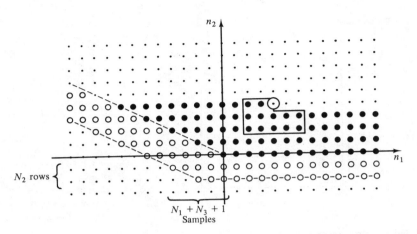

**Figure 4.6** Output mask corresponding to a recursively computable nonsymmetric half-plane filter. The open circles represent an acceptable set of initial conditions.

Although the NSHP filter is a generalization of the quadrant filter, it is also very closely related to it, since any NSHP filter can be mapped to a quadrant filter using the linear mappings introduced in Section 1.2.8. For example, the output mask shown in Figure 4.5 can be mapped into the mask of a one-quadrant filter by means of the transformation

$$m_1 = n_1 + N_3 n_2; \qquad m_2 = n_2 \tag{4.6}$$

This transformation gives the quadrant mask shown in Figure 4.7.

### 4.1.3 Boundary Conditions

Up to this point, we have said little about boundary (initial) conditions except to assume that they were always conveniently available when needed. In fact, we must be careful; we cannot choose our boundary conditions arbitrarily if we want to have a linear shift-invariant system. For example, we know that if a system is linear, it satisfies

$$T[ax(n_1, n_2)] = aT[x(n_1, n_2)] \tag{4.7}$$

for all values of the parameter $a$ including $a = 0$. This means that the response to the all-zero input must be the all-zero output. The initial conditions cannot be chosen to violate this condition and still result in a linear system. The only explicit value that can be given to the boundary conditions of a linear system is the value zero.

Given that linearity demands that all boundary values be zero, where can these samples be located? To answer this question, we must appeal to the shift invariance of the system. Consider as a simple example the (unstable) difference equation

$$y(n_1, n_2) = y(n_1 - 1, n_2) + y(n_1 + 1, n_2 - 1) + x(n_1, n_2) \tag{4.8}$$

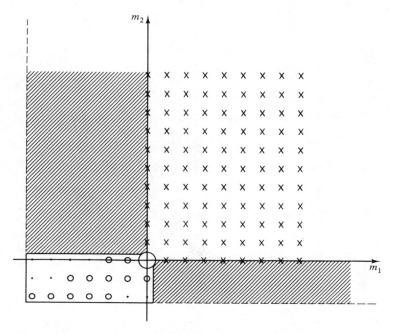

**Figure 4.7** Output mask of Figures 4.5 and 4.6 mapped to a quadrant, or quarter-plane, mask. Open circles correspond to nonzero coefficients, dots correspond to zero coefficients, and x's correspond to output samples to be computed. The initial conditions of Figure 4.6 are mapped to the shaded L-shaped region between the dashed lines and the axes.

The results of filtering a pair of identical inputs with two different sets of boundary conditions are shown in Figures 4.8 and 4.9. In part (a) of each figure, we see the boundary conditions used, in part (b) we see the response to the input $x(n_1, n_2) = \delta(n_1, n_2)$, and in part (c) we see the response to $x(n_1, n_2) = \delta(n_1 - 1, n_2 - 1)$. On the basis of these figures, we can make some very important observations. First, although the same difference equation was used, the different boundary conditions produced very different outputs. Also, the boundary conditions chosen in Figure 4.8 do not result in a shift-invariant system and those chosen in Figure 4.9 would appear to. What is the difference between these two cases?

If the overall system is linear and shift invariant, the output sequence $y(n_1, n_2)$ must be the convolution of the input sequence and the impulse response of the system. The region of support of that output, which is determined by the convolution, can be determined using methods discussed in Chapter 1. If $R_y$ denotes the region of support of $y(n_1, n_2)$, then

$$y(n_1, n_2) = 0, \qquad (n_1, n_2) \notin R_y \tag{4.9}$$

This equation tells us how the boundary conditions must be chosen. If they are chosen

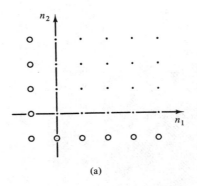

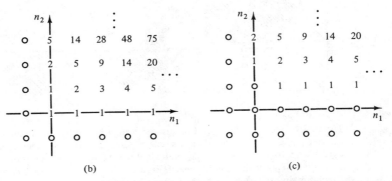

**Figure 4.8** (a) Set of boundary conditions to be used with difference equation (4.8). (b) Response to $x(n_1, n_2) = \delta(n_1, n_2)$. (c) Response to $x(n_1, n_2) = \delta(n_1 - 1, n_2 - 1)$.

to force some of the samples inside $R_y$ to be zero, as in Figure 4.8, when the convolution says they must be nonzero, something has to give; the system becomes shift-varying.

For filters of finite order, it is sufficient to specify the boundary conditions in a V-shaped band of finite width. Whereas the exact width and the orientation of the "V" depend on the shape of the output mask, the band of boundary conditions must be outside $R_y$. For a nonsymmetric half-plane output mask, shift invariance requires that the band of boundary conditions be similar to that shown in Figure 4.6, which forms an obtuse angle close to the origin. Even if we wish to filter a sequence that is confined to the first quadrant, we must nonetheless evaluate the output sequence at many samples in the second quadrant, as in Figure 4.9. This is one factor that has limited the popularity of NSHP filters.

On occasion, nonzero boundary conditions can be very useful—we are not saying that they should not be used. When they are used, however, it should be recognized that the resulting system will not be linear or shift invariant.

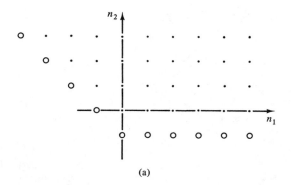

(a)

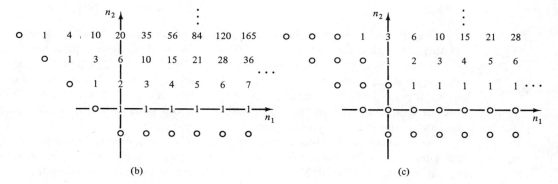

(b)                                              (c)

**Figure 4.9**  (a) Another set of boundary conditions to be used with difference equation (4.8). (b) Response to $x(n_1, n_2) = \delta(n_1, n_2)$. (c) Response to $x(n_1, n_2) = \delta(n_1 - 1, n_2 - 1)$.

### 4.1.4 Ordering the Computation of Output Samples

We saw in an earlier section that a significant dissimilarity between 1-D and $M$-dimensional difference equations concerns the order in which the output samples can be computed. For a 1-D system with a given output mask, there is at most one ordering by which the output samples can be computed; the computation is thus fully ordered. In the 2-D case, however, output samples can be computed in any one of several possible orderings; the computation is only partially ordered. This partial ordering can be represented by means of a precedence graph [3], such as the one shown in Figure 4.10 for our first quadrant example. Every output sample that must be computed corresponds to a vertex on the graph. Each sample can be computed whenever the two samples lying above it in the graph have been computed. Thus, with a first quadrant filter, the sample at (0, 0) must be evaluated first. Once its value is known, then either the sample at (1, 0) or the sample at (0, 1) can be evaluated, or if the hardware will permit it, both outputs can be computed simultaneously. Similar statements can be made at other levels in the graph.

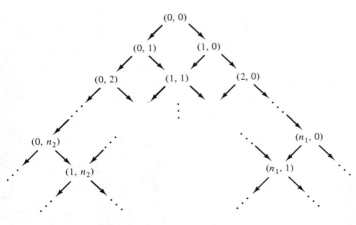

**Figure 4.10** Precedence graph for a first-quadrant recursion [3]. The indices at the nodes correspond to samples of the output array.

The precedence graph is an embodiment of the implied partial ordering of the output samples. To implement any real filter, we must specify a complete ordering. One means of doing this is through the use of an *index mapping function* or *ordering relation* [4] of the form $n = I(n_1, n_2)$. If $n' = I(n'_1, n'_2)$, then $n < n'$ implies that the output at $(n_1, n_2)$ will be computed before the output at $(n'_1, n'_2)$. The ordering relation must be consistent with the partial ordering relations. The existence of an ordering relation implies that the system is recursively computable.

As an example, consider the first quadrant filter whose partial ordering is given in Figure 4.10. Let us assume that we wish to compute the output samples only over the range $[0, N_1 - 1] \times [0, N_2 - 1]$. One candidate for the index mapping function is

$$n = I(n_1, n_2) = N_2 n_1 + n_2 \tag{4.10}$$

This corresponds to computing the output signal one column at a time, going from $(0, 0)$ to $(0, N_2 - 1)$, then from $(1, 0)$ to $(1, N_2 - 1)$, and so on. To compute the output samples row by row, we could use

$$n = n_1 + N_1 n_2 \tag{4.11}$$

As a third example, we could consider computing the output values along diagonals ($n_1 + n_2$ equal to a constant). In this case, a suitable index mapping function would be

$$n = \tfrac{1}{2}(n_1 + n_2)(n_1 + n_2 + 1) + n_1 \tag{4.12}$$

The order generated by this particular index mapping function is $(0, 0)$, $(0, 1)$, $(1, 0)$, $(0, 2)$, $(1, 1)$, $(2, 0)$, $(0, 3)$, $(1, 2)$, $(2, 1)$, . . . .

It should be obvious that several index mapping functions are often possible for the same difference equation. However, the various complete orderings may have distinct advantages and disadvantages with respect to each other. Let us clarify this statement by considering two of the most common orderings, column by column (or row by row) and diagonal by diagonal. To make the example as concrete as possible, let us assume that we want to implement a first quadrant filter with a square $3 \times 3$ output mask, a $1 \times 1$ input mask, and that we want the output to be evaluated over an $N \times N$ square region.

In the column-by-column implementation, we must store the output values in the current column which lie below the output point we are currently computing, the output values in the preceding column, and the output values in the column before that which lie above or are covered by the output mask. As illustrated in Figure 4.11, the number of output values that must be buffered is $2N + 2$. If the input samples

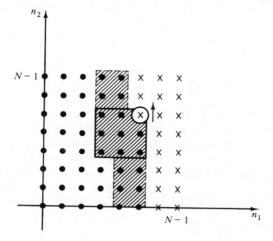

**Figure 4.11**  Illustration of the $2N + 2$ output values that must be buffered in the column-by-column implementation of a $3 \times 3$ recursive filter.

are available column by column, no additional storage will be needed to buffer the input samples. A total of $9N^2$ multiplications are required to compute all $N^2$ output values. Eight multiplications per output point take place in the output mask, and the ninth is used to scale the input values.

The diagonal-by-diagonal implementation must provide storage for the four previously computed diagonals. A detailed accounting, illustrated in Figure 4.12, indicates that in the worst case, $4N - 4$ output values must be buffered. Although this implementation requires more storage than the column-by-column implementa-

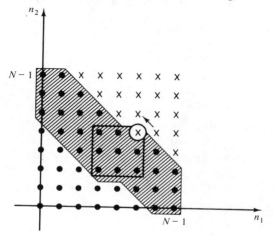

**Figure 4.12**  At most, $4N - 4$ output values must be buffered in the diagonal-by-diagonal implementation of a $3 \times 3$ recursive filter.

tion, it does have the important advantage that output samples along any diagonal can be computed independently. Consequently, if enough processing units are available, all the output samples along a particular diagonal can be computed simultaneously. Although $9N^2$ multiplications still need to be computed, if $N$ parallel processors are available, the time needed for a diagonal-by-diagonal implementation is roughly the time needed to perform $9(2N - 1)$ multiplications. Using this implemention, we trade off extra required storage for the capability of performing parallel computations.

## 4.2 MULTIDIMENSIONAL z-TRANSFORMS

In Chapter 1 we discussed the response of linear shift-invariant systems to sinusoidal excitations, which led very naturally to the Fourier transform. The z-transform is a generalization of the Fourier transform which allows us to treat exponential inputs. Because of this fact, the two transforms have much in common, but one important difference is the way in which they are used. The Fourier transform is used primarily to describe signals and to describe the actions that systems will have on them. The z-transform is used to describe systems and to provide an additional tool for manipulating difference equations. It is also invaluable for such tasks as determining filter stability. While the 2-D z-transform is related to its 1-D counterpart, the two transforms are actually quite different. For this reason, we shall try to make our discussion quite complete.

### 4.2.1 Transfer Function

Exponentials of the form $x(n_1, n_2) = z_1^{n_1} z_2^{n_2}$ are eigenfuctions of 2-D linear shift-invariant systems. This can be seen by evaluating the output array using the convolution sum.

$$y(n_1, n_2) = \sum_{k_1} \sum_{k_2} z_1^{n_1-k_1} z_2^{n_2-k_2} h(k_1, k_2) \tag{4.13}$$

$$= z_1^{n_1} z_2^{n_2} \sum_{k_1} \sum_{k_2} h(k_1, k_2) z_1^{-k_1} z_2^{-k_2} \tag{4.14}$$

$$= z_1^{n_1} z_2^{n_2} H_z(z_1, z_2) \tag{4.15}$$

where

$$H_z(z_1, z_2) \triangleq \sum_{k_1=-\infty}^{\infty} \sum_{k_2=-\infty}^{\infty} h(k_1, k_2) z_1^{-k_1} z_2^{-k_2} \tag{4.16}$$

$H_z(z_1, z_2)$ is the eigenvalue associated with the eigenfunction $z_1^{n_1} z_2^{n_2}$. It will be referred to as the *transfer function* or *system function* of the system. Like the impulse response, it constitutes a characterization of the system. In general, this summation may not converge for all values of the complex variables $z_1$ and $z_2$, but if the filter is stable, it will converge for $z_1 = e^{j\omega_1}$, $z_2 = e^{j\omega_2}$. For these values of $z_1$ and $z_2$, the transfer function becomes the frequency response since

$$H_z(e^{j\omega_1}, e^{j\omega_2}) = H(\omega_1, \omega_2) \tag{4.17}$$

### 4.2.2 The z-Transform

Let us now formally define the 2-D $z$-transform of a discrete array $x$ as

$$X_z(z_1, z_2) = \sum_{n_1=-\infty}^{\infty} \sum_{n_2=-\infty}^{\infty} x(n_1, n_2) z_1^{-n_1} z_2^{-n_2} \qquad (4.18)$$

With this definition, the transfer function is seen to be the $z$-transform of the impulse response. Letting $z_1 = e^{j\omega_1}$, $z_2 = e^{j\omega_2}$, the $z$-transform reduces to the Fourier transform. For convenience, we shall denote the surface in the $z$-domain described by $z_1 = e^{j\omega_1}$, $z_2 = e^{j\omega_2}$ as the 2-D *unit surface* or the *unit bicircle*.

The sum in (4.18) may not converge for all (or any) values of $z_1$ and $z_2$. The values of $z_1$ and $z_2$ for which the $z$-transform does converge absolutely constitute a *region of convergence* or *region of analyticity* in the $(z_1, z_2)$-hyperplane. Within the region of convergence $X_z(z_1, z_2)$ is an analytic function. The region of convergence consists of those points $(z_1, z_2)$ for which

$$\sum_{n_1} \sum_{n_2} |x(n_1, n_2)| |z_1|^{-n_1} |z_2|^{-n_2} = S_1 < \infty \qquad (4.19)$$

which in turn implies that

$$|X_z(z_1, z_2)| < \infty \qquad (4.20)$$

Whether or not a point $(z_1, z_2)$ lies in the region of convergence depends only on the magnitudes $|z_1|$, $|z_2|$ and not on the phase angles of the complex variables. Thus in the 1-D case the region of convergence of a $z$-transform is an annulus as shown in Figure 4.13. Equivalently, the region of convergence can be indicated as an interval

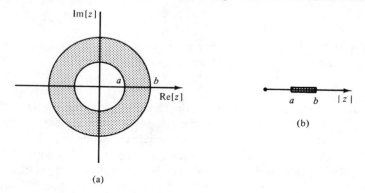

**Figure 4.13**  (a) Typical region of convergence for the $z$-transform of a 1-D sequence. (b) Equivalent representation for the annulus as a segment of a line.

on a line. The 2-D analog of the annulus is the *Reinhardt domain*. If a point $(z_1, z_2)$ lies in a Reinhardt domain, $R$, the points $(e^{j\mu} z_1, e^{j\nu} z_2)$ must also lie in $R$ for all real values of $\mu$ and $v$. Although the Reinhardt domain for a 2-D $z$-transform is a four-dimensional construct, it can be completely specified by a 2-D figure such as the one shown in Figure 4.14.

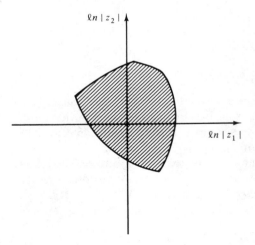

$\ell n\,|z_2|$

$\ell n\,|z_1|$

**Figure 4.14** Graphical representation for a 2-D region of convergence. The region is plotted as a function of $\ell n\,|z_1|$ and $\ell n\,|z_2|$ to simplify subsequent plots. The unit bicircle corresponds to the origin in the $(\ell n\,|z_1|, \ell n\,|z_2|)$-plane.

As in the 1-D case, the specification of the $z$-transform of a sequence is incomplete and ambiguous without the specification of the region of convergence as well. Let us consider several examples of typical regions of convergence that will be encountered. These are related to the region of support of the sequence $x(n_1, n_2)$.

**Sequences with finite support.**   For a sequence whose region of support is confined to a finite area in the $(n_1, n_2)$-plane, the $z$-transform can be expressed as

$$X_z(z_1, z_2) = \sum_{n_1=N_1}^{M_1} \sum_{n_2=N_2}^{M_2} x(n_1, n_2) z_1^{-n_1} z_2^{-n_2} \quad (4.21)$$

Since the limits on the sums are finite and the summands are also finite, this $z$-transform is seen to converge for all finite values of $z_1$ and $z_2$ except possibly for $z_1 = 0$ or $z_2 = 0$.

**Sequences with quadrant support.** Because of their analogy to 1-D causal sequences, 2-D sequences that are zero outside of the first quadrant form an important class. For sequences in this class, the 2-D $z$-transform is given by

$$X_z(z_1, z_2) = \sum_{n_1=0}^{\infty} \sum_{n_2=0}^{\infty} x(n_1, n_2) z_1^{-n_1} z_2^{-n_2} \quad (4.22a)$$

It is straightforward to show that if the point $(z_{01}, z_{02})$ lies in the region of convergence for the sum (4.22a), then all points $(z_1, z_2)$ that satisfy

$$|z_1| \geq |z_{01}|, \qquad |z_2| \geq |z_{02}| \quad (4.22b)$$

also lie in the region of convergence. This is shown graphically in Figure 4.15.

Because of this condition, we can make some statements about the boundary of the region of convergence for first-quadrant sequences. This boundary, indicated by the dashed line in Figure 4.15, must have a nonpositive slope. If it did not, we could find points that would satisfy the sufficient condition and yet lie outside the region of convergence, as shown in Figure 4.16. Because of this contradiction, we may argue that the boundary must not have a positive slope.

As a simple example, consider the 2-D first-quadrant sequence

$$x(n_1, n_2) = a^{n_1}\delta(n_1 - n_2)u(n_1, n_2) \quad (4.23)$$

which has the $z$-transform

$$X_z(z_1, z_2) = \frac{1}{1 - az_1^{-1}z_2^{-1}}$$

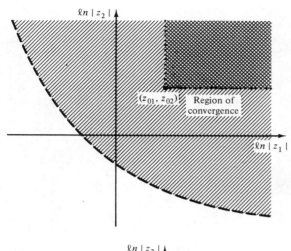

**Figure 4.15** Region of convergence of a 2-D sequence with first quadrant support lies above and to the right of the dashed line in the $(\ell n\,|z_1|,\,\ell n\,|z_2|)$-plane.

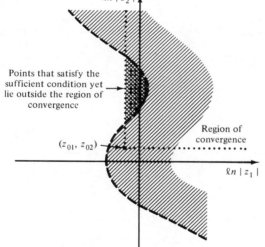

**Figure 4.16** Region of convergence with the boundary indicated cannot occur in practice for the first-quadrant sequences. Since $(z_{01}, z_{02})$ lies within the region of convergence, all points $(z_1, z_2)$ which satisfy $|z_1| \geq |z_{01}|$, $|z_2| \geq |z_{02}|$ must also lie within the region of convergence. If the boundary has a positive slope over some region, a contradiction results, indicated here by the crosshatched area.

It is straightforward to show that the region of convergence contains those points $(z_1, z_2)$ which satisfy

$$|a| < |z_1| \cdot |z_2| \tag{4.24a}$$

or equivalently

$$\ell n\,|a| < \ell n\,|z_1| + \ell n\,|z_2| \tag{4.24b}$$

The boundary of the region of convergence is thus a straight line with a slope of $-1$ in the $(\ell n\,|z_1|,\,\ell n\,|z_2|)$-plane.

For sequences with support on the second, third, or fourth quadrants, similar arguments may be made to derive restrictions on the shape of the region of convergence.

**Sequences with support on a wedge.** The region of convergence for a sequence with support on a wedge is only slightly more complicated. Suppose that a sequence has the support shown in Figure 4.17(a). The $z$-transform of such a sequence

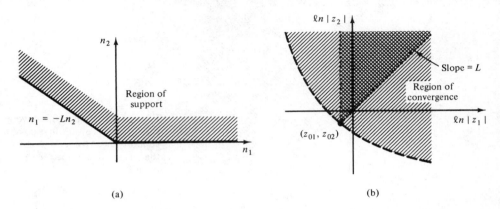

(a)                                                   (b)

**Figure 4.17** 2-D sequence with support on a wedge (a) and the region of convergence of its $z$-transform (b). If a point $(z_{01}, z_{02})$ lies within the region of support, then all points $(z_1, z_2)$ such that $\ell n\,|z_1| \geq \ell n\,|z_{01}|$ and $\ell n\,|z_2| \geq L\,\ell n\,|z_1| + \{\ell n\,|z_{02}| - L\,\ell n\,|z_{01}|\}$ (the crosshatched area) also lie within the region of convergence.

can be written as

$$X_z(z_1, z_2) = \sum_{n_2=0}^{\infty} \sum_{n_1=-Ln_2}^{\infty} x(n_1, n_2) z_1^{-n_1} z_2^{-n_2} \qquad (4.25)$$

If we define a new variable $l = n_1 + Ln_2$, we can express the sum above as

$$X_z(z_1, z_2) = \sum_{n_2=0}^{\infty} \sum_{l=0}^{\infty} x(l - Ln_2, n_2) z_1^{-l+Ln_2} z_2^{-n_2} \qquad (4.26)$$

$$= \sum_{n_2=0}^{\infty} \sum_{l=0}^{\infty} x(l - Ln_2, n_2) z_1^{-l} [z_1^{-L} z_2]^{-n_2} \qquad (4.27)$$

Clearly if the sum in (4.27) converges, the sum in (4.25) will converge. The sequence $x(l - Ln_1, n_2)$ has its support limited to the first quadrant. Thus, if the point $(z_{01}, z_{02})$ lies within the region of convergence, then $(z_1, z_2)$ will also lie within the region of convergence, provided that

$$|z_1| \geq |z_{01}| \qquad (4.28a)$$

and

$$|z_1^{-L} z_2| \geq |z_{01}^{-L} z_{02}| \qquad (4.28b)$$

These inequalities can be rewritten as

$$\ell n\,|z_1| \geq \ell n\,|z_{01}| \qquad (4.29a)$$

and

$$\ell n\,|z_2| \geq L\,\ell n\,|z_1| + \{\ell n\,|z_{02}| - L\,\ell n\,|z_{01}|\} \qquad (4.29b)$$

This region is shown in Figure 4.17(b). Again, these conditions may be used to constrain the slope of the boundary of the region of convergence.

**Sequences with support on a half-plane.**   Suppose that the sequence $x(n_1, n_2)$ has support on the upper half-plane only. Thus

$$x(n_1, n_2) = 0 \qquad \text{for } n_2 < 0$$

and

$$X_z(z_1, z_2) = \sum_{n_1=-\infty}^{\infty} \sum_{n_2=0}^{\infty} x(n_1, n_2) z_1^{-n_1} z_2^{-n_2}$$

If $(z_{01}, z_{02})$ lies within the region of convergence of $X_z$, it follows that the points $(z_1, z_2)$ which satisfy $|z_1| = |z_{01}|$ and $|z_2| \geq |z_{02}|$ also lie within the region of convergence. Consequently, the boundary of the region of convergence is constrained to be a single-valued function of $|z_1|$ (or $\ell n\,|z_1|$), as shown in Figure 4.18.

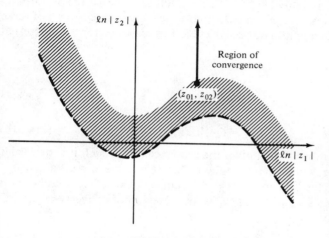

**Figure 4.18**   For 2-D sequences with support on the upper half-plane, the fact that $(z_{01}, z_{02})$ lies within the region of convergence of the z-transform implies that $(z_1, z_2)$ will also lie within the region of convergence for $|z_1| = |z_{01}|$ and $|z_2| \geq |z_{02}|$.

**Sequences with support everywhere.**   Sequences that have support over the entire $(n_1, n_2)$-plane have 2-D z-transforms whose regions of convergence may exhibit a variety of sizes and shapes. For instance, the sequence

$$x(n_1, n_2) = \exp\left(-n_1^2 - n_2^2\right)$$

has a z-transform that converges for all values of $(z_1, z_2)$. On the other hand, the z-transform of the sequence

$$x(n_1, n_2) = 2^{|n_1|} 2^{|n_2|}$$

will not converge for any value of $(z_1, z_2)$. Often, however, the z-transform of a sequence with support everywhere will converge in a region of finite area.

We can write a sequence with support everywhere as the sum of four quadrant sequences. For example,

$$x(n_1, n_2) = x_1(n_1, n_2) + x_2(n_1, n_2) + x_3(n_1, n_2) + x_4(n_1, n_2)$$

where

$$x_1(n_1, n_2) = \begin{cases} x(n_1, n_2) & \text{for } n_1 > 0, n_2 > 0 \\ \frac{1}{2}x(n_1, n_2) & \text{for } n_1 = 0, n_2 > 0 \text{ or } n_1 > 0, n_2 = 0 \\ \frac{1}{4}x(n_1, n_2) & \text{for } n_1 = n_2 = 0 \\ 0 & \text{for } n_1 < 0 \text{ or } n_2 < 0 \end{cases} \quad (4.30)$$

The sequences $x_2(n_1, n_2)$, $x_3(n_1, n_2)$, and $x_4(n_1, n_2)$ are similarly defined to have support on the second, third, and fourth quadrants, respectively. Consequently, the $z$-transform $X_z(z_1, z_2)$ can be written as the sum of four $z$-transforms of sequences with quadrant support. The region of convergence of $X_z(z_1, z_2)$ is the intersection of the regions of convergence of the four constituent $z$-transforms.

### 4.2.3 Properties of the 2-D z-Transform

The $z$-transform operation has a number of properties that can be useful in performing calculations, solving problems, and proving theorems. Below we state several of these properties. The proofs are straightforward and are left as exercises for the interested reader.

**Separable signals.**   If

$$x(n_1, n_2) = v(n_1)w(n_2) \qquad (4.31a)$$

then

$$X_z(z_1, z_2) = V_z(z_1)W_z(z_2) \qquad (4.31b)$$

Thus $X_z(z_1, z_2)$ will be separable if and only if the sequence from which it is derived is separable. In that case, $V_z(z_1)$ and $W_z(z_2)$ are the 1-D $z$-transforms of $v(n_1)$ and $w(n_2)$, respectively. A point $(z_1, z_2)$ will be in the region of convergence of $X_z$ if and only if $z_1$ is in the region of convergence of $V_z$ and $z_2$ is in the region of convergence of $W_z$.

**Linearity.**   If

$$x(n_1, n_2) = av(n_1, n_2) + bw(n_1, n_2) \qquad (4.32a)$$

then

$$X_z(z_1, z_2) = aV_z(z_1, z_2) + bW_z(z_1, z_2) \qquad (4.32b)$$

for any complex constants $a$ and $b$. The region of convergence for $X_z$ is generally the intersection of the regions of convergence for $V_z$ and $W_z$, although in rare cases it can be slightly larger. This property can be helpful in decomposing a complex system into a parallel connection of simpler systems or in building up a complex system from simpler ones.

**Shift property.**   If

$$x(n_1, n_2) = v(n_1 + m_1, n_2 + m_2) \qquad (4.33a)$$

then

$$X_z(z_1, z_2) = z_1^{m_1} z_2^{m_2} V_z(z_1, z_2) \qquad (4.33b)$$

The region of convergence for $X_z$ is the same as the region of convergence for $V_z$, except possibly for the points for which $|z_1| = 0$ or $|z_2| = 0$.

**Modulation property.**    If

$$x(n_1, n_2) = a^{n_1} b^{n_2} w(n_1, n_2) \tag{4.34a}$$

then

$$X_z(z_1, z_2) = W_z(a^{-1}z_1, b^{-1}z_2) \tag{4.34b}$$

The region of convergence for $X_z$ has the same shape as the region for $W_z$ except that it is scaled by $|a|$ in the $z_1$ variable and $|b|$ in the $z_2$ variable.

**Differentiation property.**    If

$$x(n_1, n_2) = n_1 n_2 w(n_1, n_2) \tag{4.35a}$$

then

$$X_z(z_1, z_2) = z_1 z_2 \frac{\partial^2}{\partial z_1 \, \partial z_2} W_z(z_1, z_2) \tag{4.35b}$$

The regions of convergence for $X_z$ and $W_z$ are the same.

**Conjugation properties for complex signals.**    If $x(n_1, n_2)$ is a complex signal with $z$-transform $X_z(z_1, z_2)$, then

$$x^*(n_1, n_2) \longleftrightarrow X_z^*(z_1^*, z_2^*) \tag{4.36}$$

$$\mathrm{Re}\,[x(n_1, n_2)] \longleftrightarrow \tfrac{1}{2}[X_z(z_1, z_2) + X_z^*(z_1^*, z_2^*)] \tag{4.37}$$

$$\mathrm{Im}\,[x(n_1, n_2)] \longleftrightarrow \frac{1}{2j}[X_z(z_1, z_2) - X_z^*(z_1^*, z_2^*)] \tag{4.38}$$

All of these $z$-transforms have the same region of convergence as $X_z(z_1, z_2)$.

**Reflection properties.**    If

$$x(n_1, n_2) \longleftrightarrow X_z(z_1, z_2) \tag{4.39}$$

then

$$x(-n_1, n_2) \longleftrightarrow X_z(z_1^{-1}, z_2) \tag{4.40}$$

$$x(n_1, -n_2) \longleftrightarrow X_z(z_1, z_2^{-1}) \tag{4.41}$$

$$x(-n_1, -n_2) \longleftrightarrow X_z(z_1^{-1}, z_2^{-1}) \tag{4.42}$$

**Convolution property.**    If

$$y(n_1, n_2) = \sum_{k_1} \sum_{k_2} x(n_1 - k_1, n_2 - k_2) h(k_1, k_2) \tag{4.43a}$$

then

$$Y_z(z_1, z_2) = X_z(z_1, z_2) H_z(z_1, z_2) \tag{4.43b}$$

The 2-D $z$-transform of the convolution of two sequences is the product of their $z$-transforms. The region of convergence of $Y_z$ is the intersection of the regions of convergence of $X_z$ and of $H_z$.

**Multiplication property.**    The $z$-transform of the product of two sequences is the complex convolution of their $z$-transforms. Thus

$$x(n_1, n_2)y(n_1, n_2) \longleftrightarrow \left(\frac{1}{2\pi j}\right)^2 \oint_{C_2} \oint_{C_1} X_z\left(\frac{z_1}{v_1}, \frac{z_2}{v_2}\right) Y_z(v_1, v_2) \frac{dv_1}{v_1} \frac{dv_2}{v_2} \qquad (4.44)$$

**Parseval's theorem.**    Parseval's theorem relates the inner product of two arrays to the inner product of their $z$-transforms:

$$\sum_{n_1=-\infty}^{\infty} \sum_{n_2=-\infty}^{\infty} x(n_1, n_2)y^*(n_1, n_2) = \left(\frac{1}{2\pi j}\right)^2 \oint_{C_2} \oint_{C_1} X_z(z_1, z_2)Y_z^*\left(\frac{1}{z_1^*}, \frac{1}{z_2^*}\right) \frac{dz_1}{z_1} \frac{dz_2}{z_2} \qquad (4.45)$$

The contours of integration must be closed, must encircle the origins of their respective variables in a counterclockwise fashion, and must lie totally within the region of convergence.

**Initial value theorems.**    If $x(n_1, n_2) = 0$ for $n_1 < 0$ and $n_2 < 0$, then

$$\lim_{z_1 \to \infty} X_z(z_1, z_2) = \sum_{n_2} x(0, n_2)z_2^{-n_2} \qquad (4.46)$$

$$\lim_{z_2 \to \infty} X_z(z_1, z_2) = \sum_{n_1} x(n_1, 0)z_1^{-n_1} \qquad (4.47)$$

$$\lim_{\substack{z_1 \to \infty \\ z_2 \to \infty}} X_z(z_1, z_2) = x(0, 0) \qquad (4.48)$$

**Linear mappings.**    Suppose that two 2-D arrays $x$ and $w$ are related by a linear mapping such that

$$x(n_1, n_2) = \begin{cases} w(m_1, m_2), & n_1 = Im_1 + Jm_2, \quad n_2 = Km_1 + Lm_2 \\ 0, & \text{otherwise} \end{cases} \qquad (4.49)$$

where $I, J, K$ and $L$ are integers and where $IL - KJ \neq 0$. Then

$$X_z(z_1, z_2) = W_z(z_1^I z_2^K, z_1^J z_2^L) \qquad (4.50)$$

### 4.2.4 Transfer Functions of Systems Specified by Difference Equations

Consider a 2-D linear shift-invariant (LSI) system specified by the difference equation

$$\sum_{k_1} \sum_{k_2} b(k_1, k_2)y(n_1 - k_1, n_2 - k_2) = \sum_{r_2} \sum_{r_2} a(r_1, r_2)x(n_1 - r_1, n_2 - r_2) \qquad (4.51)$$

Since the system is linear and shift invariant, the sequence $x(n_1, n_2) = z_1^{n_1} z_2^{n_2}$ is an eigenfunction and the output corresponding to this input has the form $y(n_1, n_2) = H_z(z_1, z_2)z_1^{n_1} z_2^{n_2}$. Substituting into (4.51), we see that

$$H_z(z_1, z_2) \sum_{k_1} \sum_{k_2} b(k_1, k_2)z_1^{-k_1} z_2^{-k_2} = \sum_{r_1} \sum_{r_2} a(r_1, r_2)z_1^{-r_1} z_2^{-r_2} \qquad (4.52)$$

or

$$H_z(z_1, z_2) = \frac{\sum_{r_1} \sum_{r_2} a(r_1, r_2)z_1^{-r_1} z_2^{-r_2}}{\sum_{k_1} \sum_{k_2} b(k_1, k_2)z_1^{-k_1} z_2^{-k_2}} \triangleq \frac{A_z(z_1, z_2)}{B_z(z_1, z_2)} \qquad (4.53)$$

The transfer function of a system specified by a difference equation is thus the ratio of the $z$-transforms of the coefficient arrays $a(r_1, r_2)$ and $b(k_1, k_2)$. Since each of these arrays has a finite area of support, their $z$-transforms are polynomials. For the one-quadrant mask shown in Figure 4.2, the transfer function is given by

$$H_z(z_1, z_2) = \frac{\sum_{r_1=0}^{N_1} \sum_{r_2=0}^{N_2} a(r_1, r_2) z_1^{-r_1} z_2^{-r_2}}{\sum_{k_1=0}^{N_1} \sum_{k_2=0}^{N_2} b(k_1, k_2) z_1^{-k_1} z_2^{-k_2}} \qquad (4.54)$$

under the assumption that the input mask is the same size as the output mask. A nonsymmetric half-plane filter such as the one shown in Figure 4.5 has a transfer function of the form

$$H_z(z_1, z_2) = \frac{\sum_{r_1=-N_3}^{N_1} \sum_{r_2=1}^{N_2} a(r_1, r_2) z_1^{-r_1} z_2^{-r_2} + \sum_{r_1=0}^{N_1} a(r_1, 0) z_1^{-r_1}}{\sum_{k_1=-N_3}^{N_1} \sum_{k_2=1}^{N_2} b(k_1, k_2) z_1^{-k_1} z_2^{-k_2} + \sum_{k_1=0}^{N_1} b(k_1, 0) z_1^{-k_1}} \qquad (4.55)$$

where we again have assumed that the input mask has the same shape as the output mask. This transfer function involves only positive powers of $z_2^{-1}$, but it involves both positive and negative powers of $z_1^{-1}$.

In dealing with one-dimensional transfer functions, it is extremely useful to be able to define poles and zeros. This can be done in the 2-D case as well. We will say that $H_z$ has a zero at $(z_1, z_2)$ if $A_z(z_1, z_2) = 0$ and $B_z(z_1, z_2) \neq 0$. Similarly, we will say that $H_z$ possesses a singularity at $(z_1, z_2)$ if $B_z(z_1, z_2) = 0$. Although we will often call such a singularity a pole, this should be recognized as an abuse, because multidimensional poles and zeros are quite different from their 1-D counterparts. Zeros of 1-D polynomials occur at isolated points in the $z$-plane, whereas multidimensional zeros are generally continuous surfaces. For example, consider the simple polynomial

$$B_z(z_1, z_2) = 1 - b z_1^{-1} z_2^{-1} \qquad (4.56)$$

The zeros of this polynomial are given by the relation

$$z_1 z_2 = b \qquad (4.57)$$

and they form a continuous surface or manifold in a 4-D space. Thus as $z_1$ is moved around the unit circle in the $z_1$-plane, the corresponding values of $z_2$ for which $B_z(z_1, z_2) = 0$ trace out a circle in the $z_2$-plane of radius $|b|$.

In a more general example, consider the transfer function from equation (4.54),

$$H_z(z_1, z_2) = \frac{A_z(z_1, z_2)}{B_z(z_1, z_2)} = \frac{A[z_1](z_2)}{B[z_1](z_2)} \qquad (4.58)$$

The notation used in the latter expression deserves a word of comment. The entity $B_z(z_1, z_2)$ is a polynomial in two variables with constant coefficients. Alternatively, $B_z(z_1, z_2)$ could be interpreted as a polynomial in one variable, say $z_2$, whose coefficients are themselves 1-D polynomials in the parameter $z_1$. To force the latter interpretation, we will use the notation $B[z_1](z_2)$. $B_z(z_1, z_2)$ could also be interpreted as $B[z_2](z_1)$, which is a polynomial in $z_1$ with coefficients that are polynomials in $z_2$.

If $z_1$ is fixed, the denominator polynomial $B[z_1](z_2)$ can be factored to yield

$N_2$ roots, each corresponding to a point of singularity where $B_z(z_1, z_2) = 0$. If $z_1$ is fixed at another value, $B[z_1](z_2)$ can again be factored to yield another $N_2$ points of singularity. Thus the locations of the poles and zeros of $H_z(z_1, z_2)$ with respect to the $z_2$ variable vary as a function of $z_1$, and vice versa.

The *root map* is a 2-D graph that can be quite useful for investigating the stability of an LSI system. It consists of two parts—one part shows the loci of the roots of

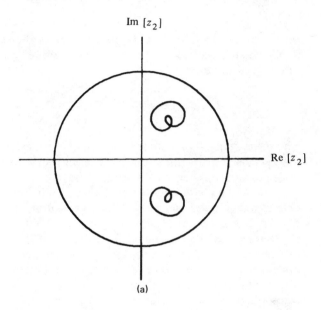

(a)

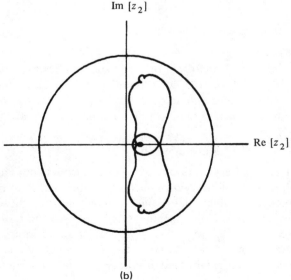

(b)

**Figure 4.19** Some examples of root maps. Each corresponds to the denominator polynomial of a lowpass filter. (a) 3 × 3. (b) 4 × 4. (c) 5 × 5 (unstable). (d) 6 × 6. (Courtesy of Gary A. Shaw and Russell M. Mersereau.)

$B[z_1](z_2)$ as the parameter $z_1$ traverses the unit circle $z_1 = e^{j\omega_1}$ for $-\pi \le \omega_1 \le \pi$. The other part shows the loci of the roots of $B[z_2](z_1)$ as the parameter $z_2$ traverses the unit circle $z_2 = e^{j\omega_2}$. If we regard the condition $B(z_1, z_2) = 0$ as an algebraic mapping between the $z_1$-plane and the $z_2$-plane, then the root map shows the image of the $z_1$ unit circle in the $z_2$-plane, and vice versa. Figure 4.19 shows some examples of root maps produced from the denominator polynomials of 2-D lowpass filters

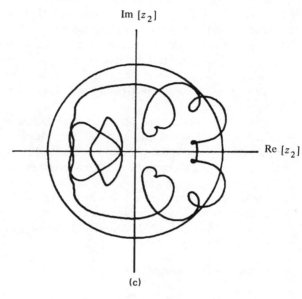

(c)

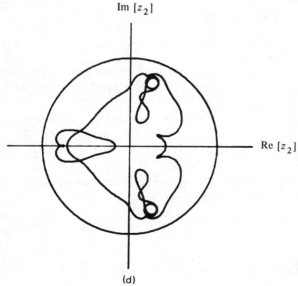

(d)

**Figure 4.19**  (*Continued*)

[17]. [Since $B(z_1, z_2) = B(z_2, z_1)$ for these filters, the two parts of each root map are identical.] The output masks varied in extent from 3 points by 3 points to 6 points by 6 points. The root maps can be quite complex even for these simple polynomials.

If there is no polynomial (other than a constant polynomial) which divides both the numerator and denominator polynomials of a rational transfer function, we say that the numerator and denominator are *coprime*. If a 2-D polynomial cannot be factored into a product of polynomials of lower order, the polynomial is said to be *irreducible*. A 2-D rational polynomial $H_z(z_1, z_2)$ is *irreducible* if its numerator and denominator polynomials are coprime. Note, however, that this does not imply that $A_z(z_1, z_2)$ and $B_z(z_1, z_2)$ are individually irreducible.

Even if a transfer function $H_z(z_1, z_2)$ is irreducible, there may be particular values of $(z_1, z_2)$ for which $A_z$ and $B_z$ are both zero. These points are termed *non-essential singularities of the second kind*. There is no 1-D counterpart to this type of 2-D singularity.

Part of the usefulness of poles in the 1-D case is due to the Fundamental Theorem of Algebra, which states that any 1-D polynomial of degree $N$ can be factored into a product of $N$ polynomials of degree 1. This allows us to look upon any high-order transfer function as a cascade of $N$ first-order systems, each of which can be described by its pole. Each pole can thus be considered in isolation. Unfortunately, there is no Fundamental Theorem of Algebra for multidimensional polynomials. It is rare that a multidimensional polynomial is factorable into factors of any order. Thus, in the 2-D case, it is generally impossible to isolate singularities unless a transfer function is constructed to be factorable.

### 4.2.5 Inverse z-Transform

As in the 1-D case, a 2-D z-transform can be inverted by means of an inversion formula which takes the form of a contour integral,

$$x(n_1, n_2) = \left(\frac{1}{2\pi j}\right)^2 \oint_{C_2} \oint_{C_1} X_z(z_1, z_2) z_1^{n_1-1} z_2^{n_2-1} \, dz_1 \, dz_2 \tag{4.59}$$

Each integral is evaluated over a contour that must be closed, must lie completely within the region of convergence of $X_z$, and must encircle the origin counterclockwise in the plane of the respective variable.

As an example, consider the computation of the inverse z-transform of the transfer function

$$H_z(z_1, z_2) = \frac{1}{1 - az_1^{-1} - bz_2^{-1}} \tag{4.60}$$

where $|a| + |b| < 1$ and the region of convergence includes the 2-D unit surface, over which we shall perform the integration. In this case,

$$h(n_1, n_2) = \left(\frac{1}{2\pi j}\right)^2 \oint_{C_2} \oint_{C_1} \frac{z_1^{n_1} z_2^{n_2}}{z_1 z_2 - az_2 - bz_1} \, dz_1 \, dz_2$$
$$= \left(\frac{1}{2\pi j}\right)^2 \oint_{C_2} \oint_{C_1} \frac{(z_2 - b)^{-1} z_1^{n_1} z_2^{n_2}}{z_1 - [az_2/(z_2 - b)]} \, dz_1 \, dz_2 \tag{4.61}$$

Let us perform the integral over $C_1$ first. In doing so, we can consider $z_2$ to be simply a parameter. This integral is the inverse 1-D z-transform of a system with a simple pole at $az_2/(z_2 - b)$. Since the contour of integration is $|z_1| = 1$, $|z_2| = 1$, it can be shown that $|az_2/(z_2 - b)| < 1$ and hence that the pole is inside the contour of integration. Applying Cauchy's residue theorem, we then observe that

$$h(n_1, n_2) = \left(\frac{1}{2\pi j}\right) a^{n_1} u(n_1) \oint_{C_2} \frac{z_2^{n_1 + n_2}}{(z_2 - b)^{n_1 + 1}} \, dz_2 \qquad (4.62)$$

[The sequence $u(n_1)$ is the 1-D step function described in Chapter 1.] The integral evaluated over $C_2$ can be recognized as the inverse z-transform of a 1-D system with an $(n_1 + 1)$st-order pole at $z_2 = b$, which lies inside the contour of integration. Applying the residue theorem one more time gives the final result:

$$h(n_1, n_2) = a^{n_1} b^{n_2} \frac{(n_1 + n_2)!}{n_1! \, n_2!} u(n_1) u(n_2) \qquad (4.63)$$

Although the example was quite simple, our solution procedure became quite involved. For more complicated transfer functions, it becomes extremely difficult, if not impossible, to compute the inverse z-transform explicitly. In the 1-D case, we could address the problem of inverting a high-order transfer function by first performing a partial fraction expansion and then expressing the inverse z-transform as a sum of simple components. This will not work in multidimensional cases since, without the ability to factor polynominals, we cannot perform partial fraction expansions. For these reasons, inverse z-transforms of multidimensional transfer functions are almost never computed analytically. If the transfer function and its region of convergence correspond to a recursively computable difference equation, it is possible to deduce a difference equation from the transfer function and drive it with a unit impulse $\delta(n_1, n_2)$ to obtain a numerical representation of the impulse response.

### 4.2.6 Two-Dimensional Flowgraphs

Earlier in this chapter we presented input and output masks to help explain how the output samples of a difference equation could be computed recursively. Masks are useful for determining whether or not a difference equation is recursively computable, what initial conditions are required to begin the recursion, and what partial orderings will preserve recursive computability. The flowgraph is another way of attempting to describe a 2-D LSI system graphically. The flowgraph can be derived from the z-transform and it is particularly useful for specifying the structure of the network. If the z-transform can be factored, for example, this has an identifiable effect on the flowgraph. Furthermore, since the flowgraph is a graph, operations can be performed on it to alter the description of an LSI system. The flowgraph is useful for analyzing the complexity of the system, analyzing the sensitivity of the system to deviations in coefficient values, and analyzing the effects of arithmetic round-off errors on the output of the system. Neither the input and output masks nor the flowgraph provides a complete characterization of the system for neither provides the ordering relation to

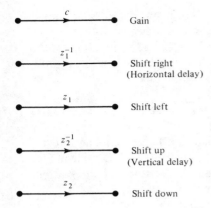

**Figure 4.20** Types of branches used for 2-D flowgraphs.

be used in realizing the system, but both are useful in trying to understand multidimensional recursive systems.

A flowgraph is a collection of *branches* that are directed connections between signal *nodes*. For any value of the ordered pair $(n_1, n_2)$, we can associate a numerical value with each node, which is determined by the branches entering the node and a few straightforward rules. There are five types of branches, which are shown diagrammatically in Figure 4.20. Each branch accepts an input value from the node at its tail and provides a value at its head. The value associated with any node is the sum of the values contributed from the branches that enter it.

The simplest branch is the gain. The value at the output of the branch is simply the value at the input multiplied by the gain of the branch. The other four types of branches are shift operators. If the input sample is $f(n_1, n_2)$, the output sample will be either $f(n_1 - 1, n_2)$, $f(n_1 + 1, n_2)$, $f(n_1, n_2 - 1)$, or $f(n_1, n_2 + 1)$, depending on the type of branch. These can be thought of as gains in the z-domain where the gains are $z_1^{-1}, z_1, z_2^{-1}$, and $z_2$, respectively, as indicated in Figure 4.20. Chan [5] has considered the use of more general shift branches, which can be obtained from our branches by means of linear transformations. This can often be useful if the output samples are computed in other than row-by-row or column-by-column fashion.

**Example 1**

As a simple example, consider the transfer function

$$H(z_1, z_2) = \frac{1}{1 - az_1^{-1} - bz_2^{-2}} \tag{4.64}$$

which corresponds to the difference equation

$$y(n_1, n_2) = x(n_1, n_2) + ay(n_1 - 1, n_2) + by(n_1, n_2 - 2) \tag{4.65}$$

This equation can be represented by the flowgraph shown in Figure 4.21.

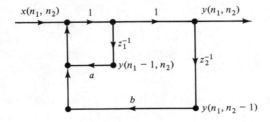

**Figure 4.21** Flowgraph representation of the system in Example 1.

**Example 2**

As a slightly more complicated example, consider the filter with transfer function

$$H(z_1, z_2) = \frac{1}{1 - \sum\limits_{\substack{n_1 = 0 \\ (n_1, n_2) \neq 0}}^{2} \sum\limits_{n_2 = 0}^{2} c(n_1, n_2) z_1^{-n_1} z_2^{-n_2}} \tag{4.66}$$

which corresponds to the difference equation

$$y(n_1, n_2) = x(n_1, n_2) + \sum_{\substack{k_1=0 \\ (k_1, k_2) \neq 0}}^{2} \sum_{k_2=0}^{2} c(k_1, k_2) y(n_1 - k_1, n_2 - k_2) \qquad (4.67)$$

With a little thought, this can be seen to be represented by the flowgraph shown in Figure 4.22.

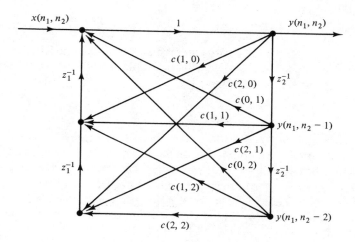

**Figure 4.22**  Flowgraph representation of the system in Example 2.

In the case of one-dimensional causal filters, flowgraphs can be used directly for implementation, since the shift operator can be realized as a single word of storage. In two dimensions, however, the shift operators are more difficult to implement. Consider the flowgraph shown in Figure 4.22. If the order of processing is chosen to be column by column, the output values will be computed in the order $y(n_1, n_2 - 2)$, $y(n_1, n_2 - 1)$, $y(n_1, n_2)$. Consequently, the $z_2^{-1}$ shift operators can be realized as a single word of storage. The $z_1^{-1}$ shift operators, however, need to buffer an entire column's worth of signal values, since the output of a $z_1^{-1}$ branch is $w(n_1 - 1, n_2)$ if its input is $w(n_1, n_2)$. The flowgraph, in its simplicity, obscures the problem of implementing both horizontal and vertical shift operators, and for that reason, its use is somewhat limited.

## 4.3 STABILITY OF RECURSIVE SYSTEMS

A system is stable if its output is well behaved for all reasonable inputs. Any transient behavior due to abrupt changes in the input array should be localized and the steady-state behavior of the system (i.e., the response for large values of the indices $n_1$ and $n_2$) should be predictable. To make this concept more formal, researchers have defined several different types of stability, but for LSI systems all of these essentially bound the impulse response in some fashion. The most extensively studied stability criterion is the bounded-input bounded-output (BIBO) criterion, which is the one that we will consider in this book in some detail. A system is stable in the BIBO sense if every bounded input sequence produces a bounded output sequence. That is, if the input signal satisfies

$$|x(n_1, n_2)| < P \qquad (4.68)$$

then the output signal must satisfy

$$|y(n_1, n_2)| < Q \qquad (4.69)$$

where $P$ and $Q$ are finite positive numbers. As we stated in Chapter 1, this implies that a linear shift-invariant system will be stable if and only if its impulse response is absolutely summable; that is,

$$\sum_{n_1=-\infty}^{\infty} \sum_{n_2=-\infty}^{\infty} |h(n_1, n_2)| = S_1 < \infty \qquad (4.70)$$

Although (4.70) is both correct and fundamental, it is not particularly useful. If it is to be used as a stability test, an infinite sum must be evaluated. Merely truncating the sum is unsatisfactory because a truncated sum will always be finite. Furthermore, (4.70) requires that the impulse response be available. Filter design algorithms usually provide either the coefficients of a difference equation or a filter transfer function. Thus we would like to be able to determine the stability of a filter directly from its transfer function.

### 4.3.1 Stability Theorems

If a 2-D sequence is absolutely summable, its $z$-transform will be analytic on the unit surface $|z_1| = |z_2| = 1$. The converse of this statement is also true: If the $z$-transform of a sequence is analytic on the unit surface, then the sequence is absolutely summable and the filter is stable. For the practical case of rational transfer functions

$$H_z(z_1, z_2) = \frac{A_z(z_1, z_2)}{B_z(z_1, z_2)} \qquad (4.71)$$

guaranteeing the analyticity of $H_z$ for $|z_1| = |z_2| = 1$ is equivalent to ensuring that $B_z \neq 0$ for $|z_1| = |z_2| = 1$. This, in turn, requires that the unit surface $|z_1| = |z_2| = 1$ lie in the region of convergence of the transfer function.

Although this condition is easy to test, the region of convergence for a transfer function is rarely given explicitly. Instead, we are more typically given the functional form of the transfer function and the region of support of the impulse response. The impulse response of a general, recursively computable filter has support only on a wedge-shaped sector. Earlier we saw that any sequence with wedge support could be linearly mapped into another sequence with first-quadrant support. Since the mapping is reversible, the original sequence will be absolutely summable if and only if the mapped sequence is absolutely summable [6]. The stability of any filter whose impulse response has wedge support can thus be equated to the stability of a filter with first-quadrant support. Consequently, we will restrict our attention for the moment to the stability of LSI systems with rational transfer functions whose impulse responses have support on the first quadrant of the $(n_1, n_2)$-plane.

The stability of a one-dimensional recursive filter is related to the locations of its poles. In the multidimensional case, stability is likewise related to the zero set of $B_z(z_1, z_2)$, the denominator polynominal. In some cases, however, the numerator polynominal can affect stability as well. This can happen when nonessential singularities of the second kind occur on the unit bicircle, a topic we defer until Section 4.3.3. For the moment, however, we can avoid the issue by assuming that $A_z(z_1, z_2) = 1$.

The earliest stability theorem, which is due to Shanks [7, 8], extends the idea of examining pole locations to the 2-D case.

**Theorem** (Shanks).    Let $H_z(z_1, z_2) = 1/B_z(z_1, z_2)$ be a first-quadrant recursive filter. This filter is stable if and only if $B_z(z_1, z_2) \neq 0$ for every point $(z_1, z_2)$ such that $|z_1| \geq 1$ or $|z_2| \geq 1$.

Although straightforward to understand, Shanks' theorem is difficult to use. It requires that the whole exterior of the unit bicircle be searched for points of singularity. An equivalent, but less well-known, result was implied by Shanks [7]. (See also [17].)

**Theorem** (Shanks).    Let $H_z(z_1, z_2) = 1/B_z(z_1, z_2)$ be a first-quadrant recursive filter. Then $H_z(z_1, z_2)$ is stable if and only if the following conditions are true:

(a) $B_z(z_1, z_2) \neq 0, |z_1| \geq 1, |z_2| = 1$
(b) $B_z(z_1, z_2) \neq 0, |z_1| = 1, |z_2| \geq 1$

A similar theorem was stated by Huang [9]. Correct proofs can be found in [10, 11].

**Theorem** (Huang).    Let $H_z(z_1, z_2) = 1/B_z(z_1, z_2)$ be a first-quadrant recursive filter. This filter is stable if and only if $B_z(z_1, z_2)$ satisfies the following two conditions:

(a) $B_z(z_1, z_2) \neq 0, |z_1| \geq 1, |z_2| = 1$
(b) $B_z(a, z_2) \neq 0, |z_2| \geq 1$ for any $a$ such that $|a| \geq 1$

The second condition of Huang's theorem is a 1-D stability condition; the first condition is 2-D, but $z_2$ is confined to its unit circle. The roles of $z_1$ and $z_2$ in this theorem can be interchanged. The majority of the implementations of stability tests between 1972 and 1977 were based on Huang's test. DeCarlo et al. [12] and Strintzis [13], however, independently showed that Huang's test could also be simplified. This third test is stated below.

**Theorem** (DeCarlo–Strintzis).    Let $H_z(z_1, z_2) = 1/B_z(z_1, z_2)$ be a first-quadrant recursive filter. This filter is stable if and only if $B(z_1, z_2)$ satisfies the following three conditions:

(a) $B_z(z_1, z_2) \neq 0$ for $|z_1| = 1$ and $|z_2| = 1$

(b) $B_z(a, z_2) \neq 0$ for $|z_2| \geq 1$ for any $a$ such that $|a| = 1$

(c) $B_z(z_1, b) \neq 0$ for $|z_1| \geq 1$ for any $b$ such that $|b| = 1$

Here again (b) and (c) correspond to 1-D stability conditions and (a) is a 2-D condition. DeCarlo *et al.* [12] and O'Connor [14] have also provided some alternative stability criteria in which conditions (b) and (c) are changed, but these are straightforward modifications. In particular, we can pick $a = b = 1$ for the DeCarlo–Strintzis theorem. It should be noted that condition (a) simply requires evaluation of the $z$-transform over the unit bicircle, which is the evaluation of a Fourier transform. This test can thus be interpreted as a generalization of the 1-D Nyquist stability criterion.

Although some reasonably sophisticated mathematics has been used to prove this theorem, we can get a heuristic feeling for how it works by simply looking at a root map of $B_z(z_1, z_2)$. Recall that a root map is the image of the unit circle of one variable (say $z_1$) in the plane of the other variable under the implicit mapping $B_z(z_1, z_2) = 0$. Every polynomial has two root maps—one which is the image in the $z_1$-plane and one which is the image in the $z_2$-plane. The conditions of Shanks's theorem (in the second form) will be satisfied if all the root images of a polynomial in each of its root maps lie within the unit circle.

In Figure 4.23 we have shown the two possible forms that a root map can assume and have categorized the corresponding denominator polynomials accordingly as class A or class B. For a polynomial of class A, at least one of the root images intersects the unit circle. For a polynomial of class B, none of the images intersects the unit circle; the root images (which must be closed curves) are thus either completely enclosed by or completely outside the unit circle. Clearly, root maps of class A correspond to unstable systems. For a system to be stable, it must be of class B and all the root images must lie inside the unit circle. Condition (a) of the DeCarlo–Strintzis theorem requires that the denominator polynomial belong to class B. In

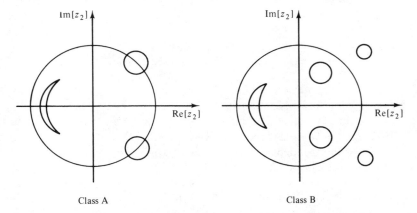

Class A                            Class B

**Figure 4.23**   Root maps for the two classes of 2-D polynomials.

addition, conditions (b) and (c) require that the root images be inside the respective unit circles of both maps. If a root image does not intersect the unit circle, all the points on the root image are either inside or outside the unit circle. Consequently, we need to examine only one point on each root image to determine if the entire root image is inside the unit circle. This can be done, for example, by examining the roots of the two 1-D polynomials $B_z(z_1, 1)$ and $B_z(1, z_2)$.

**Example 3**

As an example, consider the first-quadrant filter with the transfer function

$$H_z(z_1, z_2) = \frac{1}{1 - az_1^{-1} - bz_2^{-1}} \tag{4.72}$$

In Section 4.2 we saw that this filter has the impulse response

$$h(n_1, n_2) = \frac{(n_1 + n_2)!}{n_1! \, n_2!} a^{n_1} b^{n_2} u(n_1) u(n_2) \tag{4.73}$$

For this filter, $B(\omega_1, \omega_2) = 1 - ae^{-j\omega_1} - be^{-j\omega_2}$. The first condition of the DeCarlo–Strintzis theorem requires that

$$|1 - ae^{-j\omega_1} - be^{-j\omega_2}| > 0 \tag{4.74}$$

The second and third conditions require that the $z$-transforms $B_z(1, z_2)$ and $B_z(z_1, 1)$ have no roots outside the unit circle. This gives

$$|b| < |1 - a| \tag{4.75}$$

$$|a| < |1 - b| \tag{4.76}$$

A little manipulation of these inequalities reveals that (4.75) and (4.76) are special cases of (4.74) and that (4.74) is equivalent to requiring that

$$|a| + |b| < 1 \tag{4.77}$$

## *4.3.2  Stability Testing

In the preceding section we presented four stability theorems. These theorems by themselves have marginal utility if they cannot be used to generate numerical algorithms for testing stability. O'Connor [14] and Jury [15] discuss the implementation of stability tests in depth. Here we will confine our attention to the implementation of a test based on the DeCarlo–Strintzis theorem. In this respect, our discussion will follow closely the work of Shaw [16, 17].

The stability test we will examine in detail has three parts, corresponding to conditions (a), (b), and (c) of the DeCarlo–Strintzis theorem. Condition (b) consists of examining the root distribution of the 1-D polynomial

$$P_z(z_2) \triangleq B_z(1, z_2) = \sum_{n_2} \left[ \sum_{n_1} b(n_1, n_2) \right] z_2^{-n_2} \tag{4.78}$$

to ensure that $P_z(z_2) \neq 0$ for $|z_2| \leq 1$. This can be accomplished by using a Marden–Jury table [6, 18] or by applying the Argument Principle [16, 17]. The Argument Principle, stated loosely, says that the net change in the argument (phase) of $P_z(z_2)$ as $z_2$ moves around the unit circle once in a counterclockwise direction is equal to zero if and only if all of the roots of $P_z(z_2)$ are inside the unit circle. [We are assuming, as before, that $b(n_1, n_2)$ has first-quadrant support.] Condition (c) of the DeCarlo–Strintzis theorem consists of applying a similar test to the 1-D polynomial

$$Q_z(z_1) \triangleq B_z(z_1, 1) = \sum_{n_1} [\sum_{n_2} b(n_1, n_2)] z_1^{-n_1} \qquad (4.79)$$

Finally, assuming that $B_z(z_1, z_2)$ has passed these two 1-D tests, we must search for zeros of $B_z$ on the unit bicircle, or equivalently, we must search for zeros of the Fourier transform $B(\omega_1, \omega_2)$. One way to do this is to use the Argument Principle to look for root loci that cross the $z_2$ unit circle in the root map. Let us consider $z_1 = e^{j\omega_1}$ to be fixed and examine the 1-D parametric polynomial $B[z_1](z_2)$. Since $B[1](z_2) = P_z(z_2)$, we know that all the roots of $B[1](z_2)$ are inside the unit circle. Now, we can let $\omega_1$ (and consequently $z_1$) take on discrete values and check the distribution of zeroes of $B[z_1](z_2)$ by applying the Argument Principle. Conceptually, we are sampling the loci of the roots on the $z_2$ root map. If the net change in the phase of $B[z_1](z_2)$ is nonzero after $z_2$ has moved around the unit circle once, then at least one root has moved outside the unit circle. To get outside, it must have crossed the unit circle at some point, meaning that $B(\omega_1, \omega_2) = 0$ at some point. Consequently, the filter is unstable. If we thoroughly test $B[z_1](z_2)$ in this manner, there is no need to also test $B[z_2](z_1)$ since an intersection between a root locus and the $z_2$ unit circle implies and is implied by the intersection of a root locus and the $z_1$ unit circle.

In practice, $\omega_1$ is allowed to take on the values $2\pi k_1 / N_1$ so that the coefficients of the parametric polynomial $B[z_1](z_2)$ can be computed with a fast Fourier transform. $N_1$ must be chosen large enough, implying that the root map is sampled often enough, to avoid missing a root locus that sneaks over the unit circle and back. With this test, there is still a potential problem of a root locus becoming tangent to the unit circle, thus making $B(\omega_1, \omega_2) = 0$ at some point, without actually crossing the unit circle [16, 17].

To apply the Argument Principle, we must compute the phase of a 1-D polynominal $P_z(z_2)$ as $z_2$ moves around the unit circle, or equivalently, the phase of the 1-D Fourier transform $P(\omega_2)$. Ordinarily, the phase function can be defined as

$$\text{ARG} \, [P(\omega_2)] \triangleq \tan^{-1} \frac{\text{Im} \, [P(\omega_2)]}{\text{Re} \, [P(\omega_2)]} \qquad (4.80a)$$

or

$$\text{ARG} \, [P(\omega_2)] \triangleq \text{Im} \, [\ell n \, P(\omega_2)] \qquad (4.80b)$$

In both of the expressions above, the principal value is usually taken since the inverse tangent and complex logarithm functions are multivalued. Use of the principal value, however, can introduce artificial discontinuities of $2\pi$ in ARG $[P(\omega_2)]$, making it impossible to measure the net change in the phase of $P(\omega_2)$ as $\omega_2$ goes from 0 to $2\pi$.

We can circumvent this problem by defining an *unwrapped* phase function $\phi(\omega_2)$ as the integral of the phase derivative. For notational convenience, let us define

$$P_R(\omega_2) \triangleq \text{Re} \, [P(\omega_2)]; \qquad P_I(\omega_2) \triangleq \text{Im} \, [P(\omega_2)]$$

Then, by formally computing the derivative of the right side of (4.80a) or (4.80b) and equating it to the derivative of $\phi(\omega_2)$, we get

$$\frac{d\phi(\omega_2)}{d\omega_2} = \frac{P_R(\omega_2)\dfrac{dP_I(\omega_2)}{d\omega_2} - P_I(\omega_2)\dfrac{dP_R(\omega_2)}{d\omega_2}}{P_R^2(\omega_2) + P_I^2(\omega_2)} \tag{4.81}$$

The unwrapped phase function $\phi(v_2)$ at a particular frequency $v_2$ is computed by integration.

$$\phi(v_2) = \phi(0) + \int_0^{v_2} \frac{d\phi(\omega_2)}{d\omega_2}\, d\omega_2 \tag{4.82}$$

Because $P(\omega_2)$ is a trigonometric polynomial, the derivative of $\phi(\omega_2)$, and consequently $\phi(\omega_2)$ itself, will be continuous as long as $P(\omega_2) \neq 0$. Tribolet [19] has developed a computationally efficient algorithm for performing this calculation.

Now, the Argument Principle can be applied to test the stability of $1/B_z(z_1, z_2)$. Condition (b) of the DeCarlo–Strintzis theorem will be true if the unwrapped phase of the polynomial $P(\omega_2) = B(0, \omega_2)$ shows no net change after $\omega_2$ has gone from zero to $2\pi$. This is equivalent to saying that the unwrapped phase function is periodic. Similarly, condition (c) will be satisfied if the unwrapped phase of the polynomial $Q(\omega_1) = B(\omega_1, 0)$ is periodic. Finally, condition (a) will be satisfied if the unwrapped phase of the parametric trigonometric polynomial $B[\omega_1](\omega_2)$ is periodic for every real value of $\omega_1$.

In cases where $B(\omega_1, \omega_2)$ comes very close to zero, the 1-D phase unwrapping algorithm may fail. In this case, the stability testing algorithm developed by Shaw [16, 17] performs a localized search for the minimum of $|B(\omega_1, \omega_2)|$ to see if it actually attains zero. If it does not, the filter is theoretically stable although, in practice, it may be sensitive to coefficient errors or other finite-wordlength problems that render it unusable.

This series of 1-D tests for phase periodicity can be interpreted as computing a 2-D unwrapped phase function for $B(\omega_1, \omega_2)$ and testing to see that it is continuous and doubly periodic [6, 14]. We can define the 2-D unwrapped phase function

$$\phi(v_1, v_2) = \phi(0, 0) + \int_0^{v_2} \frac{\partial\phi(0, \omega_2)}{\partial\omega_2}\, d\omega_2 + \int_0^{v_1} \frac{\partial\phi(\omega_1, v_2)}{\partial\omega_1}\, d\omega_1 \tag{4.83}$$

where the partial derivatives are given by formulas analogous to (4.81). This function will be continuous as long as $B(\omega_1, \omega_2) \neq 0$. If it is also periodic; that is, if

$$\phi(\omega_1 + 2\pi, \omega_2) = \phi(\omega_1, \omega_2) \tag{4.84a}$$

$$\phi(\omega_1, \omega_2 + 2\pi) = \phi(\omega_1, \omega_2) \tag{4.84b}$$

then condition (b) of the DeCarlo–Strintzis theorem is satisfied since it is equivalent to

$$\phi(0, \omega_2 + 2\pi) = \phi(0, \omega_2) \tag{4.85}$$

Similarly, condition (c) is satisfied since it is equivalent to

$$\phi(\omega_1 + 2\pi, 0) = \phi(\omega_1, 0) \tag{4.86}$$

Condition (a) is satisfied directly by the continuity of $\phi(\omega_1, \omega_2)$, which requires that $B(\omega_1, \omega_2) \neq 0$. This continuity, in turn, is checked by applying the Argument Principle to either of the parametric polynomials $B[\omega_1](\omega_2)$ or $B[\omega_2](\omega_1)$ to look for violations of (4.84a) or (4.84b). Consequently, the stability theorem can be stated in terms of the phase function as follows [6, 14]:

**Theorem** (O'Connor)   A 2-D recursive filter with a frequency response $H(\omega_1, \omega_2) = 1/B(\omega_1, \omega_2)$ is stable if the unwrapped phase function $\phi(\omega_1, \omega_2)$ is continuous and doubly periodic.

This theorem does not require that $b(n_1, n_2)$ have support only on the first quadrant. As long as $b(n_1, n_2)$ can be mapped to the first quadrant by a linear one-to-one mapping, the theorem will hold since a linear mapping will not affect the continuity or periodicity of the unwrapped phase.

This implementation of the stability test represents a set of necessary conditions. It is possible that an unstable filter could pass this test if the root loci are not sampled finely enough to detect excursions across the unit circle, or if a root locus intersects but does not cross the unit circle. However, the degree of assurance that a filter is, in fact, stable rests with the user who picks the sampling density in the stability test.

Stability tests that use algebraic methods [15] to test for positive definiteness can theoretically determine the stability of a filter without question. In practice, however, these methods are complicated to implement and can suffer from computational noise due to finite-wordlength arithmetic which can call their conclusions into question [6, 14, 16, 17].

### 4.3.3 Effect of the Numerator Polynomial on Stability

In the 1-D case, if the numerator and denominator polynomials of a transfer function have no common factors, the stability of a filter depends only on the locations of its poles. In the case of 2-D filters, however, the numerator can affect the stability of the filter. As an example, consider the following three transfer functions (from Goodman [20]):

$$F_z(z_1, z_2) = \frac{1}{1 - \frac{1}{2}z_1^{-1} - \frac{1}{2}z_2^{-1}} \tag{4.87}$$

$$G_z(z_1, z_2) = \frac{(1 - z_1^{-1})^8(1 - z_2^{-1})^8}{1 - \frac{1}{2}z_1^{-1} - \frac{1}{2}z_2^{-1}} \tag{4.88}$$

$$H_z(z_1, z_2) = \frac{(1 - z_1^{-1})(1 - z_2^{-1})}{1 - \frac{1}{2}z_1^{-1} - \frac{1}{2}z_2^{-1}} \tag{4.89}$$

$F_z$ is unstable, from the example of the preceding section. $G_z$ and $H_z$ have the same denominator polynomial as $F_z$, yet $G_z$ is stable and $H_z$ is not. If we look a little more closely we see that the only point on or inside the bicircle where the denominator polynomials are zero is at the point $z_1 = z_2 = 1$. The numerator polynomials of both $G_z$ and $H_z$ are also zero at this point. Thus $G_z$ and $H_z$ both possess a nonessential

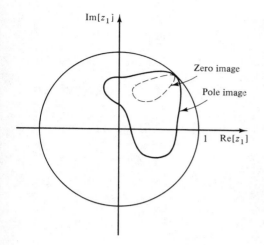

**Figure 4.24** Nonessential singularity of the second kind on the unit bicircle.

singularity of the second kind on the unit bicircle. This situation is depicted graphically in Figure 4.24. One of the pole images and one of the zero images of the pole and zero root maps are tangent at a point which is on the unit bicircle. Heuristically, filter stability is related to the degree of tangency. At this time, there is no general, straightforward means of determining whether a transfer function which has no poles outside the unit bicircle and which has a nonessential singularity of the second kind on the unit bicircle is stable or not. Goodman [20], however, did establish the stability of (4.88) and the instability of (4.89).

The stability theorems of the preceding sections are simply sufficient conditions for stability when there is a nontrivial numerator polynomial. They are necessary conditions for stability only if there are no nonessential singularities of the second kind on the unit bicircle.

### 4.3.4 Multidimensional Stability Theorems

All of the stability theorems from the preceding sections can be generalized to the multi-dimensional case. Although the statement of these theorems is straightforward, their implementation is not. The volume of computation and the necessity for preserving numerical accuracy make the implementation of these higher-dimensional stability theorems a true test of the programmer's art. As we saw in the 2-D case, it is only necessary to consider the equivalent of first-quadrant tests, since any filter with $M$-dimensional wedge support can be mapped into a filter with support on the $M$-dimensional analog of the first quadrant.

**Theorem** (Justice and Shanks [8]).    Let $H_z(z_1, z_2, \ldots, z_M) = 1/B_z(z_1, z_2, \ldots, z_M)$ be a "first-quadrant" recursive filter; that is, $h(n_1, n_2, \ldots, n_M) = 0$ unless $n_1 \geq 0$, $n_2 \geq 0, \ldots, n_M \geq 0$. This filter is stable if and only if $B(z_1, z_2, \ldots, z_M) \neq 0$ for any point $(z_1, z_2, \ldots, z_M)$ such that $|z_1| \geq 1$ or $|z_2| \geq 1$ or $\ldots$ or $|z_M| \geq 1$.

Huang's 2-D stability theorem was generalized to the multidimensional case by Anderson and Jury [21].

**Theorem** (Anderson and Jury).    Let $H_z(z_1, z_2, \ldots, z_M) = 1/B_z(z_1, z_2, \ldots, z_M)$ be a "first-quadrant" recursive filter. This filter is stable if and only if the following hold:

$$B_z(z_1, 0, \ldots, 0) \neq 0 \quad \text{when } |z_1| \geq 0$$
$$B_z(z_1, z_2, \ldots, 0) \neq 0 \quad \text{when } |z_1| = 1 \text{ and } |z_2| \geq 0$$

.
.
.

$$B_z(z_1, z_2, \ldots, z_{M-1}, 0) \neq 0 \quad \text{when } |z_1| = \cdots = |z_{M-2}| = 1$$
$$\text{and } |z_{M-1}| \geq 0$$
$$B_z(z_1, z_2, \ldots, z_{M-1}, z_M) \neq 0 \quad \text{when } |z_1| = \cdots = |z_{M-1}| = 1$$
$$\text{and } |z_M| \geq 0$$

The $M$-dimensional version of the DeCarlo–Strintzis test is very similar to the 2-D version. It was discovered by both DeCarlo *et al.* [12] and Strintzis [13].

**Theorem** (DeCarlo–Strintzis).   Let $H_z(z_1, z_2, \ldots, z_M) = 1/B_z(z_1, z_2, \ldots, z_M)$ be a "first-quadrant" recursive filter. This filter is stable if and only if the following hold:

(a) $B_z(z_1, z_2, \ldots, z_M) \neq 0$ for $|z_1| = 1, |z_2| = 1, \ldots, |z_M| = 1$

(b) $B_z(1, 1, \ldots, 1, z_k, 1, \ldots, 1) \neq 0$ for $|z_k| \geq 1$ for each $k = 1, 2, \ldots, M$. Equivalently, the unwrapped phase function $\phi(\omega_1, \ldots, \omega_M)$ must be periodic and continuous.

This theorem suggests a stability test consisting of $M$ 1-D stability tests plus a search for zeroes over the $M$-dimensional unit surface $|z_1| = |z_2| = \cdots = |z_M| = 1$.

If $H_z(z_1, z_2, \ldots, z_M)$ has a numerator polynomial $A_z(z_1, z_2, \ldots, z_M)$ which is coprime with $B(z_1, z_2, \ldots, z_M)$, then these three theorems give sufficient conditions for stability. They are necessary if there are no nonessential singularities of the second kind on the $M$-dimensional unit surface.

## 4.4 TWO-DIMENSIONAL COMPLEX CEPSTRUM

At this point it is appropriate to discuss the multidimensional complex cepstrum. Like its one-dimensional counterpart [22], the multidimensional complex cepstrum is a sequence obtained by performing an inverse $z$-transform on the complex logarithm of the $z$-transform of a sequence. The term "cepstrum" (pronounced kep'strm) was coined [23] to indicate that it is the inverse of a logarithmic *spec*trum. The complex cepstrum (or simply "cepstrum" for short) is potentially useful in multidimensional filtering and deconvolution problems, but more important, it provides an analytic tool that has been found useful in studying the stability and factorability of multidimensional transfer functions.

### 4.4.1 Definition of the Complex Cepstrum

Let $x(n_1, n_2)$ be a 2-D sequence with a $z$-transform $X_z(z_1, z_2)$ which converges in some region of convergence $R$. Then the 2-D complex cepstrum, denoted $\hat{x}(n_1, n_2)$, is defined as the inverse $z$-transform of $\ell n\, X_z(z_1, z_2)$; that is,

$$\hat{x}(n_1, n_2) = \frac{1}{(2\pi j)^2} \oint \oint \ell n \, X_z(z_1, z_2) z_1^{n_1-1} z_2^{n_2-1} \, dz_1 \, dz_2 \qquad (4.90)$$

Note that the expression $\ell n \, X_z(z_1, z_2)$ involves a complex logarithm, which is multi-valued because of the ambiguity of $2\pi$ in its imaginary part. In order to be able to compute the inverse $z$-transform, the function $\ell n \, X_z(z_1, z_2)$ must be analytic in some region. This implies that we must define the complex logarithm function so that $\ell n \, X_z(z_1, z_2)$ is single-valued, continuous, and differentiable. We shall attack this problem in Section 4.4.2.

The complex cepstrum of two signals combined by convolution is the sum of the complex cepstra of the two signals. For example, let

$$y(n_1, n_2) = \sum_{k_1} \sum_{k_2} h(k_1, k_2) x(n_1 - k_1, n_2 - k_2) \qquad (4.91)$$

Then

$$Y_z(z_1, z_2) = H_z(z_1, z_2) X_z(z_1, z_2) \qquad (4.92)$$

so that

$$\ell n \, Y_z(z_1, z_2) = \ell n \, H_z(z_1, z_2) + \ell n \, X_z(z_1, z_2) \qquad (4.93)$$

Then, by applying the inverse $z$-transform, we see that

$$\hat{y}(n_1, n_2) = \hat{h}(n_1, n_2) + \hat{x}(n_1, n_2) \qquad (4.94)$$

Because of this property, the complex cepstrum is useful in studying filter transfer functions that are products of factors. For example, if

$$H_z(z_1, z_2) = \frac{A_z(z_1, z_2) C_z(z_1, z_2)}{B_z(z_1, z_2) D_z(z_1, z_2)} \qquad (4.95)$$

then

$$\hat{h}(n_1, n_2) = \hat{a}(n_1, n_2) - \hat{b}(n_1, n_2) + \hat{c}(n_1, n_2) - \hat{d}(n_1, n_2) \qquad (4.96)$$

For a separable transfer function of the form

$$H_z(z_1, z_2) = F_z(z_1) G_z(z_2) \qquad (4.97)$$

it can be shown that the complex cepstrum has the form

$$\hat{h}(n_1, n_2) = \hat{f}(n_1) \delta(n_2) + \hat{g}(n_2) \delta(n_1) \qquad (4.98)$$

### 4.4.2 Existence of the Complex Cepstrum [24]

Not every 2-D sequence $x(n_1, n_2)$ will possess a complex cepstrum. For the complex cepstrum to exist, the function $\ell n \, X_z(z_1, z_2)$ must be analytic in some region of convergence $\hat{R}$. This implies that $\ell n \, X_z(z_1, z_2)$ must be continuous, differentiable, and be periodic as $z_1$ and $z_2$ trace out closed contours. Let us look at the special case where the contour integrals in equation (4.90) are taken around the unit bicircle $|z_1| = |z_2| = 1$. Then (4.90) becomes

$$\hat{x}(n_1, n_2) = \frac{1}{(2\pi)^2} \int_0^{2\pi} \int_0^{2\pi} \ell n \, X(\omega_1, \omega_2) \exp(j\omega_1 n_1 + j\omega_2 n_2) \, d\omega_1 \, d\omega_2 \qquad (4.99)$$

The function $\hat{X}(\omega_1, \omega_2) = \ell n\, X(\omega_1, \omega_2)$ must be continuous, differentiable, and doubly periodic in $\omega_1$ and $\omega_2$. By writing $X(\omega_1, \omega_2)$ in polar form, we see that

$$\hat{X}(\omega_1, \omega_2) = \ell n\, |X(\omega_1, \omega_2)| + j\phi(\omega_1, \omega_2) \qquad (4.100)$$

$\hat{X}(\omega_1, \omega_2)$ will be continuous and differentiable if $\phi(\omega_1, \omega_2)$ is taken to be the un-wrapped phase function (Section 4.3.2) and if $X(\omega_1, \omega_2)$ is finite and not equal to zero. Requiring $\hat{X}(\omega_1, \omega_2)$ to be doubly periodic is thus equivalent to requiring that the unwrapped phase function $\phi(\omega_1, \omega_2)$ be doubly periodic. We shall show that if it is not periodic, it can be written as the sum of a periodic phase component and a linear phase component. The linear phase component can be removed by creating a new sequence $y(n_1, n_2)$, which is simply a shifted version of $x(n_1, n_2)$. Since the unwrapped phase function for the sequence $y(n_1, n_2)$ is periodic and continuous, we can define a complex cepstrum for $y(n_1, n_2)$.

For simplicity, let $x(n_1, n_2)$ be a finite-extent array so that $X(\omega_1, \omega_2)$ is simply a trigonometric polynomial. We shall assume that $X(\omega_1, \omega_2) \neq 0$. Then, by considering the parametric function $X[\omega_2](\omega_1)$, we can ask what the net change in the unwrapped phase is as $\omega_1$ varies from zero to $2\pi$ for a fixed value of $\omega_2$. The Argument Principle tells us that the net change will be $2\pi K_1(\omega_2)$, where $K_1(\omega_2)$ is an integer related to the number of roots of $X_z[e^{j\omega_2}](z_1)$ inside the $z_1$ unit circle. Thus

$$\phi_X(2\pi, \omega_2) = \phi_X(0, \omega_2) + 2\pi K_1(\omega_2) \qquad (4.101)$$

Similarly, we can derive the expression

$$\phi_X(\omega_1, 2\pi) = \phi_X(\omega_1, 0) + 2\pi K_2(\omega_1) \qquad (4.102)$$

If $K_1(\omega_2)$ is not constant as a function of $\omega_2$, then the number of roots of $X_z[e^{j\omega_2}](z_1)$ inside the $z_1$ unit circle must have changed as $\omega_2$ was varied. This could happen only if a root moved from inside to outside the unit circle, or vice versa. In either case, the root would have to lie on the unit circle at some point since the roots move in a continuous manner as $\omega_2$ is varied. This, however, would violate our original assumption that $X(\omega_1, \omega_2) \neq 0$. Consequently, $K_1(\omega_2)$ and, by analogy, $K_2(\omega_1)$ are integer constants which we shall denote simply by $K_1$ and $K_2$. Now, let us consider the sequence

$$y(n_1, n_2) \triangleq x(n_1 - K_1, n_2 - K_2) \qquad (4.103)$$

which is simply a shifted version of $x(n_1, n_2)$. The Fourier transform of $y(n_1, n_2)$ is given by

$$Y(\omega_1, \omega_2) = X(\omega_1, \omega_2) \exp{(-j\omega_1 K_1 - j\omega_2 K_2)} \qquad (4.104)$$

Consequently, the unwrapped phase of $Y(\omega_1, \omega_2)$ is given by

$$\phi_Y(\omega_1, \omega_2) = \phi_X(\omega_1, \omega_2) - \omega_1 K_1 - \omega_2 K_2$$

so that $\phi_Y(\omega_1, \omega_2)$ satisfies

$$\phi_Y(2\pi, \omega_2) = \phi_Y(0, \omega_2) \qquad (4.105a)$$

$$\phi_Y(\omega_1, 2\pi) = \phi_Y(\omega_1, 0) \qquad (4.105b)$$

More generally, $\phi_Y(\omega_1, \omega_2)$ can be shown to be continuous and doubly periodic, so that $Y(\omega_1, \omega_2)$ satisfies the conditions necessary to define the complex cepstrum $\hat{y}(n_1, n_2)$.

$$\hat{y}(n_1, n_2) = \frac{1}{4\pi^2} \int_{-\pi}^{\pi} \int_{-\pi}^{\pi} \{\ell n \mid Y(\omega_1, \omega_2)\mid + j\phi_Y(\omega_1, \omega_2)\} \exp{(j\omega_1 n_1 + j\omega_2 n_2)}\, d\omega_1\, d\omega_2$$

(4.106)

### 4.4.3 Causality, Minimum Phase, and the Complex Cepstrum

In one-dimensional signal processing, the concept of the minimum-phase (or more descriptively, minimum-delay [22]) signal is often useful. These signals have a number of interesting properties: they are causal; they have all their poles and zeros inside the unit circle; they have most of their energy close to the origin (minimum-delay); they are absolutely summable; and they have inverses that are both causal and absolutely summable. A 1-D minimum-phase signal can be shown to have a complex cepstrum which is also causal and absolutely summable.

Maximum-phase signals can also be defined. Simply, they are minimum-phase signals reversed in time. A maximum-phase signal is anticausal with an anticausal inverse and an anticausal complex cepstrum.

In general, any absolutely summable signal, if appropriately shifted in time, can be written as the convolution of a minimum-phase signal with a maximum-phase signal. Consequently, its complex cepstrum will be the sum of a causal part and an anticausal part, and it will be absolutely summable.

Causality and anticausality can be interpreted as 1-D regions of support. It can be stated that, in the 1-D case, if an absolutely summable signal has an absolutely summable inverse that occupies the same region of support, the complex cepstrum will also occupy that same region of support. Conversely, an absolutely summable cepstrum with support in some region corresponds to an absolutely summable signal and an absolutely summable inverse with support in the same region. For minimum-phase signals, this region is the nonnegative part of the time axis; for maximum-phase signals, it is the nonpositive part of the time axis.

The relationship between regions of support of a signal, its inverse, and its cepstrum are straightforwardly extended to the 2-D case [25]. Let $\hat{X}(\omega_1, \omega_2) \triangleq \ell n\, X(\omega_1, \omega_2)$. Then

$$\frac{\partial \hat{X}(\omega_1, \omega_2)}{\partial \omega_1} = \frac{1}{X(\omega_1, \omega_2)} \frac{\partial X(\omega_1, \omega_2)}{\partial \omega_1}$$

(4.107)

By taking the inverse Fourier transform of (4.107), we see that

$$n_1 \hat{x}(n_1, n_2) = \sum_{k_1} \sum_{k_2} k_1 x(k_1, k_2) v(n_1 - k_1, n_2 - k_2)$$

(4.108)

where $v(n_1, n_2)$ is the inverse of $x(n_1, n_2)$ obtained by evaluating the inverse Fourier transform of $1/X(\omega_1, \omega_2)$. Similarly, we can derive the expression

$$n_2 \hat{x}(n_1, n_2) = \sum_{k_1} \sum_{k_2} k_2 x(k_1, k_2) v(n_1 - k_1, n_2 - k_2)$$

(4.109)

By examining equations (4.108) and (4.109) closely, we see that the region of support of $\hat{x}(n_1, n_2)$ is determined by the regions of support of $x(n_1, n_2)$ and its inverse $v(n_1, n_2)$. For example, if both $x(n_1, n_2)$ and $v(n_1, n_2)$ have support only in the first quadrant, then so will $\hat{x}(n_1, n_2)$ since a first-quadrant sequence convolved with another first-quadrant sequence must yield a first-quadrant sequence. Similar geometric arguments about the region of support of the convolution of two signals can be made for the other quadrants and half-planes, as well as for wedge-shaped regions of support.

It is also simple to demonstrate that the region of support of the complex cepstrum implies the region of support of the signal and its inverse [2, 25]. Because

$$X(\omega_1, \omega_2) = \exp[\hat{X}(\omega_1, \omega_2)] \tag{4.110}$$

we can use the series expansion for the exponential function to derive

$$X(\omega_1, \omega_2) = \sum_{k=0}^{\infty} \frac{1}{k!} \hat{X}^k(\omega_1, \omega_2) \tag{4.111}$$

Then, taking the inverse Fourier transform of both sides, we get

$$x(n_1, n_2) = \sum_{k=0}^{\infty} \frac{1}{k!} \hat{c}_k(n_1, n_2) \tag{4.112}$$

where $\hat{c}_k(n_1, n_2)$ is the sequence obtained by convolving $\hat{x}(n_1, n_2)$ with itself $k$ times. Consequently, the region of support of $\hat{x}(n_1, n_2)$ convolved with itself an infinite number of times is the region of support of $x(n_1, n_2)$. Since

$$V(\omega_1, \omega_2) = \frac{1}{X(\omega_1, \omega_2)} = \exp[-\hat{X}(\omega_1, \omega_2)] \tag{4.113}$$

the region of support of $v(n_1, n_2)$ can be expressed in the same way.

We can define a multidimensional minimum-phase signal as one that is absolutely summable and whose inverse and complex cepstrum are absolutely summable and have the same region of support. This region of support should be a convex region such as a quadrant, wedge, nonsymmetric half-plane, or proper half-plane. In multidimensional signal processing, the region of support must be a part of any definition of minimum phase.

### 4.4.4 Spectral Factorization [2]

The cepstrum is useful in providing one approach to the difficult problem of 2-D spectral factorization. This problem arises in the design and implementation of IIR filters (which we shall study in detail in Chapter 5). The spectral factorization problem may be stated in several ways. In the signal domain, we may describe it in terms of a real, symmetric finite-extent autocorrelation sequence $r(n_1, n_2)$ whose Fourier transform $R(\omega_1, \omega_2)$ is positive definite. Ideally, we seek a real, finite-extent, minimum-phase sequence $b(n_1, n_2)$ such that

$$r(n_1, n_2) = b(n_1, n_2) ** b(-n_1, -n_2) \tag{4.114}$$

Equivalently, in the Fourier domain, we wish to find $B(\omega_1, \omega_2)$ such that

$$R(\omega_1, \omega_2) = |B(\omega_1, \omega_2)|^2 \qquad (4.115)$$

Were it not for the restriction that $b(n_1, n_2)$ be a minimum-phase sequence, we could set

$$B(\omega_1, \omega_2) = \sqrt{R(\omega_1, \omega_2)} \exp[j\phi(\omega_1, \omega_2)] \qquad (4.116)$$

where $\phi(\omega_1, \omega_2)$ is an arbitrary phase function. To satisfy the restriction, however, we must choose the minimum-phase phase function $\phi_{mp}(\omega_1, \omega_2)$ and even that, in general, will not guarantee that $b(n_1, n_2)$ will have finite extent. We can also state the spectral factorization problem in the $z$-domain by taking $z$-transform of equation (4.114) to get the equation

$$R_z(z_1, z_2) = B_z(z_1, z_2)B_z(z_1^{-1}, z_2^{-1}) \qquad (4.117)$$

for which we seek a solution $B_z(z_1, z_2)$ which is a minimum-phase 2-D polynomial.

The 1-D spectral factorization problem may be stated in a similar way, but unlike its 2-D counterpart, the 1-D problem may be solved by applying the Fundamental Theorem of Algebra. Let us dwell on this point briefly, since it represents a significant difference between 1-D and 2-D digital signal processing. In the 1-D case, we are given a symmetric polynomial of the form

$$R_z(z) = \sum_{n=-N+1}^{N-1} r(n)z^{-n} \qquad (4.118)$$

with the properties $R_z(z) = R_z(z^{-1})$ and $R_z(e^{j\omega}) > 0$. We seek a minimum-phase polynomial $B_z(z)$ of the form

$$B_z(z) = \sum_{n=0}^{N-1} b(n)z^{-n} \qquad (4.119)$$

such that

$$R_z(z) = B_z(z)B_z(z^{-1}) \qquad (4.120)$$

The Fundamental Theorem of Algebra permits us to write $R_z(z)$ in the factored form

$$R_z(z) = A \prod_{i=1}^{N-1} (1 - q_i z^{-1})(1 - q_i z) \qquad (4.121)$$

where $A$ is a positive real constant and the roots $\{q_i\}$ satisfy

$$|q_i| < 1 \qquad (4.122)$$

The desired 1-D minimum-phase spectral factor $B_z(z)$ is obtained simply choosing the roots of (4.121) that lie inside the unit circle $|z| = 1$. Thus

$$B_z(z) = \sqrt{A} \prod_{i=1}^{N-1} (1 - z^{-1}q_i) \qquad (4.123)$$

which satisfies (4.120). Note that the sequence $b(n)$ corresponding to the inverse $z$-transform of (4.123) will be causal, minimum-phase, and have finite extent as desired.

Because of the general inability to factor 2-D polynomials and obtain polynomial factors, the foregoing spectral factorization procedure cannot be extended to two dimensions. For example, consider the 2-D positive real function

$$R(\omega_1, \omega_2) = 5 + 2\cos\omega_1 + 2\cos\omega_2 > 0 \tag{4.124}$$

or the corresponding $z$-domain polynomial

$$R_z(z_1, z_2) = 5 + z_1 + z_1^{-1} + z_2 + z_2^{-1} \tag{4.125}$$

It is simply not possible to find a 2-D polynomial of the form

$$B_z(z_1, z_2) = a + bz_1^{-1} + cz_2^{-1} + dz_1^{-1}z_2^{-1} \tag{4.126}$$

(or any other finite form for that matter) such that

$$R_z(z_1, z_2) = B_z(z_1, z_2)B_z(z_1^{-1}, z_2^{-1}) \tag{4.127}$$

However, there does exist a *transcendental factorization* of $R_z(z_1, z_2)$ that satisfies (4.127) but results in a factor $B_z(z_1, z_2)$ that has an infinite number of terms.

The transcendental factorization of $R_z(z_1, z_2)$ is derived by expressing (4.114) in the cepstral domain and making use of the properties of the complex cepstrum described earlier. We see that the cepstrum of $r(n_1, n_2)$ may be expressed as

$$\hat{r}(n_1, n_2) = \hat{b}(n_1, n_2) + \hat{b}(-n_1, -n_2) \tag{4.128}$$

where $\hat{r}(n_1, n_2)$ may be computed by

$$\hat{r}(n_1, n_2) = \frac{1}{4\pi^2} \int_{-\pi}^{\pi} \int_{-\pi}^{\pi} \ell n \left[ R(\omega_1, \omega_2) \right] \exp\left( j\omega_1 n_1 + j\omega_2 n_2 \right) d\omega_1 \, d\omega_2 \tag{4.129}$$

Since $R(\omega_1, \omega_2)$ is real and positive, there is no ambiguity in specifying its logarithm; the problem of phase unwrapping associated with the complex logarithm does not arise in this context.

We saw in Section 4.4.3 that a 2-D sequence $b(n_1, n_2)$ with quadrant support will be a minimum-phase sequence if its cepstrum $\hat{b}(n_1, n_2)$ has quadrant support. A similar result holds for a sequence with nonsymmetric half-plane support. For example, if

$$\hat{b}(n_1, n_2) = 0 \quad\quad \text{for } n_2 < 0 \text{ and}$$
$$\text{for } n_2 = 0, n_1 < 0 \tag{4.130}$$

then $b(n_1, n_2)$ is a minimum-phase sequence such that

$$b(n_1, n_2) = 0 \quad\quad \text{for } n_2 < 0 \text{ and}$$
$$\text{for } n_2 = 0, n_1 < 0 \tag{4.131}$$

If we require that $\hat{b}(n_1, n_2)$ have the nonsymmetric half-plane form (4.130), $\hat{b}(n_1, n_2)$ and $\hat{b}(-n_1, -n_2)$ will overlap only at the origin. From equation (4.128), we see that

$$\hat{b}(n_1, n_2) = \begin{cases} \hat{r}(n_1, n_2) & \text{for } n_2 > 0 \text{ and} \\ & \text{for } n_2 = 0, n_1 > 0 \\ \frac{1}{2}\hat{r}(0, 0) & \text{for } n_1 = n_2 = 0 \\ 0 & \text{otherwise} \end{cases} \tag{4.132}$$

The sequence $b(n_1, n_2)$, corresponding to the cepstrum $\hat{b}(n_1, n_2)$ given by (4.132), will be a minimum-phase sequence with nonsymmetric half-plane support, but there is no guarantee that $b(n_1, n_2)$ will have finite extent. On the other hand, if $r(n_1, n_2)$ were constructed by computing the autocorrelation function of a minimum-phase finite-extent sequence $b(n_1, n_2)$, the 2-D spectral factorization procedure outlined above will reproduce $b(n_1, n_2)$ exactly from $r(n_1, n_2)$.

One further point is worth noting. If $b(n_1, n_2)$ is a minimum-phase finite-extent sequence with support on the first quadrant only, the cepstrum $\hat{r}(n_1, n_2)$ of the auto-correlation of $b(n_1, n_2)$ will be identically zero in the second and fourth quadrants.

$$\hat{r}(n_1, n_2) = 0 \quad \text{for } n_2 > 0 \text{ and } n_1 < 0$$
$$\text{and for } n_1 > 0 \text{ and } n_2 < 0 \tag{4.133}$$

Consequently, equation (4.133) represents a necessary condition for the existence of a spectral factor corresponding to a $b(n_1, n_2)$ with finite extent and first-quadrant support. Since wedge-shaped regions of support may be mapped to the first quadrant with a linear transformation, a similar necessary condition exists for $b(n_1, n_2)$ to be a minimum-phase finite-extent sequence with support on a wedge.

### *4.4.5 Computing the 2-D Complex Cepstrum [26]

We shall discuss two methods of computing the complex cepstrum of a 2-D discrete signal $x(n_1, n_2)$. The first method uses the discrete Fourier transform implemented by an FFT algorithm to approximate the necessary forward and inverse z-transforms using the unit bicircle $|z_1| = |z_2| = 1$ as the contour of integration. In addition, a phase unwrapping algorithm is necessary to avoid the artificial discontinuities that would be introduced by using the principal value of the phase. Because of the use of the DFT, this method yields a spatially aliased approximation to the true complex cepstrum.

The second method we discuss uses a recursive formula to compute the complex cepstrum exactly when $x(n_1, n_2)$ is a minimum-phase sequence with first-quadrant support. If the sequence $x(n_1, n_2)$ does not have minimum phase, the result of the recursive computation will grow without bound.

The first method of computing the 2-D complex cepstrum is a straightforward attempt to implement equation (4.99) numerically. Implementation of the necessary forward and inverse Fourier transforms requires the 2-D FFT, which will be computed in the row–column fashion. We shall assume that the sequence $x(n_1, n_2)$ whose complex cepstrum we wish to compute has finite extent. Its region of support is the rectangle $0 \le n_1 < M_1$, $0 \le n_2 < M_2$. By padding $x(n_1, n_2)$ with additional zero-valued samples, we can create a sequence whose region of support is $0 \le n_1 < N_1$, $0 \le n_2 < N_2$. We take the $(N_1 \times N_2)$-point DFT of this augmented sequence to obtain $N_1 N_2$ samples of the Fourier transform of $x(n_1, n_2)$. As before, we shall denote these samples by

$$X(k_1, k_2) = X(\omega_1, \omega_2)\Big|_{\substack{\omega_1 = 2\pi k_1/N_1 \\ \omega_2 = 2\pi k_2/N_2}} \tag{4.134}$$

Next, we must compute the unwrapped phase of $X(k_1, k_2)$ so that we can evaluate $\ell n \, X(k_1, k_2)$ in a consistent manner. Should the sampled, unwrapped phase function, denoted $\phi_x(k_1, k_2)$, contain any linear phase terms, they must be removed.

Samples of the unwrapped phase function can be computed in the following manner. First, the initial value of the phase function $\phi_x(0, 0)$ is computed by

$$\phi_x(0, 0) = \tan^{-1} \left\{ \frac{\text{Im} \, [X(0, 0)]}{\text{Re} \, [X(0, 0)]} \right\} \tag{4.135}$$

where, of course,

$$X(0, 0) = \sum_{n_1=0}^{M_1-1} \sum_{n_2=0}^{M_2-1} x(n_1, n_2) $$

If $x(n_1, n_2)$ is a purely real sequence, $\phi_x(0, 0)$ is usually taken to be zero. [There will be an overall phase ambiguity since the principal value of the inverse tangent function is used in (4.135), but this ambiguity is unimportant.]

Next, a 1-D phase unwrapping algorithm such as Tribolet's method [19] is applied to unwrap the 2-D phase function. Tribolet's algorithm can be considered to be a "black box" which accepts a 1-D sequence $x(n)$ and an initial phase value $\phi_x(0)$ as its inputs and delivers the 1-D DFT $X(k)$ and the sampled, unwrapped phase function $\phi_x(k)$ as its outputs. We can use Tribolet's algorithm to unwrap the 2-D phase function along the line $k_2 = 0$ by supplying the 1-D sequence

$$x_1(n_1) = \sum_{n_2=0}^{M_2-1} x(n_1, n_2) \tag{4.136}$$

and the initial phase value $\phi_x(0, 0)$ as inputs. The resulting outputs will be the $N_1$-point DFT $X(k_1, 0)$ and the sampled, unwrapped phase function $\phi_x(k_1, 0)$, including the sample $\phi_x(N_1, 0)$. At this point, we can deduce the amount of linear phase in the $k_1$ direction by computing the phase difference

$$2\pi K_1 = \phi_x(N_1, 0) - \phi_x(0, 0) \tag{4.137}$$

After that, we form a set of 1-D sequences by computing the DFTs of the rows of $x(n_1, n_2)$. This gives us

$$S(k_1, n_2) \triangleq \sum_{n_1=0}^{M_1-1} x(n_1, n_2) W_{N_1}^{n_1 k_1} \tag{4.138}$$

For each $k_1$ from zero through $N_1 - 1$, we use $S(k_1, n_2)$ (as a function of $n_2$) and the initial phase value $\phi_x(k_1, 0)$ as inputs to the 1-D phase unwrapping algorithm. The outputs will be $X(k_1, k_2)$ and the sampled, unwrapped phase $\phi_x(k_1, k_2)$, including the samples for $k_2 = N_2$. The unwrapped phase is examined to determine the amount of linear phase in the $k_2$ direction. The phase difference

$$2\pi K_2 = \phi_x(k_1, N_2) - \phi_x(k_1, 0) \tag{4.139}$$

should be independent of $k_1$ if the continuous Fourier transform $X(\omega_1, \omega_2)$ does not go to zero anywhere.

Now we can compute the phase function

$$\phi_Y(k_1, k_2) = \phi_x(k_1, k_2) - \frac{2\pi k_1}{N_1} K_1 - \frac{2\pi k_2}{N_2} K_2 \tag{4.140}$$

to remove the linear phase components and we can compute the log-magnitude function

$$\ell \mathrm{n}\,|\,Y(k_1, k_2)| = \ell \mathrm{n}\,|\,X(k_1, k_2)| \qquad (4.141)$$

Finally, using an inverse 2-D DFT, we can implement the discrete version of equation (4.106).

$$\hat{y}_{sa}(n_1, n_2) = \frac{1}{N_1 N_2} \sum_{k_1=0}^{N_1-1} \sum_{k_2=0}^{N_2-1} \{\ell \mathrm{n}\,|\,Y(k_1, k_2)| + j\phi_Y(k_1, k_2)\} W_{N_1}^{-n_1 k_1} W_{N_2}^{-n_2 k_2} \qquad (4.142)$$

The subscript in $\hat{y}_{sa}(n_1, n_2)$ serves to remind the reader that this sequence is a spatially aliased version of the complex cepstrum $\hat{y}(n_1, n_2)$ given by equation (4.106). The two sequences are related by

$$\hat{y}_{sa}(n_1, n_2) = \sum_{j_1=-\infty}^{\infty} \sum_{j_2=-\infty}^{\infty} \hat{y}(n_1 + j_1 N_1, n_2 + j_2 N_2) \qquad (4.143)$$

Even though $x(n_1, n_2)$ and therefore $y(n_1, n_2) = x(n_1 - K_1, n_2 - K_2)$ is a sequence with a finite region of support, the complex cepstrum $\hat{y}(n_1, n_2)$ will have an infinite region of support in general. Obviously, $\hat{y}_{sa}(n_1, n_2)$ will be a good approximation to $\hat{y}(n_1, n_2)$ only if $\hat{y}(n_1, n_2)$ becomes small as $n_1 \rightarrow N_1$ and $n_2 \rightarrow N_2$.

When $x(n_1, n_2)$ is a minimum-phase signal with support in the first quadrant, the complex cepstrum can be computed exactly by a recursive formula. We can derive the recursion by rewriting equation (4.107) to get

$$X(\omega_1, \omega_2) \frac{\partial \hat{X}(\omega_1, \omega_2)}{\partial \omega_1} = \frac{\partial X(\omega_1, \omega_2)}{\partial \omega_1} \qquad (4.144)$$

Taking the inverse 2-D Fourier transform of both sides of this equation gives us

$$\sum_{k_1=-\infty}^{\infty} \sum_{k_2=-\infty}^{\infty} x(n_1 - k_1, n_2 - k_2) k_1 \hat{x}(k_1, k_2) = n_1 x(n_1, n_2) \qquad (4.145)$$

Since $x(n_1, n_2)$ has minimum phase and first-quadrant support and since, by implication, $\hat{x}(n_1, n_2)$ has first-quadrant support, the infinite sums in (4.145) can be replaced by finite sums to give us

$$\sum_{k_1=0}^{n_1} \sum_{k_2=0}^{n_2} x(n_1 - k_1, n_2 - k_2) k_1 \hat{x}(k_1, k_2) = n_1 x(n_1, n_2) \qquad (4.146)$$

To simplify the equations, let us assume for the moment that $x(n_1, n_2)$ is normalized so that $x(0, 0) = 1$. [Scaling $x(n_1, n_2)$ affects only the value of $\hat{x}(0, 0)$ in the complex cepstrum.] Then the $(k_1, k_2) = (n_1, n_2)$ term can be pulled out of the summation and the remaining terms subtracted from both sides to yield

$$\hat{x}(n_1, n_2) = x(n_1, n_2) - \frac{1}{n_1} \sum_{\substack{k_1=0 \\ (k_1, k_2) \neq (n_1, n_2)}}^{n_1} \sum_{k_2=0}^{n_2} x(n_1 - k_1, n_2 - k_2) k_1 \hat{x}(k_1, k_2) \qquad (4.147)$$

$$\text{for } n_1 \neq 0$$

Similarly, we can derive

$$\hat{x}(n_1, n_2) = x(n_1, n_2) - \frac{1}{n_2} \sum_{\substack{k_1=0 \\ (k_1, k_2) \neq (n_1, n_2)}}^{n_1} \sum_{k_2=0}^{n_2} x(n_1 - k_1, n_2 - k_2) k_2 \hat{x}(k_1, k_2) \qquad (4.148)$$

$$\text{for } n_2 \neq 0$$

The value of $\hat{x}(0, 0)$ can be derived by applying the 2-D initial value theorem to get

$$\hat{x}(0, 0) = \ell n\, x(0, 0) \qquad (4.149)$$

Because of the use of the complex logarithm, the imaginary part of $\hat{x}(0, 0)$ will have an ambiguity of $2\pi$ times an integer. This is the same overall ambiguity encountered in computing the initial phase sample $\phi_x(0, 0)$ in the DFT method of computing the complex cepstrum.

After $\hat{x}(0, 0)$ has been computed, the values of $\hat{x}(0, n_2)$ can be computed from recursion (4.148) and the values of $\hat{x}(n_1, 0)$ can be computed from recursion (4.147). After that, the remaining values of $\hat{x}(n_1, n_2)$ for $n_1 > 0$ and $n_2 > 0$ can be computed using either recursive formula.

The recursive method of computing the 2-D complex cepstrum has the advantage that it yields a theoretically exact result, not an aliased approximation as the DFT method does. It also has the obvious disadvantage that it works only if a signal has minimum phase. In general, it is not known beforehand if a signal does, in fact, have minimum phase. If we blindly apply the recursion to a non-minimum-phase signal that has support only on the first quadrant, the computed cepstrum will grow without bound.

## PROBLEMS

**4.1.** A 2-D system is described by the difference equation

$$y(n_1, n_2) - 0.9y(n_1, n_2 - 1) + 0.5y(n_1 - 1, n_2 - 1) = x(n_1, n_2)$$

Find $y(n_1, n_2)$ for $0 \le n_1 \le 3$, $0 \le n_2 \le 3$ if $x(n_1, n_2) = \delta(n_1, n_2)$ under the assumption that $y(n_1, n_2) = 0$ for $n_1 < 0$ or $n_2 < 0$.

**4.2.** Consider a recursively computable digital filter with the output mask shown in Figure P4.2. This mask corresponds to the difference equation

$$y(n_1, n_2) + ay(n_1 - 1, n_2 - 1) + by(n_1, n_2 - 1) + cy(n_1 + 1, n_2 - 1) = x(n_1, n_2)$$

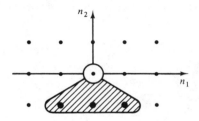

**Figure P4.2**

(a) Determine all possible recursion directions that are allowable with this mask.
(b) Determine the region of support of the impulse response of the filter.
(c) Find a sufficient set of initial conditions which will guarantee that the system is linear, shift invariant, and recursively computable.

**4.3.** Repeat Problem 4.2 for the filter with the output mask shown in Figure P4.3.

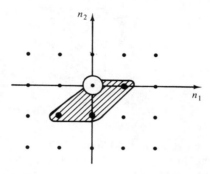

**Figure P4.3**

**4.4.** Consider a recursively computable 2-D filter for which the output mask is confined to a sector, with the hole of the mask at the vertex as shown in Figure P4.4.

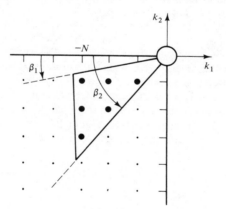

**Figure P4.4**

(a) Find the region of support of the resulting filter assuming a trivial input mask.
(b) Find all possible directions of recursion.
(c) What effect, if any, does the shape of the rear boundary of the mask have on your answers to parts (a) and (b)?

**4.5.** In Figure P4.5 we have a nonsymmetric half-plane output mask.
(a) Find a transformation of variables such that in the new coordinate system this filter can be implemented with a one-quadrant mask.

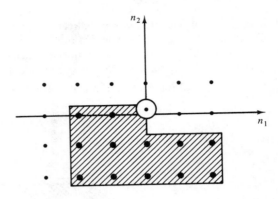

**Figure P4.5**

(b) Sketch the transformed output mask.
(c) Find the inverse transformation that will convert the mask from part (b) into the mask of Figure P4.5.
(d) Using the result above, determine which samples in the transformed output plane need to be computed and which can be ignored.

**4.6.** Consider the NSHP filter with output mask shown in Figure P4.5. The initial conditions that accompany this mask are shown in Figure 4.6 in the text.
(a) Draw the precedence graph that is appropriate for this recursion. Your result should be similar to Figure 4.10 except for the labeling of the nodes.
(b) The precedence graph that you derived in part (a) can be obtained from Figure 4.10 with a transformation of variables. Determine that transformation and show that it is the inverse of a transformation that maps the NSHP output mask into a quarter-plane mask.
(c) Using the results of part (b), show that the precedence graph of any filter with sector support is equivalent to Figure 4.10.
(d) For the original mask of Figure P4.5, determine an ordering relation that will allow the maximum possible number of output samples to be computed simultaneously.

**4.7.** Find the $z$-transform of the following 2-D arrays and give the associated regions of convergence
(a) $\delta(n_1, n_2)$
(b) $a^{(n_1+n_2)} u(n_1, n_2)$
(c) $x(n_1, n_2) = \begin{cases} \delta(n_1), & n_2 = 0, 1 \ldots, N-1 \\ 0, & \text{otherwise} \end{cases}$

**4.8.** Consider a sequence $x(n_1, n_2)$ with support in the shaded region of the $(n_1, n_2)$-plane shown in Figure P4.8. Let $X_z(z_1, z_2)$ be the 2-D $z$-transform of $x(n_1, n_2)$.
(a) If the point $(z_{01}, z_{02})$ is within the region of convergence of $X_z(z_1, z_2)$ what other points are guaranteed to be in the region of convergence?
(b) What can you say about the slope of the boundary of the region of convergence?

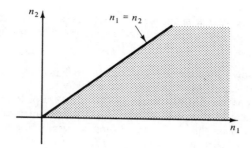

**Figure P4.8**

**4.9.** The 1-D sequence $h(n)$ has a z-transform, $H_z(z)$, which converges for $a < |z| < b$. Determine the z-transform and region of convergence of the sequence

$$g(n_1, n_2) = h(n_1 + n_2)h(n_1 - n_2)$$

**4.10.** In this chapter we stated a number of properties of the z-transform without proof. These properties are derived in this problem.

(a) Derive the following properties of the z-transform given that $X_z(z_1, z_2)$ is the z-transform of $x(n_1, n_2)$.

(1) $x^*(n_1, n_2) \longleftrightarrow X_z^*(z_1^*, z_2^*)$

(2) $\text{Re}\,[x(n_1, n_2)] \longleftrightarrow \frac{1}{2}[X_z(z_1, z_2) + X_z^*(z_1^*, z_2^*)]$

(3) $\text{Im}\,[x(n_1, n_2)] \longleftrightarrow \frac{1}{2j}[X_z(z_1, z_2) - X_z^*(z_1^*, z_2^*)]$

(4) $x(-n_1, n_2) \longleftrightarrow X_z(z_1^{-1}, z_2)$

(5) $x(n_1, -n_2) \longleftrightarrow X_z(z_1, z_2^{-1})$

(6) $x(-n_1, -n_2) \longleftrightarrow X_z(z_1^{-1}, z_2^{-1})$

(7) $x(n_2, n_1) \longleftrightarrow X_z(z_2, z_1)$

(b) Derive Parseval's theorem

$$\sum_{n_1=-\infty}^{\infty} \sum_{n_2=-\infty}^{\infty} x(n_1, n_2)y^*(n_1, n_2) = \frac{1}{(2\pi j)^2} \oint_{C_2} \oint_{C_1} X_z(z_1, z_2) Y_z^*\left(\frac{1}{z_1^*}, \frac{1}{z_2^*}\right) z_1^{-1} z_2^{-1}\, dz_1\, dz_2$$

**4.11.** Consider the 2-D transfer function

$$H_z(z_1, z_2) = \frac{1}{1 - az_1^{-1} - bz_2^{-1} - cz_1^{-1}z_2^{-1}}$$

where $a$, $b$, and $c$ are real coefficients.

(a) Write an algebraic expression for the root locus in the $z_2$-plane; that is, find an expression of the form $z_2 = f(e^{-j\omega_1})$ such that the denominator of $H_z(e^{j\omega_1}, z_2)$ equals zero.

(b) Show that in order for $H_z(z_1, z_2)$ to be stable it is necessary that

$$\left|\frac{b+c}{1-a}\right| < 1 \qquad \text{and} \qquad \left|\frac{b-c}{1+a}\right| < 1$$

(c) Are there other necessary conditions you can derive? (*Hint:* Consider the root locus in the $z_1$-plane.)

**4.12.** Suppose that a sequence $h(n_1, n_2)$ has a 2-D $z$-transform $H_z(z_1, z_2) = 1/B_z(z_1, z_2)$. Then the root maps for $H(z_1, z_2)$ are given by the algebraic relations

$$B_z(z_1, e^{j\omega_2}) = 0; \qquad B_z(e^{j\omega_1}, z_2) = 0$$

   **(a)** Now define a new sequence $g(n_1, n_2) \triangleq h(n_1, n_2)a^{n_1}b^{n_2}$ where $a$ and $b$ are complex constants. Write an expression for $G_z(z_1, z_2)$ in terms of $B_z(z_1, z_2)$.

   **(b)** Suppose that $H_z(z_1, z_2) \triangleq 1/(1 + z_1 + z_2)$. Sketch the root maps for $H_z(z_1, z_2)$. Indicate the point on the $z_2$-plane root map corresponding to $\omega_1 = 0$ and the point on the $z_1$-plane root map corresponding to $\omega_2 = 0$.

   **(c)** As in part (b), assume that $H_z(z_1, z_2) = 1/(1 + z_1 + z_2)$. Draw the root maps for $G_z(z_1, z_2)$ for the following values of $a$ and $b$, indicating the points where $\omega_1 = 0$ ($z_2$-plane) and $\omega_2 = 0$ ($z_1$-plane) as you did in part (b).

   (1) $a = 1, b = \frac{1}{2}$
   (2) $a = 2, b = j$
   (3) $a = \frac{1}{2}e^{j\pi/4}, b = -3j/4$

**4.13.** Determine the inverse $z$-transforms of the following transfer functions. In all cases, assume that the region of convergence includes the 2-D unit bicircle.

   **(a)** $H_z(z_1, z_2) = \dfrac{z_1^{-1}}{1 - az_1^{-3} - bz_2^{-2}}$

   **(b)** $H_z(z_1, z_2) = \dfrac{1}{1 - az_1^{-1}z_2^{-2}}$

   **(c)** $H_z(z_1, z_2) = \dfrac{1}{1 - az_1^{-1}z_2^{-2} - bz_1^{-1}}$

**4.14.** One means of determining a transfer function for a system from a flowgraph is to define a sequence at the output of each summing node, write a $z$-domain equation at each node, and then, by combining equations, eliminate all variables but the input and output. Using this approach, find the transfer functions for the two systems shown in Figure P4.14.

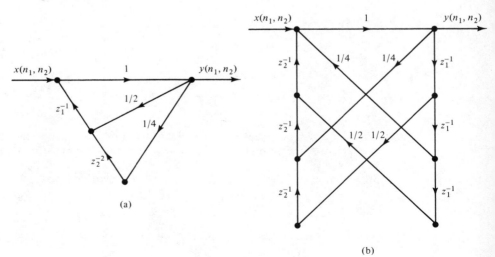

**Figure P4.14**

**4.15.** Every flowgraph represents a difference equation or family of difference equations. One can find the difference equations corresponding to a given flowgraph by defining sequences at the output of each summing node and by writing a difference equation at each summing node. Using this procedure, determine the difference equations corresponding to the two networks in Problem 4.14.

**4.16.** Prove that a 2-D separable first-quadrant recursive filter $H_z(z_1, z_2) = H_1(z_1)H_2(z_2)$ is stable if and only if $H_1(z_1)$ and $H_2(z_2)$ correspond to the transfer functions of stable causal 1-D filters.

**4.17.** State whether or not the following filters are stable.

(a) $H_a(z_1, z_2) = \dfrac{1}{1 + 0.9z_1^{-1} - 0.9z_2^{-2} - 0.81z_1^{-1}z_2^{-2}}$

(b) $H_b(z_1, z_2) = \dfrac{1}{1 + 0.9z_1^{-1} - 0.81z_1^{-1}z_2^{-2}}$

(c) $H_c(z_1, z_2) = \dfrac{1}{1 - 0.9z_2^{-1} - 0.81z_1^{-1}z_2^{-2}}$

**4.18.** Suppose that the impulse response $h(n_1, n_2)$ of a recursive filter has support only on the first quadrant and that $H_z(z_1, z_2) = 1/B_z(z_1, z_2)$. For $h(n_1, n_2)$ to represent the impulse response of a stable filter, it is necessary and sufficient that

$$B_z(e^{j\omega_1}, e^{j\omega_2}) \neq 0, \quad B_z(z_1, 1) \neq 0 \quad \text{for } |z_1| \geq 1 \quad \text{and} \quad B_z(1, z_2) \neq 0 \quad \text{for } |z_2| \geq 1$$

(a) Suppose that a new impulse response $g(m_1, m_2) \triangleq h(n_1, n_2)$ is created by the linear mapping

$$\begin{bmatrix} m_1 \\ m_2 \end{bmatrix} = \begin{bmatrix} 1 & -1 \\ 0 & 1 \end{bmatrix} \begin{bmatrix} n_1 \\ n_2 \end{bmatrix}$$

Sketch the region of support of $g(m_1, m_2)$.

(b) Let $G_z(w_1, w_2) \triangleq 1/C_z(w_1, w_2)$. Relate $C_z(w_1, w_2)$ and $B_z(z_1, z_2)$.

(c) What conditions must $C_z(w_1, w_2)$ satisfy to ensure that $g(m_1, m_2)$ is the impulse response of a stable filter?

**4.19.** Suppose that $x(n_1, n_2) \triangleq a^{n_1}b^{n_2}u(n_1, n_2)$ where $|a| < 1$ and $|b| < 1$.

(a) Write an expression for $X_z(z_1, z_2)$. Sketch the region of convergence for $X_z(z_1, z_2)$ in the $(\ell n\, |z_1|, \ell n\, |z_2|)$-plane.

(b) Now define $y(m_1, m_2) \triangleq x(n_1, n_2)$ where the coordinates $(m_1, m_2)$ and $(n_1, n_2)$ are related by the linear mapping

$$\begin{bmatrix} m_1 \\ m_2 \end{bmatrix} = \begin{bmatrix} 2 & 1 \\ 1 & 1 \end{bmatrix} \begin{bmatrix} n_1 \\ n_2 \end{bmatrix}$$

Sketch the region of support of $y(m_1, m_2)$.

(c) Let

$$Y_z(w_1, w_2) = \sum_{m_1} \sum_{m_2} y(m_1, m_2) w_1^{-m_1} w_2^{-m_2}$$

Relate $Y_z(w_1, w_2)$ to $X_z(z_1, z_2)$. Sketch the region of convergence of $Y_z(w_1, w_2)$ in the $(\ell n\, |w_1|, \ell n\, |w_2|)$-plane.

(d) Let $X(\omega_1, \omega_2) \triangleq X_z(e^{j\omega_1}, e^{j\omega_2})$ and let $\phi_x(\omega_1, \omega_2)$ be the unwrapped phase function. Show that $\phi_x(\omega_1, \omega_2)$ can be written in the form $\phi_1(\omega_1) + \phi_2(\omega_2)$. Demonstrate that $\phi_x(\omega_1, \omega_2)$ is continuous, odd, and doubly periodic.

(e) Let $Y(v_1, v_2) \triangleq Y_z(e^{jv_1}, e^{jv_2})$ and let $\phi_y(v_1, v_2)$ be the unwrapped phase function. Relate $\phi_y(v_1, v_2)$ and $\phi_x(\omega_1, \omega_2)$. Demonstrate that $\phi_y(v_1, v_2)$ is continuous, odd, and periodic.

**4.20.** In Section 4.4.1 we claimed that if

$$h(n_1, n_2) = f(n_1)g(n_2)$$

then the cepstrum of $h(n_1, n_2)$ was given by

$$\hat{h}(n_1, n_2) = \hat{f}(n_1)\delta(n_2) + \hat{g}(n_2)\delta(n_1)$$

where $\hat{f}(n_1)$ is the 1-D complex cepstrum of $f(n_1)$ and $\hat{g}(n_2)$ is the 1-D complex cepstrum of $g(n_2)$. Show that this result is correct.

**4.21.** Let $\hat{x}(n_1, n_2)$ denote the complex cepstrum of $x(n_1, n_2)$. Determine the complex cepstrum of $g(m_1, m_2)$ where

$$g(m_1, m_2) = x(n_1, n_2), \qquad m_1 = An_1 + Bn_2; \quad m_2 = Cn_1 + Dn_2$$

and $AD - BC = 1$.

**4.22.** In this chapter we have not specifically discussed recursive systems for processing hexagonally sampled sequences. Such systems, however, can be defined and they are quite similar to their rectangular counterparts. In Figure P4.22 we show two output masks for hexagonally sampled recursive systems. The one in P4.22(a) is called a one-third plane mask and the one in P4.22(b) is called a one-sixth plane mask.

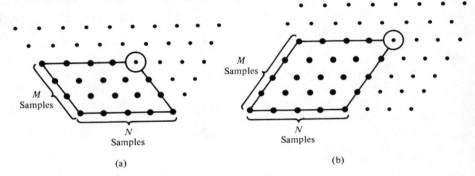

**Figure P4.22**

**(a)** Determine all possible recursion directions that are allowable with the one-third plane mask.

**(b)** Determine the region of support of the impulse response of the filter assuming that the system is to be linear and shift invariant and that the input mask sees only a single sample at the origin (i.e., the numerator of the system function is a nonzero constant).

**(c)** Repeat parts (a) and (b) for the one-sixth plane output mask.

**(d)** Find a linear transformation of coordinates that will convert the one-third plane mask of Figure P4.22(a) into a one-sixth plane filter.

**4.23.** The $z$-transform of a hexagonally sampled signal can be defined as

$$X_z(z_1, z_2) = \sum_{n_1=-\infty}^{\infty} \sum_{n_2=-\infty}^{\infty} x(n_1, n_2) z_1^{-n_1} z_2^{-n_2}$$

(a) With this definition, how are the $z$-transform and the Fourier transform, $X_V(\Omega_1, \Omega_2)$, of a hexagonally sampled signal related? (See Section 1.5.)

(b) Two LSI systems, one rectangular and one hexagonal, both have the transfer function

$$H_z(z_1, z_2) = \frac{A_z(z_1, z_2)}{B_z(z_1, z_2)}$$

How are the frequency responses of these two systems related?

(c) The rectangular system in part (b) is known to be stable. Is this enough to guarantee the stability of the hexagonal system? Explain your answer.

**4.24.** Consider the first quadrant signal:

$$x(n_1, n_2) = \frac{(n_1 + n_2)!}{n_1! \, n_2!} a^{n_1} b^{n_2} u(n_1, n_2)$$

where $|a| + |b| < 1$. Show that the complex cepstrum $\hat{x}(n_1, n_2)$ is given by

$$\hat{x}(n_1, n_2) = \begin{cases} \dfrac{(n_1 + n_2 - 1)!}{n_1! \, n_2!} a^{n_1} b^{n_2} u(n_1, n_2), & n_1 \geq 0, \, n_2 \geq 0, \, (n_1, n_2) \neq (0, 0) \\ 0, & (n_1, n_2) = (0, 0), \, n_1 < 0, \text{ or } n_2 < 0 \end{cases}$$

by using the following steps:

(a) Derive a closed-form expression for $X_z(z_1, z_2)$.

(b) Show that $n_1 s(n_1, n_2) \longleftrightarrow -z_1 \dfrac{\partial}{\partial z_1} S_z(z_1, z_2)$ for an arbitrary sequence $s(n_1, n_2)$.

(c) Derive an expression for $-z_1 \dfrac{\partial}{\partial z_1} \hat{X}_z(z_1, z_2)$ using the fact that $\hat{X}_z(z_1, z_2) = \ell\mathrm{n} \, [X_z(z_1, z_2)]$.

(d) Compute $n_1 \hat{x}(n_1, n_2)$ and $n_2 \hat{x}(n_1, n_2)$ and finally $\hat{x}(n_1, n_2)$.

# REFERENCES

1. Russell M. Mersereau and Dan E. Dudgeon, "Two-Dimensional Digital Filtering," *Proc. IEEE*, 63, no. 4 (Apr. 1975), 610–23.

2. Michael P. Ekstrom and John W. Woods, "Two-Dimensional Spectral Factorization with Applications in Recursive Digital Filtering," *IEEE Trans. Acoustics, Speech, and Signal Processing*, ASSP-24, no. 2 (Apr. 1976), 115–28.

3. David S. K. Chan, "A Novel Framework for the Description of Realization Structures for 1-D and 2-D Digital Filters," *1976 IEEE Electronics and Space Convention Record*, pp. 157(A-H).

4. Michael T. Manry and J. K. Aggarwal, "Picture Processing Using One-Dimensional Implementations of Discrete Planar Filters," *IEEE Trans. Acoustics, Speech, and Signal Processing*, ASSP-22, no. 3 (June 1974), 164–73.

5. David S. K. Chan, "Theory and Implementation of Multidimensional Discrete Systems for Signal Processing," Ph.D. thesis, Department of Electrical Engineering and Computer Science, Massachusetts Institute of Technology (1978).

6. Brian T. O'Connor and Thomas S. Huang, "Stability of General Two-Dimensional Recursive Digital Filters," *IEEE Trans. Acoustics, Speech, and Signal Processing*, ASSP-26, no. 6 (Dec. 1978), 550–60.

7. John L. Shanks, Sven Treitel, and James H. Justice, "Stability and Synthesis of Two-Dimensional Recursive Filters," *IEEE Trans. Audio and Electroacoustics*, AU-20, no. 2 (June 1972), 115–28.

8. James H. Justice and John L. Shanks, "Stability Criterion for $N$-Dimensional Digital Filters," *IEEE Trans. Automatic Control*, AC-18, no. 3 (June 1973), 284–86.

9. Thomas S. Huang, "Stability of Two-Dimensional Recursive Filters," *IEEE Trans. Audio and Electroacoustics*, AU-20, no. 2 (June 1972), 158–63.

10. Dennis Goodman, "An Alternate Proof of Huang's Stability Theorem," *IEEE Trans. Acoustics, Speech, and Signal Processing*, ASSP-24, no. 5 (Oct. 1976), 426–27.

11. Daniel L. Davis, "A Correct Proof of Huang's Theorem on Stability," *IEEE Trans. Acoustics, Speech, and Signal Processing*, ASSP-24, no. 5 (Oct. 1976), 425–26.

12. R. DeCarlo, J. Murray, and R. Saeks, "Multivariate Nyquist Theory," *Int. J. Control*, 25 (1977), 657–75.

13. Michael G. Strintzis, "Test of Stability of Multidimensional Filters," *IEEE Trans. Circuits and Systems*, CAS-24, no. 8 (Aug. 1977), 432–37.

14. Brian T. O'Connor, "Techniques for Determining the Stability of Two-Dimensional Recursive Filters and Their Application to Image Restoration," Ph.D. thesis, School of Electrical Engineering, Purdue University (May 1978).

15. E. I. Jury, "Stability of Multidimensional Scalar and Matrix Polynominals," *Proc. IEEE*, 66, no. 9 (Sept. 1978), 1018–48.

16. G. A. Shaw, "An Algorithm for Testing Stability of Two-Dimensional Digital Recursive Filters," *Proc. IEEE Int. Conf. Acoustics, Speech, and Signal Processing* (Apr. 1978), 769–72.

17. G. A. Shaw and R. M. Mersereau, "Design, Stability and Performance of Two-Dimensional Recursive Digital Filters," Technical Report E21-B05-1, Georgia Institute of Technology School of Electrical Engineering (Dec. 1979).

18. E. I. Jury, *Theory and Application of the z-Transform Method* (New York: John Wiley & Sons, Inc., 1964).

19. Jose M. Tribolet, "A New Phase Unwrapping Algorithm," *IEEE Trans. Acoustics, Speech, and Signal Processing*, ASSP-25, no. 2 (Apr. 1977), 170–77.

20. Dennis Goodman, "Some Stability Properties of Two-Dimensional Linear Shift-Invariant Digital Filters," *IEEE Trans. Circuits and Systems*, CAS-24, no. 4 (Apr. 1977), 201–8.

21. B. D. O. Anderson and E. I. Jury, "Stability of Multidimensional Digital Filters," *IEEE Trans. Circuits and Systems*, CAS-21, no. 2 (Mar. 1974), 300–304.

22. Alan V. Oppenheim and Ronald W. Schafer, *Digital Signal Processing* (Englewood Cliffs, N.J.: Prentice-Hall, Inc., 1975).

23. B. P. Bogert, M. J. R. Healy, and J. W. Tukey, "The Quefrency Alanysis of Time Series for Echoes: Cepstrum, Pseudo-autocovariance, Cross-Cepstrum, and Saphe Cracking," *Time Series Analysis* (Proc. Symp.), M. Rosenblatt, ed. (New York: John Wiley & Sons, Inc., 1963), 209–43.

24. Dan E. Dudgeon, "The Existence of Cepstra for Two-Dimensional Rational Polynomials," *IEEE Trans. Acoustics, Speech, and Signal Processing*, ASSP-23, no. 2 (Apr. 1975), 242–43.

25. Dan E. Dudgeon, "Two-Dimensional Recursive Filtering," Sc.D. thesis, Department of Electrical Engineering and Computer Science, Massachusetts Institute of Technology (May 1974).

26. Dan E. Dudgeon, "The Computation of Two-Dimensional Cepstra," *IEEE Trans. Acoustics, Speech, and Signal Processing*, ASSP-25, no. 6 (Dec. 1977), 476–84.

# 5

# DESIGN AND IMPLEMENTATION
# OF TWO-DIMENSIONAL IIR FILTERS

In this chapter we discuss implementation techniques for 2-D IIR (infinite-extent impulse response) digital filters and some algorithms for determining the parameters of filters that satisfy desired specifications. As we saw in Chapter 4, the input and output signals of a 2-D IIR filter obey a constant-coefficient linear partial difference equation from which the value of an output sample can be computed using the input samples and previously computed output samples. Because the values of the output samples are fed back, the 2-D IIR filter, like its 1-D counterpart, can be unstable.

The promise of IIR filters is a potential reduction in computation compared to FIR filters when performing comparable filtering operations. By feeding back output samples, we can use a filter with fewer coefficients (hence less computation) to implement a desired operation. On the other hand, IIR filters pose some potentially significant implementation and stabilization problems not encountered with FIR filters. For this reason, we examine 2-D IIR implementation methods before considering design algorithms. It makes no sense to design what cannot be implemented, and, as we shall shortly see, implementation can be a problem in two dimensions.

## 5.1 CLASSICAL 2-D IIR FILTER IMPLEMENTATIONS

There are a variety of 1-D IIR filter implementations that have been studied, including direct form structures, cascade structures, and parallel structures [1]. The classical 2-D IIR implementations that we examine in this section are derived from these 1-D implementations, but they also exhibit properties not found in the 1-D case. For example, in Section 4.1.4, we examined the question of ordering the computation of the output sample values. For 1-D difference equations, the ordering is generally totally specified. In the 2-D case, we have some latitude in deciding which output sample will be computed next.

This problem also manifests itself when we try to draw block diagrams for filter structures. Block diagrams usually imply that serial data flow through them in a well-defined order, but for 2-D flowgraphs, the order is not completely defined, merely constrained by precedence relations. This point is explored in detail in Section 5.3.

### 5.1.1 Direct Form Implementations

An IIR filter may be implemented in direct form by rearranging its difference equation to express one output sample in terms of the input samples and previously computed output samples. In Section 4.1 we studied difference equations in some detail and encountered some of the issues associated with the direct form implementation, such as ordering the computations, recursive computability, and the use of input and output masks. In this section we continue to study direct form implementations and examine the case where the input mask has nontrivial extent.

For a first-quadrant filter, the input signal $x(n_1, n_2)$ and the output signal $y(n_1, n_2)$ are related by

$$y(n_1, n_2) = \sum_{l_1=0}^{L_1-1} \sum_{l_2=0}^{L_2-1} a(l_1, l_2) x(n_1 - l_1, n_2 - l_2)$$
$$- \sum_{\substack{k_1=0 \\ (k_1,k_2) \neq (0,0)}}^{K_1-1} \sum_{k_2=0}^{K_2-1} b(k_1, k_2) y(n_1 - k_1, n_2 - k_2) \tag{5.1}$$

[With no loss of generality, we have set $b(0, 0) = 1$.] Since the response of the filter to an impulse $\delta(n_1, n_2)$ is by definition the impulse response $h(n_1, n_2)$, we can derive the relationship

$$h(n_1, n_2) = a(n_1, n_2) - \sum_{\substack{k_1=0 \\ (k_1,k_2) \neq (0,0)}}^{K_1-1} \sum_{k_2=0}^{K_2-1} b(k_1, k_2) h(n_1 - k_1, n_2 - k_2) \tag{5.2}$$

By taking the 2-D $z$-transform of both sides, we can solve for the system function $H_z(z_1, z_2)$, which is given by

$$H_z(z_1, z_2) = \frac{\displaystyle\sum_{l_1=0}^{L_1-1} \sum_{l_2=0}^{L_2-1} a(l_1, l_2) z_1^{-l_1} z_2^{-l_2}}{\displaystyle\sum_{k_1=0}^{K_1-1} \sum_{k_2=0}^{K_2-1} b(k_1, k_2) z_1^{-k_1} z_2^{-k_2}}$$
$$\triangleq \frac{A_z(z_1, z_2)}{B_z(z_1, z_2)} \tag{5.3}$$

This ratio may be viewed as resulting from the cascade of two filters, an FIR filter with a system function equal to $A_z(z_1, z_2)$ and a purely recursive filter with a system function equal to $1/B_z(z_1, z_2)$, as shown in Figure 5.1.

Suppose that the size of the input mask is $L_1 \times L_2$ points, that the implementation is column by column, and that we desire $N_2$ output points in each column. Then the number of input samples we must be able to store is $(L_1 - 1)N_2 + L_2 - 1$, which is slightly more than $L_1 - 1$ columns. This amount of storage is necessary just to accommodate the input mask, which corresponds to the FIR stage in the cascade shown in Figure 5.1.

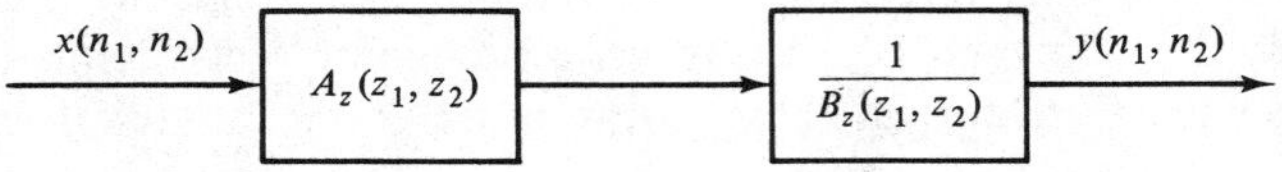

**Figure 5.1** Representation for the filter with the system function $H_z(z_1, z_2) = A_z(z_1, z_2)/B_z(z_1, z_2)$.

The second stage of the cascade is an IIR filter with a constant numerator. If its output mask is $K_1 \times K_2$ points with the output hole at the northeast corner, a column-by-column realization will require $(K_1 - 1)N_2 + K_2 - 1$ storage elements. Thus the total storage requirement for a direct form column-by-column realization is $(L_1 + K_1 - 2)N_2 + L_2 + K_2 - 2$ samples.

Since we have interpreted $H_z(z_1, z_2)$ as a cascade of two LSI filters, we can reverse the order of the two filters, as shown in Figure 5.2, without altering $H_z(z_1, z_2)$. The

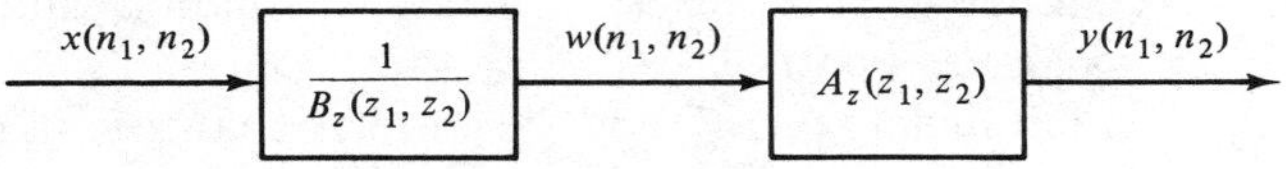

**Figure 5.2** Alternative realization of $H_z(z_1, z_2)$, called the "direct form II" realization.

resulting realization, which we shall call the 2-D *direct-form II* realization, is analogous to the 1-D direct form II structure [1]. The 2-D direct form II structure reduces the amount of storage required by allowing storage to be shared between the feedback and feedforward loops. The system in Figure 5.2 corresponds to implementing the equation

$$w(n_1, n_2) = x(n_1, n_2) - \sum_{\substack{k_1=0 \\ (k_1,k_2) \neq (0,0)}}^{K_1-1} \sum_{k_2=0}^{K_2-1} b(k_1, k_2)w(n_1 - k_1, n_2 - k_2) \tag{5.4}$$

and then forming the output $y(n_1, n_2)$ by filtering $w(n_1, n_2)$ with the FIR filter $A_z(z_1, z_2)$.

$$y(n_1, n_2) = \sum_{l_1=0}^{L_1-1} \sum_{l_2=0}^{L_2-1} a(l_1, l_2)w(n_1 - l_1, n_2 - l_2) \tag{5.5}$$

The entire intermediate output array $w(n_1, n_2)$ need not be stored simultaneously. Only the part of $w(n_1, n_2)$ needed to compute subsequent values of $w(n_1, n_2)$ and $y(n_1, n_2)$ by equations (5.4) and (5.5) must be retained.

Consider the $K_1 \times K_2$ output mask associated with equation (5.4) shown in Figure 5.3. It is swept over the $w(n_1, n_2)$ array column by column to generate the output values $w(n_1, n_2)$ according to equation (5.4). The input mask, which is associated with the filter transfer function $A_z(z_1, z_2)$, covers a rectangular region $L_1$ points long by $L_2$ points high. For the moment, let us assume $L_1 \leq K_1$ and $L_2 \leq K_2$. If we have just finished computing the output value $w(n_1, n_2)$ for a particular point $(n_1, n_2)$, we can immediately compute $y(n_1, n_2)$ according to equation (5.5) since all of the

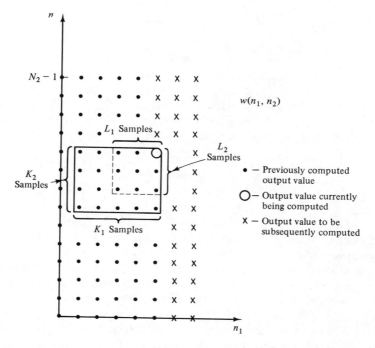

**Figure 5.3**  $K_1 \times K_2$ output mask (solid line) and an $L_1 \times L_2$ input mask (dashed line) can be overlapped in the direct form II structure to save storage. The storage needed for a column-by-column recursion is shown by the shaded region. Dots, previously computed output values; circle, output value currently being computed; crosses, output values to be computed.

required samples of $w(n_1, n_2)$ are available, as shown in Figure 5.3. In this case, there is no additional storage needed by the filter to implement the numerator of the transfer function.

In the more general case, $L_1$ may be greater than $K_1$. In this case, it is necessary to save $L_1 - K_1$ additional columns of $w(n_1, n_2)$ in order to implement equation (5.5). The amount of storage needed for the column-by-column direct form II structure is thus $\max[(L_1 - 1)N_2 + L_2 - 1,\ (K_1 - 1)N_2 + K_2 - 1)]$ or approximately $N_2 \max(K_1, L_1)$.

### 5.1.2 Cascade and Parallel Implementations

As we saw in Chapter 1, LSI systems may be constructed from the cascade and parallel interconnections of simpler LSI systems. Thus by interconnecting simple 2-D IIR filters in cascade or in parallel, we may construct a more complicated 2-D IIR filter. For example, consider the filter, shown in Figure 5.4, which is a cascade of $N$ 2-D IIR filters. If we let

$$H_z^{(i)}(z_1, z_2) \triangleq \frac{A_z^{(i)}(z_1, z_2)}{B_z^{(i)}(z_1, z_2)} \qquad (5.6)$$

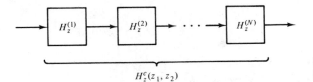

**Figure 5.4** Cascade of $N$ simple 2-D IIR filters.

$H_z^c(z_1, z_2)$

denote the transfer function of the $i$th filter in the cascade, then the overall transfer function $H_z^c(z_1, z_2)$ is given by the product

$$H_z^c(z_1, z_2) \triangleq \prod_{i=1}^{N} H_z^{(i)}(z_1, z_2) = \frac{\prod_{i=1}^{N} A_z^{(i)}(z_1, z_2)}{\prod_{i=1}^{N} B_z^{(i)}(z_1, z_2)} \qquad (5.7)$$

Since $H_z^c(z_1, z_2)$ has the form of a ratio of 2-D polynomials, it represents the transfer function of a 2-D IIR filter. However, because of the factored form of the numerator and denominator polynomials, it should be clear that the form of $H_z^c(z_1, z_2)$ cannot be used to represent the transfer function of an arbitrary 2-D IIR filter (except for the trivial case where $N = 1$). In contrast, any 1-D rational transfer function may be expressed in a factored form corresponding to a cascade of simple first-order filters.

Although we cannot generally decompose an arbitrary 2-D IIR filter into a cascade of simpler filters, we can synthesize 2-D IIR filters by cascading simpler filters together. For example, consider the case illustrated in Figure 5.5. A circularly symmetric lowpass filter is cascaded with a highpass filter to form a bandpass filter. The passbands of the lowpass and highpass filters are indicated by slanted lines, and their

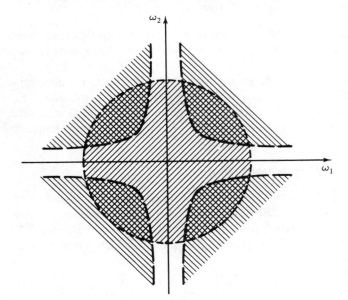

**Figure 5.5** Bandpass filter resulting from the cascade of a lowpass filter and a highpass filter.

intersection represents the passband of the resulting bandpass filter. Costa and Venetsanopoulos [2] used a more refined version of this basic approach to design circularly symmetric 2-D IIR lowpass filters by cascading rotated replicas of 1-D filters.

The cascade filter architecture of Figure 5.4 can be interpreted in two ways. If each box represents a physically separate filtering device, the data being filtered see a processing *pipeline*, analogous to a factory assembly line. The throughput rate of the pipeline can be very high because computation is occurring simultaneously within each filtering device in the pipeline. The total storage required is the sum of the storage needed by each subfilter $H_z^{(i)}(z_1, z_2)$.

Alternatively, Figure 5.4 could be interpreted as an algorithmic flowchart. First, the input signal is filtered by $H_z^{(1)}(z_2, z_2)$; then the resulting signal is filtered by $H_z^{(2)}(z_1, z_2)$; and so on. A single subroutine, properly written, could be used to implement a general subfilter $H_z^{(i)}(z_1, z_2)$. When called repeatedly and supplied with the proper coefficients and data, this subroutine could be used to implement the cascade filter of Figure 5.4. Naturally, the throughput rate would be less than that of the pipeline implementation since the computations proceed sequentially rather than simultaneously. On the other hand, the primary storage required to implement the cascade filter by sequential processing is less than that needed by the pipeline. The subroutine needs to have only enough primary storage to implement the subfilter $H_z^{(i)}(z_1, z_2)$ with the largest storage requirement. [The entire signal, however, would have to be kept in secondary storage (e.g., on a disk).]

We can also build up complicated 2-D IIR filters by the parallel interconnection of subfilters, as shown in Figure 5.6. In this case, the overall transfer function becomes

$$H_z^p(z_1, z_2) \triangleq \sum_{i=1}^{N} H_z^{(i)}(z_1, z_2) \tag{5.8}$$

Using equation (5.6) and putting the sum in (5.8) over a common denominator, we get the expanded form

$$H_z^p(z_1, z_2) \triangleq \frac{A_z^p(z_1, z_2)}{B_z^p(z_1, z_2)} = \frac{\sum_{j=1}^{N} \prod_{i \neq j} A_z^{(j)}(z_1, z_2) B_z^{(i)}(z_1, z_2)}{\prod_{i=1}^{N} B_z^{(i)}(z_1, z_2)} \tag{5.9}$$

As in the cascade filter case, $H_z^p(z_1, z_2)$ has a denominator polynomial which can be written in factored form. Consequently, the parallel form cannot be used to implement an arbitrary 2-D rational system function. Nevertheless, we can synthesize interesting 2-D IIR filters which can be implemented by a parallel architecture. For example, the parallel form may be advantageous when designing a filter with multiple passbands.

The parallel implementation can also be useful for implementing a 2-D IIR filter whose impulse response is not confined to a single quadrant, such as a symmetric filter. In this case, an impulse response that has support on the entire $(n_1, n_2)$-plane can be partitioned into four separate impulse responses, one for each quadrant. A filter corresponding to each of the quadrant impulse responses can then be designed

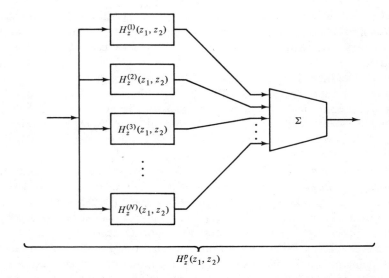

**Figure 5.6**   Parallel interconnection of $N$ simple 2-D IIR filters to form a more complex 2-D IIR filter $H_z^p(z_1, z_2)$.

and the four of them connected in parallel to realize the desired full-plane impulse response. We shall see an example of this later in Section 5.4 when we discuss space-domain filter design techniques.

Like the cascade architecture, the parallel architecture of Figure 5.6 can be interpreted in two ways. If each box is said to represent a physically separate filtering device, the output data are duplicated and sent to each of the $N$ parallel subfilters. Computation takes place simultaneously in each filter and the total amount of storage is the sum of storage needed by each of the subfilters. Alternatively, we could use a subroutine to implement a general subfilter as we did in the cascade case. The parallel implementation then consists of calling the subroutine $N$ times, using the same input data but different coefficients, and adding up the $N$ output signals. In this case, computation occurs sequentially since each subfilter output signal is computed in turn, but the total amount of primary storage required can be less.

## 5.2 ITERATIVE IMPLEMENTATIONS FOR 2-D IIR FILTERS [3, 4]

In this section we explore a different approach to the problem of implementing a 2-D digital filter with a rational transfer function. In many applications such as image processing, all the signal sample values are available simultaneously; that is, the entire signal is stored in a computer memory and is available for processing. (In contrast, when processing 1-D real-time signals, only the sample values corresponding to the past are available.) In this case it seems plausible to consider examining the 2-D output signal and using it in conjunction with the input signal to generate a "better" version of the output signal. The iterative implementation described in this section uses this familiar concept of feedback to generate successively better approximations to the desired output signal.

### 5.2.1 Basic Iterative Implementation

One of the motivations behind the iterative implementation is to attempt to realize IIR filters with impulse responses which are not recursively computable. In the 1-D case, the transfer function of an IIR filter can be factored into causal and anticausal parts which can be implemented separately using difference equations. In two or more dimensions, however, the lack of a factorization theorem precludes this approach, forcing us to use either an approximate factorization or to design an explicitly factorable transfer function. The iterative implementation offers a third alternative and, as we shall see later, also gives us a means of incorporating boundary conditions (see Section 5.2.3).

To make the notation a little less cumbersome, let us derive the iterative implementation in terms of the IIR frequency response $H(\omega_1, \omega_2)$ rather than the transfer function $H_z(z_1, z_2)$. In general, $H(\omega_1, \omega_2)$ can be represented as

$$H(\omega_1, \omega_2) \triangleq \frac{A(\omega_1, \omega_2)}{B(\omega_1, \omega_2)} = \frac{\sum_{l_1} \sum_{l_2} a(l_1, l_2) \exp(-j\omega_1 l_1 - j\omega_2 l_2)}{\sum_{k_1} \sum_{k_2} b(k_1, k_2) \exp(-j\omega_1 k_1 - j\omega_2 k_2)} \tag{5.10}$$

where the arrays $a(l_1, l_2)$ and $b(k_1, k_2)$ are finite-extent arrays. As before, we shall assume that the ratio is normalized so that $b(0, 0) = 1$.

Now, let us define a new trigonometric polynomial $C(\omega_1, \omega_2)$ as follows:

$$C(\omega_1, \omega_2) \triangleq 1 - B(\omega_1, \omega_2) \tag{5.11}$$

Then we can write

$$H(\omega_1, \omega_2) = A(\omega_1, \omega_2)/[1 - C(\omega_1, \omega_2)] \tag{5.12}$$

If we let $X(\omega_1, \omega_2)$ represent the spectrum of the input signal $x(n_1, n_2)$ and $Y(\omega_1, \omega_2)$ represent the spectrum of the output signal $y(n_1, n_2)$ and we assume that $|Y(\omega_1, \omega_2)|$ and $|X(\omega_1, \omega_2)|$ are bounded, then

$$Y(\omega_1, \omega_2) = H(\omega_1, \omega_2)X(\omega_1, \omega_2)$$
$$= \frac{A(\omega_1, \omega_2)X(\omega_1, \omega_2)}{1 - C(\omega_1, \omega_2)} \tag{5.13}$$

Multiplying both sides of this equation by $[1 - C(\omega_1, \omega_2)]$ and then rearranging terms, we get the implicit relation

$$Y(\omega_1, \omega_2) = A(\omega_1, \omega_2)X(\omega_1, \omega_2) + C(\omega_1, \omega_2)Y(\omega_1, \omega_2) \tag{5.14}$$

In the signal domain, this equation becomes

$$y(n_1, n_2) = a(n_1, n_2) ** x(n_1, n_2) + c(n_1, n_2) ** y(n_1, n_2) \tag{5.15}$$

where, as before, the double asterisk denotes 2-D convolution.

Equations (5.14) and (5.15) represent implicit relationships among the input signal, the output signal, and the coefficients of the IIR filter, but they do not necessarily represent a convenient, or even feasible, way of computing $y(n_1, n_2)$ given $x(n_1, n_2)$. Suppose, however, that we make an educated guess at the output signal. Then by substituting this guess into the right side of equation (5.15) for $y(n_1, n_2)$, we can compute a better approximation to $y(n_1, n_2)$. Naturally, this process can be continued. Let $y_{i-1}(n_1, n_2)$ represent the $(i-1)$st approximation to the correct output signal $y(n_1, n_2)$. Then we can compute the $i$th approximation by

$$y_i(n_1, n_2) = a(n_1, n_2) ** x(n_1, n_2) + c(n_1, n_2) ** y_{i-1}(n_1, n_2) \qquad (5.16)$$

In the frequency domain, this equation corresponds to

$$Y_i(\omega_1, \omega_2) = A(\omega_1, \omega_2)X(\omega_1, \omega_2) + C(\omega_1, \omega_2)Y_{i-1}(\omega_1, \omega_2) \qquad (5.17)$$

The question we must now address is whether or not the sequence of approximations $\{y_i(n_1, n_2)\}$ converges to the correct output signal $y(n_1, n_2)$. This question can be more easily approached in the frequency domain. We shall assume for convenience that $Y_{-1}(\omega_1, \omega_2) \triangleq 0$, so that

$$Y_0(\omega_1, \omega_2) = A(\omega_1, \omega_2)X(\omega_1, \omega_2) \qquad (5.18a)$$

$$Y_1(\omega_1, \omega_2) = [A(\omega_1, \omega_2) + C(\omega_1, \omega_2)A(\omega_1, \omega_2)]X(\omega_1, \omega_2) \qquad (5.18b)$$

and so on. In general, $Y_I(\omega_1, \omega_2)$ is given by

$$Y_I(\omega_1, \omega_2) = \left[A(\omega_1, \omega_2)\sum_{i=0}^{I} C^i(\omega_1, \omega_2)\right]X(\omega_1, \omega_2) \qquad (5.19)$$

Since

$$\sum_{i=0}^{I} C^i(\omega_1, \omega_2) = \frac{1 - C^{I+1}(\omega_1, \omega_2)}{1 - C(\omega_1, \omega_2)} \qquad (5.20)$$

we see that

$$Y_I(\omega_1, \omega_2) = A(\omega_1, \omega_2)\frac{1 - C^{I+1}(\omega_1, \omega_2)}{1 - C(\omega_1, \omega_2)}X(\omega_1, \omega_2) \qquad (5.21)$$

Now, if we assume that

$$|C(\omega_1, \omega_2)| < 1 \qquad (5.22)$$

then

$$\lim_{I\to\infty} Y_I(\omega_1, \omega_2) = \frac{A(\omega_1, \omega_2)X(\omega_1, \omega_2)}{1 - C(\omega_1, \omega_2)} \qquad (5.23)$$

$$= Y(\omega_1, \omega_2)$$

as desired. In essence, each iteration of equation (5.17) generates another term in an infinite geometric series. The series will converge if condition (5.22) is satisfied. In this case it is possible to show that $Y_i(\omega_1, \omega_2)$ converges uniformly to $Y(\omega_1, \omega_2)$ and $y_i(n_1, n_2)$ converges uniformly to $y(n_1, n_2)$ as $i$ goes to infinity.

Because $a(n_1, n_2)$ and $b(n_1, n_2)$ are arrays of finite extent, the computations that occur during each iteration are FIR filtering operations. Consequently, the rational frequency response $H(\omega_1, \omega_2)$ of a 2-D IIR filter can be realized exactly by an infinite number of FIR filtering operations.

The use of the iterative computation (5.16) can be visualized as a simple, first-order digital filter that processes a sequence of signals rather than a sequence of sample values. To help describe this analogy in a little more detail, let us generalize the input signal slightly to make it a function of the iteration variable $i$. Thus the input signal becomes $x_i(n_1, n_2)$. For the particular case described above, we can think of $x_i(n_1, n_2)$ as a 1-D step function in $i$ multiplied by $x(n_1, n_2)$.

$$x_i(n_1, n_2) = x(n_1, n_2)u(i) \qquad (5.24)$$

With this generalization in the form of the input signal, the iteration (5.16) becomes

$$y_i(n_1, n_2) = a(n_1, n_2) ** x_i(n_1, n_2) + c(n_1, n_2) ** y_{i-1}(n_1, n_2) \qquad (5.25)$$

This equation can be represented by the block diagram shown in Figure 5.7. During

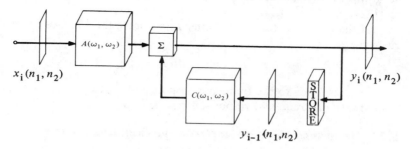

**Figure 5.7**    Block diagram for the iterative implementation of $H(\omega_1, \omega_2)$. (Courtesy of Dan E. Dudgeon, *IEEE Trans. Acoustics, Speech, and Signal Processing,* © 1980 IEEE.)

each iteration, estimates of entire signals are being fed around the loop. The STORE operator stores the result of the preceding iteration; it is analogous to the 1-D shift operator usually represented by $z^{-1}$ in block diagrams of 1-D digital filters. In this case, however, it stores an entire signal rather than a single sample value.

The condition for convergence of the iteration, $|C(\omega_1, \omega_2)| < 1$, is fairly restrictive. (In the next subsection, we shall examine some approaches for reducing or eliminating this restriction.) Not all stable 2-D IIR filters satisfy the condition for convergence. An IIR filter is BIBO stable as long as $C(\omega_1, \omega_2) \neq 1$. Obviously, the condition for convergence of the iteration, $|C(\omega_1, \omega_2)| < 1$, is much more restrictive than the condition for stability.

To compute an infinite number of iterations requires an infinite amount of computation. In practice, the number of iterations will be limited, giving a signal $y_I(n_1, n_2)$ that is only an approximation to the desired $y(n_1, n_2)$. It is appropriate to consider the size of the approximation error. This problem is more easily approached in the frequency domain. We can relate the Fourier transforms of $y_I(n_1, n_2)$ and $y(n_1, n_2)$ by the formula

$$Y_I(\omega_1, \omega_2) = Y(\omega_1, \omega_2)[1 - C^{I+1}(\omega_1, \omega_2)] \qquad (5.26)$$

Ideally, the complex ratio $Y_I(\omega_1, \omega_2)/Y(\omega_1, \omega_2)$ should be unity; thus we can take the absolute value of the difference between this ratio and unity as one measure of the error introduced by terminating the computation after $I$ iterations. Letting $E(\omega_1, \omega_2)$ represent the spectral error as a function of frequency, we can write

$$E(\omega_1, \omega_2) \triangleq \left| \frac{Y_I(\omega_1, \omega_2)}{Y(\omega_1, \omega_2)} - 1 \right| = |C(\omega_1, \omega_2)|^{I+1} \qquad (5.27)$$

If we specify a tolerable degree of spectral error over some band of frequencies $\Omega$ by writing

$$E(\omega_1, \omega_2) < \epsilon \qquad \text{for } (\omega_1, \omega_2) \in \Omega \qquad (5.28)$$

($\epsilon$ is a small positive constant), we can use equation (5.27) to determine how many iterations are needed to achieve the specification. Conversely, if the number of iterations $I$ is fixed, equation (5.27) can be used to specify the restrictions to be placed on $|C(\omega_1, \omega_2)|$.

If the process is terminated after $I$ iterations, the effective frequency response is given by

$$H_I(\omega_1, \omega_2) = A(\omega_1, \omega_2) \sum_{i=0}^{I} C^i(\omega_1, \omega_2) \qquad (5.29)$$

which corresponds to the frequency response of an FIR filter that approximates the desired rational frequency response $H(\omega_1, \omega_2)$.

### 5.2.2 Generalizations of the Iterative Implementation

Perhaps the most troublesome limitation of the iterative implementation described in the preceding subsection is that there exist stable, well-behaved IIR filters that do not satisfy the convergence condition (5.22). Hence these filters cannot be realized by the iterative computation. In this subsection we discuss some generalizations that allow any stable IIR filter to be implemented iteratively.

To make the derivations easier, let us look at the special case where the denominator polynomial $B(\omega_1, \omega_2)$ is purely real and strictly positive.

$$B(\omega_1, \omega_2) > 0 \qquad (5.30)$$

We can write the desired IIR frequency response as

$$H(\omega_1, \omega_2) = \frac{A(\omega_1, \omega_2)}{B(\omega_1, \omega_2)} = \frac{\lambda A(\omega_1, \omega_2)}{\lambda B(\omega_1, \omega_2)} \qquad (5.31)$$

where $\lambda$ is a constant parameter to be determined. Now we can redefine $C(\omega_1, \omega_2)$ to be

$$C(\omega_1, \omega_2) \triangleq 1 - \lambda B(\omega_1, \omega_2) \qquad (5.32)$$

Deriving an iterative computation as before for the frequency response $\lambda A(\omega_1, \omega_2) / \lambda B(\omega_1, \omega_2)$, we get

$$Y_i(\omega_1, \omega_2) = \lambda A(\omega_1, \omega_2) X(\omega_1, \omega_2) + C(\omega_1, \omega_2) Y_{i-1}(\omega_1, \omega_2) \qquad (5.33)$$

The condition for convergence, $|C(\omega_1, \omega_2)| < 1$, remains as before, but now $C(\omega_1, \omega_2)$ contains the free parameter $\lambda$. Since we have assumed that $B(\omega_1, \omega_2)$ is strictly positive, we see that $C(\omega_1, \omega_2) < 1$ for positive values of $\lambda$. All that remains is to guarantee that $C(\omega_1, \omega_2) > -1$ to satisfy the condition for convergence. This can be done by setting $\lambda$ to a constant in the range

$$0 < \lambda < \frac{2}{\max_{(\omega_1, \omega_2)} B(\omega_1, \omega_2)} \tag{5.34}$$

thus assuring that $|C(\omega_1, \omega_2)| < 1$. The case where $B(\omega_1, \omega_2)$ is real and positive has some practical significance, since it includes some symmetric 2-D IIR filters which are important in image processing.

In the more general case, $B(\omega_1, \omega_2)$ is a complex trigonometric polynomial function which satisfies

$$B(\omega_1, \omega_2) \neq 0 \tag{5.35}$$

for a stable 2-D IIR filter. For this general case, $H(\omega_1, \omega_2)$ can be rewritten as

$$H(\omega_1, \omega_2) = \frac{A(\omega_1, \omega_2)}{B(\omega_1, \omega_2)} = \frac{\lambda B^*(\omega_1, \omega_2)A(\omega_1, \omega_2)}{\lambda |B(\omega_1, \omega_2)|^2} \tag{5.36}$$

Now the denominator function $|B(\omega_1, \omega_2)|^2$ is real and positive, so we can proceed as before. We redefine $C(\omega_1, \omega_2)$ to be

$$C(\omega_1, \omega_2) \triangleq 1 - \lambda |B(\omega_1, \omega_2)|^2 \tag{5.37}$$

and the new iterative computation becomes

$$Y_i(\omega_1, \omega_2) = \lambda B^*(\omega_1, \omega_2)A(\omega_1, \omega_2)X(\omega_1, \omega_2) + C(\omega_1, \omega_2)Y_{i-1}(\omega_1, \omega_2) \tag{5.38}$$

As before, $C(\omega_1, \omega_2)$ must satisfy the convergence condition $|C(\omega_1, \omega_2)| < 1$. To ensure this, we set $\lambda$ to a constant in the range

$$0 < \lambda < \frac{2}{\max_{(\omega_1, \omega_2)} |B(\omega_1, \omega_2)|^2} \tag{5.39}$$

Aside from permitting the realization of any stable 2-D rational frequency response, the iterative computation (5.38) has some interesting properties. For instance, it is easy to demonstrate that the phase of the spectrum of the estimate converges to the correct phase after a single iteration. Notationally,

$$\phi[Y(\omega_1, \omega_2)] = \phi[Y_i(\omega_1, \omega_2)] \qquad \text{for } i \geq 0 \tag{5.40}$$

with

$$Y_{-1}(\omega_1, \omega_2) \triangleq 0$$

The explanation for this interesting observation follows from the fact that the phase of $B^*(\omega_1, \omega_2)$ is equal to the phase of $1/B(\omega_1, \omega_2)$. Thus

$$\begin{aligned}
\phi[Y_0(\omega_1, \omega_2)] &= \phi[A(\omega_1, \omega_2)] + \phi[B^*(\omega_1, \omega_2)] + \phi[X(\omega_1, \omega_2)] \\
&= \phi[A(\omega_1, \omega_2)] - \phi[B(\omega_1, \omega_2)] + \phi[X(\omega_1, \omega_2)] \\
&= \phi[Y(\omega_1, \omega_2)]
\end{aligned} \tag{5.41}$$

Succeeding iterations serve only to improve the magnitude of the spectrum of the estimate.

The iteration (5.38) is more powerful than the simpler iteration (5.17) in that it permits a larger class of frequency responses to be realized. However, it also requires more computation to be implemented. The input signal $x(n_1, n_2)$ must be prefiltered by the cascade of $\lambda B^*(\omega_1, \omega_2)$ and $A(\omega_1, \omega_2)$ rather than $A(\omega_1, \omega_2)$ alone. The FIR filter corresponding to $C(\omega_1, \omega_2)$ is also more complicated, involving roughly twice as much computation as before when implemented by direct convolution.

Because of the way we have redefined $C(\omega_1, \omega_2)$, the relation between the error $E(\omega_1, \omega_2)$ and $C(\omega_1, \omega_2)$ is still valid. We repeat it here for convenience:

$$E(\omega_1, \omega_2) = |C(\omega_1, \omega_2)|^{I+1} \tag{5.42}$$

If $|C(\omega_1, \omega_2)|$ is close to zero in some particular band of frequencies, the error $E(\omega_1, \omega_2)$ will rapidly go to zero in that band as the number of iterations increases. Conversely, if $|C(\omega_1, \omega_2)|$ is close to 1 for some spectral region, the error will remain significant in that region even after several iterations.

We might wish to have greater control over the distribution of the errors in the frequency plane in order to spread the errors more evenly across the entire frequency band or to reduce errors in an important frequency band at the expense of a relatively unimportant frequency band. We can gain this greater control by making one further generalization. Instead of using the factor $\lambda B^*(\omega_1, \omega_2)$ in equations (5.36) and (5.37), we can use a general frequency-dependent relaxation function $\lambda(\omega_1, \omega_2)$. Thus we can write

$$H(\omega_1, \omega_2) = \frac{\lambda(\omega_1, \omega_2)A(\omega_1, \omega_2)}{\lambda(\omega_1, \omega_2)B(\omega_1, \omega_2)} \tag{5.43}$$

and redefine $C(\omega_1, \omega_2)$ once again.

$$C(\omega_1, \omega_2) \triangleq 1 - \lambda(\omega_1, \omega_2)B(\omega_1, \omega_2) \tag{5.44}$$

Then the iterative computation becomes

$$Y_i(\omega_1, \omega_2) = \lambda(\omega_1, \omega_2)A(\omega_1, \omega_2)X(\omega_1, \omega_2) + C(\omega_1, \omega_2)Y_{i-1}(\omega_1, \omega_2) \tag{5.45}$$

The objective now becomes to pick $\lambda(\omega_1, \omega_2)$ so that $|C(\omega_1, \omega_2)| \simeq 0$ over an important frequency band. [In theory, we could pick $\lambda(\omega_1, \omega_2) = B^{-1}(\omega_1, \omega_2)$ and be done in one iteration, but we cannot implement $B^{-1}(\omega_1, \omega_2)$ directly.] For practical reasons, we constrain $\lambda(\omega_1, \omega_2)$ to be the frequency response of an FIR filter. By making the spatial extent of the inverse Fourier transform of $\lambda(\omega_1, \omega_2)$ relatively large, we can force $|C(\omega_1, \omega_2)|$ to be small. However, a large spatial extent requires more computation during each iteration of equation (5.45); consequently, there is a trade-off between the amount of computation done at each iteration and the number of iterations needed to drive the error $E(\omega_1, \omega_2)$ down to some prescribed value across the frequency band of interest.

### *5.2.3 Truncation, Boundary Conditions, and Signal Constraints [4]

In practice, the STORE operator in Figure 5.7, which must buffer an entire signal, has finite size. As the iterative computation continues, however, the extent of $y_i(n_1, n_2)$ will grow. Consequently, at some point $y_i(n_1, n_2)$ will exceed the available storage capacity, and information will be lost to further iterations.

A natural question to ask, therefore, is how truncation of the signal estimate at each iteration will affect the final result. For simplicity, let us consider the simple iteration of equation (5.16). If we have an input $x(n_1, n_2)$, we can use the iteration to find (in principle) a signal $y(n_1, n_2)$ that satisfies the implicit relation (5.15), repeated here.

$$y(n_1, n_2) = a(n_1, n_2) ** x(n_1, n_2) + c(n_1, n_2) ** y(n_1, n_2) \qquad (5.46)$$

Because of the finite capacity of the STORE operator, the iteration which is actually implemented may be written

$$\tilde{y}_i(n_1, n_2) = T[a(n_1, n_2) ** x(n_1, n_2) + c(n_1, n_2) ** \tilde{y}_{i-1}(n_1, n_2)] \qquad (5.47)$$

where $T[f(n_1, n_2)]$ is equal to $f(n_1, n_2)$ in some region I and equal to zero outside of $I$. It can be proven [4] that there exists a signal $\tilde{y}(n_1, n_2)$ which satisfies the following implicit relationship:

$$\tilde{y}(n_1, n_2) = T[a(n_1, n_2) ** x(n_1, n_2) + c(n_1, n_2) ** \tilde{y}(n_1, n_2)] \qquad (5.48)$$

The difference between $y(n_1, n_2)$ and $\tilde{y}(n_1, n_2)$ represents the error caused by truncation. We can define an error signal

$$e(n_1, n_2) \triangleq y(n_1, n_2) - \tilde{y}(n_1, n_2) \qquad (5.49)$$

and then derive an implicit relation for $e(n_1, n_2)$, using the fact that the truncation operation is linear, distributive with respect to addition, and idempotent ($T[T[x]] = T[x]$).

$$\begin{aligned} T[e(n_1, n_2)] &= T[y(n_1, n_2) - \tilde{y}(n_1, n_2)] \\ &= T[a(n_1, n_2) ** x(n_1, x_2) + c(n_1, n_2) ** y(n_1, n_2)] \\ &\quad - T[a(n_1, n_2) ** x(n_1, n_2) + c(n_1, n_2) ** \tilde{y}(n_1, n_2)] \\ &= T[c(n_1, n_2) ** y(n_1, n_2)] - T[c(n_1, n_2) ** \tilde{y}(n_1, n_2)] \\ &= T[c(n_1, n_2) ** \{y(n_1, n_2) - \tilde{y}(n_1, n_2)\}] \\ &= T[c(n_1, n_2) ** e(n_1, n_2)] \end{aligned} \qquad (5.50)$$

Thus, in the region $I$, the error signal satisfies

$$e(n_1, n_2) = c(n_1, n_2) ** e(n_1, n_2) \qquad \text{for } (n_1, n_2) \in I \qquad (5.51)$$

Outside $I$ the error signal is simply $y(n_1, n_2)$ since $\tilde{y}(n_1, n_2)$ is zero.

The error signal $e(n_1, n_2)$ can be obtained from an iterative computation which incorporates boundary conditions, specifically

$$e_i(n_1, n_2) = \begin{cases} c(n_1, n_2) ** e_{i-1}(n_1, n_2), & (n_1, n_2) \in I \\ y(n_1, n_2), & (n_1, n_2) \notin I \end{cases} \qquad (5.52)$$

The iterative computation can be viewed as a special case of the more general iteration

$$f_i(n_1, n_2) = \begin{cases} a(n_1, n_2) ** x(n_1, n_2) + c(n_1, n_2) ** f_{i-1}(n_1, n_2), & (n_1, n_2) \in I \\ bc(n_1, n_2), & (n_1, n_2) \notin I \end{cases} \qquad (5.53)$$

where $f_i(n_1, n_2)$ is the $i$th estimate of the output signal, $x(n_1, n_2)$ is the input signal as before, and $bc(n_1, n_2)$ represents the boundary conditions. The boundary conditions

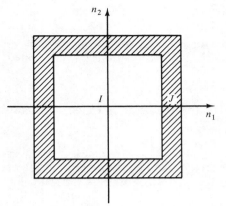

**Figure 5.8** In general, the region $J$ surrounds the region $I$ in the $(n_1, n_2)$-plane. The thickness of $J$ depends on the extent of the coefficient array $c(n_1, n_2)$.

are specified in a region $J$, which is generally taken to surround the region $I$ as shown in Figure 5.8. The thickness of the region $J$ depends on the spatial extent of the coefficient array $c(n_1, n_2)$. Basically, $J$ must be wide enough to cover the array $c(n_1, n_2)$ when $f_i(n_1, n_2)$ is being computed for $(n_1, n_2)$ on the edge of the region $I$.

Since $y(n_1, n_2) = \tilde{y}(n_1, n_2) + e(n_1, n_2)$, it can be shown that $y(n_1, n_2)$ is a solution to the iteration

$$y_i(n_1, n_2) = \begin{cases} a(n_1, n_2) ** x(n_1, n_2) + c(n_1, n_2) ** \\ \qquad\qquad y_{i-1}(n_1, n_2), & (n_1, n_2) \in I \\ y(n_1, n_2), & (n_1, n_2) \notin I \end{cases}$$

$$(5.54)$$

Thus, despite the effects of truncation, the correct signal can still be computed if its values are known outside the region $I$. [If the answer is desired only inside the region $I$, then $y(n_1, n_2)$ must be known only in the region $J$, not outside $I + J$.]

To demonstrate the use of boundary conditions to mitigate the effects of truncation, we present a very simple 1-D example. Let us consider the iteration

$$y_i(n) = A\delta(n) + by_{i-1}(n - 1) + by_{i-1}(n + 1) \qquad (5.55)$$

which has the solution

$$y(n) = a^{|n|} \qquad (5.56)$$

The parameters $A$ and $b$ are related to the constant $a$ as follows:

$$A = \frac{1 - a^2}{1 + a^2}, \qquad b = \frac{a}{1 + a^2} \qquad (5.57)$$

Figure 5.9(a) shows $y(n)$ for the case $a = 0.5$. Now, suppose the iteration is truncated so that (5.55) becomes

$$\tilde{y}_i(n) = \begin{cases} A\delta(n) + b\tilde{y}_{i-1}(n - 1) + b\tilde{y}_{i-1}(n + 1), & |n| < 4 \\ 0, & |n| \geq 4 \end{cases} \qquad (5.58)$$

Figure 5.9(b) shows a plot of $\tilde{y}(n)$ obtained by iterating (5.58) until convergence was achieved. The difference $y(n) - \tilde{y}(n)$ is shown in Figure 5.9(c).

We can generate the correct $y(n)$ using the truncated iteration if we apply the correct boundary conditions at $n = \pm 4$, namely, $y(\pm 4) = a^4 = 0.0625$. Figure 5.10 shows the output signal after $i = 2, 10, 20,$ and $50$ iterations to demonstrate convergence to the correct solution.

Although we shall not discuss it in detail here, signal constraints can be incorporated into the iterative computation as well. The imposition of boundary conditions may be regarded as one type of constraint placed on the output signal. For some

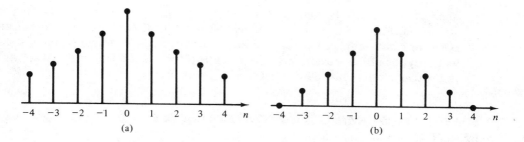

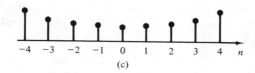

**Figure 5.9**   Correct solution (a), the solution to the truncated iteration (b), and the difference signal (c) for a simple iterative implementation.

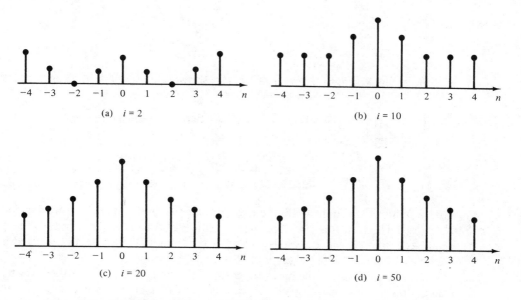

**Figure 5.10**   Generating the correct solution $y(n)$ with a truncated interation through the use of boundary conditions. (a) 2 iterations. (b) 10 iterations. (c) 20 iterations. (d) 50 iterations.

applications, it may be known that the correct output signal must obey other constraints such as being strictly positive or bandlimited. In these cases, a constraint operator can be applied at each iteration to ensure that the output signal, if it exists, will obey the proper constraints. (See Chapter 7 for a discussion of signal constraints in the context of iterative reconstruction algorithms.)

## 5.3  SIGNAL FLOWGRAPHS AND STATE-VARIABLE REALIZATIONS

In this section, we shall briefly look at signal flowgraphs for 2-D IIR filters. Like their 1-D counterparts, 2-D signal flowgraphs attempt to depict in a circuit-like manner the transformations undergone by 2-D signals as they flow through 2-D filters. We discussed flowgraphs and their relation to the $z$-transform in Section 4.2.6. Shortly, we shall examine the circuit elements typically used in 2-D signal flowgraphs and their realizations. Later in this section we relate the signal flowgraphs to 2-D state-variable realizations for IIR filters.

### 5.3.1  Circuit Elements and Their Realization

Let us consider the simple 2-D IIR filter transfer function given by

$$H_z(z_1, z_2) = \frac{a_{00} + a_{10}z_1^{-1} + a_{01}z_2^{-1} + a_{11}z_1^{-1}z_2^{-1} + a_{21}z_1^{-2}z_2^{-1}}{1 - c_{10}z_1^{-1} - c_{01}z_2^{-1} - c_{11}z_1^{-1}z_2^{-1} - c_{20}z_1^{-2} - c_{21}z_1^{-2}z_2^{-1}}$$

$$= \frac{A_z(z_1, z_2)}{1 - C_z(z_1, z_2)} \qquad (5.59)$$

A simple block diagram for $H_z(z_1, z_2)$ is shown in Figure 5.11. The input signal

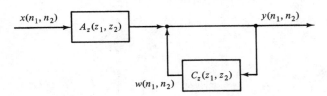

**Figure 5.11**  Block diagram for the transfer function $H_z(z_1, z_2) = A_z(z_1, z_2)/[1 - C_z(z_1, z_2)]$. Circuit nodes with more than one signal path entering are considered summing nodes. All signals paths leaving a common circuit node contain the same signal.

$x(n_1, n_2)$ flows through a filter corresponding to the numerator transfer function $A_z(z_1, z_2)$. The resulting signal is added to the signal $w(n_1, n_2)$ to produce the output signal $y(n_1, n_2)$. The denominator transfer function $1 - C_z(z_1, z_2)$ is realized by the feedback loop containing $C_z(z_1, z_2)$.

To be truly useful, the block diagram of Figure 5.11 must be broken down into a signal flowgraph consisting of the canonic circuit elements introduced in Section 4.2.6—summers, gains, and shifts. Since we are dealing with two dimensions, there are two fundamental shift operators which may occur along a signal flow path, the horizontal shift indicated by $z_1^{-1}$ and the vertical shift indicated by $z_2^{-1}$. [We shall omit from consideration the inverse-shift operators $z_1$ and $z_2$. In most cases of practical interest, they can be eliminated by multiplying both the numerator and denominator polynomials of $H_z(z_1, z_2)$ by the appropriate powers of $z_1^{-1}$ and $z_2^{-1}$.]

Let us look at a signal flowgraph representing the numerator polynomial

$$A_z(z_1, z_2) = a_{00} + a_{10}z_1^{-1} + a_{01}z_2^{-1} + a_{11}z_1^{-1}z_2^{-1} + a_{21}z_1^{-2}z_2^{-1} \tag{5.60}$$

which is shown in Figure 5.12. Note the chain of two $z_1^{-1}$ operators descending on the

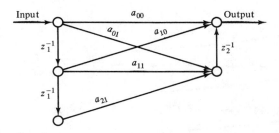

**Figure 5.12**  Signal flowgraph depicting the transfer function $A_z(z_1, z_2)$.

left and the single $z_2^{-1}$ operator ascending on the right. The nodes along these two vertical paths are connected by branches with the appropriate gains. If we label the nodes in both the $z_1^{-1}$ chain and the $z_2^{-1}$ chain 0, 1, 2, and so on, from the top down, the $i$th node in the $z_1^{-1}$ chain is connected to the $j$th node in the $z_2^{-1}$ chain by a branch with a gain factor of $a_{ij}$.

Similarly, the signal flowgraph for the polynomial $C_z(z_1, z_2)$ is shown in Figure 5.13. Since there is no $c_{00}$ term ($c_{00} = 0$), there is no direct connection between the

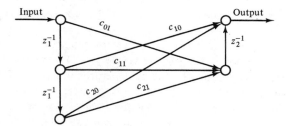

**Figure 5.13**  Signal flowgraph depicting the polynomial $C_z(z_1, z_2)$.

input and output nodes of this signal flowgraph. Thus any path from the input node to the output node will encounter at least one $z_1^{-1}$ or $z_2^{-1}$ shift operator. This fact becomes important when $C_z(z_1, z_2)$ is placed in a feedback loop as in Figure 5.11.

At this point it is appropriate to discuss realizations for the two shift operators $z_1^{-1}$ and $z_2^{-1}$. At their simplest level, the shift operators merely select the "previous" sample value in the horizontal or vertical direction. When the input to a $z_1^{-1}$ operator is the sample $x(n_1, n_2)$, the output will be $x(n_1 - 1, n_2)$. Similarly, for a $z_2^{-1}$ operator, the output will be $x(n_1, n_2 - 1)$ when the input is $x(n_1, n_2)$. Consequently, a realization of either shift operator must embody the appropriate amount of memory to retain the "previous" sample in the appropriate direction.

The flowgraph, like the input and output masks, does not explicitly indicate the order in which the signal indices $(n_1, n_2)$ assume their integer values or the range of values taken on by $n_1$ and $n_2$. Unfortunately, the particular ordering relation chosen (see Section 4.1.4) has an impact on the particular realizations of the $z_1^{-1}$ and $z_2^{-1}$ operators [5]. For example, let us assume that we will cycle through the indices in a row-by-row manner and that the length of interest of each row is $N$ samples. Thus the ordered pair $(n_1, n_2)$ takes on the values $(0, 0)$, $(1, 0)$, . . . , $(N - 1, 0)$, $(0, 1)$, $(1, 1)$, . . . , $(N - 1, 1)$, $(0, 2)$, as shown in Figure 5.14. In this case, the $z_1^{-1}$ operator is

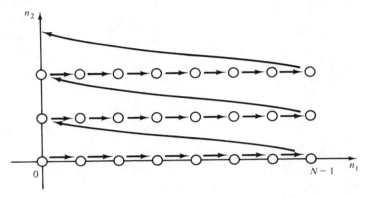

**Figure 5.14**   Row-by-row ordering of samples with $N$ samples in each row.

implemented simply as a single storage register, releasing the value $x(n_1 - 1, n_2)$ when the new value $x(n_1, n_2)$ is clocked into it. At the end of each row, however, the $z_1^{-1}$ register must be reset to the initial condition of the next row, so that the sample value "preceeding" $x(0, n_2)$ is the initial value $x(-1, n_2)$ and not the actual previous value $x(N - 1, n_2 - 1)$.

The $z_2^{-1}$ operator for this example is even more complicated. Because of the row-by-row ordering we have imposed, the sample $x(n_1, n_2 - 1)$ precedes the sample $x(n_1, n_2)$ by exactly $N$ samples. Consequently, the $z_2^{-1}$ operator must be able to buffer $N$ sample values in a first-in, first-out manner like a shift register memory. To begin, the shift register must be loaded with the initial values $x(n_1, -1)$ for $0 \leq n_1 < N$. Thereafter, sample values simply enter and exit the $z_2^{-1}$ shift register without regard to row boundaries.

It should be apparent to the reader that the simple elegance of the 1-D filter implementation based on signal flowgraphs has been lost in the 2-D case. The 1-D shift operator $z^{-1}$ always corresponds to a single storage register, one unit of memory. As we have just seen, the $z_1^{-1}$ operator can also be implemented with a single unit of memory in the row-by-row case (if we are careful to reset it at the end of each row), but the $z_2^{-1}$ operator requires $N$ units of storage. Naturally, in a column-by-column implementation the storage requirements of the $z_1^{-1}$ and $z_2^{-1}$ operators are reversed. Finally, the amount of storage needed to implement the $z_2^{-1}$ operator depends on the number of samples in each row that we are interested in computing. Consequently, a realization of a $z_2^{-1}$ operator for the case $N = 64$ may be of little value for the case $N = 75$.

We have been considering the flowgraph to be similar to a circuit schematic, telling us which components are to be connected together. However, it may be fruitful to consider an alternative interpretation of the flowgraph, namely as an algorithm for computing the desired output sample values. For example, let us return to the flowgraph for $A_z(z_1, z_2)$ given in Figure 5.12 and write an equation for computing the sample value at each node. We shall let $x(n_1, n_2)$ denote the input signal and $y(n_1, n_2)$ denote the output signal. The node signals in the $z_1^{-1}$ chain will be labeled $g_0(n_1, n_2)$, $g_1(n_1, n_2)$, and $g_2(n_1, n_2)$ for nodes 0, 1, and 2, respectively (top to bottom). Similarly, the node signals in the $z_2^{-1}$ chain will be labeled $v_0(n_1, n_2)$ and $v_1(n_1, n_2)$. Then we can write

$$g_2(n_1, n_2) = g_1(n_1 - 1, n_2) \tag{5.61a}$$

$$g_1(n_1, n_2) = g_0(n_1 - 1, n_2) \tag{5.61b}$$

$$g_0(n_1, n_2) = x(n_1, n_2) \tag{5.61c}$$

$$v_0(n_1, n_2) = a_{00}g_0(n_1, n_2) + a_{10}g_1(n_1, n_2) + v_1(n_1, n_2 - 1) \tag{5.61d}$$

$$v_1(n_1, n_2) = a_{01}g_0(n_1, n_2) + a_{11}g_1(n_1, n_2) + a_{21}g_2(n_1, n_2) \tag{5.61e}$$

$$y(n_1, n_2) = v_0(n_1, n_2) \tag{5.61f}$$

(The reason for the peculiar ordering of the equations will become apparent in a moment.) As we cycle through the desired set of ordered pairs $(n_1, n_2)$, we can use the equations (5.61) to generate the output sample values. Naturally, the ordering of the values $(n_1, n_2)$ is important to avoid the situation where a sample of a node signal which has not yet been computed is needed in another computation.

As an example, let us explore a software implementation of equations (5.61) for the case where the input samples arrive row by row, as in Figure 5.14, and the values of the output signal in an $N \times N$ region are desired. Figure 5.15 shows one

```
; compute equations (5.61) for row-by-row input stream
array v₁(0:N−1); allocate storage for z₂⁻¹ shift
scalar v₀, h₀; output and input nodes
scalar h₁, h₂; allocate storage for z₁⁻¹ shifts
index n₁, n₂; index variables
constants a₀₀, a₁₀, a₀₁, a₁₁, a₂₁; filter coefficients
{ instructions to initialize the array v₁ to zero}
loop for n₂ = 0 to N−1
{
       h₀ ← 0
       h₁ ← 0
       loop for n₁ = 0 to N−1
       {
              h₂ ← h₁                           ; equation (5.61a)
              h₁ ← h₀                           ; equation (5.61b)
              h₀ ← next_input_sample            ; equation (5.61c)
              v₀ ← a₀₀h₀ + a₁₀h₁ + v₁(n₁)       ; equation (5.61d)
              v₁(n₁) ← a₀₁h₀ + a₁₁h₁ + a₂₁h₂    ; equation (5.61e)
              next_output_sample ← v₀           ; equation (5.61f)

       }
}
```

**Figure 5.15**  Possible software implementation for equations (5.61).

possible software implementation of this process (in an admittedly loose form) using a generic high-level language. Take a minute to study Figure 5.15. By examining the computer code in detail, we see the reason for the ordering of equations (5.61a–f), namely, to avoid using an extra, temporary variable to hold sample values when the node signals are updated. Because of the row-by-row sequencing of input samples, only one storage variable is necessary to represent each of the node signals in the $z_1^{-1}$ chain ($h_0$, $h_1$, and $h_2$). However, $N$ storage elements are needed to implement a $z_2^{-1}$ shift operator; hence an array, $v_1(n_1)$, is used to store the sample values that will be needed to compute the output samples in the *next* row. The array $v_1(n_1)$ is initialized with the horizontal initial conditions, usually zeros. At the beginning of each row, $h_0$ and $h_1$ are also reinitialized with zeros. This corresponds to the resetting of the $z_1^{-1}$ shift operators mentioned previously.

Given this age of microprocessors and very large scale integrated circuits, it is appropriate to mention one possible implementation scheme for flowgraphs involving several processors rather than a single processor. Stated simply, the idea is to use a separate processor to compute the sample values of each node signal in the flowgraph. Naturally, all of the nodal processors must be informed of the indexing order for $(n_1, n_2)$ and when to reset any storage elements.

Basically, the scheme would involve two alternating processing epochs, a communication epoch and a computation epoch. During the communications epoch each nodal processor makes available the current sample value of its node signal to any other processor that needs the information. During the computation epoch, each processor computes the next value of its node signal based on the information it gathered during the communications epoch. Each nodal processor is responsible for buffering the sample values, obtained from other processors, which will be needed for subsequent computations. In this way, the $z_1^{-1}$ and $z_2^{-1}$ shift operators are realized.

### 5.3.2 Minimizing the Number of Shift Operators

Let us return to the original problem of this section, drawing a detailed signal flow-graph for the system function $H_z(z_1, z_2)$ given by equation (5.59). We can do this simply enough by combining the flowgraphs in Figures 5.11, 5.12 and 5.13 to get the flowgraph shown in Figure 5.16. This flowgraph can be made even simpler. Because the shift operation is distributive over addition, we can combine the two $z_2^{-1}$ operators

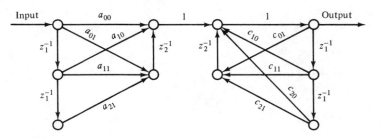

**Figure 5.16**   Detailed signal flowgraph for the system function $H_z(z_1, z_2)$.

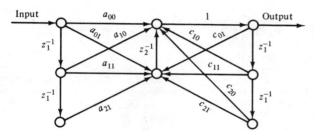

**Figure 5.17**   Improved signal flowgraph for $H_z(z_1, z_2)$.

into a single one, yielding the flowgraph of Figure 5.17. Doing so reduces the number of shift operators that need to be implemented and consequently the amount of storage necessary.

There are other signal flowgraphs which give rise to the desired system function $H_z(z_1, z_2)$. For example, we could invert the order of the $A_z(z_1, z_2)$ filter and the feedback loop containing $C_z(z_1, z_2)$ to obtain the block diagram shown in Figure 5.18.

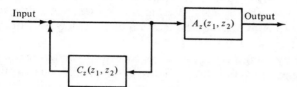

**Figure 5.18**   Alternative block diagram for $H_z(z_1, z_2)$.

Then, when we substitute Figures 5.12 and 5.13 for the blocks as before, the two $z_1^{-1}$ chains will contain the same data and can be merged to yield the signal flowgraph shown in Figure 5.19. This flowgraph has a total of four shift operators, and it minimizes the number of $z_1^{-1}$ operators.

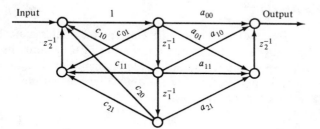

**Figure 5.19**   Detailed signal flowgraph for $H_z(z_1, z_2)$ which minimizes the number of $z_1^{-1}$ shift operators.

Another signal flowgraph that minimizes the number of $z_1^{-1}$ operators may be obtained from Figure 5.19 by the 2-D transposition theorem to obtain a transposed network. Like its 1-D counterpart [1], the 2-D transposition theorem states that the transposed network, which is obtained by reversing the directions of all the arrows in a signal flowgraph, will have the same system function as the original network. If we reverse the direction of all the arrows in Figure 5.19 and then redraw the flowgraph with the input port on the left and the output port on the right, we get the flowgraph shown in Figure 5.20. This transposed flowgraph may be preferred in implementations

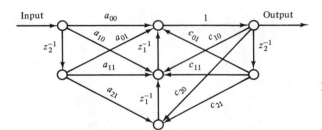

**Figure 5.20** Another signal flowgraph representing $H_z(z_1, z_2)$. It was obtained by transposing the flowgraph in Figure 5.19.

with limited wordlengths since the attenuation due to the "zeros" of $H_z(z_1, z_2)$ occurs before the gain due to the "poles," thus lessening somewhat the possibility of arithmetic overflow in the intermediate computations.

Using the notion of transposition at both the flowgraph level and the block diagram level (note that Figure 5.18 is the transpose of Figure 5.11), the flowgraph can be manipulated to yield a realization that minimizes the total number of shift operators. As we saw earlier, however, a $z_2^{-1}$ operator will require substantially more storage than a $z_1^{-1}$ operator for a row-by-row ordering of input samples. Consequently, it may be more economical to minimize not the total number of shift operators (as in the 1-D case) but the number of $z_2^{-1}$ operators.

If the filter is realized by using a separate microprocessor to compute samples of each node signal, storage may be less of an issue. In this case, we may want to minimize the total number of nodes in a flowgraph in order to reduce the number of microprocessors in an implementation.

As digital technology progresses, the relative costs of storage, computation, and interconnectivity keep changing. In the future, digital systems designers may have radically different criteria for optimizing a filter realization.

### 5.3.3 State-Variable Realizations

In 1-D linear systems theory and control theory, the concept of a filter state has played an important role. Basically, the filter state at any point in time contains all the information necessary to compute the remainder of the filter output signal given the input signal. One-dimensional single-input, single-output filter realizations based on a state-variable model can be written in the form

$$\mathbf{s}(n + 1) = \mathbf{As}(n) + \mathbf{B}x(n) \qquad (5.62a)$$

$$y(n) = \mathbf{Cs}(n) + Dx(n) \qquad (5.62b)$$

This form relates the input $x(n)$ and the output $y(n)$ through a state vector $\mathbf{s}(n)$. The state vector evolves in time according to equation (5.62a). The matrices $\mathbf{A}$, $\mathbf{B}$, $\mathbf{C}$, and the $1 \times 1$ matrix $D$ govern the exact form of the input–output relationship. (In general, these matrices may vary with the index $n$ and the input and output signals may be vectors as well.) Quite often the components of the state vector are taken to be the outputs of the $z^{-1}$ delay operators in a flowgraph representation of the 1-D filter. A classic problem in state-variable theory is to find the matrices $\mathbf{A}$, $\mathbf{B}$, $\mathbf{C}$, and

$D$ which will realize a particular system function $H_z(z)$ with a minimum number of state variables.

A similar approach may be taken to develop a 2-D state-variable model. We shall briefly outline the model developed by Roesser [6] and studied further by Kung *et al.* [7]. Roesser's model makes use of two kinds of state variables: horizontally propagating state variables, which we shall denote as $g_0, g_1, \ldots$, and vertically propagating state variables, which we shall denote as $v_0, v_1, \ldots$. The horizontally propagating state variables, taken together, form a vector $\mathbf{g}(n_1, n_2)$; similarly, a vector $\mathbf{v}(n_1, n_2)$ can also be formed. The state-variable model for a 2-D LSI single-input, single-output filter then takes the form

$$\begin{bmatrix} \mathbf{g}(n_1 + 1, n_2) \\ \mathbf{v}(n_1, n_2 + 1) \end{bmatrix} = \begin{bmatrix} \mathbf{A}_1 & \mathbf{A}_2 \\ \mathbf{A}_3 & \mathbf{A}_4 \end{bmatrix} \begin{bmatrix} \mathbf{g}(n_1, n_2) \\ \mathbf{v}(n_1, n_2) \end{bmatrix} + \begin{bmatrix} \mathbf{B}_1 \\ \mathbf{B}_2 \end{bmatrix} x(n_1, n_2) \qquad (5.63\text{a})$$

$$y(n_1, n_2) = [\mathbf{C}_1 \quad \mathbf{C}_2] \begin{bmatrix} \mathbf{g}(n_1, n_2) \\ \mathbf{v}(n_1, n_1) \end{bmatrix} + Dx(n_1, n_2) \qquad (5.63\text{b})$$

Note in equation (5.63a) that $\mathbf{g}(n_1, n_2)$ is propagated horizontally to get $\mathbf{g}(n_1 + 1, n_2)$ and $\mathbf{v}(n_1, n_2)$ is propagated vertically to get $\mathbf{v}(n_1, n_2 + 1)$.

We can compute a system function

$$F_z(z_1, z_2) \triangleq \frac{Y_z(z_1, z_2)}{X_z(z_1, z_2)}$$

for this class of filters by taking the 2-D $z$-transforms of equations (5.63) and solving for the ratio $Y_z / X_z$. Doing so gives us

$$F_z(z_1, z_2) = [\mathbf{C}_1 \quad \mathbf{C}_2] \left( \begin{bmatrix} \mathbf{Z}_1 & 0 \\ 0 & \mathbf{Z}_2 \end{bmatrix} - \begin{bmatrix} \mathbf{A}_1 & \mathbf{A}_2 \\ \mathbf{A}_3 & \mathbf{A}_4 \end{bmatrix} \right)^{-1} \begin{bmatrix} \mathbf{B}_1 \\ \mathbf{B}_2 \end{bmatrix} + D \qquad (5.64)$$

The submatrix $\mathbf{Z}_1$ is simply $z_1$ times an identity matrix of the appropriate size. Similarly, $\mathbf{Z}_2$ is $z_2$ times an identity matrix. The objective of the state-variable realization problem is to find the matrices $\mathbf{A}, \mathbf{B}, \mathbf{C}$, and $D$ which yield an $F_z(z_1, z_2)$ that equals or approximates a desired system function $H_z(z_1, z_2)$. In essence, equations (5.63) represent an implementation for which a design algorithm must be found. Space does not permit us to pursue an extended discussion of the design problem; the interested reader is referred to Kung et al. [7] and, in general, the control theory literature [8]. Below we shall discuss state-variable realizations that are derived from signal flowgraphs.

One choice for the state variables is the output signals from the shift operators. Thus $\mathbf{g}(n_1, n_2)$ is a vector containing the output signals from the $z_1^{-1}$ operators and $\mathbf{v}(n_1, n_2)$ contains the output signals from the $z_2^{-1}$ operators. (Note that the output signal of a shift-operator signal path is not necessarily the same as the nodal signal at the node to which the signal path points.) If a state variable corresponds to the output of a shift operator, the next value of that state variable must correspond to the input of the shift operator. To obtain the submatrices $\mathbf{A}_1, \mathbf{A}_2, \mathbf{A}_3$, and $\mathbf{A}_4$ in equation (5.63a), we write the input signal of each shift operator in terms of the outputs of all the shift operators, taking care to include all shift-free paths from output to input.

Let us return to the signal flowgraph in Figure 5.20 and develop a state-variable implementation from it. We shall call the output of the top $z_1^{-1}$ operator $g_0(n_1, n_2)$, the output of the lower $z_1^{-1}$ operator $g_1(n_1, n_2)$, the output of the left $z_2^{-1}$ operator $v_0(n_1, n_2)$, and the output of the right $z_2^{-1}$ operator $v_1(n_1, n_2)$ as indicated in Figure 5.21.

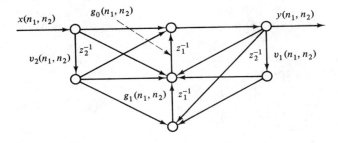

**Figure 5.21** Flowgraph of Figure 5.20 showing signal labeling conventions. Note that labels such as $g_0(n_1, n_2)$ correspond to the outputs of the signal paths containing shift operators and not to the nodal signals.

For this example, equations (5.63) become

$$
\begin{bmatrix} g_0(n_1 + 1, n_2) \\ g_1(n_1 + 1, n_2) \\ v_0(n_1, n_2 + 1) \\ v_1(n_1, n_2 + 1) \end{bmatrix} = \begin{bmatrix} c_{10} & 1 & a_{11} + a_{01}c_{10} & c_{11} + c_{01}c_{10} \\ c_{20} & 0 & a_{21} + a_{01}c_{20} & c_{21} + c_{01}c_{20} \\ 0 & 0 & 0 & 0 \\ 1 & 0 & a_{01} & c_{01} \end{bmatrix} \begin{bmatrix} g_0(n_1, n_2) \\ g_1(n_1, n_2) \\ v_0(n_1, n_2) \\ v_1(n_1, n_2) \end{bmatrix}
$$

$$
+ \begin{bmatrix} a_{10} + a_{00}c_{10} \\ a_{00}c_{20} \\ 1 \\ a_{00} \end{bmatrix} x(n_1, n_2) \tag{5.65a}
$$

$$
y(n_1, n_2) = \begin{bmatrix} 1 & 0 & a_{01} & c_{01} \end{bmatrix} \begin{bmatrix} g_0(n_1, n_2) \\ g_1(n_1, n_2) \\ v_0(n_1, n_2) \\ v_1(n_1, n_2) \end{bmatrix} + [a_{00}]x(n_1, n_2) \tag{5.65b}
$$

These two vector-matrix equations represent an algorithm for computing the samples of the output signal from the samples of the input signal. Just as in the preceding subsection, the amount of memory required to store the state variables depends on the order in which the output samples are to be computed. It is possible to envision a multiprocessor architecture for computing equation (5.65a) by assigning each processor the responsibility of computing the next value of a particular state variable given the current input value and the current state-variable values. Equation (5.65b) could be implemented by a fifth microprocessor to generate the desired output signal values. In such an architecture, minimization of the number of microprocessors corresponds to the minimization of the number of state variables, a problem studied thoroughly in the control theory literature [7, 8]. Other state-variable forms with the same number of state variables can also be found that will realize the same system function $H_z(z_1, z_2)$ and may exhibit lower coefficient sensitivity or round-off noise [7, 9].

For the special case of "all-pole" 2-D IIR filters, that is, filters with a system function of the form

$$H_z(z_1, z_2) = \frac{a_{00}}{B_z(z_1, z_2)} \tag{5.66}$$

where $a_{00}$ is a constant and $B_z(z_1, z_2)$ is a 2-D polynomial, it can be shown that state-variable realizations based on signal flowgraphs, using the output of the shift operators as the state variables, require the minimum number of state variables. They are *minimal* realizations [7].

Interestingly enough, in the more general case where the numerator polynomial $A_z(z_1, z_2)$ is not constant, state-variable realizations based on conventional signal flowgraphs like that of Figure 5.20 may not be minimal. Consider the following example. Let

$$H_z(z_1, z_2) = \frac{a_{10}z_1^{-1} + a_{01}z_2^{-1} + a_{11}z_1^{-1}z_2^{-1}}{1 - c_{10}z_1^{-1} - c_{01}z_2^{-1} - c_{11}z_1^{-1}z_2^{-1}} \tag{5.67}$$

A signal flowgraph representation for this system function using three shift operators is shown in Figure 5.22.

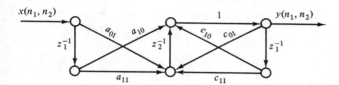

Figure 5.22 Signal flowgraph representation of a system function. Note that there is no $a_{00}$ branch.

Kung *et al.* [7] have shown that the following state-variable equations, which use only two shift operators, will also realize $H_z(z_1, z_2)$.

$$\begin{bmatrix} g(n_1 + 1, n_2) \\ v(n_1, n_2 + 1) \end{bmatrix} = \begin{bmatrix} c_{10} & p \\ q & c_{01} \end{bmatrix} \begin{bmatrix} g(n_1, n_2) \\ v(n_1, n_2) \end{bmatrix} + \begin{bmatrix} 1 \\ 1 \end{bmatrix} x(n_1, n_2) \tag{5.68a}$$

$$y(n_1, n_2) = [a_{10} \quad a_{01}] \begin{bmatrix} g(n_1, n_2) \\ v(n_1, n_2) \end{bmatrix} \tag{5.68b}$$

The constants $p$ and $q$ are determined by solving the simultaneous nonlinear equations

$$pq = c_{11} + c_{10}c_{01} \tag{5.69a}$$

$$a_{10}p^2 - (a_{01}c_{10} + a_{10}c_{01} + a_{11})p + (a_{01}c_{11} + a_{01}c_{10}c_{01}) = 0 \tag{5.69b}$$

Using equations (5.64) and (5.69), it is straightforward but tedious to show that

$$F_z(z_1, z_2) = [a_{10} \quad a_{01}] \begin{bmatrix} z_1 - c_{10} & -p \\ -q & z_2 - c_{01} \end{bmatrix}^{-1} \begin{bmatrix} 1 \\ 1 \end{bmatrix} \tag{5.70}$$

does, in fact, equal the $H_z(z_1, z_2)$ given in equation (5.67). From equations (5.68), we can construct a signal flowgraph with only two shift operators (Figure 5.23).

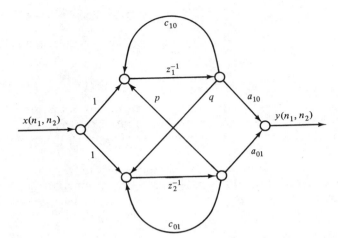

**Figure 5.23** Alternative signal flowgraph for $H_z(z_1, z_2)$ using only two shift operators. The gains $p$ and $q$ are complex numbers in general.

Kung et al. [7] have shown that state-variable realizations of the form of equations (5.68) may be generalized to any system function $H_z(z_1, z_2)$ which satisfies the following three conditions. The constant term in the numerator, $a_{00}$, must be zero, the largest powers of $z_1^{-1}$ in the numerator and denominator polynomials must be equal, and the largest powers of $z_2^{-1}$ in the numerator and denominator polynomials must be equal.

There is one potential difficulty with state-variable realizations of this type. The nonlinear equations defining $p$ and $q$ may result in complex values for these constants. For example, when $a_{10} = a_{01} = 1$, $a_{11} = 0$, $c_{10} = c_{01} = 2$, and $c_{11} = 1$, we can compute that $p = q^* = 2 \pm j$. The introduction of complex arithmetic and storage for complex signals may offset any hardware, computation, or storage savings gained by using the two-state-variable form in equations (5.68). The filter designer must decide which approach is more suitable to his or her particular application.

## 5.4 SPACE-DOMAIN DESIGN TECHNIQUES

In this section we examine several techniques for designing 2-D IIR filters based on minimizing error functionals in the space domain. The aim, in general, is to determine the coefficient arrays $a(n_1, n_2)$ and $b(n_1, n_2)$ of a 2-D recursive filter so that the filter response $y(n_1, n_2)$ to a specified input signal $x(n_1, n_2)$ is a good approximation to some specified, desired output signal $d(n_1, n_2)$. Usually, the specified input signal is taken to be the unit sample $x(n_1, n_2) = \delta(n_1, n_2)$ so that $d(n_1, n_2)$ represents the desired impulse response. For a particular application, however, it may be advantageous to choose $x(n_1, n_2)$ to be some other signal, such as a unit step or unit ramp function.

Space-domain design techniques are useful when the application calls for a filter whose response to a particular input approximates a well-defined signal. In other situations, where the desired characteristics of the filter to be designed are specified in the frequency domain, other techniques, such as those described in Section 5.5, may be more useful.

Although other error norms are certainly possible, the mean-squared error norm and its relatives are the most widely used in space-domain approximations because of their mathematical tractability. Typically, the mean-squared error $e_2$ is given by

$$e_2 \triangleq \sum_{n_1=-\infty}^{\infty} \sum_{n_2=-\infty}^{\infty} [y(n_1, n_2) - d(n_1, n_2)]^2 \qquad (5.71)$$

(We are assuming that $y$ and $d$ are real-valued signals.) Using Parseval's theorem, one can show that $e_2$ is also equal to the frequency-domain mean-squared error given by

$$e_2 = E_2 \triangleq \frac{1}{4\pi^2} \int_{-\pi}^{\pi} \int_{-\pi}^{\pi} |Y(\omega_1, \omega_2) - D(\omega_1, \omega_2)|^2 \, d\omega_1 \, d\omega_2 \qquad (5.72)$$

This relation is important because the dependence of $y$ on the filter parameters $a(n_1, n_2)$ and $b(n_1, n_2)$ is more easily written in the frequency domain.

$$Y(\omega_1, \omega_2) = \frac{A(\omega_1, \omega_2) X(\omega_1, \omega_2)}{B(\omega_1, \omega_2)} \qquad (5.73)$$

where

$$A(\omega_1, \omega_2) = \sum_{n_1} \sum_{n_2} a(n_1, n_2) \exp(-j\omega_1 n_1 - j\omega_2 n_2) \qquad (5.74a)$$

and

$$B(\omega_1, \omega_2) = \sum_{n_1} \sum_{n_2} b(n_1, n_2) \exp(-j\omega_1 n_1 - j\omega_2 n_2) \qquad (5.74b)$$

In most derivations of the design algorithms, it is assumed that $a(n_1, n_2)$, $b(n_1, n_2)$, $x(n_1, n_2)$, $y(n_1, n_2)$, and $d(n_1, n_2)$ have their support confined to the first quadrant. Although we will make a similar assumption, it should be clear that other assumed regions of support could be used. A potentially more serious assumption concerns the limits of summation for equation (5.71). This summation must have finite limits for computational tractability. If the extent of the summation is sufficiently large, however, it can be argued that errors outside this region should be small if the designed filter is stable.

The error, $e_2$, can be minimized in theory by setting its partial derivatives with respect to the parameters $\{a(n_1, n_2), b(n_1, n_2)\}$ equal to zero. To begin, we define the error signal

$$e(n_1, n_2) \triangleq y(n_1, n_2) - d(n_1, n_2) \qquad (5.75)$$

Then

$$e_2 = \sum_{n_1} \sum_{n_2} e^2(n_1, n_2) \qquad (5.76)$$

so that

$$\frac{\partial e_2}{\partial a(p_1, p_2)} = \sum_{n_1} \sum_{n_2} 2e(n_1, n_2) \frac{\partial e(n_1, n_2)}{\partial a(p_1, p_2)}$$
$$= \sum_{n_1} \sum_{n_2} 2e(n_1, n_2) \frac{\partial y(n_1, n_2)}{\partial a(p_1, p_2)} \qquad (5.77a)$$

Similarly,

$$\frac{\partial e_2}{\partial b(q_1, q_2)} = \sum_{n_1} \sum_{n_2} 2e(n_1, n_2) \frac{\partial y(n_1, n_2)}{\partial b(q_1, q_2)} \qquad (5.77b)$$

The partial derivatives on the right side of the equations (5.77) can be obtained as follows. Assuming that all the signals and coefficient arrays of interest have support on the first quadrant only, we can express $y(n_1, n_2)$ by the recursion

$$y(n_1, n_2) = \sum_{p_1=0}^{N_1-1} \sum_{p_2=0}^{N_2-1} a(p_1, p_2) x(n_1 - p_1, n_2 - p_2)$$
$$- \sum_{\substack{q_1=0 \\ (q_1,q_2) \neq (0,0)}}^{M_1-1} \sum_{q_2=0}^{M_2-1} b(q_1, q_2) y(n_1 - q_1, n_2 - q_2) \tag{5.78}$$

[assuming that $b(0, 0) = 1$]. Then it is straightforward [10] to derive recursive relationships for $\partial y(n_1, n_2)/\partial a(p_1, p_2)$ and $\partial y(n_1, n_2)/\partial b(q_1, q_2)$.

$$\frac{\partial y(n_1, n_2)}{\partial a(p_1, p_2)} = x(n_1 - p_1, n_2 - p_2)$$
$$- \sum_{\substack{m_1=0 \\ (m_1,m_2) \neq (0,0)}}^{M_1-1} \sum_{m_2=0}^{M_2-1} b(m_1, m_2) \frac{\partial y(n_1 - m_1, n_2 - m_2)}{\partial a(p_1, p_2)} \tag{5.79a}$$

$$\frac{\partial y(n_1, n_2)}{\partial b(q_1, q_2)} = -y(n_1 - q_1, n_2 - q_2)$$
$$- \sum_{\substack{m_1=0 \\ (m_1,m_2) \neq (0,0)}}^{M_1-1} \sum_{m_2=0}^{M_2-1} b(m_1, m_2) \frac{\partial y(n_1 - m_1, n_2 - m_2)}{\partial b(q_1, q_2)} \tag{5.79b}$$

These recursions are started with initial conditions equal to zero. Furthermore, it is possible to show that the partial derivatives are related to one another by

$$\frac{\partial y(n_1, n_2)}{\partial a(p_1, p_2)} = \frac{\partial y(n_1 - p_1, n_2 - p_2)}{\partial a(0, 0)} \tag{5.80a}$$

and

$$\frac{\partial y(n_1, n_2)}{\partial b(q_1, q_2)} = \frac{\partial y(n_1 - q_1 + 1, n_2 - q_2)}{\partial b(1, 0)} \tag{5.80b}$$

$$= \frac{\partial y(n_1 - q_1, n_2 - q_2 + 1)}{\partial b(0, 1)} \tag{5.80c}$$

Thus the necessary partial derivatives can be computed numerically with two recursion relations.

It is generally not possible to solve analytically for the coefficient values $\{a(p_1, p_2), b(q_1, q_2)\}$ which minimize $e_2$. Consequently, we shall be forced to consider algorithmic methods of minimizing $e_2$. The subsections below detail three algorithms that have been applied to the problem of designing 2-D IIR filters using a space-domain error criterion.

### 5.4.1 Shanks's Method

One of the earliest methods used to design 2-D IIR filters was developed by Shanks *et al.* [11]. To avoid the highly nonlinear problem described above, they minimized a modified error function. If we let $E(\omega_1, \omega_2)$ represent the 2-D Fourier transform of the error signal $e(n_1, n_2)$, then

$$E(\omega_1, \omega_2) \triangleq \frac{A(\omega_1, \omega_2)X(\omega_1, \omega_2)}{B(\omega_1, \omega_2)} - D(\omega_1, \omega_2) \tag{5.81}$$

The modified error spectrum $E'(\omega_1, \omega_2)$ is obtained by multiplying $E(\omega_1, \omega_2)$ by $B(\omega_1, \omega_2)$.

$$E'(\omega_1, \omega_2) \triangleq A(\omega_1, \omega_2)X(\omega_1, \omega_2) - B(\omega_1, \omega_2)D(\omega_1, \omega_2) \tag{5.82}$$

This corresponds to the error signal

$$e'(n_1, n_2) = a(n_1, n_2) ** x(n_1, n_2) - b(n_1, n_2) ** d(n_1, n_2) \tag{5.83}$$

which is *linear* in the filter coefficients $\{a(n_1, n_2), b(n_1, n_2)\}$.

The total error $e'_2$ is computed by summing the squared modified error values.

$$e'_2 \triangleq \sum_{n_1} \sum_{n_2} [e'(n_1, n_2)]^2 \tag{5.84}$$

Shanks minimized $e'_2$ in the following fashion. First, he let the input signal $x(n_1, n_2)$ be the unit sample $\delta(n_1, n_2)$. Then, since the numerator array $a(n_1, n_2)$ is zero outside the rectangle $0 \leq n_1 < N_1, 0 \leq n_2 < N_2$, equation (5.83) becomes simply

$$e'(n_1, n_2) = -\sum_{q_1=0}^{M_1-1} \sum_{q_2=0}^{M_2-1} b(q_1, q_2)d(n_1 - q_1, n_2 - q_2) \qquad \text{for } n_1 \geq N_1 \text{ or } n_2 \geq N_2 \tag{5.85}$$

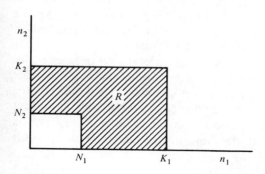

**Figure 5.24**  Region $R$ over which the square of the modified error signal is summed in Shanks's design method.

Now we can substitute this result into equation (5.84) provided that the sum in equation (5.84) is over the region $R$ shown in Figure 5.24. (The upper limits of $K_1$ and $K_2$ are imposed for computational tractability.)

If we differentiate $e'_2$ with respect to the denominator coefficients $b(q_1, q_2)$, remembering that $b(0, 0) = 1$, we get the following set of linear equations:

$$\sum_{m_1=0}^{M_1-1} \sum_{m_2=0}^{M_2-1} b(m_1, m_2)r(m_1, m_2; q_1, q_2) = 0$$

$$\text{for } 0 \leq m_1 < M_1, \quad 0 \leq m_2 < M_2 \tag{5.86}$$

$$\text{but } (m_1, m_2) \neq (0, 0)$$

where the function $r(m_1, m_2; q_1, q_2)$ is given by

$$r(m_1, m_2; q_1, q_2) = \sum_{(n_1, n_2) \in R} \sum d(n_1 - m_1, n_2 - m_2)\, d(n_1 - q_1, n_2 - q_2) \tag{5.87}$$

and the sum is over the region $R$ shown in Figure 5.24. Since $b(0, 0) = 1$, we can rewrite these equations to obtain the set of $M_1M_2 - 1$ normal equations in the $M_1M_2 - 1$ unknowns $\{b(m_1, m_2): (m_1, m_2) \neq (0, 0)\}$ given below.

$$\sum_{\substack{m_1=0 \\ (m_1, m_2) \neq (0,0)}}^{M_1-1} \sum_{m_2=0}^{M_2-1} b(m_1, m_2)r(m_1, m_2; q_1, q_2) = -r(0, 0; q_1, q_2)$$

$$\tag{5.88}$$

$$\text{for } 0 \leq q_1 < M_1, \quad 0 \leq q_2 < M_2 \qquad \text{but } (q_1, q_2) \neq (0, 0)$$

Solving (5.88) gives us the denominator coefficients $\{b(n_1, n_2)\}$. To find the numerator coefficients $\{a(n_1, n_2)\}$, we can make use of the fact that

$$a(n_1, n_2) ** x(n_1, n_2) = b(n_1, n_2) ** y(n_1, n_2) \qquad (5.89)$$

Since $x(n_1, n_2) = \delta(n_1, n_2)$ and $y(n_1, n_2) \simeq d(n_1, n_2)$, we can write

$$a(n_1, n_2) \simeq b(n_1, n_2) ** d(n_1, n_2) \qquad (5.90)$$

In general, the extent of $a(n_1, n_2)$ computed this way will exceed the extent of the rectangle $0 \le n_1 < N_1, 0 \le n_2 < N_2$. In this case, $a(n_1, n_2)$ can be truncated or windowed to the proper extent. This technique works reasonably well if the desired response $d(n_1, n_2)$ can be approximated well with a 2-D IIR filter [11].

The major advantage of Shanks's design technique is that it requires only the solution of linear equations in finding the coefficients for a direct form implementation. On the other hand, this technique does not minimize the true mean-squared error between $y(n_1, n_2)$ and $d(n_1, n_2)$ as desired. Furthermore, an IIR filter designed with this method may not be stable.

### 5.4.2 Descent Methods for Space-Domain Design

The general techniques of nonlinear optimization may be applied to the problem of designing 2-D IIR filters. Space does not permit us to make a general comparison and evaluation of all available techniques, but [10] and [12] provide a starting point for the interested reader. In this subsection, we will outline several optimization strategies frequently used to design both 1-D and 2-D IIR filters.

For convenience, let us define the parameter vector $\mathbf{p}$ as the set of filter coefficients $\{a(n_1, n_2), b(n_1, n_2)\}$. Perturbations of the parameter vector will be denoted $\Delta\mathbf{p}$. In general, the nonlinear optimization algorithms are iterative in nature. At each iteration, they attempt to reduce the approximation error $e_2$ by perturbing the current values of the parameters. We can indicate the functional dependence of the error on the parameter vector as $e_2(\mathbf{p})$. Then at each iteration, we seek an incremental vector $\Delta\mathbf{p}$ such that

$$e_2(\mathbf{p} + \Delta\mathbf{p}) < e_2(\mathbf{p}) \qquad (5.91)$$

(In practice, $\Delta\mathbf{p}$ may be multiplied by a positive scalar $\lambda$. Then, $\Delta\mathbf{p}$ can be thought of as providing a search direction with $\lambda$ controlling the step size.)

In the method of steepest descent, the direction vector $\Delta\mathbf{p}$ is taken to be the negative of the gradient of the error functional.

$$\Delta\mathbf{p} = -\nabla e_2(\mathbf{p})$$
$$= \left[ -\frac{\partial e_2}{\partial a(n_1, n_2)}, \ldots, -\frac{\partial e_2}{\partial b(n_1, n_2)}, \ldots \right]' \qquad (5.92)$$

The partial derivatives of $e_2$ are related to the partial derivatives of the output signal $y(n_1, n_2)$ by equations (5.77).

Although it is relatively simple to use, the method of steepest descent converges rather slowly [10, 13]. Consequently, it is often used only during the first few iterations of an attempted minimization.

The Newton method approximates the error functional $e_2(\mathbf{p} + \Delta\mathbf{p})$ by a second-order Taylor series of the form

$$e_2(\mathbf{p} + \Delta\mathbf{p}) \simeq e_2(\mathbf{p}) + \nabla e_2 \cdot \Delta\mathbf{p} + \tfrac{1}{2}\Delta\mathbf{p}'\mathbf{H}\,\Delta\mathbf{p} \tag{5.93}$$

where $\mathbf{H}$ is the Hessian matrix whose $ij$th component is the second-order partial derivative $\partial^2 e_2 / \partial p_i \partial p_j$ and $p_i$ and $p_j$ are the $i$th and $j$th components of the parameter vector $\mathbf{p}$. To find the perturbation vector $\Delta\mathbf{p}$, equation (5.93) is differentiated with respect to $\Delta\mathbf{p}$ and set equal to zero. This yields the linear set of equations

$$\mathbf{H}\,\Delta\mathbf{p} = -\nabla e_2 \tag{5.94}$$

to be solved for $\Delta\mathbf{p}$ at each iteration.

Theoretically, this method exhibits quadratic convergence when the parameter vector $\mathbf{p}$ is sufficiently close to optimum. However, to use this method it is necessary to compute second-order partial derivatives to obtain $\mathbf{H}$. This, of course, represents an additional computational burden. Despite its potential for rapid convergence, the Newton method often fails to converge in space-domain minimizations because the Hessian matrix can exhibit negative eigenvalues when $\mathbf{p}$ is close to the optimum [10]. Nevertheless, the Newton method has been used with some success in frequency-domain minimizations, as we shall see in Section 5.6.1.

The Fletcher–Powell minimization algorithm [14] has been used successfully to design direct form 2-D IIR filters [13]. This algorithm has been called "a steepest-descent algorithm with accelerated convergence" since it uses first-order partial derivatives plus information from previous iterations to estimate second-order partial derivatives. The Fletcher–Powell algorithm, along with the steepest descent and Newton algorithms, requires an initial estimate for the parameter vector $\mathbf{p}$. Judicious choice of this initial estimate can speed convergence as well as avoid local minima in searching for the global minimum. Often Shanks's algorithm [11] is used to provide a good starting point.

A linearization approach has been used successfully to design 1-D and 2-D IIR filters [10, 15, 16]. In this method, we use a first-order Taylor series expansion of the variation of the output signal $y(n_1, n_2)$ as a function of the parameter vector $\mathbf{p}$. To emphasize this functional dependence and to simplify the notation somewhat, we shall use a subscript $\mathbf{p}$ and suppress the dependence on $(n_1, n_2)$. Thus we define

$$y_{\mathbf{p}} \triangleq y(n_1, n_2)|_{\mathbf{p}} \tag{5.95}$$

Now, we form the first-order Taylor series

$$y_{\mathbf{p}+\Delta\mathbf{p}} \simeq y_{\mathbf{p}} + \nabla y_{\mathbf{p}} \cdot \Delta\mathbf{p} \tag{5.96}$$

where the components of the gradient vector $\nabla y_{\mathbf{p}}$ are the partial derivatives of $y_{\mathbf{p}}$ with respect to the filter coefficients comprising the parameter vector $\mathbf{p}$. The error functional can now be approximated by

$$e_2(\mathbf{p} + \Delta\mathbf{p}) = \sum_{n_1} \sum_{n_2} (y_{\mathbf{p}+\Delta\mathbf{p}} - d)^2$$

$$\simeq \sum_{n_1} \sum_{n_2} (y_{\mathbf{p}} + \nabla y_{\mathbf{p}} \cdot \Delta\mathbf{p} - d)^2 \qquad (5.97)$$

If we differentiate this approximation with respect to the perturbation vector $\Delta\mathbf{p}$ and set the resulting equations equal to zero, we get

$$0 = 2 \sum_{n_1} \sum_{n_2} (y_{\mathbf{p}} + \nabla y_{\mathbf{p}} \cdot \Delta\mathbf{p} - d)\nabla y_{\mathbf{p}} \qquad (5.98)$$

These normal equations are to be solved for $\Delta\mathbf{p}$. They can be written more concisely by using the relation

$$\nabla e_2(\mathbf{p}) = 2 \sum_{n_1} \sum_{n_2} (y_{\mathbf{p}} - d)\nabla y_{\mathbf{p}} \qquad (5.99)$$

Combining equations (5.99) and (5.98), we get the linear set of equations

$$\nabla e_2(\mathbf{p}) + 2\mathbf{Q}_{\mathbf{p}} \, \Delta\mathbf{p} = \mathbf{0} \qquad (5.100)$$

where $\mathbf{Q}_{\mathbf{p}}$ is a matrix whose $ij$th entry is

$$Q_{ij} = \sum_{n_1} \sum_{n_2} \frac{\partial y_{\mathbf{p}}}{\partial p_i} \frac{\partial y_{\mathbf{p}}}{\partial p_j} \qquad (5.101)$$

Since $\mathbf{p}$ is known at the start of the iteration, the quantities $y_{\mathbf{p}} - d$ and $\nabla y_{\mathbf{p}}$ can be computed. From these, $\nabla e_2(\mathbf{p})$ and $\mathbf{Q}_{\mathbf{p}}$ are easily computed to get the normal equations in the form (5.100), which are then solved for $\Delta\mathbf{p}$.

Strictly speaking, none of these techniques can guarantee BIBO $(L_1)$ stability or mean-squared $(L_2)$ stability since, in practice, the squared error is summed only over a finite region in the $(n_1, n_2)$-plane. However, if the desired signal $d(n_1, n_2)$ is square-summable and the region over which the error is summed is chosen large enough, the designed filter will tend to be stable in both the mean-squared and BIBO senses.

In the interest of completeness, let us mention the Levenberg–Marquardt minimization algorithm [17], which has been used for 2-D IIR filter design in the frequency domain [18], This algorithm can be viewed as a compromise between the Newton method and the steepest-descent method. It is thus iterative in nature and solves a set of linear equations at each iteration for the direction vector $\Delta\mathbf{p}$. The version used in [18] for filter design computed the necessary partial derivatives numerically rather than using explicit formulas.

### 5.4.3 Iterative Prefiltering Design Method

Shaw and Mersereau [13] have developed an alternative method for minimum mean-squared error design of 2-D IIR filters. The technique, which is an extension of a 1-D system identification procedure [19], involves rewriting the error norm to yield a set of linear equations to be solved for the filter parameters $\{a(n_1, n_2), b(n_1, n_2)\}$ at each iteration.

Equations (5.72) can be rewritten as

$$
\begin{aligned}
e_2 &= \frac{1}{4\pi^2} \int_{-\pi}^{\pi} \int_{-\pi}^{\pi} \left| \frac{A(\omega_1, \omega_2)X(\omega_1, \omega_2)}{B(\omega_1, \omega_2)} - D(\omega_1, \omega_2) \right|^2 d\omega_1 \, d\omega_2 \\
&= \frac{1}{4\pi^2} \int_{-\pi}^{\pi} \int_{-\pi}^{\pi} \left| \frac{1}{B(\omega_1, \omega_2)} \right|^2 \left| A(\omega_1, \omega_2)X(\omega_1, \omega_2) \right. \\
&\qquad \left. - D(\omega_1, \omega_2)B(\omega_1, \omega_2) \right|^2 d\omega_1 \, d\omega_2 \\
&= \frac{1}{4\pi^2} \int_{-\pi}^{\pi} \int_{-\pi}^{\pi} |G(\omega_1, \omega_2)|^2 \left| A(\omega_1, \omega_2)X(\omega_1, \omega_2) \right. \\
&\qquad \left. - D(\omega_1, \omega_2)B(\omega_1, \omega_2) \right|^2 d\omega_1 \, d\omega_2
\end{aligned}
\tag{5.102}
$$

where we have simply defined

$$
G(\omega_1, \omega_2) \triangleq \frac{1}{B(\omega_1, \omega_2)}
\tag{5.103}
$$

In the space domain, equation (5.102) can be written

$$
e_2 = \sum_{n_1} \sum_{n_2} e^2(n_1, n_2)
\tag{5.104}
$$

with

$$
e(n_1, n_2) = g(n_1, n_2) ** [a(n_1, n_2) ** x(n_1, n_2) - d(n_1, n_2) ** b(n_1, n_2)]
\tag{5.105}
$$

So far, we have not really altered the minimization problem since $g(n_1, n_2)$ is a function of the denominator coefficients $\{b(n_1, n_2)\}$. If, however, $g(n_1, n_2)$ did not depend on $b(n_1, n_2)$, then minimization of equation (5.104) would result in a set of linear equations to be solved. In the context of an iterative optimization, we can use $b(n_1, n_2)$ from the previous iteration to get $g(n_1, n_2)$ for the current iteration. Thus $g(n_1, n_2)$ will not be variable during the current iteration. Then a set of linear equations may be solved to get the new values of $\{a(n_1, n_2), b(n_1, n_2)\}$.

Setting the partial derivatives of (5.104) equal to zero gives us

$$
0 = \sum_{n_1} \sum_{n_2} e(n_1, n_2) \frac{\partial e(n_1, n_2)}{\partial a(p_1, p_2)}
\tag{5.106a}
$$

$$
0 = \sum_{n_1} \sum_{n_2} e(n_1, n_2) \frac{\partial e(n_1, n_2)}{\partial b(q_1, q_2)}
\tag{5.106b}
$$

It is straightforward to verify that the partial derivatives of equation (5.105), which are needed in (5.106), are given by

$$
\frac{\partial e(n_1, n_2)}{\partial a(p_1, p_2)} = g(n_1 - p_1, n_2 - p_2) ** x(n_1 - p_1, n_2 - p_2)
\tag{5.107a}
$$

$$
\frac{\partial e(n_1, n_2)}{\partial b(q_1, q_2)} = -g(n_1 - q_1, n_2 - q_2) ** d(n_1 - q_1, n_2 - q_2)
\tag{5.107b}
$$

under the assumption that $g(n_1, n_2)$ is not variable during the current iteration. The sequence $g(n_1, n_2)$ for the current iteration, denoted $g_i(n_1, n_2)$, is computed from $b_{i-1}(n_1, n_2)$ using the recursion

$$g_i(n_1, n_2) = \delta(n_1, n_2) - \sum_{q_1=0}^{M_1-1} \sum_{q_2=0}^{M_2-1} b_{i-1}(q_1, q_2) g_i(n_1 - q_1, n_2 - q_2) \qquad (5.108)$$

At convergence, the error signals computed by (5.105) and (5.75) will be identical. However, this error may not be the true minimum [13, 19] since the true partial derivatives of $e(n_1, n_2)$ with respect to the filter coefficients $\{b(n_1, n_2)\}$ are not equal to (5.107b), where $g(n_1, n_2)$ was assumed to be constant. Consequently, when (5.106b) equals zero, the true gradient may be nonzero, pointing the way for a further reduction of $e_2$.

The true partial derivative of $e(n_1, n_2)$ with respect to $b(q_1, q_2)$ is given by [13]

$$\frac{\partial e(n_1, n_2)}{\partial b(q_1, q_2)} = v(n_1 - q_1, n_2 - q_2) \qquad (5.109)$$

where $v(n_1, n_2)$ is the inverse Fourier transform of

$$V(\omega_1, \omega_2) \triangleq \frac{-A(\omega_1, \omega_2)X(\omega_1, \omega_2)}{B^2(\omega_1, \omega_2)} \qquad (5.110)$$

After the iteration based on (5.106) and (5.107) has converged, we can take the resulting filter parameters $\{a(n_1, n_2), b(n_1, n_2)\}$ as the starting point for a secondary iteration which uses (5.109) instead of (5.107b). The equations to be solved at each iteration will still be linear [13].

The iterative prefiltering design method converges much faster in practice than more general optimization techniques such as Fletcher–Powell [14]. On occasion, however, the filter at any particular iteration can become unstable, usually resulting in severe numerical problems when computing $g(n_1, n_2)$ for the next iteration. Although stability cannot be proven analytically, if the method converges in practice, it seems to always yield a stable filter [13].

### Example 1 [13]

The iterative prefiltering design method was used to design a 2-D lowpass filter. The desired response $d(n_1, n_2)$ was the first quadrant of a $23 \times 23$ circularly symmetric FIR filter, designed by the window method, with a nominal cutoff frequency of $0.5\pi$ and a transition bandwidth of $0.225\pi$. The nominal passband gain was 1.0 with a maximum ripple of 0.017. The maximum stopband ripple was 0.0076. Because only the first quadrant of the impulse response was used, $d(n_1, n_2)$ was nonzero only in a $12 \times 12$ region in the $(n_1, n_2)$-plane. The final circularly symmetric response is constructed by using four quadrant filters in parallel.

Using the iterative prefiltering design method, a 2-D IIR filter of order $4 \times 4$ in both the numerator and denominator was designed to approximate $d(n_1, n_2)$. The total mean-squared error over 144 samples in $(n_1, n_2)$ was $7.33 \times 10^{-6}$ at convergence, which took only four iterations to achieve (two of each type). The final, composite filter's passband ripple was 0.015, its stopband ripple was 0.0072, and its transition bandwidth was $0.258\pi$.

For comparison, the general Fletcher–Powell [14] optimization algorithm was used to design essentially the same filter. It required approximately 80 iterations to reach convergence. For this example, the iterative prefiltering design method was much more efficient computationally.

## 5.5 FREQUENCY-DOMAIN DESIGN TECHNIQUES

In the preceding section, we examined some techniques for designing 2-D IIR filters using space-domain error criteria. In this section we study design techniques that use frequency-domain error criteria. Because of Parseval's theorem, the mean-squared error is the same in both domains; that is,

$$\sum_{n_1} \sum_{n_2} [y(n_1, n_2) - d(n_1, n_2)]^2 = \frac{1}{4\pi^2} \int_{-\pi}^{\pi} \int_{-\pi}^{\pi} |Y(\omega_1, \omega_2) - D(\omega_1, \omega_2)|^2 \, d\omega_1 \, d\omega_2$$

$$(5.111)$$

This relationship will provide a convenient bridge between the two types of error criteria.

Other error measurements are possible, of course, as we saw in Chapter 3. The $L_\infty$ (or Chebyshev) frequency-domain error is given by

$$E_\infty \triangleq \max_{(\omega_1, \omega_2)} |Y(\omega_1, \omega_2) - D(\omega_1, \omega_2)| \qquad (5.112)$$

and the $L_p$ error norm is given by

$$E_p \triangleq \left[ \frac{1}{4\pi^2} \int_{-\pi}^{\pi} \int_{-\pi}^{\pi} |Y(\omega_1, \omega_2) - D(\omega_1, \omega_2)|^p \, d\omega_1 \, d\omega_2 \right]^{1/p} \qquad (5.113)$$

If $p$ is chosen to be relatively large, say 20, the $L_p$ norm becomes a good approximation to the $L_\infty$ norm.

Frequency-domain design methods are popular for several reasons. First, the approximating function $Y(\omega_1, \omega_2)$ is easily written in closed form as a function of the filter parameters $\{a(n_1, n_2), b(n_1, n_2)\}$, allowing simple derivation of any partial derivatives. Second, it is often the case that the desired response is not fully specified. For example, we may be interested in only approximating a certain magnitude response without specifying a particular phase response. Such a partial specification is much easier to impose in the frequency domain than in the space domain.

### 5.5.1 General Minimization Procedures

The general minimization techniques mentioned in Section 5.4 can also be applied to the frequency-domain error minimization problem. Usually, the error is summed over a finite number of samples in the frequency domain rather than being integrated over the square $-\pi \leq \omega_1, \omega_2 < \pi$. In addition, the sample frequencies are usually spaced evenly so that 2-D FFTs may be used to compute $Y(\omega_1, \omega_2)$ from $\{a(n_1, n_2), b(n_1, n_2)\}$. In addition, a nonnegative weighting function can be included if the filter designer wishes to specify frequencies where low approximation error is important. Thus, for the mean-squared error case, the problem becomes one of minimizing the approximation error functional

$$J_a \triangleq \sum_k W(\omega_{1k}, \omega_{2k}) \left[ \frac{A(\omega_{1k}, \omega_{2k})}{B(\omega_{1k}, \omega_{2k})} - D(\omega_{1k}, \omega_{2k}) \right]^2 \qquad (5.114)$$

where $W(\omega_1, \omega_2)$ is the weighting function, $(\omega_{1k}, \omega_{2k})$ are the frequency samples selected for the minimization, and as before we have tactily assumed that $X(\omega_1, \omega_2) \triangleq$ 1 for convenience.

The linearization approach discussed in the preceding section can be applied to this problem. Again, we shall use the vector **p** to denote the filter parameters and the vector $\Delta\mathbf{p}$ to denote an incremental change in **p**. Using the notation

$$Y_\mathbf{p} = \frac{A_\mathbf{p}}{B_\mathbf{p}} \triangleq \frac{A(\omega_1, \omega_2)}{B(\omega_1, \omega_2)}\bigg|_\mathbf{p} \tag{5.115}$$

to represent the filter responses when the parameter values are given by **p**, we can form the linearization

$$Y_{\mathbf{p}+\Delta\mathbf{p}} \simeq Y_\mathbf{p} + \nabla Y_\mathbf{p} \cdot \Delta\mathbf{p} \tag{5.116}$$

where $\nabla Y_\mathbf{p}$ is the gradient vector consisting of the partial derivatives

$$\frac{\partial Y(\omega_1, \omega_2)}{\partial a(p_1, p_2)} = \frac{\exp(-j\omega_1 p_1 - j\omega_2 p_2)}{B(\omega_1, \omega_2)} \tag{5.117a}$$

$$\frac{\partial Y(\omega_1, \omega_2)}{\partial b(q_1, q_2)} = \frac{-A(\omega_1, \omega_2)\exp(-j\omega_1 q_1 - j\omega_2 q_2)}{B^2(\omega_1, \omega_2)} \tag{5.117b}$$

Now the value of $J_a$ (equation 5.114) for the parameter setting $\mathbf{p} + \Delta\mathbf{p}$ can be written approximately as

$$J_a(\mathbf{p} + \Delta\mathbf{p}) \simeq \sum_k W[Y_\mathbf{p} + \nabla Y_\mathbf{p} \cdot \Delta\mathbf{p} - D]^2 \tag{5.118}$$

If we differentiate this expression with respect to $\Delta\mathbf{p}$ and set the result equal to zero, we get a set of linear equations of the form

$$\nabla J_a(\mathbf{p}) + 2\mathbf{Q}_\mathbf{p} \Delta\mathbf{p} = 0 \tag{5.119}$$

to be solved for $\Delta\mathbf{p}$. In this case, $\nabla J_a(\mathbf{p})$ denotes the gradient of (5.114) and the matrix $\mathbf{Q}_\mathbf{p}$ has as its $ij$th entry

$$Q_{ij} \triangleq \sum_k \frac{\partial Y_\mathbf{p}(\omega_{1k}, \omega_{2k})}{\partial p_i} \frac{\partial Y_\mathbf{p}(\omega_{1k}, \omega_{2k})}{\partial p_j} \tag{5.120}$$

where $p_i$ and $p_j$ represent the $i$th and $j$th components of the parameter vector **p** respectively. At each iteration, equation (5.119) is solved for a new incremental vector $\Delta\mathbf{p}$. [In practice, $\Delta\mathbf{p}$ is multiplied by a positive scalar step size parameter $\lambda$ adjusted to ensure that $J_a(\mathbf{p} + \lambda\Delta\mathbf{p}) < J_a(\mathbf{p})$.] Other optimization approaches can be similarly applied to minimize the functional $J_a$.

In some cases it may be advantageous to perform minimizations over subsets of the filter parameters in sequence during each iteration. For example, within an iteration, one could hold the denominator coefficients $\{b(n_1, n_2)\}$ constant while varying the numerator coefficients $\{a(n_1, n_2)\}$ to reduce $J_a$ and then hold the $\{a(n_1, n_2)\}$ constant while varying the $\{b(n_1, n_2)\}$. This reduces the number of parameters being varied at any one time and may result in a reduction of computation and computational inaccuracies.

With these frequency-domain design methods, there is no guarantee that a stable filter will result. If the designed filter is unstable, either it must be stabilized using the techniques described in Section 5.7, or it must be discarded. To avoid this problem, filter design algorithms have been developed that incorporate a stability constraint, and we shall examine one of these techniques in Section 5.5.3.

### 5.5.2  Magnitude and Magnitude-Squared Design Algorithms

The functional $J_a$ in equation (5.114) is a measure of the difference between two complex functions, the desired response $D(\omega_1, \omega_2)$ and the actual filter response $Y(\omega_1, \omega_2)$. In some applications, however, we may care only that the magnitude (or magnitude-squared) of the actual filter response approximate some real, desired response $D(\omega_1, \omega_2)$. By stating the approximation problem in this way, we are relinquishing control of the filter's phase or group delay response, reasoning that the phase response is not important for a particular application or that we intend to use a zero-phase implementation.

In general, we could specify some function of $Y(\omega_1, \omega_2)$ to approximate a desired response $D(\omega_1, \omega_2)$. Then the optimization problem becomes one of minimizing some functional of the difference $f(Y) - D$. If the $L_2$ error norm is to be used in the minimization, we could construct an error measure of the form

$$J_a = \sum_k W(\omega_{1k}, \omega_{2k})\left[ f\left(\frac{A(\omega_{1k}, \omega_{2k})}{B(\omega_{1k}, \omega_{2k})}\right) - D(\omega_{1k}, \omega_{2k})\right]^2 \qquad (5.121)$$

In this case, general optimization algorithms, such as Fletcher–Powell [14] and Levenberg–Marquardt [17], or techniques described in the preceding subsection, such as linearization, can be applied to find the filter coefficients $\{a(n_1, n_2), b(n_1, n_2)\}$ that minimize $J_a$. This formulation ensures that the designed filter will have the appropriate form to be implemented by a finite-order difference equation; however, there is no guarantee that the designed filter will be stable and, depending on the function $f$, the necessary partial derivatives $\partial f/\partial a(p_1, p_2)$ and $\partial f/\partial b(q_1, q_2)$ may be tedious or difficult to compute.

### 5.5.3  Magnitude Design with a Stability Constraint

It is possible to formulate a filter design procedure which includes a "stability error," $J_s$, to be minimized together with the usual approximation error [18]. This stability error, which is a crude measure of how unstable a filter is, is a type of penalty function. It should be zero for stable filters and large for unstable ones. The filter can then be designed by minimizing

$$J \triangleq J_a + \alpha J_s \qquad (5.122)$$

where the positive constant $\alpha$ weights the relative importance of $J_a$ and $J_s$. Ekstrom et al. [18] used nonlinear optimization techniques to minimize $J$. Their stability error was based on the difference between the denominator coefficient array and the minimum-phase array with the same autocorrelation function.

The minimum-phase array in question may be determined by first computing the autocorrelation function

$$r_b(n_1, n_2) = \sum_{q_1} \sum_{q_2} b(q_1, q_2) b(q_1 + n_1, q_2 + n_2) \tag{5.123}$$

of the NSHP denominator coefficient array $b(n_1, n_2)$. Then the Fourier transform of $r_b$, denoted $R_b(\omega_1, \omega_2)$, must be split into its minimum- and maximum-phase components. This is accomplished by spectral factorization using the complex cepstrum (Section 4.4.4).

To do this, we form the cepstrum $\hat{r}_b(n_1, n_2)$ of the autocorrelation function and multiply it by a nonsymmetric half-plane window $w(n_1, n_2)$ [see equation (4.132)] to obtain the cepstrum

$$\hat{b}_{\text{mp}}(n_1, n_2) = \hat{r}_b(n_1, n_2) w(n_1, n_2) \tag{5.124}$$

The subscript "mp" serves to remind us that this cepstrum corresponds to a minimum-phase sequence $b_{\text{mp}}(n_1, n_2)$.

If our designed filter is stable, its denominator coefficient array $b(n_1, n_2)$ is a minimum-phase sequence with nonsymmetric half-plane support. In this case, $b(n_1, n_2)$ is equal to $b_{\text{mp}}(n_1, n_2)$; otherwise, it is not. Consequently, the functional

$$J_s = \sum_{n_1} \sum_{n_2} [b(n_1, n_2) - b_{\text{mp}}(n_1, n_2)]^2 \tag{5.125}$$

may be used as the stability error.

In practice, $J_s$ is rarely driven to zero because of numerical errors in computing the cepstrum $\hat{r}_b(n_1, n_2)$. In general, $\hat{r}_b(n_1, n_2)$ has infinite extent, and spatially aliasing results when the FFT is used to compute it. As discussed in Section 4.4.5, the degree of aliasing can be controlled by increasing the size of the FFT.

### 5.5.4 Zero-Phase IIR Frequency-Domain Design Methods

Often, especially in applications such as image processing, one may want to filter a signal with a filter whose impulse response is symmetric. Such filters will have a real-valued, or zero-phase, frequency response. Historically, zero-phase IIR filters could be implemented in two ways, cascade or parallel.

In the cascade approach, a filter whose impulse response is $h(n_1, n_2)$ is cascaded with a filter whose impulse response is $h(-n_1, -n_2)$. The overall impulse response of the cascade is $h(n_1, n_2) ** h(-n_1, -n_2)$ and the overall frequency response is the real, nonnegative function

$$C(\omega_1, \omega_2) = |H(\omega_1, \omega_2)|^2 \tag{5.126}$$

As this equation shows, the frequency response of the cascade is limited to nonnegative functions of $(\omega_1, \omega_2)$. In addition, the cascade suffers from some computational problems due to transient effects. The output samples of second filter in the cascade are computed by a recursion which runs in the opposite direction from that of the first filter. If $h(n_1, n_2)$ is an IIR filter, its output has infinite extent, and in theory an infinite number of its output samples must be evaluated before filtering with $h(-n_1, -n_2)$ can begin, even if the ultimate output is desired only over a limited region. Truncating the computations from the first filter can introduce errors. As a practical matter, the output from the first filter must be computed far enough out in space so that any initial transients from the second filter will have effectively died out in the region of interest of the final output.

In the parallel approach, the outputs of two nonsymmetric half-plane (or four quarter-plane) IIR filters are added to form the final output signal. As in the cascade case, the second filter is a space-reversed version of the first, so the overall frequency response is given by

$$P(\omega_1, \omega_2) = H(\omega_1, \omega_2) + H^*(\omega_1, \omega_2)$$
$$= 2 \operatorname{Re}\left[H(\omega_1, \omega_2)\right] \tag{5.127}$$

This approach avoids the problems of the cascade approach for zero-phase implementation, but it is best suited to 2-D IIR filters designed in the space domain, where the desired filter response $d(n_1, n_2)$ can be partitioned into the proper regions of support.

A relatively new method of implementing zero-phase IIR filters, the iterative implementation, was discussed in detail in Section 5.2. The frequency-domain design methods described below will produce filters that can be implemented by this approach.

The frequency response of a zero-phase 2-D IIR filter can be written as

$$H(\omega_1, \omega_2) = \frac{A(\omega_1, \omega_2)}{B(\omega_1, \omega_2)}$$
$$= \frac{\displaystyle\sum_{n_1=-N_1+1}^{N_1-1} \sum_{n_2=-N_2+1}^{N_2-1} a(n_1, n_2) \exp\left(-j\omega_1 n_1 - j\omega_2 n_2\right)}{\displaystyle\sum_{m_1=-M_1+1}^{M_1-1} \sum_{m_2=-M_2+1}^{M_2-1} b(m_1, m_2) \exp\left(-j\omega_1 m_1 - j\omega_2 m_2\right)} \tag{5.128}$$

(As before, we assume that $b(0, 0) = 1$.) Since

$$a(n_1, n_2) = a(-n_1, -n_2) \tag{5.129a}$$
$$b(n_1, n_2) = b(-n_1, -n_2) \tag{5.129b}$$

$A(\omega_1, \omega_2)$ and $B(\omega_1, \omega_2)$ can be written more concisely as

$$A(\omega_1, \omega_2) = \sum_{n_1} \sum_{n_2} a'(n_1, n_2) \cos\left(\omega_1 n_1 + \omega_2 n_2\right) \tag{5.130a}$$
$$B(\omega_1, \omega_2) = \sum_{m_1} \sum_{m_2} b'(m_1, m_2) \cos\left(\omega_1 m_1 + \omega_2 m_2\right) \tag{5.130b}$$

where

$$a'(0, 0) \triangleq a(0, 0)$$
$$a'(n_1, n_2) \triangleq 2a(n_1, n_2) \quad \text{for } (n_1, n_2) \neq (0, 0) \tag{5.131a}$$
$$b'(0, 0) \triangleq b(0, 0) = 1$$
$$b'(m_1, m_2) \triangleq 2b(m_1, m_2) \quad \text{for } (m_1, m_2) \neq (0, 0) \tag{5.131b}$$

[The sums in (5.130) are taken over the appropriate finite-extent NSHP regions.] At this point, we could formulate a mean-squared error functional that could be minimized by the techniques described earlier. The result of the minimization would yield the zero-phase filter coefficients $\{a'(n_1, n_2), b'(n_1, n_2)\}$ which are easily related to the polynomial coefficients $a\{(n_1, n_2), b(n_1, n_2)\}$ through equations (5.131). Then the techniques of Section 5.2 could be used to implement the designed filter.

Alternatively, we may want to minimize an $L_\infty$ error functional of the form

$$E = \left\| D(\omega_1, \omega_2) - \frac{A(\omega_1, \omega_2)}{B(\omega_1, \omega_2)} \right\|$$
$$\triangleq \max_{(\omega_1, \omega_2)} \left| D(\omega_1, \omega_2) - \frac{A(\omega_1, \omega_2)}{B(\omega_1, \omega_2)} \right| \tag{5.132}$$

This is highly a nonlinear optimization problem, but there is an iterative technique called differential correction [20–24] which can minimize $E$ by solving a linear programming problem at each iteration. Note that $A$, $B$, and $D$ are real-valued functions.

Let us assume that after $k$ iterations we have an approximation to the desired real frequency response $D(\omega_1, \omega_2)$. We shall denote this approximation

$$H_k(\omega_1, \omega_2) \triangleq \frac{A_k(\omega_1, \omega_2)}{B_k(\omega_1, \omega_2)} \tag{5.133}$$

Next we compute the $L_\infty$ error for this approximation:

$$E_k = \| D(\omega_1, \omega_2) - H_k(\omega_1, \omega_2) \|$$
$$= \max_{(\omega_1, \omega_2)} \left| D(\omega_1, \omega_2) - \frac{A_k(\omega_1, \omega_2)}{B_k(\omega_1, \omega_2)} \right| \tag{5.134}$$

Now we can define the differential correction functional

$$\delta_k \triangleq \max_{(\omega_1, \omega_2)} \frac{|D(\omega_1, \omega_2)B(\omega_1, \omega_2) - A(\omega_1, \omega_2)| - E_k B(\omega_1, \omega_2)}{B_k(\omega_1, \omega_2)} \tag{5.135}$$

This functional depends on the filter parameters $\{a'(n_1, n_2), b'(n_1, n_2)\}$ through $A(\omega_1, \omega_2)$ and $B(\omega_1, \omega_2)$ in the numerator of (5.135). By adjusting these parameters, we can minimize $\delta_k$, which in general is less than zero. The iteration continues until $\delta_{k+1} \geq 0$, at which time the best approximation to $D(\omega_1, \omega_2)$ is given by $H_k(\omega_1, \omega_2)$. The minimization of $\delta_k$ at each iteration can be accomplished using the techniques of linear programming (see [25], for example). The differential correction algorithm, although mathematically elegant, can require a substantial amount of computation owing to the solution of a linear programming problem at each iteration.

We can also design 2-D zero-phase IIR filters by applying the McClellan transformation discussed in Section 3.5.3 to the numerator and denominator polynomials of 1-D zero-phase IIR filters. For example, suppose that

$$H(\omega) = \frac{A(\omega)}{B(\omega)}$$

$$= \frac{\sum_{n=0}^{N-1} a'(n) \cos(\omega n)}{\sum_{m=0}^{M-1} b'(m) \cos(\omega m)} \tag{5.136}$$

$$= \frac{\sum_{n=0}^{N-1} a'(n) T_n [\cos \omega]}{\sum_{m=0}^{M-1} b'(m) T_m [\cos \omega]}$$

where $T_n[x]$ is the $n$th Chebyshev polynomial. Then we can make the substitution of a low-order zero-phase 2-D trigonometric polynomial $F(\omega_1, \omega_2)$ for $\cos\omega$ to obtain the 2-D frequency response

$$H(\omega_1, \omega_2) = \frac{\displaystyle\sum_{n=0}^{N-1} a'(n) T_n[F(\omega_1, \omega_2)]}{\displaystyle\sum_{m=0}^{M-1} b'(m) T_m[F(\omega_1, \omega_2)]} \tag{5.137}$$

$$= \frac{A(\omega_1, \omega_2)}{B(\omega_1, \omega_2)}$$

For example, we could take the magnitude-squared response of the 1-D digital Butterworth lowpass filter as our symmetric 1-D filter. This frequency response is given by

$$H(\omega) = \frac{1}{1 + \left[\dfrac{\tan(\omega/2)}{\tan(\omega_c/2)}\right]^{2N}} \tag{5.138}$$

where $\omega_c$ is the cutoff frequency [1]. Using trigonometric identities we can rewrite $H(\omega)$ in terms of $\cos\omega$:

$$H(\omega) = \frac{(1 + \cos\omega)^N}{(1 + \cos\omega)^N + \alpha(1 - \cos\omega)^N} \tag{5.139}$$

where

$$\alpha \triangleq \left(\cotan\frac{\omega_c}{2}\right)^{2N} = \left(\frac{1 + \cos\omega_c}{1 - \cos\omega_c}\right)^N \tag{5.140}$$

Then we can substitute for $\cos\omega$ the 2-D function

$$F(\omega_1, \omega_2) = \tfrac{1}{2}[-1 + \cos\omega_1 + \cos\omega_2 + (\cos\omega_1)(\cos\omega_2)] \tag{5.141}$$

to obtain the frequency response of a nearly circularly symmetric IIR lowpass filter.

At each stage of the iterative implementation, it is necessary to filter the signal with a frequency response of the form

$$C(\omega_1, \omega_2) = 1 - \lambda B(\omega_1, \omega_2) \tag{5.142}$$

for real, positive-valued functions $B(\omega_1, \omega_2)$. As discussed in Section 5.2, this is a 2-D FIR filtering operation which may be carried out using the methods of Chapter 3. In particular, when $B(\omega_1, \omega_2)$ has been designed using the McClellan transformation, the frequency response $C(\omega_1, \omega_2)$ may be implemented by using a modified version of the specialized structure described in Section 3.5.4.

### 5.5.5 Frequency Transformations

In the interest of completeness, we shall briefly discuss some simple frequency-domain transformations which can map both 1-D and 2-D IIR filters into other 2-D IIR filters. These transformations can be useful for designing lowpass, highpass, and bandpass filters as well as multiple passband filters. The reader interested in greater detail is directed to [26, 27].†

These transformations are most conveniently discussed using the system function notation

$$H_z(z_1, z_2) \triangleq \frac{A_z(z_1, z_2)}{B_z(z_1, z_2)} \qquad (5.143)$$

where $A_z(z_1, z_2)$ and $B_z(z_1, z_2)$ are 2-D polynomials. The objective of the frequency transformations is to map a stable, rational system function into another stable, rational system function. In general, the filter designer also wants to preserve some of the characteristics of the prototype filter, such as its stopband attenuation and passband ripple, while altering other characteristics, such as the location and number of the passbands.

For transformations from a 1-D IIR prototype $H_z(z)$ to a 2-D IIR filter $G_z(z_1, z_2)$, we can make the substitution

$$z^{-1} = F_z(z_1, z_2) \qquad (5.144)$$

Thus

$$G_z(z_1, z_2) = H_z(F_z^{-1}(z_1, z_2)) \qquad (5.145)$$

To ensure that $G_z$ is a stable filter, both $H_z$ and $F_z$ must be stable [26, 27]. In addition, we want the unit circle $z = e^{j\omega}$ to be mapped to the frequency plane $(z_1, z_2) = (e^{j\omega_1}, e^{j\omega_2})$. This forces the magnitude of $F_z(z_1, z_2)$ to be unity.

$$|F_z(z_1, z_2)| = 1 \qquad (5.146)$$

Thus the mapping function $F_z$ must be a stable, all-pass filter.

Chakrabarti and Mitra [27] have stated that the only admissible transformation for mapping 1-D IIR filters into 2-D IIR filters has the form

$$F_z(z_1, z_2) = z_1^{-p} z_2^{-q} \qquad (5.147)$$

where $p$ and $q$ are positive rational numbers. For example, consider the simple 1-D IIR filter

$$H_z(z) = \frac{1}{1 - az^{-1}}, \qquad |a| < 1 \qquad (5.148)$$

which has the impulse response

$$h(n) = a^n u(n) \qquad (5.149)$$

where $u(n)$ is the 1-D unit step function.

Now let us apply the transformation

$$F_z(z_1, z_2) = z_1^{-1} z_2^{-2} \qquad (5.150)$$

to get the 2-D IIR filter

$$G_z(z_1, z_2) = \frac{1}{1 - az_1^{-1} z_2^{-2}} \qquad (5.151)$$

---

†The reader should be aware that these references define the 2-D z-transform in terms of positive powers of $z_1$ and $z_2$. Consequently, their statements are formulated differently than ours.

Taking the 2-D inverse $z$-transform of $G_z$ gives us the impulse response

$$g(n_1, n_2) = a^{n_1}\delta(n_2 - 2n_1)u(n_1, n_2) \tag{5.152}$$

Figure 5.25 shows the impulse responses $h(n)$ and $g(n_1, n_2)$.

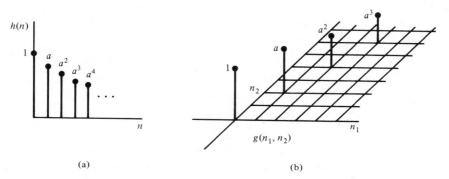

(a)                                           (b)

**Figure 5.25**   Transformation $F_z(z_1, z_2) = z_1^{-1}z_2^{-2}$ maps $h(n)$ into $g(n_1, n_2)$.

The general 2-D-to-2-D transformation is characterized by two mapping functions

$$z_1^{-1} \longleftarrow F_1(z_1, z_2)$$
$$z_2^{-1} \longleftarrow F_2(z_1, z_2) \tag{5.153}$$

Both $F_1$ and $F_2$ must be stable, 2-D all-pass filters. First quadrant filters will have the general form

$$F_i(z_1, z_2) = \pm \frac{\sum_{n_1=0}^{N_1-1} \sum_{n_2=0}^{N_2-1} f_i(n_1, n_2)z_1^{-n_1}z_2^{-n_2}}{z_1^{-N_1}z_2^{-N_2} \sum_{n_1=0}^{N_1-1} \sum_{n_2=0}^{N_2-1} f_i(n_1, n_2)z_1^{n_1}z_2^{n_2}} \qquad \text{for } i = 1, 2 \tag{5.154}$$

Designing $F_1$ and $F_2$ to effect a desired frequency transformation is a difficult problem, even when $N_1$ and $N_2$ are small [26].

The general transformation can be simplified and specialized by making $F_1$ independent of $z_2$ and $F_2$ independent of $z_1$. Thus the transformation becomes

$$.z_1^{-1} \longleftarrow F_1(z_1)$$
$$z_2^{-1} \longleftarrow F_2(z_2)$$

This simplified transformation scales the frequency axes independently. By mapping a small number of points on the frequency plane (usually one or two, depending on the order of the transformation) to other points, one can define $F_1$ and $F_2$ [26]. In general, this transformation can be viewed as placing a window, whose size is dictated by the order of $F_1$ and $F_2$, over the doubly periodic frequency response $H(\omega_1, \omega_2)$. Depending on the parameter values of the all-pass functions $F_1$ and $F_2$, $H(\omega_1, \omega_2)$ may be locally stretched or compressed, and the window may be centered at $(0, 0)$, $(0, \pi)$, $(\pi, 0)$, or

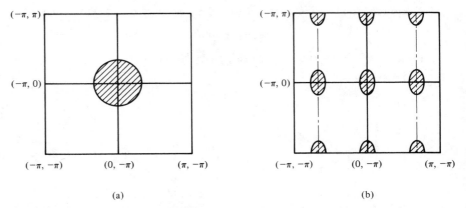

**Figure 5.26**  Circularly symmetric 2-D lowpass filter (a) is transformed into multiband filter (b) using a transformation in which $F_1$ is a third-order all-pass function and $F_2$ is a second-order all-pass function. Shaded areas denote pass-bands.

$(\pi, \pi)$. Figure 5.26 shows a 2-D circularly symmetric lowpass filter which has been transformed to a multiband filter using a third-order all-pass function for $F_1$ and a second-order all-pass function for $F_2$.

## 5.6 DESIGN TECHNIQUES FOR SPECIALIZED STRUCTURES

In this section we examine three design techniques geared to three specialized implementations—cascade filters, separable denominator filters, and lattice filters. Because of the absence of a fundamental theorem of algebra for multidimensional polynomials, it is usually necessary to tailor a design algorithm to suit a particular implementation. In the subsections below, we shall see three examples of specialized design algorithms.

### 5.6.1 Cascade Designs

As we saw in Chapter 3, cascade implementations must be considered specialized structures in two dimensions. For a cascade implementation, it is necessary to formulate the design problem in cascade form. For example, we could force the frequency response of the designed filter to have the form

$$H(\omega_1, \omega_2) = \beta \prod_{i=1}^{N} H_i(\omega_1, \omega_2) \tag{5.155}$$

where $\beta$ is an overall gain constant and

$$H_i(\omega_1, \omega_2) = \frac{\sum_{n_1=0}^{N_1-1} \sum_{n_2=0}^{N_2-1} a_i(n_1, n_2) \exp(-j\omega_1 n_1 - j\omega_2 n_2)}{\sum_{m_1=0}^{M_1-1} \sum_{m_2=0}^{M_2-1} b_i(m_1, m_2) \exp(-j\omega_1 m_1 - j\omega_2 m_2)} \tag{5.156}$$

Usually, $N_1$, $N_2$, $M_1$, and $M_2$ are small (equal to 2 or 3) and $a_i(0, 0)$ and $b_i(0, 0)$ are normalized to unity [28]. Consequently, each $H_i(\omega_1, \omega_2)$ will have no more than 16 free parameters and the full frequency response will have no more than $16N + 1$ free parameters. Symmetry or other constraints, such as a constant numerator, may also be imposed by the designer to reduce the number of parameters.

With $H(\omega_1, \omega_2)$ in the form (5.155), it is straightforward to set up an $L_2$ or $L_p$ minimization problem which can be solved using the optimization techniques described in the preceding section. For example, Maria and Fahmy [28] use a Newton-type optimization to solve an approximation problem in which the error functional is given by

$$J_a = \sum_{k=1}^{K} [|H(\omega_{1k}, \omega_{2k})| - D(\omega_{1k}, \omega_{2k})]^p \qquad (5.157)$$

The cascade formulation has several advantages which may be important for practical applications. First, like its 1-D counterpart, the frequency response of a 2-D IIR filter will be less sensitive to coefficient perturbations in a cascade structure than in a direct-form structure. Second, a stability check can be incorporated into the design procedure since checking the stability of the low-order subfilters in the cascade is relatively easy [28]. Finally, the optimization may be performed in stages. At each iteration, the coefficients of all but one of the subfilters can be held constant while the coefficients of the remaining subfilter are varied to reduce $J_a$. On the next iteration, the coefficients of another subfilter are varied.

Cascade filters may be implemented by the methods discussed in Section 5.1.2.

### 5.6.2 Separable Denominator Designs

Separable filters have a number of advantages, both in design and implementation, since they are an outer product of two 1-D filters. Because of their simplicity, however, they offer a very limited range of possible impulse and frequency responses. An interesting compromise approach has been developed [13, 29] which combines a nonseparable numerator polynomial with a separable denominator polynomial. These separable denominator designs retain much of the flexibility of nonseparable designs and yet offer the implementation advantages of separable IIR filters.

The frequency response of a 2-D separable denominator filter is given by

$$H(\omega_1, \omega_2) = \frac{A(\omega_1, \omega_2)}{B_1(\omega_1)B_2(\omega_2)} \qquad (5.158)$$

We may view $H(\omega_1, \omega_2)$ as the cascade of a 2-D nonseparable FIR filter $A(\omega_1, \omega_2)$ and a 2-D separable "all-pole" IIR filter $1/B_1(\omega_1)B_2(\omega_2)$. Since $A(\omega_1, \omega_2)$ by itself can approximate some desired frequency response $D(\omega_1, \omega_2)$ arbitrarily closely as the number of free parameters is allowed to increase, it follows that $H(\omega_1, \omega_2)$ can also be used to approximate $D(\omega_1, \omega_2)$ arbitrarily closely. The use of $1/B_1(\omega_1)B_2(\omega_2)$ can be viewed as an attempt to improve the approximation in a manner that results in a computationally more efficient implementation than simply increasing the number of free parameters in $A(\omega_1, \omega_2)$.

Separable denominator filters have several implementation advantages. If we neglect the numerator, which may be implemented separately as an FIR filter, then the remaining part of the filter is a separable filter whose frequency response has the form $1/B_1(\omega_1)B_2(\omega_2)$. The implementation of this part of the filter can be structured as a set of 1-D convolutions (implemented by a difference equation) on the rows of the signal array followed by another set of 1-D convolutions on the columns of the resulting signal array. The entire implementation of the separable denominator filter, as shown in Figure 5.27, takes the form

$$f(n_1, n_2) \triangleq \sum_{p_1} \sum_{p_2} a(p_1, p_2)x(n_1 - p_1, n_2 - p_2) \tag{5.159a}$$

$$s(n_1, n_2) = -\sum_{q_1=1}^{M_1-1} b_1(q_1)s(n_1 - q_1, n_2) + f(n_1, n_2) \tag{5.159b}$$

$$y(n_1, n_2) = -\sum_{q_2=1}^{M_2-1} b_2(q_2)y(n_1, n_2 - q_2) + s(n_1, n_2) \tag{5.159c}$$

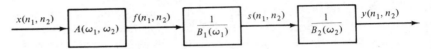

**Figure 5.27** Cascade that implements the separable denominator filter $A(\omega_1, \omega_2)/B(\omega_1)B(\omega_2)$.

Equation (5.159a) represents the implementation of the numerator response $A(\omega_1, \omega_2)$. Although it is written in the form of a direct convolution, any appropriate FIR implementation strategy (Chapter 3) may be used. Equation (5.159b) represents the set of 1-D filtering operations on the rows and, similarly, equation (5.159c) represents the set of 1-D filtering operations on the columns. The 1-D filtering operations in each set may be computed in parallel, using multiple processors, to obtain a high throughput rate.

Since equations (5.159) and Figure 5.27 can be interpreted as a cascade of three filters, the order of the three filters may be permuted without altering the overall frequency response. For practical reasons such as round-off noise considerations, one permutation may be preferred over the others.

In terms of the filter parameters $\{a(n_1, n_2), b_1(n_1), b_2(n_2)\}$, the frequency response of the separable denominator filter has the form

$$H(\omega_1, \omega_2) = \frac{\sum_{n_1} \sum_{n_2} a(n_1, n_2) \exp(-j\omega_1 n_1 - j\omega_2 n_2)}{\sum_{m_1=0}^{M_1-1} \sum_{m_2=0}^{M_2-1} b_1(m_1)b_2(m_2) \exp(-j\omega_1 m_1 - j\omega_2 m_2)} \tag{5.160}$$

where $b_1(0) = b_2(0) = 1$. Naturally, the denominator polynomial can be separated into the product of two 1-D polynomials, but we have written (5.160) as it stands to illustrate the following point. Because of the requirement that the denominator be separable, the number of denominator free parameters in the design problem is reduced from $(M_1 M_2 - 1)$ to $(M_1 + M_2 - 2)$. However, the design problem becomes more nonlinear because the free parameters $\{b_1(m_1), b_2(m_2)\}$ multiply each other in

(5.160). Consequently, the optimization algorithm we use will be faced with a smaller, but more nonlinear, minimization problem.

Shaw and Mersereau [13] have developed one design approach for separable denominator filters which minimizes an $L_2$ space-domain error. Their design algorithm is a version of the iterative prefiltering approach discussed in Section 5.4.3 modified to account for the additional nonlinearity introduced by the separable denominator. Recall that in the iterative prefiltering design method, we seek to minimize the space domain error

$$e_2 = \sum_{n_1} \sum_{n_2} e^2(n_1, n_2) \tag{5.161}$$

where

$$e(n_1, n_2) = g(n_1, n_2) ** [a(n_1, n_2) ** x(n_1, n_2) - d(n_1, n_2) ** b(n_1, n_2)] \tag{5.162}$$

The sequence $g(n_1, n_2)$ is the inverse Fourier transform of $1/B(\omega_1, \omega_2)$. For the separable denominator case, $B(\omega_1, \omega_2) = B_1(\omega_1)B_2(\omega_2)$, so that $b(n_1, n_2) = b_1(n_1)b_2(n_2)$ and $g(n_1, n_2) = g_1(n_1)g_2(n_2)$. To simplify matters, we shall assume that the input signal $x(n_1, n_2)$ is a 2-D impulse. Then (5.162) becomes

$$e(n_1, n_2) = [g_1(n_1)g_2(n_2)] ** \{a(n_1, n_2) - d(n_1, n_2) ** [b_1(n_1)b_2(n_2)]\} \tag{5.163}$$

If we pretend, as in Section 5.4.3, that $g_1$ and $g_2$ are constant, then the partial derivatives of $e(n_1, n_2)$ with respect to the numerator coefficients $\{a(p_1, p_2)\}$ are given by

$$\frac{\partial e(n_1, n_2)}{\partial a(p_1, p_2)} = g_1(n_1 - p_1)g_2(n_2 - p_2) \tag{5.164}$$

and the partial derivatives with respect to the denominator parameters are given by

$$\frac{\partial e(n_1, n_2)}{\partial b_1(q_1)} = -\sum_{q_2=0}^{M_2-1} b_2(q_2)\{[g_1(n_1 - q_1)g_2(n_2 - q_2)] ** d(n_1 - q_1, n_2 - q_2)\}$$
$$\tag{5.165a}$$

$$\frac{\partial e(n_1, n_2)}{\partial b_2(q_2)} = -\sum_{q_1=0}^{M_1-1} b_1(q_1)\{[g_1(n_1 - q_1)g_2(n_2 - q_2)] ** d(n_1 - q_1, n_2 - q_2)\}$$
$$\tag{5.165b}$$

Because $\partial e/\partial b_1$ and $\partial e/\partial b_2$ depend on $b_1$ and $b_2$, respectively, the equations generated by differentiating $e_2$ and setting the resulting derivatives equal to zero will still be nonlinear. Shaw and Mersereau [13] avoid this problem by using the values of $\{b_1(q_1), b_2(q_2)\}$ from the previous iteration to compute the partial derivatives (5.165). The sequences $g_1(n_1)$ and $g_2(n_2)$ are also computed from the previous values of $\{b_1(q_1), b_2(q_2)\}$ by the 1-D iterations

$$g_1(n_1) = \delta(n_1) - \sum_{q_1=1}^{M_1-1} b_1(q_1)g_1(n_1 - q_1) \tag{5.166a}$$

$$g_2(n_2) = \delta(n_2) - \sum_{q_2=1}^{M_2-1} b_2(q_2)g_2(n_2 - q_2) \tag{5.166b}$$

As we discussed in Section 5.4.3, computing the partial derivatives under the assumption that $g(n_1, n_2)$, or in this case $g_1(n_1)g_2(n_2)$, is constant does not force the error gradient to zero. As in the nonseparable case, a second mode of iterations may be entered after the primary iteration has converged. This second mode replaces the partial derivatives (5.165) by

$$\frac{\partial e(n_1, n_2)}{\partial b_1(q_1)} = v_1(n_1 - q_1, n_2) \tag{5.167a}$$

$$\frac{\partial e(n_1, n_2)}{\partial b_2(q_2)} = v_2(n_1, n_2 - q_2) \tag{5.167b}$$

where $v_1(n_1, n_2)$ is the inverse Fourier transform of

$$V_1(\omega_1, \omega_2) \triangleq \frac{-A(\omega_1, \omega_2)}{B_1^2(\omega_1)B_2(\omega_2)} \tag{5.168a}$$

and $v_2(n_1, n_2)$ is the inverse Fourier transform of

$$V_2(\omega_1, \omega_2) \triangleq \frac{-A(\omega_1, \omega_2)}{B_1(\omega_1)B_2^2(\omega_2)} \tag{5.168b}$$

Because the denominator polynomial is separable, verification that the designed filter is stable becomes simply a matter of testing the stability of the two 1-D IIR filters $1/B_1(\omega_1)$ and $1/B_2(\omega_2)$.

**Example 2 [13]**

A quadrant filter with a symmetric separable denominator polynomial $B(\omega_1, \omega_2) = B_1(\omega_1)B_1(\omega_2)$ was designed to approximate the FIR lowpass filter in Example 1 at the end of Section 5.4.3. The one-quadrant filter numerator had 25 independently variable parameters and the denominator was restricted to 4 independent parameters. The error after 13 primary iterations was $1.90 \times 10^{-5}$; after 3 primary iterations and 8 secondary iterations, it was slightly lower at $1.78 \times 10^{-5}$. It is conjectured that the error is only slightly different because only 4 of the 29 parameters were affected by using (5.165) for the partial derivatives instead of (5.167) [13]. The full, four-quadrant frequency response had a passband ripple of 0.0034, a stopband ripple of 0.021, and a transition bandwidth of $0.233\pi$.

## *5.6.3 Lattice Structures

Lattice structures exist for both 1-D and 2-D all-zero (FIR) and all-pole digital filters. We begin by considering the 1-D case and will later generalize these results to the two-dimensional case.

Let $b(n)$, $n = 1, 2, \ldots, N$ be a sequence of length $N$ and define

$$B_z^{(N)}(z) = 1 + \sum_{n=1}^{N} b(n)z^{-n} \tag{5.169}$$

$B_z^{(N)}(z)$ can be thought of either as the transfer function of an FIR filter whose tap weights are $\{b(n)\}$ or as the denominator of the transfer function of an all-pole IIR filter. We can now define a reversible mapping between $\{b(n)\}$ and another set of $N$ numbers $\{k(p), p = 1, 2, \ldots, N\}$, known as *reflection coefficients* [30], by means of the following iterative procedure. For $p = N, N - 1, \ldots, 1$ set

$$k(p) = b^{(p)}(p) \qquad (5.170a)$$

$$B_z^{(p-1)}(z) = \frac{1}{1 - k^2(p)}[B_z^{(p)}(z) - k(p)z^{-p}B_z^{(p)}(z^{-1})] \qquad (5.170b)$$

As one proceeds through the iteration the orders of the polynomials are reduced. The number $b^{(p)}(p)$ is the coefficient of $z^{-p}$ in $B_z^{(p)}(z)$, the last coefficient in this polynomial. The recursion in (5.170) can also be driven the other way so that $\{b(n), n = 1, 2, \ldots, N\}$ is obtained from $\{k(p), p = 1, 2, \ldots, N\}$. This yields the recursion

$$B_z^{(0)}(z) = 1 \qquad (5.171a)$$

$$B_z^{(p)}(z) = B_z^{(p-1)}(z) + k(p)z^{-p}B_z^{(p-1)}(z^{-1}) \qquad (5.171b)$$

Equation (5.171b) is repeated for $p = 1, 2, \ldots, N$.

Either the $\{b(n)\}$ or the $\{k(p)\}$ can be used to implement the filter. In the former case, we have a direct form realization. In the latter case, we have a lattice filter. In Figure 5.28(a) we show a lattice realization of a filter with transfer function $B_z^{(N)}(z)$, and in Figure 5.28(b) we show a lattice filter with transfer function $1/B_z^{(N)}(z)$.

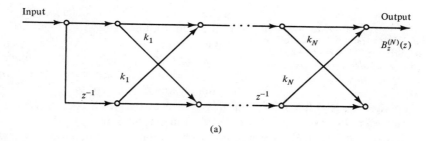

(a)

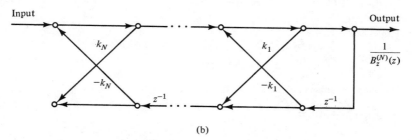

(b)

**Figure 5.28**  (a) Lattice form for an $N$th-order 1-D FIR prediction filter. (b) Lattice form for the inverse filter.

The 1-D lattice filters have two nice features which we would like to see in a multidimensional generalization. First, they have good coefficient quantization properties. The frequency response is not particularly sensitive to perturbations in $\{k(p)\}$. Second, if $|k(p)| < 1$ for $1 \leq p \leq N$, we are guaranteed that $B_z^{(N)}(z)$ will be a minimum-phase polynomial.

Both Marzetta [31, 32] and Harris [33] have extended these ideas to the two-dimensional case, but they have done it in very different ways. Marzetta's approach was to define a 2-D sequence of reflection coefficients; Harris' was to define a 1-D sequence of reflection functions. We will begin with Marzetta's formulation.

Let $\{b(n_1, n_2), 0 \leq n_1 \leq N_1, 0 \leq n_2 \leq N_2, (n_1, n_2) \neq (0, 0)\}$ be a 2-D FIR array and let

$$B_z^{(N_1, N_2)}(z_1, z_2) = 1 + \sum_{\substack{n_1=0 \\ (n_1, n_2) \neq (0,0)}}^{N_1} \sum_{n_2=0}^{N_2} b(n_1, n_2) z_1^{-n_1} z_2^{-n_2}$$

(5.172)

We would like to define a set of $N_1 \times N_2$ reflection coefficients $\{k(p_1, p_2)\}$ which provide an alternative representation for the filter such that the condition $|k(p_1, p_2)| < 1$ for $0 \leq p_1 \leq N_1, 0 \leq p_2 \leq N_2$ implies that $B_z^{(N_1, N_2)}(z_1, z_2)$ is a 2-D minimum-phase polynomial. Unfortunately, Marzetta [31, 32] showed that this cannot be done in general. He further showed that to preserve the minimum-phase property required an infinite set of reflection coefficients $\{k(p_1, p_2)\}$ defined on the "continuous support" region $R_{N_1 N_2}$, which is shown as solid dots in Figure 5.29.

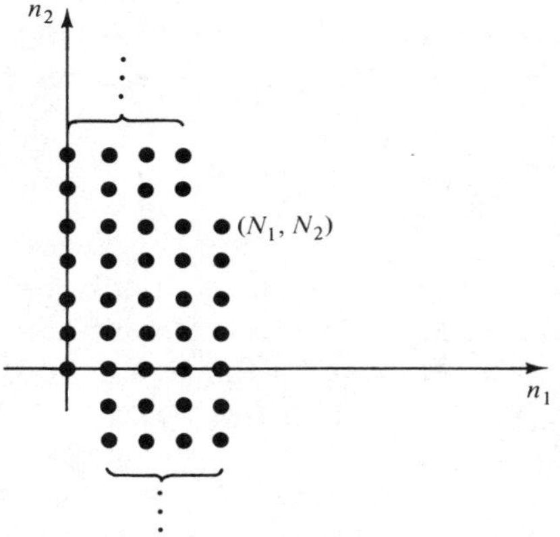

**Figure 5.29** Region of "continuous support" for 2-D prediction filters, denoted $R_{N_1 N_2}$. (Courtesy of Thomas L. Marzetta, *IEEE Trans. Acoustics, Speech, and Signal Processing*, © 1980 IEEE.)

Although a finite-extent sequence $\{b(n_1, n_2)\}$ results in an infinite sequence of reflection coefficients, fortunately the converse of this result is not true. If a polynomial is defined by a finite sequence of reflection coefficients $\{k(p_1, p_2)\}$, that polynomial will still have finite degree. Furthermore, if $|k(p_1, p_2)| < 1$ for all $(p_1, p_2)$, that polynomial will have minimum phase. The polynomial can be found from the $\{k(p_1, p_2)\}$ by means of the following recursion, which is a generalization of (5.171).

$$B_z^{(0, 0)}(z_1, z_2) = 1$$

(5.173a)

$$B_z^{(p_1, p_2)}(z_1, z_2) = B_z^{(p_1, p_2-1)}(z_1, z_2) + k(p_1, p_2) z_1^{-p_1} z_2^{-p_2} B_z^{(p_1, p_2-1)}(z_1^{-1}, z_2^{-1})$$

(5.173b)

$$B_z^{(p_1, -\infty)}(z_1, z_2) = B_z^{(p_1-1, \infty)}(z_1, z_2)$$

(5.173c)

This recursion is repeated for all values of $(p_1, p_2)$ for which $k(p_1, p_2) \neq 0$.

**Example 3 [32]**

Let the reflection coefficient sequence $k(p_1, p_2)$ be given by

$$k(0, 1) = 0.8$$
$$k(1, -1) = -0.6$$
$$k(1, 0) = 0.2$$
$$k(1, 1) = 0.1$$

(5.174)

$$k(p_1, p_2) = 0 \quad \text{for all other values of } (p_1, p_2)$$

Then equations (5.173) can be used to generate the polynomial representation

$$B_z^{(1,1)}(z_1, z_2) = 1 + 0.7z_2^{-1} - 0.14z_2^{-2} - 0.048z_2^{-3} - 0.48z_1^{-1}z_2^2$$
$$- 0.4496z_1^{-1}z_2 + 0.268z_1^{-1} + 0.1z_1^{-1}z_2^{-1} \tag{5.175}$$

Note that the region of support of the polynomial coefficients is larger than that of the reflection coefficient sequence.

Because of the equivalence of the reflection coefficient representation and the polynomial representation of 2-D minimum-phase filters with quarter-plane or nonsymmetric half-plane support, we can design filters of the form $1/B_z(z_1, z_2)$ by determining their reflection coefficient sequences. Furthermore, if the reflection coefficients have magnitudes which are less than 1, the stability of the filter is guaranteed. The design procedure is complicated by the fact that the relationship between the design parameters $\{k(p_1, p_2)\}$ and the denominator function $B_z(z_1, z_2)$ is iterative. Although some attempts have been made to design 2-D IIR filters using the reflection coefficients representation [34], the problem is not yet solved.

2-D FIR and IIR filters given in terms of a reflection coefficient array $k(p_1, p_2)$ may be implemented using a 2-D lattice structure. The 2-D lattice structure, basically a straightforward generalization of the 1-D lattice structure, can be derived from the recursive formulation (5.173).

Harris [33] developed a different lattice structure by treating the problem as a 1-D vector filtering problem. This approach is particularly well suited to array processing problems where the number of samples in one dimension (space) is small compared to the number of samples in the other dimension (time).

To describe this approach, we must introduce the notion of a *symmetric half-plane (SHP) filter*. A SHP filter is a 2-D IIR filter whose output mask is symmetric with respect to one of its variables, has a finite number of coefficients, and has its output hole on an edge. Thus if the finite-extent array $b(n_1, n_2)$ satisfies

$$b(n_1, n_2) = 0 \qquad \text{for } n_2 < 0$$

and

$$b(n_1, n_2) = b(-n_1, n_2) \qquad \text{for } n_2 \geq 0 \tag{5.176}$$

then the filter $1/B_z(z_1, z_2)$ is a SHP filter. (For simplicity, we shall assume that the numerator polynomial is simply a constant.) An example of a SHP filter output mask is shown in Figure 5.30.

A SHP filter, such as the one whose output mask is depicted in Figure 5.30, is not recursively computable in the sense used in Chapter 4. However, we can visualize the SHP filter as a 1-D vector processor which computes an entire row of output values as a function of the input and the previous $N$ rows of the output values. If the boundary conditions on the right and left sides of Figure 5.30 are given, then the usual 2-D difference equation relating the input signal $x$ to the output signal $y$ can be written in matrix-vector form as

$$\sum_{m=0}^{N} \mathbf{b}(m)\mathbf{y}(n - m) = \mathbf{x}(n) \tag{5.177}$$

or, equivalently,

$$\mathbf{b}(0)\mathbf{y}(n) = \mathbf{x}(n) - \sum_{m=1}^{N} \mathbf{b}(m)\mathbf{y}(n - m) \tag{5.178}$$

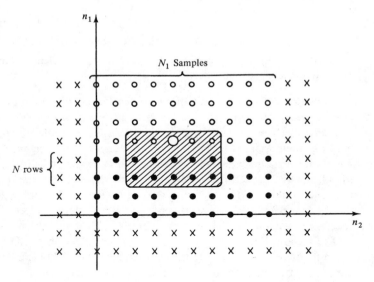

**Figure 5.30** SHP filter output mask. Dots, computed output values; circles, output values yet to be computed; ×'s, boundary conditions.

In this notation, $\mathbf{x}(n)$ represents the $n$th row of the input signal, $\mathbf{y}(n)$ represents the $n$th row of the output signal, and, under the assumption of zero boundary conditions, $\mathbf{b}(n)$ is a matrix that incorporates the effect of the $n$th row of the denominator coefficient array. It has the form of a Toeplitz matrix. The $ij$th element of the matrix $\mathbf{b}(n)$ is given by

$$[\mathbf{b}(n)]_{ij} = b(i - j, n) \qquad \text{for } 0 \le i, j \le N \tag{5.179}$$

When $\mathbf{b}(0)$ is equal to the identity matrix $\mathbf{I}$, we can use the 1-D vector recursion (5.178) to solve for $\mathbf{y}(n)$. From (5.179), we see that this condition occurs when

$$b(n_1, 0) = \delta(n_1) \tag{5.180}$$

which corresponds to the recursively computable output mask shown in Figure 5.31.

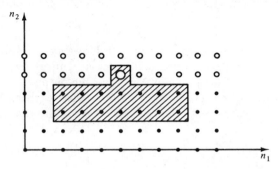

**Figure 5.31** Example of an output mask for a recursively computable SHP filter.

In the more general case, $\mathbf{y}(n)$ must be determined by inverting the $\mathbf{b}(0)$ matrix and multiplying it with the vector on the right side of (5.178). Harris [33] advocates the use of a triangular factorization of $\mathbf{b}(0)$ in computing the inverse since it preserves the banded structure of $\mathbf{b}(0)$ and leads to a relatively efficient implementation.

It can be shown that every minimum-phase SHP polynomial $B_z(z_1, z_2)$ of finite extent can be represented by a sequence of reflection coefficient functions $\{K_p(z)\}$ [33]. Each reflection coefficient function satisfies the constraint

$$|K_p(z)| < 1 \qquad \text{for } |z| = 1 \tag{5.181}$$

Given a reflection coefficient function representation $\{K_p(z)\}$, the corresponding SHP polynomial can be constructed through the generalized recursion

$$B_z^{(p)}(z_1, z_2) = B_z^{(p-1)}(z_1, z_2) + K_p(z_1)z_2^{-p}B_z^{(p-1)}(z_1, z_2^{-1}), \quad p = 1, \ldots, N \tag{5.182}$$

The recursion is initiated by a 1-D nonzero polynomial in the variable $z_1$.

$$B_z^{(0)}(z_1, z_2) = Q_z(z_1) \tag{5.183}$$

where

$$Q_z(z) \neq 0 \qquad \text{for } |z| = 1 \tag{5.184}$$

The reflection coefficient functions can be derived from a backward recursion similar in spirit to equation (5.170) [33].

In most practical filter design problems, the $K_p(z)$ are chosen to be symmetric polynomials of the form

$$K_p(z) = \sum_{n=-M}^{M} k_p(n)z^{-n} = K_p(z^{-1}) \tag{5.185}$$

with $k_p(n) = k_p(-n)$, but they may also be rational polynomials in $z$ when derived from an arbitrary SHP polynomial $B_z(z_1, z_2)$. In either case, if the reflection coefficient functions obey the constraint given in (5.181), the filter $1/B_z(z_1, z_2)$ is guaranteed to be BIBO stable.

Let us assume that we have designed a set of reflection coefficient functions $\{K_p(z); p = 1, N\}$ that represent an SHP polynomial $B_z(z_1, z_2)$ whose inverse we wish to implement in lattice form. To develop the structure of the lattice, we shall define the auxiliary functions [33]

$$F_p(z_1, z_2) \triangleq \frac{B_z^{(p)}(z_1, z_2)}{B_z(z_1, z_2)} \tag{5.186a}$$

$$G_p(z_1, z_2) \triangleq \frac{z_2^{-p}B_z^{(p)}(z_1, z_2^{-1})}{B_z(z_1, z_2)} \tag{5.186b}$$

From equations (5.182) and (5.186), the following recursive relationships may then be derived:

$$F_{p-1}(z_1, z_2) = F_p(z_1, z_2) - K_p(z_1)z_2^{-1}G_{p-1}(z_1, z_2) \tag{5.187a}$$

$$G_p(z_1, z_2) = z_2^{-1}G_{p-1}(z_1, z_2) + K_p(z_1)F_{p-1}(z_1, z_2) \tag{5.187b}$$

The desired overall transfer function $1/B_z(z_1, z_2)$ may be written as

$$\frac{1}{B_z(z_1, z_2)} = \frac{F_0(z_1, z_2)}{B_z^{(0)}(z_1, z_2)} \tag{5.188}$$

Using equations (5.187) and (5.188), Harris [33] developed the 1-D vector lattice structure shown in Figure 5.32, to implement the SHP filter $1/B_z(z_1, z_2)$.

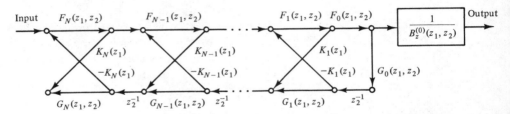

**Figure 5.32**     1-D vector lattice structure for implementing the 2-D SHP filter $1/B_z(z_1, z_2)$.

Figure 5.32 is most easily interpreted if we consider the $n_1$ variable to be a spatial variable and the $n_2$ variable to be a time variable. The input and output signals are then a sequence of vectors, each vector representing all the spatial samples at a given instant of time. Similarly, the internal signals present at the nodes of the lattice structure are also sequences of vectors. The shift operators $(z_2^{-1})$ along the lower rail of the lattice represent a delay of one time unit, and consequently they embody enough storage to buffer one entire row of spatial samples (i.e., one vector).

The gains on the cross-links are actually time-independent 1-D spatial filters. They are represented as $z_1$-domain multiplications which, of course, correspond to the convolution of the elements of a vector with the inverse $z$-transform of the reflection coefficient function. The last filter before the output node, $1/B_z^{(0)}(z_1, z_2)$, is also a time-independent 1-D spatial filter by virtue of equation (5.183). The node labels in Figure 5.32 [e.g., $F_{N-1}(z_1, z_2)$ and $G_{N-1}(z_1, z_2)$] represent the transfer function from the input node to the labeled node.

In some respects, the 1-D vector lattice implementation is similar in spirit to the iterative implementation of Section 5.2 (see Figure 5.7). In the iterative implementation, the iteration variable could be considered a time variable and the various filtering operations could be considered time-independent spatial filters, like the reflection coefficient functions in the vector lattice structure. For certain applications, especially the processing of signals from an array of spatially distributed sensors, the representation of multidimensional filtering operations as 1-D vector operations is a reasonable approach. Harris [33] has used his vector lattice structure to design and implement 2-D digital filters, such as fan filters and wave migration filters, which have applications in geophysical signal processing.

Because of limitations of space, we shall not endeavor to describe in detail the design algorithms developed by Harris [33] for the vector lattice structure. Instead, we shall given an overview of one design approach and mention some of its interesting aspects.

Harris's approach to the design problem uses an iterative optimization algorithm, similar to the method of steepest descent, to minimize the error functional

$$J = J_a + J_s \tag{5.189}$$

The functional $J_a$ is a measure of the approximation error in the frequency response and $J_s$ is a stability error.

He was interested in the simple case of designing a filter whose magnitude response approximates unity gain in some passband PB, zero gain in some stopband SB, and is arbitrary over the remaining regions of the frequency plane. His approximation error is interesting because it treats stopband and passband errors in different ways. Specifically, it has the form

$$
J_a = \frac{1}{4\pi^2} \iint\limits_{(\omega_1,\omega_2)\in\mathrm{PB}} [1 - |H(\omega_1,\omega_2)|^2]^2 \, d\omega_1 \, d\omega_2
$$

$$
+ \frac{\alpha}{4\pi^2} \iint\limits_{(\omega_1,\omega_2)\in\mathrm{SB}} |H(\omega_1,\omega_2)|^2 \, d\omega_1 \, d\omega_2
\tag{5.190}
$$

The parameter $\alpha$ provides a weight to control the relative importance of stopband and passband errors. This error functional, which is used to ensure that $J_a$ is differentiable [33], can be rationalized by observing that a stopband value of $\Delta$ contributes $\Delta^2$ to the integrand while a passband value of $1 + \Delta$ contributes approximately $4\Delta^2$ to the integrand. Thus, except for a proportionality constant which can be controlled by the parameter $\alpha$, both stopband and passband errors of order $\Delta$ make contributions to $J_a$ of order $\Delta^2$.

The stability error $J_s$ may assume the form of either a penalty function or a barrier function [35]. A penalty function is a function that is zero when the filter parameters correspond to a stable filter and grows as the filter becomes more and more unstable. The stability error discussed in Section 5.5.3 is one example of a penalty function. In the current context, one penalty function for reflection coefficient functions whose magnitude exceeds unity is given by [33]

$$
H_s = \mu \sum_{p=1}^{N} \frac{1}{\pi} \int_0^\pi \{\max[0, |K_p(e^{j\omega})| - 1]\}^2 \, d\omega
\tag{5.191}
$$

where $\mu$ is a proportionality constant.

A barrier function, on the other hand, is a function that is defined over the class of stable filters but grows to infinity as the boundary between the classes of stable and unstable filters is approached. For the current problem, an appropriate barrier function is given by [33]

$$
J_s = \frac{1}{\mu} \sum_{p=1}^{N} \frac{1}{\pi} \int_0^\pi \frac{d\omega}{1 - |K_p(e^{j\omega})|^2}
\tag{5.192}
$$

If any of the reflection coefficient functions $K_p(e^{j\omega})$ approach unity at any frequency, the stability error will grow. When using a barrier function, one must guarantee that the initial estimate of the filter parameter values results in a stable filter. The barrier function will then prevent subsequent estimates from resulting in an unstable filter.

The minimization of $J$ in (5.189) by most gradient algorithms requires that the partial derivatives of $J$ with respect to the parameters of $\{k_p(n)\}$ be evaluated. This calculation is straightforward but tedious due to the fact that the $\{k_p(n)\}$ affect the denominator of the transfer function through a recursion. The details of this calculation are given in [33].

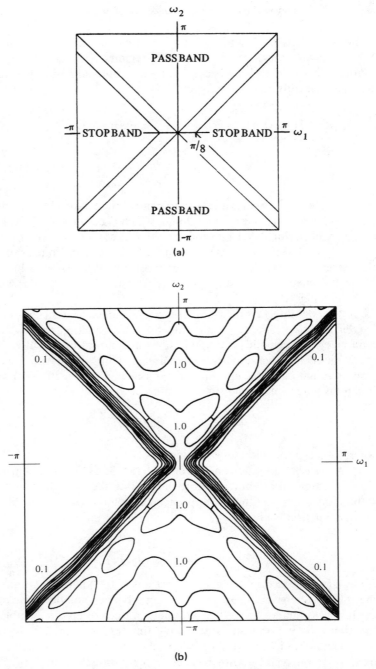

**Figure 5.33** (a) 90° fan filter specifications. (b) Magnitude of 55-coefficient fan filter linear contour plot (contour interval 0.1). (c) Perspective plot. (d) Reflection coefficient functions for 55-parameter fan filter. (Courtesy of David B. Harris.)

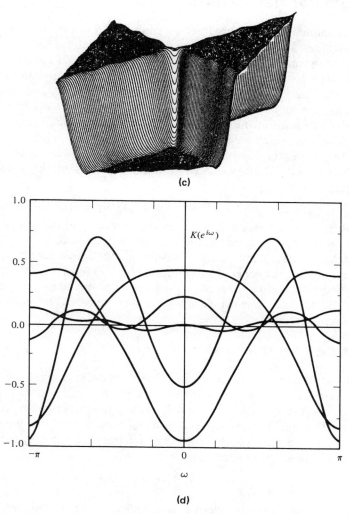

(c)

(d)

**Figure 5.33**    (*Continued*)

Harris [33] used this design method to design the 90° fan filter whose specifications are shown in Figure 5.33(a). The filter had 55 independent parameters, 25 for the quadrant symmetric numerator polynomial $A(\omega_1, \omega_2)$ and 30 for the SHP denominator polynomial $B(\omega_1, \omega_2)$. The transition bandwidth was $\pi/8$, the passband ripple was less than 0.01, and the stopband attenuation was approximately 34 dB. Figure 5.33(b) and (c) shows a contour plot and a perspective plot of the filter's frequency-response magnitude, and Figure 5.33(d) shows the five reflection coefficient functions used in the filter.

## *5.7 STABILIZATION TECHNIQUES

In this section we describe briefly two techniques for attempting to stabilize an unstable 2-D IIR filter. In the best of all possible worlds, methods for stabilization would not be necessary since all IIR filter design algorithms would generate only stable filters. Even in this less than ideal world, some of the algorithms discussed in this chapter incorporate stability checks or stability error minimizations to ensure the stability of their designed filters. Historically, however, the early 2-D IIR filter design algorithms did not control the stability or instability of the resulting filter designs, and stabilization techniques were needed to convert an unstable design into a useful, stable design.

In general, the objective of any stabilization technique is to convert an unstable IIR filter $H_u(\omega_1, \omega_2)$ into a stable IIR filter $H(\omega_1, \omega_2)$ such that the spectral magnitude is preserved.†

$$|H_u(\omega_1, \omega_2)| = |H(\omega_1, \omega_2)| \qquad (5.193)$$

Except for the rare case of a nonessential singularity of the second kind (see Chapter 4), the numerator polynomial does not affect the stability of an IIR filter. Thus, in order that an IIR filter be stable, its denominator coefficient array $b(n_1, n_2)$ must be a minimum-phase sequence. Furthermore, $b(n_1, n_2)$ must be a finite-extent array with an appropriate region of support (such as quarter-plane or nonsymmetric half-plane) if the filter is to be implemented as a recursively computable finite-order 2-D difference equation.

### 5.7.1 Cepstral Stabilization

The cepstral stabilization technique is based on the idea that the complex cepstrum of a minimum-phase signal has the same support as the signal itself (Section 4.4.3). We have already seen in Section 4.4.4 how a minimum-phase sequence can be constructed from an arbitrary sequence in a manner that preserves the spectral magnitude. In the notation of this section, we first construct the autocorrelation function of the unstable denominator array $b_u(n_1, n_2)$.

$$r_b(m_1, m_2) = \sum_{n_1} \sum_{n_2} b_u(n_1, n_2) b_u(n_1 - m_1, n_2 - m_2) \qquad (5.194)$$

Then we compute the Fourier transform, $R_b(\omega_1, \omega_2)$, of $r_b(m_1, m_2)$, which is real and nonnegative. We then take the logarithm of $R_b$ (if it is strictly positive), and compute the inverse Fourier transform to get the cepstrum, $\hat{r}_b(n_1, n_2)$, which has the property that

$$\hat{r}_b(n_1, n_2) = \hat{r}_b(-n_1, -n_2) \qquad (5.195)$$

We can then get the cepstrum of the desired minimum-phase sequence by multiplying $\hat{r}_b$ by a NSHP window function

---

†It is possible to formulate other stabilization problems in which other attributes, such as the spectral phase or group delay, are preserved. To the authors' knowledge, no practical application requiring the solution of such a stabilization problem has arisen.

$$\hat{b}(n_1, n_2) = w(n_1, n_2)\hat{r}_b(n_1, n_2) \qquad (5.196)$$

Finally, $b(n_1, n_2)$ is computed from $\hat{b}(n_1, n_2)$ by inverting the cepstral computations.

Ideally, $b(n_1, n_2)$ will be a minimum-phase sequence and will therefore represent the denominator array of a stable 2-D IIR filter. In practice, there are several potential problems. First, if $B_u(\omega_1, \omega_2)$ is zero at some frequency, this method will not be able to stabilize the filter. Second, $b(n_1, n_2)$ will generally consist of an infinite number of nonzero values, meaning that the filter cannot be implemented with a finite amount of computation. Third, using DFTs to calculate the necessary Fourier transforms means that spatial aliasing is present in the computed values of $\hat{r}_b$, $\hat{b}$, and $b$. Consequently, the computed array $b(n_1, n_2)$ may actually correspond to an unstable filter!

In spite of these potential problems, the cepstral stabilization technique has proved useful. The problem of spatial aliasing can be controlled somewhat by increasing the size of the DFTs used. This, of course, requires more computation. The spatial extent of $b(n_1, n_2)$ can be controlled by windowing [36] and, if $b(n_1, n_2)$ still results in an unstable filter, the window can be given an exponential shape to force any errant roots of $B_z(z_1, z_2)$ inside the unit bicircle. The expedient of windowing $b(n_1, n_2)$ will mean that the spectral magnitude will not be preserved, but in many cases the perturbations introduced are inconsequential.

### 5.7.2 Shaw's Stabilization Technique [13]

This stabilization technique is based on interpreting the 2-D denominator polynomial $B_z(z_1, z_2)$ as a parametric 1-D polynomial $B[z_1](z_2)$, as in Chapter 4. The stability theorems of Chapter 4 can be written in the following form for filters with support on the half-plane $n_2 \geq 0$.

**Theorem [13].**    The filter $1/B_z(z_1, z_2)$ is BIBO stable if and only if

(a) $B_z(z_1, z_2) = 0$    for $|z_1| = |z_2| = 1$ $\qquad (5.197a)$
(b) $B_z(1, z_2) \neq 0$    for $|z_2| > 1$ $\qquad (5.197b)$
(c) $B_z(z_1, 1)$ has continuous, odd, and periodic phase for $z_1 = \exp[j\omega_1]$. (There is no linear-phase component.)

The region of support $\{(n_1, n_2): n_2 \geq 0\}$ includes as special cases the first quadrant, the second quadrant, two nonsymmetric half-plane regions, and, of course, the half-plane region.

We shall follow the development of Shaw [13] by first considering an unstable, first-quadrant denominator array $b_u(n_1, n_2)$ which satisfies (5.197a) and (5.197b) but violates condition (c) of the theorem. If we simply shift $b_u(n_1, n_2)$ by the appropriate number of samples (say $m_1$) along the $n_1$-axis, we can remove the linear phase component to yield the stable denominator array

$$b(n_1, n_2) = b_u(n_1 - m_1, n_2) \qquad (5.198)$$

with a half-plane region of support in general. (It is possible to implement a 2-D IIR filter whose denominator coefficient array has half-plane support using the iterative implementation of Section 5.2 or a generalization of the half-plane implementation of Section 5.6.3. In this section, however, we are assuming that the user wishes to implement the filter with the classical 2-D difference equation computation.)

The array $b(n_1, n_2)$ can be altered to obtain the stable array $b_{ns}(n_1, n_2)$ with nonsymmetric half-plane support. First, consider the first row of $b(n_1, n_2)$ to be the 1-D sequence $b[0](n_1)$. Now we can compute the 1-D $z$-transform of $b[0](n_1)$, reflect the roots outside the unit circle back inside the unit circle, and perform an inverse $z$-transform to obtain a causal, minimum-phase sequence $b_{ns}[0](n_1)$. This operation corresponds to generating the 1-D all-pass filter

$$\text{AP}_z(z_1) = \frac{\text{MP}_z(z_1^{-1})}{\text{MP}_z(z_1)} \tag{5.199}$$

where $\text{MP}_z(z_1)$ is the maximum-phase polynomial whose roots are the roots of the $z$-transform of $b[0](n_1)$ which are outside the unit circle. In particular, we have for the first row

$$b_{ns}[0](n_1) = \text{ap}(n_1) * b[0](n_1) \tag{5.200}$$

and in general

$$b_{ns}(n_1, n_2) = \text{ap}(n_1) * b(n_1, n_2) \tag{5.201}$$

The array $b_{ns}(n_1, n_2)$ corresponds to a stable filter with the correct spectral magnitude, and it has nonsymmetric half-plane support. Unfortunately, $b_{ns}(n_1, n_2)$ may have infinite extent in the $n_1$ variable. Consequently, some sort of windowing must be applied to obtain a useful denominator array.

In the general case, an unstable denominator polynomial may violate conditions (5.197a) or (5.197b) as well. Let us now consider the case where (5.197a) is satisfied but (5.197b) is not. In this case, we can write the unstable denominator polynomial $B_z^{(u)}(z_1, z_2)$ as the 1-D parameterized polynomial $B_z^{(u)}[e^{j\omega_1}](z_2)$. Then, in theory, for every value of $\omega_1$ between 0 and $2\pi$, we may factor $B_z^{(u)}[e^{j\omega_1}](z_2)$ into the product of a minimum-phase and a maximum-phase polynomial. The maximum-phase polynomial can be converted into a minimum-phase polynomial by reflecting its roots inside the $z_2$ unit circle. The stable denominator array $B_z[e^{j\omega_1}](z_2)$ is formed as the product of these two minimum-phase polynomials. At this point, the corresponding $b(n_1, n_2)$ array will have symmetric half-plane support and can be subjected to the process described above to obtain a nonsymmetric half-plane array.

When $B_z^{(u)}(z_1, z_2)$ violates (5.197a) as well, the foregoing procedure will not quite work because of the presence of zeros on the $z_2$ unit circle. The roots of $B_z^{(u)}[e^{j\omega_1}](z_2)$ that are strictly outside the $z_2$ unit circle may still be reflected inside to obtain the polynomial

$$B_z'[e^{j\omega_1}](z_2) \neq 0 \qquad \text{for all } \omega_1 \text{ and } |z_2| > 1 \tag{5.202}$$

Then, by applying the weighting

$$w(n_2) = a^{n_2}, \qquad 0 < a < 1 \tag{5.203}$$

to the corresponding coefficients $b'(n_1, n_2)$, the roots of $B_z'$ may be moved inside the $z_2$ unit circle.

Thus for this case

$$b(n_1, n_2) = a^{n_2}b'(n_1, n_2) \tag{5.204}$$

Again, it will be necessary in general to subject $b(n_1, n_2)$ to the all-pass filtering to obtain a stable array with nonsymmetric half-plane support.

In practice, of course, $B_z^{(u)}[e^{j\omega_1}](z_2)$ is evaluated using a DFT for only a finite number of values of $\omega_1$. This, as well as other numerical effects, will introduce some spatial aliasing into the stabilized denominator array, which may be reduced by using a larger DFT. Shaw and Mersereau [13] discuss in detail the consequences of a numerical implementation of this stabilization strategy and give several examples.


## PROBLEMS

**5.1.** Consider a 3-D IIR filter which has the transfer function

$$H_z(z_1, z_2, z_3) = \frac{\sum_{k_1=0}^{K_1-1} \sum_{k_2=0}^{K_2-1} \sum_{k_3=0}^{K_3-1} a(k_1, k_2, k_3) z_1^{-k_1} z_2^{-k_2} z_3^{-k_3}}{1 - \sum_{\substack{k_1=0 \\ (k_1,k_2,k_3)\neq(0,0,0)}}^{K_1-1} \sum_{k_2=0}^{K_2-1} \sum_{k_3=0}^{K_3-1} c(k_1, k_2, k_3) z_1^{-k_1} z_2^{-k_2} z_3^{-k_3}}$$

The output of this filter is desired for values of $(n_1, n_2, n_3)$ in the range $0 \leq n_1 \leq N_1 - 1; 0 \leq n_2 \leq N_2 - 1; 0 \leq n_3 \leq N_3 - 1$ and the output samples are computed in an order such that the index $n_1$ is incremented most rapidly, and $n_3$ most slowly. That is, outputs are computed in the order $y(0, 0, 0)$, $y(1, 0, 0)$, $y(2, 0, 0)$, ..., $y(N_1 - 1, 0, 0)$, $y(0, 1, 0)$, $y(1, 1, 0)$, ....

(a) If a direct form I implementation of the filter is used, how many words of storage are necessary to hold all the samples of the input and output arrays that are needed for future computations? Assume that samples of the input arrive in the same order in which samples of the output are computed.

(b) If instead a direct form II implementation is used, how many words of storage are needed?

(c) If $N_1$, $N_2$, and $N_3$ are unequal and storage is the only concern, how should the computations be ordered?

**5.2.** Suppose that we have a cascaded 2-D IIR filter with the system function

$$H_z(z_1, z_2) = \frac{\alpha}{B_z(z_1, z_2)C_z(z_1, z_2)}$$

where

$$B_z(z_1, z_2) = \sum_{n_1=0}^{N_1-1} \sum_{n_2=0}^{N_2-1} b(n_1, n_2) z_1^{-n_1} z_2^{-n_2}$$

$$b(0, 0) \triangleq 1$$

$$C_z(z_1, z_2) = \sum_{m_1=0}^{M_1-1} \sum_{m_2=0}^{M_2-1} c(m_1, m_2) z_1^{-m_1} z_2^{-m_2}$$

$$c(0, 0) \triangleq 1$$

and $\alpha$ is a real number. [The coefficients $\{b(n_1, n_2), c(m_1, m_2)\}$ are assumed to be real also.]

(a) Write a set of two difference equations that may be used to compute the result of passing an input signal $x(n_1, n_2)$ through the cascade of two subfilters to realize the overall filter $H_z(z_1, z_2)$. Call the output signal from the first subfilter $s(n_1, n_2)$ and the overall output signal $y(n_1, n_2)$.

(b) Assuming the signal samples as well as the filter coefficients are real-valued, how many multiplications and additions are required to compute each sample of $s(n_1, n_2)$? How many for each sample of $y(n_1, n_2)$?

(c) Now assume that the difference equations derived in part (a) are implemented in a row-by-row fashion. How much storage is required to compute $s(n_1, n_2)$ for $0 \le n_1 < K_1$, $0 \le n_2 < K_2$? (Assume that $K_1 \gg N_1$, $K_1 \gg M_1$, $K_2 \gg N_2$, $K_2 \gg M_2$ and neglect storage needed for boundary conditions.) How much storage is required to compute $y(n_1, n_2)$ for the same values of $(n_1, n_2)$?

(d) Let $D_z(z_1, z_2) \triangleq B_z(z_1, z_2)C_z(z_1, z_2)$. Then $D_z(z_1, z_2)$ can be written in the form

$$D_z(z_1, z_2) = \sum_{n_1=0}^{N_1+M_1-2} \sum_{n_2=0}^{N_2+M_2-2} d(n_1, n_2) z_1^{-n_1} z_2^{-n_2}$$

Write an equation relating $d(n_1, n_2)$, $b(n_1, n_2)$, and $c(n_1, n_2)$. What value does $d(0, 0)$ have? Write a difference equation relating $x(n_1, n_2)$ and $y(n_1, n_2)$ using the coefficients $\{d(n_1, n_2)\}$.

(e) How many multiplications and additions are required to compute each sample of $y(n_1, n_2)$ using the difference equation in part (d)? How much storage is necessary to compute $y(n_1, n_2)$ for $0 \le n_1 < K_1$, $0 \le n_2 < K_2$ using the difference equation in part (d)? (Again, neglect storage needed for boundary conditions.) Is the cascade or direct form implementation for $H_z(z_1, z_2)$ more efficient?

**5.3.** Although the iterative implementation discussed in Section 5.2 is intended primarily for realizing IIR filters which are not recursively computable, we can nonetheless use it to realize causal 1-D and 2-D IIR filters to understand better how it works. Let us start with a simple 1-D case.

(a) Suppose that $H_z(z) = 1/(1 - az^{-1})$. Write an implicit relation between the $z$-transform of the input $X_z(z)$ and the $z$-transform of the output $Y_z(z)$ with the form

$$Y_z(z) = A_z(z)X_z(z) + C_z(z)Y_z(z)$$

If we write

$$C_z(z) = \sum_n c(n)z^{-n}$$

what is the sequence $c(n)$?

(b) Write the time-domain iteration corresponding to the implicit $z$-transform relation in part (a). Evaluate all the convolution operations; don't merely indicate a convolution using "$*$."

(c) Let $y_i(n)$ represent the output after $i$ iterations of the formula in part (b). Assuming $y_0(n) = x(n) = \delta(n)$, evaluate $y_1(n)$, $y_3(n)$, $y_i(n)$, and $y(n) = y_\infty(n)$. Derive an expression for the error

$$e_2 = \sum_n [y(n) - y_i(n)]^2$$

**5.4.** Let $H_z(z_1, z_2) = 1/(1 - \frac{1}{4}z_1^{-1} - \frac{1}{2}z_2^{-1})$. Derive the corresponding iterative implementation in the signal domain.

(a) What does $c(n_1, n_2)$ equal?

**(b)** Let $y_i(n_1, n_2)$ denote the output signal after the $i$th iteration and assume that $y_0(n_1, n_2) = x(n_1, n_2) = \delta(n_1, n_2)$. Evaluate $y_1(n_1, n_2)$ and $y_3(n_1, n_2)$.

**(c)** The correct solution $y(n_1, n_2) = y_\infty(n_1, n_2)$ is given by

$$y(n_1, n_2) = \frac{(n_1 + n_2)!}{n_1! \, n_2!} \left(\frac{1}{4}\right)^{n_1} \left(\frac{1}{2}\right)^{n_2} u(n_1, n_2)$$

If we use the iterative implementation to generate the estimate $y_I(n_1, n_2)$, for which points in the first quadrant does $y_I(n_1, n_2) = y(n_1, n_2)$? Draw a sketch of this region.

**5.5.** We saw in Section 5.2.1 that the iterative implementation of a recursive filter could be interpreted as a first-order feedback system in which a complete multidimensional signal is fed back. We can consider higher-order feedback systems as well.

**(a)** By substituting the iteration equation (5.16) into itself, derive an iteration formula in which $y_i(n_1, n_2)$ is obtained directly from $y_{i-2}(n_1, n_2)$. This recursion will require only one-half the number of iterations of the original recursion to reach a given level of convergence.

**(b)** Generalize the result of part (a) to find a recursion that will require only $1/N$ times as many iterations as the original recursion.

**(c)** How does the overall complexity of the recursion you derived in part (b) compare with the original recursion of equation (5.16)? Use as a measure of complexity the number of multiplications required to compute the equivalent of $N$ iterations of the original algorithm.

**5.6.** Let $H_z(z_1, z_2) = 1/B_z(z_1, z_2)$ where

$$B_z(z_1, z_2) = 1 - c_1 z_1^{-1} - c_2 z_2^{-1} + c_1 c_2 z_1^{-1} z_2^{-1}$$

**(a)** For what values of $c_1$ and $c_2$ is $H_z(z_1, z_2)$ stable, assuming that $h(n_1, n_2)$ has support on the first quadrant only?

**(b)** Assume that $C_z(z_1, z_2) \triangleq 1 - B_z(z_1, z_2)$. Can you find any values of $c_1$ and $c_2$ such that $H_z(z_1, z_2)$ is stable but $C_z(z_1, z_2)$ violates the convergence criterion $|C_z(z_1, z_2)| < 1$ for $|z_1| = |z_2| = 1$? (For simplicity you might wish to examine the case where $c_1 = c_2$.)

**(c)** Can you find a real, positive value for $\lambda$ such that the redefined function

$$C_z(z_1, z_2) \triangleq 1 - \lambda B_z(z_1, z_2)$$

satisfies the convergence criterion for all values of $c_1$ and $c_2$ such that $H_z(z_1, z_2)$ is stable?

**5.7.** In Section 5.2.2 the use of a constant relaxation parameter $\lambda$ was shown to broaden the class of filters for which a convergent iteration could be obtained. It can also affect the rate of convergence. As before, let

$$H(\omega_1, \omega_2) = \frac{A(\omega_1, \omega_1)}{B(\omega_1, \omega_2)}$$

and set $C(\omega_1, \omega_2) = 1 - \lambda B(\omega_1, \omega_2)$ for a constant $\lambda$. For simplicity, assume that $B(\omega_1, \omega_2)$ is purely real and bounded so that it lies in the range

$$0 < B_0 \leq B(\omega_1, \omega_2) \leq B_1$$

**(a)** Derive an expression for the error on the $I$th iteration in terms of $\lambda$ and $B(\omega_1, \omega_2)$.

**(b)** If $B(\omega_1, \omega_2)$ satisfies the bounds above, how should $\lambda$ be chosen to minimize the error?

**5.8.** Consider a separable FIR system with the transfer function

$$H_z(z_1, z_2) = (b_0 + b_1 z_1^{-1} + b_2 z_1^{-2})(1 + c_1 z_2^{-1} + c_2 z_2^{-2} + c_3 z_2^{-3})$$

Find a flowgraph that represents this system and involves only the coefficients $b_0$, $b_1$, $b_2$, $c_1$, $c_2$, and $c_3$.

**5.9. (a)** Write a set of difference equations corresponding to the flowgraph of Figure 5.20. Write the equations in an order that will permit the necessary variables to be computed with a minimum amount of intermediate storage. Assume that the outputs are computed rowwise.
   **(b)** Will these equations also permit the output to be computed columnwise?

**5.10. (a)** Write a series of difference equations that will realize the network shown in Figure P5.10.

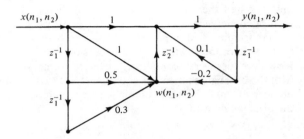

**Figure P5.10**

   **(b)** What is the transfer function of this network?
   **(c)** Draw the flowgraph of another network with the same transfer function.

**5.11.** Consider the flowgraph shown in Figure P5.11.

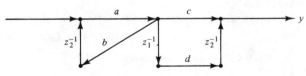

**Figure P5.11**

   **(a)** Derive an expression for the system function

$$H_z(z_1, z_2) = \frac{Y_z(z_1, z_2)}{X_z(z_1, z_2)}$$

   **(b)** Write the corresponding 2-D difference equation that relates the input signal $x(n_1, n_2)$ to the output signal $y(n_1, n_2)$.
   **(c)** Derive a set of state-variable equations for this flowgraph. Let $g_0(n_1, n_2)$ denote the horizontal state variable which is the output of the $z_1^{-1}$ branch in the middle of the flowgraph. Let $v_0(n_1, n_2)$ and $v_1(n_1, n_2)$ denote the two vertical state variables: $v_0$ corresponding to the output of the left-hand $z_2^{-1}$ branch and $v_1$ corresponding to the output of the right-hand $z_2^{-1}$ branch.

**(d)** Using equation (5.64), show that $F_z(z_1, z_2)$ computed from the matrices derived in the state-variable representation [part (c)] is in fact equal to the system function $H_z(z_1, z_2)$ [part (a)].

**(e)** Can you find an alternative state-variable realization for $H_z(z_1, z_2)$ which uses only two state variables? If so, derive it. (*Hint:* Consider the given flowgraph as an example of a direct form implementation. What does the corresponding direct form II flowgraph look like?) Verify that $F_z(z_1, z_2)$ for this two-variable form does indeed equal $H_z(z_1, z_2)$.

**5.12.** Suppose that we have a two-input, two-output 2-D system which is specified in the following state-space form:

$$\begin{bmatrix} g(n_1 + 1, n_2) \\ v(n_1, n_2 + 1) \end{bmatrix} = \begin{bmatrix} A_1 & A_2 \\ A_3 & A_4 \end{bmatrix} \begin{bmatrix} g(n_1, n_2) \\ v(n_1, n_2) \end{bmatrix} + \begin{bmatrix} B_1 & B_2 \\ B_3 & B_4 \end{bmatrix} \begin{bmatrix} x_1(n_1, n_2) \\ x_2(n_1, n_2) \end{bmatrix}$$

$$\begin{bmatrix} y_1(n_1, n_2) \\ y_2(n_1, n_2) \end{bmatrix} = \begin{bmatrix} C_1 & C_2 \\ C_3 & C_4 \end{bmatrix} \begin{bmatrix} g(n_1, n_2) \\ v(n_1, n_2) \end{bmatrix} + \begin{bmatrix} D_1 & D_2 \\ D_3 & D_4 \end{bmatrix} \begin{bmatrix} x_1(n_1, n_2) \\ x_2(n_1, n_2) \end{bmatrix}$$

where $x_1$ and $x_2$ are the two input signals, $y_1$ and $y_2$ are the two output signals, and $g$ and $v$ are the states. Assume that the coefficients $A_1, \ldots, A_4, B_1, \ldots, B_4, C_1, \ldots,$ $C_4, D_1, \ldots, D_4$ are real-valued scalars. Now, let us define $X_1(z_1, z_2)$, $X_2(z_1, z_2)$, $Y_1(z_1, z_2)$, and $Y_2(z_1, z_2)$ to be the z-transforms of $x_1(n_1, n_2)$, $x_2(n_1, n_2)$, $y_1(n_1, n_2)$, and $y_2(n_1, n_2)$, respectively. We can also define a system function matrix $\mathbf{F}(z_1, z_2)$ as

$$\mathbf{F}(z_1, z_2) \triangleq \begin{bmatrix} F_{11}(z_1, z_2) & F_{12}(z_1, z_2) \\ F_{21}(z_1, z_2) & F_{22}(z_1, z_2) \end{bmatrix}$$

where

$$F_{ij}(z_1, z_2) \triangleq \frac{Y_i(z_1, z_2)}{X_j(z_1, z_2)}; \qquad i, j = 1, 2$$

Derive an expression for $\mathbf{F}(z_1, z_2)$ in terms of the constants $\{A_1, A_2, \ldots, D_4\}$ used in the state-space equations above and the complex variables $z_1^{-1}$ and $z_2^{-1}$.

**5.13.** Most of the design algorithms discussed in this chapter assume that the numerator array $a(n_1, n_2)$ and the denominator array $b(n_1, n_2)$ have their support on the first quadrant. In this problem we will show that any filter with support on a wedge can be designed using a first-quadrant design algorithm.

Assume that the ideal response $i(n_1, n_2)$ is to be approximated by the impulse response $h(n_1, n_2)$ which has support on the wedge $W$. We then use the following procedure:

1. Find a linear transformation that will map the wedge $W$ onto the first quadrant.
2. Use this transformation to map $i(n_1, n_2)$ into a new sequence $j(m_1, m_2)$.
3. Use a first-quadrant design algorithm to find an approximation $g(m_1, m_2)$ to $j(m_1, m_2)$.
4. Use the inverse of the mapping in step 1 to map $g(m_1, m_2)$ onto $h(n_1, n_2)$.

**(a)** If

$$j(m_1, m_2) = \begin{cases} i(n_1, n_2), & m_1 = An_1 + Bn_2; \quad m_2 = Cn_1 + Dn_2 \\ 0, & \text{elsewhere} \end{cases}$$

where $A, B, C, D$ are integers and $AD \neq BC$, how is $J(\omega_1, \omega_2)$ related to $I(\omega_1, \omega_2)$?

**(b)** If the output mask of the filter has the shape shown in Figure P5.13, determine $A$, $B$, $C$, and $D$.

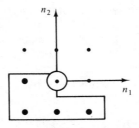

**Figure P5.13**

**5.14.** In Section 5.4 we showed that

$$\frac{\partial y(n_1, n_2)}{\partial a(p_1, p_2)} = x(n_1 - p_1, n_2 - p_2) - \sum_{\substack{m_1=0 \\ (m_1, m_2) \neq (0, 0)}}^{M_1-1} \sum_{m_2=0}^{M_2-1} b(m_1, m_2) \frac{\partial y(n_1 - m_1, n_2 - m_2)}{\partial a(p_1, p_2)}$$

$$\frac{\partial y(n_1, n_2)}{\partial b(q_1, q_2)} = -y(n_1 - q_1, n_2 - q_2) - \sum_{\substack{m_1=0 \\ (m_1, m_2) \neq (0, 0)}}^{M_1-1} \sum_{m_2=0}^{M_2-1} b(m_1, m_2) \frac{\partial y(n_1 - m_1, n_2 - m_2)}{\partial b(q_1, q_2)}$$

**(a)** Show that

$$\frac{\partial y(n_1, n_2)}{\partial a(p_1, p_2)} = \frac{\partial y(n_1 - p_1, n_2 - p_2)}{\partial a(0, 0)}$$

**(b)** Show that

$$\frac{\partial y(n_1, n_2)}{\partial b(q_1, q_2)} = \frac{\partial y(n_1 - q_1 + 1, n_2 - q_2)}{\partial b(1, 0)} = \frac{\partial y(n_1 - q_1, n_2 - q_2 + 1)}{\partial b(0, 1)}$$

**5.15.** **(a)** Generalize Shanks's design procedure to permit the design of nonsymmetric half-plane IIR filters which have a transfer function of the form

$$H_z(z_1, z_2) = \frac{\displaystyle\sum_{k_1=1}^{N_1} \sum_{k_2=-N_2}^{N_2} a(k_1, k_2) z_1^{-k_1} z_2^{-k_2} + \sum_{k_2=0}^{N_2} a(0, k_2) z_2^{-k_2}}{1 + \displaystyle\sum_{k_1=1}^{N_1} \sum_{k_2=-N_2}^{N_2} b(k_1, k_2) z_1^{-k_1} z_2^{-k_2} + \sum_{k_2=1}^{N_2} b(0, k_2) z_2^{-k_2}}$$

Sketch a possible choice for the region $R$ over which the error must be minimized. Is this choice for the region $R$ unique?

**5.16.** A filter $h(n_1, n_2)$ is designed to minimize the error

$$E = \sum_{n_1} \sum_{n_2} [h(n_1, n_2) - d(n_1, n_2)]^2$$

where

$$\sum_{n_1} \sum_{n_2} d^2(n_1, n_2)$$

is finite. If the sums are performed over the whole of the $(n_1, n_2)$-plane, show that $h(n_1, n_2)$ will be stable in the mean-squared sense.

[*Hint:* It may be helpful to make use of the Schwartz inequality

$$\left| \sum_{n_1} \sum_{n_2} x(n_1, n_2) y(n_1, n_2) \right| \leq \left\{ \sum_{n_1} \sum_{n_2} x^2(n_1, n_2) \right\}^{1/2} \left\{ \sum_{n_1} \sum_{n_2} y^2(n_1, n_2) \right\}^{1/2}$$

**5.17.** Often we wish to design a recursive filter that is symmetric. One approach is simply to use a symmetric ideal response and use an unconstrained algorithm. This approach usually works, but it involves performing a larger than necessary optimization. Another approach is to apply the constraints in the beginning, reducing the number of degrees of freedom in the design.

Consider a frequency-domain design chosen to minimize the functional

$$J_a = \sum_k W(\omega_{1k}, \omega_{2k})\left[\frac{A(\omega_{1k}, \omega_{2k})}{B(\omega_{1k}, \omega_{2k})} - D(\omega_{1k}, \omega_{2k})\right]^2$$

using the linearization approach of Section 5.5.1. Determine the gradient of $J_a$ if we require that

$$a(k_1, k_2) = a(k_2, k_1), \qquad 0 \le k_1, k_2 \le N - 1$$
$$b(k_1, k_2) = b(k_2, k_1), \qquad 0 \le k_1, k_2 \le N - 1$$

**5.18.** The filter $h(n_1, n_2)$ has support on the first quadrant. A composite response, which is realized by implementing in parallel four rotations of $h(n_1, n_2)$, is given by

$$\tilde{h}(n_1, n_2) = h(n_1, n_2) + h(-n_1, n_2) + h(-n_1, -n_2) + h(n_1, -n_2)$$

**(a)** Determine the composite frequency response $\tilde{H}(\omega_1, \omega_2)$ in terms of $H(\omega_1, \omega_2)$. Assume that $h(n_1, n_2)$ is real.

**(b)** Show that the resulting implementation is zero-phase.

**5.19.** A filter $H_z(z_1, z_2)$ has the lowpass characteristic shown in Figure P5.19(a). Find frequency transformations $F_1(z_1, z_2) \longrightarrow z_1^{-1}$, $F_2(z_1, z_2) \longrightarrow z_2^{-1}$ that can be applied to this filter to yield the responses shown in Figure P5.19(b–d).

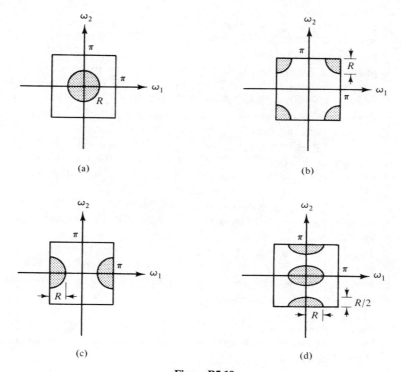

Figure P5.19

**5.20.** Let a 2-D reflection coefficient sequence $k(p_1, p_2)$ be given by

$$k(0, 1) = 0.7$$
$$k(1, -1) = 0.5$$
$$k(1, 0) = -0.3$$
$$k(1, 1) = 0.1$$
$$k(p_1, p_2) = 0 \qquad \text{for all other values of } (p_1, p_2)$$

Determine the polynomial $B_z(z_1, z_2)$ that corresponds to $k(p_1, p_2)$.

# REFERENCES

1. Alan V. Oppenheim and Ronald W. Schafer, *Digital Signal Processing* (Englewood Cliffs, N.J.: Prentice-Hall, Inc., 1975).
2. José M. Costa and Anastasios N. Venetsanopoulos, "Design of Circularly Symmetric Two-Dimensional Recursive Filters," *IEEE Trans. Acoustics, Speech, and Signal Processing*, ASSP-22, no. 6 (Dec. 1974), 26–35.
3. Dan E. Dudgeon, "An Iterative Implementation for 2-D Digital Filters," *IEEE Trans. Acoustics, Speech, and Signal Processing*, ASSP-28, no. 6 (Dec. 1980), 666–71.
4. Thomas F. Quatieri and Dan E. Dudgeon, "Implementation of 2-D Digital Filters by Iterative Methods," *IEEE Trans. Acoustics, Speech, and Signal Processing*, ASSP-30, no. 3 (June 1982), 473–87.
5. Michael T. Manry and J. K. Aggarwal, "Picture Processing Using One-Dimensional Implementations of Discrete Planar Filters," *IEEE Trans. Acoustics, Speech, and Signal Processing*, ASSP-22, no. 3 (June 1974), 164–73.
6. Robert P. Roesser, "A Discrete State–Space Model for Linear Image Processing," *IEEE Trans. Automatic Control*, AC-20, no. 1 (Feb. 1975), 1–10.
7. Sun-Yuan Kung, Bernard C. Lévy, Martin Morf, and Thomas Kailath, "New Results in 2-D Systems Theory, Part II: 2-D State–Space Models—Realization and the Notions of Controllability, Observability, and Minimality," *Proc. IEEE*, 65, no. 6 (June 1977), 945–61.
8. Alan S. Willsky, *Digital Signal Processing and Control and Estimation Theory: Points of Tangency, Areas of Intersection, and Parallel Directions* (Cambridge, Mass.: The MIT Press, 1979).
9. David S. K. Chan, "A Simple Derivation of Minimal and Near-Minimal Realizations of 2-D Transfer Functions," *Proc. IEEE*, 66, no. 4 (Apr. 1978), 515–16.
10. James A. Cadzow, "Recursive Digital Filter Synthesis via Gradient Based Algorithms," *IEEE Trans. Acoustics, Speech, and Signal Processing*, ASSP-24, no. 5 (Oct. 1976), 349–55.
11. John L. Shanks, Sven Treitel, and James H. Justice, "Stability and Synthesis of Two-Dimensional Recursive Filters," *IEEE Trans. Audio and Electroacoustics*, AU-20, no. 2 (June 1972), 115–28.

12. J. M. Ortega and W. C. Rheinboldt, *Iterative Solution of Nonlinear Equations in Several Variables* (New York: Academic Press, Inc., 1970).

13. G. A. Shaw and R. M. Mersereau, "Design, Stability, and Performance of Two-Dimensional Recursive Digital Filters," Technical Report E21-B05-1, Georgia Institute of Technology School of Electrical Engineering (Dec. 1979).

14. R. Fletcher and M. J. D. Powell, "A Rapidly Convergent Descent Method for Minimization," *Computer J.*, 6, no. 2 July (1963), 163–68.

15. Miguel S. Bertrán, "Approximation of Digital Filters in One and Two Dimensions," *IEEE Trans. Acoustics, Speech, and Signal Processing*, ASSP-23, no. 5 (Oct. 1975), 438–43.

16. B. Nowrowzian, M. Ahmadi, and R. A. King, "On the Space-Domain Design Techniques of $N$-Dimensional Recursive Digital Filters," *Proc. IEEE Int. Conf. Acoustics, Speech, and Signal Processing* (Apr. 1977), 531–34.

17. K. M. Brown and J. E. Dennis, Jr., "Derivative-Free Analogues of the Levenberg-Marquardt and Gauss Algorithm for Non-linear Least-Squares Approximation," *Numerische Mathematik*, 18 (1972), 289–97.

18. Michael P. Ekstrom, Richard F. Twogood, and John W. Woods, "Two-Dimensional Recursive Filter Design—A Spectral Factorization Approach," *IEEE Trans. Acoustics, Speech, and Signal Processing*, ASSP-28, no. 1 (Feb. 1980), 16–26.

19. K. Steiglitz and L. E. McBride, "A Technique for the Identification of Linear Systems," *IEEE Trans. Automatic Control*, AC-10 (Oct. 1965), 461–64.

20. E. W. Cheney, *Introduction to Approximation Theory* (New York: McGraw-Hill Book Company, 1966).

21. I. Barrodale, M. J. D. Powell, and F. D. K. Roberts, "The Differential Correction Algorithm for Rational $l_\infty$ Approximation", *SIAM J. Numerical Analysis*, 9 (Sept. 1972), 493–504.

22. Dan E. Dudgeon, "Two-Dimensional Recursive Filtering," Sc.D. thesis, Department of Electrical Engineering and Computer Science, Massachusetts Institute of Technology (May 1974).

23. Dan E. Dudgeon, "Recursive Filter Design Using Differential Correction," *IEEE Trans. Acoustics, Speech, and Signal Processing*, ASSP-22, no. 6 (Dec. 1974), 443–48.

24. Dan E. Dudgeon, "Two-Dimensional Recursive Filter Design Using Differential Correction," *IEEE Trans. Acoustics, Speech, and Signal Processing*, ASSP-23, no. 3 (June 1975), 264–67.

25. W. W. Garvin, *Introduction to Linear Programming* (New York: McGraw-Hill Book Company, 1960).

26. Neil A. Pendergrass, Sanjit K. Mitra, and Ely I. Jury, "Spectral Transformations for Two-Dimensional Digital Filters," *IEEE Trans. Circuits and Systems*, CAS-23, no. 1 (Jan. 1976), 26–35.

27. Satyabrata Chakrabarti and Sanjit K. Mitra, "Design of Two-Dimensional Digital Filters via Spectral Transformations," *Proc. IEEE*, 65, no. 6 (June 1977), 905–14.

28. Gamal A. Maria and Moustafa M. Fahmy, "An $l_p$ Design Technique for Two-Dimensional Digital Recursive Filters," *IEEE Trans. Acoustics, Speech, and Signal Processing*, ASSP-22, no. 1 (Feb. 1974), 15–21.

29. J. F. Abramatic, F. Germain, and E. Rosencher, "Design of 2-D Recursive Filters with Separable Denominator Transfer Functions," *Proc. IEEE Int. Conf. Acoustics, Speech, and Signal Processing* (Apr. 1979), 24–27.

30. John D. Markel and Augustine H. Gray, *Linear Prediction of Speech* (Berlin: Springer-Verlag, 1976).

31. Thomas L. Marzetta, "A Linear Prediction Approach to Two-Dimensional Spectral Factorization and Spectral Estimation," Ph.D. thesis, Department of Electrical Engineering and Computer Science, Massachusetts Institute of Technology (Feb. 1978).

32. Thomas L. Marzetta, "Two-Dimensional Linear Prediction: Autocorrelation Arrays, Minimum-Phase Prediction Error Filters, and Reflection Coefficient Arrays," *IEEE Trans. Acoustics, Speech, and Signal Processing*, ASSP-28, no. 6 (Dec. 1980), 725–33.

33. David B. Harris, "Design and Implementation of Rational 2-D Digital Filters," Ph.D. thesis, Department of Electrical Engineering and Computer Science, Massachusetts Institute of Technology (Nov. 1979).

34. Thomas L. Marzetta, "The Design of 2-D Recursive Filters in the 2-D Reflection Coefficient Domain," *Proc. IEEE Int. Conf. Acoustics, Speech, and Signal Processing* (Apr. 1979), 32–35.

35. David G. Luenberger, *Optimization by Vector Space Methods* (New York: John Wiley & Sons, Inc., 1969).

36. Michael P. Ekstrom and John W. Woods, "Two-Dimensional Spectral Factorization with Applications in Recursive Digital Filtering," *IEEE Trans. Acoustics, Speech, and Signal Processing*, ASSP-24, no. 2 (Apr. 1976), 115–28.

# 6

# PROCESSING SIGNALS CARRIED
# BY PROPAGATING WAVES

We receive signals carried by propagating waves constantly, giving us information about events happening at some distance from us. Our ears receive acoustic waves propagating past them and our eyes receive signals from a small region of the electromagnetic spectrum. These are senses that allow us to perceive and interpret events from which we are physically removed.

In order to learn more about our physical environment, humankind has strived to extend its sensory capabilities. In modern times, radar, sonar, and seismic signal processing are familiar examples of sensory extension. They include the reception and processing of propagating signals by passive as well as active systems. A passive receiver receives signals emitted by the distant event and carried to it by propagating waves; it simply "looks" or "listens." An active system, on the other hand, produces some sort of radiation—usually acoustic or electromagnetic—which may be beamed in a particular direction. This radiation is reflected by objects (actually by discontinuities in the propagation medium) and returns to the receiver, where it is analyzed.

The principal focus of this chapter is the digital processing of signals received by an array of spatially distributed sensors. We shall discuss some fundamental aspects of this topic, primarily as an example of the application of multidimensional digital filtering and spectral analysis to the problem of extracting information from propagating energy. We shall not attempt to provide a comprehensive guide to the use of multidimensional signal processing techniques for sonar, radar, geophysical, or other applications [1].

Often the processing of signals carried by propagating waves attempts to separate a signal from noise, interference, or even other signals. Consequently, this task can be thought of as the localization of signal energy in time, frequency, direction of propagation, or some other variable. Multidimensional digital filtering can be applied to the problem of extracting information from a signal carried by a propagating wave since it provides a mechanism for segregating signals with a particular set of parameter values from other signals.

In Section 6.2 we study the operation of beamforming. Beamforming is the reception of energy propagating in a particular direction while rejecting energy propagating in other directions. Signals in the beam are passed and those out of the beam are attenuated. Beamforming is thus analogous to bandpass filtering.

Multidimensional spectral estimation can also be applied to problems of segregating signal energy. By estimating the multidimensional spectrum, we can detect signal components in different frequency bands and estimate their strength. This, in effect, performs the segregation of signal energy that is important in several applications. In later sections we explore in more detail how multidimensional signal estimation can be applied to problems of this type.

## 6.1 ANALYSIS OF SPACE–TIME SIGNALS

Propagating waves and the signals they carry can be modeled as functions of space and time, and they can be analyzed by using multidimensional Fourier methods. Thus if $s(\mathbf{x}, t)$ represents a signal that is a function of spatial position $\mathbf{x}$ and time $t$, we can use a continuous 4-D Fourier transform to obtain the 4-D wavenumber–frequency spectrum $S(\mathbf{k}, \omega)$ [2].

$$S(\mathbf{k}, \omega) \triangleq \int_{-\infty}^{\infty} \int_{-\infty}^{\infty} s(\mathbf{x}, t) \exp\left[-j(\omega t - \mathbf{k}'\mathbf{x})\right] d\mathbf{x}\, dt \qquad (6.1)$$

[Notice the signs in the exponential. $S(\mathbf{k}, \omega)$ has been deliberately defined in this fashion so that our later equations will reflect our intuitive understanding of propagating waves.] The variable $\omega$ represents the familiar temporal frequency. Analogously, the wavenumber vector $\mathbf{k}$ represents the spatial frequency. It is a vector quantity since it represents the number of waves per unit distance in each of the three orthogonal spatial directions. The scalar term $\mathbf{k}'\mathbf{x}$ in the exponent of equation (6.1) is the dot product of the wavenumber vector $\mathbf{k} \triangleq (k_x, k_y, k_z)'$ and the position vector $\mathbf{x}$.

### 6.1.1 Elemental Signals

Elemental signals of the form

$$e(\mathbf{x}, t) \triangleq \exp\left[j(\omega_o t - \mathbf{k}_0'\mathbf{x})\right] \qquad (6.2)$$

represent propagating plane waves. From the inverse 4-D Fourier transform,

$$s(\mathbf{x}, t) = \frac{1}{(2\pi)^4} \int_{-\infty}^{\infty} \int_{-\infty}^{\infty} S(\mathbf{k}, \omega) \exp\left[j(\omega t - \mathbf{k}'\mathbf{x})\right] d\mathbf{k}\, d\omega \qquad (6.3)$$

we observe that any signal $s(\mathbf{x}, t)$ can be decomposed into a superposition of propagating plane waves. By defining the vector $\boldsymbol{\alpha}_0$ as

$$\boldsymbol{\alpha}_0 \triangleq \frac{\mathbf{k}_0}{\omega_0} \tag{6.4}$$

we can write equation (6.2) as

$$e(\mathbf{x}, t) = \exp[j\omega_0(t - \boldsymbol{\alpha}_0'\mathbf{x})] \tag{6.5}$$

Thus $e(\mathbf{x}, t)$ can be interpreted as a plane wave propagating in the $\boldsymbol{\alpha}_0$ direction with a speed equal to $1/|\boldsymbol{\alpha}_0|$. Since the magnitude $|\boldsymbol{\alpha}_0|$ is equal to the inverse of the propagation speed, $\boldsymbol{\alpha}_0$ is sometimes called the *slowness vector*.

If we take the Fourier transform of the elemental signal $e(\mathbf{x}, t)$ using (6.1), we get

$$E(\mathbf{k}, \omega) = \delta(\mathbf{k} - \mathbf{k}_0)\delta(\omega - \omega_0) \tag{6.6}$$

which is a 4-D impulse (a Dirac delta function) in $(\mathbf{k}, \omega)$-space at the point $\mathbf{k} = \mathbf{k}_0$ and $\omega = \omega_0$. Each point in $(\mathbf{k}, \omega)$-space thus corresponds to a plane wave in $(\mathbf{x}, t)$-space with a particular orientation and frequency [2, 3].

Consider the simplified plots of $(\mathbf{k}, \omega)$-space in Figure 6.1. The variable $\omega$ is represented by the vertical axis and $k_x$ and $k_y$ are represented by the horizontal plane. (We are neglecting the $k_z$ variable for the moment to simplify the figures.) In Figure 6.1(a) we see that all signal components at the same frequency $\omega$ lie in a plane parallel to the $(k_x, k_y)$-plane. Signal components with the same speed of propagation $c$ will lie on the surface of a cone, as shown in Figure 6.1(b), since $c = \omega/|\mathbf{k}|$. Signal components propagating in the same direction fall along a half-plane which is perpendicular to the $(k_x, k_y)$-plane since direction of propagation is indicated by the direction of the $\mathbf{k}$-vector, as shown in Figure 6.1(c). In some instances, we will also be interested in signals that lie at the intersections of these surfaces. For example, signal components with the same speed and direction of propagation lie along a line formed by the intersection of the cone in Figure 6.1(b) and the half-plane in Figure 6.1(c).

### 6.1.2 Filtering in Wavenumber–Frequency Space

In processing signals that are a function of space and time, we are often interested in extracting signal components at certain frequencies and certain velocities of propagation (both speed and direction). This problem can be stated as a multidimensional filtering problem [3]; it is analogous to separating the frequency components of a 1-D signal by using bandpass filters. For example, suppose that a signal $s(\mathbf{x}, t)$ has a wavenumber–frequency spectrum $S(\mathbf{k}, \omega)$ given by equation (6.1). We can conceive of sending this signal through a 4-D linear, shift-invariant filter with an impulse response $h(\mathbf{x}, t)$ to yield the output signal $f(\mathbf{x}, t)$. The filter impulse response is designed to pass signal components of interest and to reject those, such as additive noise, which are not of interest. The input and output signals are related through the 4-D continuous convolution integral

$$f(\mathbf{x}, t) = \int_{-\infty}^{\infty} \int_{-\infty}^{\infty} h(\mathbf{x} - \boldsymbol{\xi}, t - \tau)s(\boldsymbol{\xi}, \tau)\, d\boldsymbol{\xi}\, d\tau \tag{6.7}$$

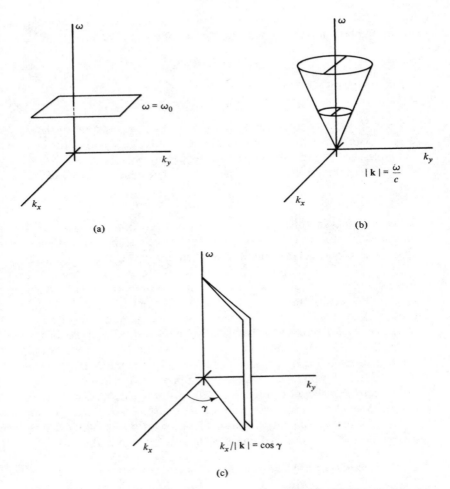

**Figure 6.1** Loci of points in $(\mathbf{k}, \omega)$-space corresponding to: (a) signals with a common frequency $\omega_0$; (b) signals with a common propagation speed; (c) signals with a common propagation direction.

In the wavenumber–frequency domain, the output spectrum is equal to the product of the input spectrum and the filter's wavenumber–frequency response.

$$F(\mathbf{k}, \omega) = H(\mathbf{k}, \omega) S(\mathbf{k}, \omega) \qquad (6.8)$$

To extract the desired signal components, we must design the filter's wavenumber–frequency response $H(\mathbf{k}, \omega)$ so that it is near unity in the desired regions of $(\mathbf{k}, \omega)$-space and near zero everywhere else. Thus, if we want to pass signal components in some narrow band around the frequency $\omega_0$ regardless of the speed or direction of propagation, $H(\mathbf{k}, \omega)$ would take on the form of a 1-D bandpass frequency response which does not depend on $\mathbf{k}$. If we want to extract the signal components with a particular frequency, speed, and direction of propagation, $H(\mathbf{k}, \omega)$ would take the form of a 4-D bandpass filter whose passband is centered on the particular values of $(\mathbf{k}_0, \omega_0)$.

This general filtering approach suffers from two major problems: The signal $s(\mathbf{x}, t)$ is usually not known for all positions $\mathbf{x}$, and the filtered signal $f(\mathbf{x}, t)$ is usually needed at only one position. Nonetheless, this general approach is useful as a framework for studying systems that process signals carried by propagating waves.

## 6.2 BEAMFORMING

Beamforming is one type of filtering that can be conveniently applied to signals carried by propagating waves. The objective of a beamforming system is to isolate signal components that are propagating in a particular direction. It is generally assumed that the waves all propagate with the same speed $c$, so that the signals of interest lie on the surface of the cone $\omega = c|\mathbf{k}|$ in $(\mathbf{k}, \omega)$-space. Ideally, the passband of the beamformer is the intersection of this cone with the plane containing the desired direction vector, as shown in Figure 6.2.

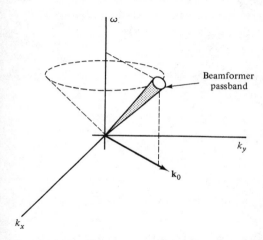

### 6.2.1 Weighted Delay-and-Sum Beamformer

Beamforming can be accomplished by applying signal processing operations to the signals received by an array of receivers. Let us assume that we have an array of $N$ receivers distributed in space, such that the $i$th receiver is located at the position $\mathbf{x}_i$ ($i = 0, 1, \ldots, N-1$) and the signal received by it is denoted $r_i(t)$. Because of the fixed positions of the receivers, they sample the signal $s(\mathbf{x}, t)$ spatially. If we assume that

**Figure 6.2**   Passband of an ideal beamformer lies at the intersection of the cone $\omega = c|\mathbf{k}|$ and the plane containing the desired propagation direction $\mathbf{k}_0$.

this sampling is ideal, the $i$th receiver signal $r_i(t)$ will be given by

$$r_i(t) = s(\mathbf{x}_i, t) \qquad (6.9)$$

One of the simplest beamforming systems is called the *weighted delay-and-sum beamformer*. The beamformer output $bf(t)$ is formed by averaging weighted and delayed versions of the receiver signals

$$bf(t) \triangleq \frac{1}{N} \sum_{i=0}^{N-1} w_i r_i(t - \tau_i) \qquad (6.10)$$

The weight and the relative delay for the $i$th receiver are given by $w_i$ and $\tau_i$, respectively. The delays $\tau_i$ are chosen to center the beamformer's passband along some particular orientation in $(\mathbf{k}, \omega)$-space. (This is frequently called "steering the beam.") For example, if all the receivers lie in the same plane and we want to steer the beam perpendicular to that plane, $\tau_i$ should be zero for all the receivers. Plane waves approaching from the perpendicular direction will be added together in phase while those approaching from other directions will be added with different phases and will tend to cancel themselves out.

To pass plane waves propagating in a particular direction with a slowness vector equal to $\boldsymbol{\alpha}_0$, the delays should be set equal to

$$\tau_i = -\boldsymbol{\alpha}_0' \mathbf{x}_i \tag{6.11}$$

### 6.2.2 Array Pattern

Ideally, the beamformer should pass signal components propagating with a slowness vector $\boldsymbol{\alpha}_0$ and reject all other components. In practice, this ideal cannot be achieved. The array pattern, derived below, indicates how closely this ideal can be approximated. Suppose the beam is steered toward $\boldsymbol{\alpha}_0$ so that the receiver delays are given by (6.11), and let the space–time signal $s(\mathbf{x}, t)$ be a plane wave propagating in a different direction $\boldsymbol{\alpha}$. Then

$$s(\mathbf{x}, t) = \exp[j\omega(t - \boldsymbol{\alpha}'\mathbf{x})] \tag{6.12}$$

The beamformer output will be given by

$$
\begin{aligned}
bf(t) &= \frac{1}{N} \sum_{i=0}^{N-1} w_i r_i(t - \tau_i) \\
&= \frac{1}{N} \sum_{i=0}^{N-1} w_i s(\mathbf{x}_i, t + \boldsymbol{\alpha}_0' \mathbf{x}_i) \\
&= \frac{1}{N} \sum_{i=0}^{N-1} w_i \exp[j\omega(t + \boldsymbol{\alpha}_0' \mathbf{x}_i - \boldsymbol{\alpha}' \mathbf{x}_i)] \\
&= \left\{ \frac{1}{N} \sum_{i=0}^{N-1} w_i \exp[-j\omega(\boldsymbol{\alpha} - \boldsymbol{\alpha}_0)' \mathbf{x}_i] \right\} \exp(j\omega t) \\
&= W(\omega(\boldsymbol{\alpha} - \boldsymbol{\alpha}_0)) \exp(j\omega t) \\
&= W(\mathbf{k} - \mathbf{k}_0) \exp(j\omega t)
\end{aligned}
\tag{6.13}
$$

The function

$$W(\mathbf{k}) \triangleq \frac{1}{N} \sum_{i=0}^{N-1} w_i \exp(-j\mathbf{k}'\mathbf{x}_i) \tag{6.14}$$

is called the *array pattern*. It is essentially the Fourier transform of the receiver weighting function $w_i$, taking into account the positions $\mathbf{x}_i$ of the receivers. The array pattern $W(\mathbf{k} - \mathbf{k}_0)$ indicates the attenuation suffered by a plane wave propagating with a slowness vector $\boldsymbol{\alpha}$, or equivalently a wavenumber vector $\mathbf{k} = \omega\boldsymbol{\alpha}$, when the beam is steered in the direction parallel to the vector $\mathbf{k}_0$.

More generally, the signal $s(\mathbf{x}, t)$ is composed of many plane-wave components with different temporal frequencies and different directions of propagation. We can use the wavenumber–frequency spectrum $S(\mathbf{k}, \omega)$ to represent this composition as before by writing

$$s(\mathbf{x}, t) = \frac{1}{(2\pi)^4} \int_{-\infty}^{\infty} \int_{-\infty}^{\infty} S(\mathbf{k}, \omega) \exp\left[j(\omega t - \mathbf{k}'\mathbf{x})\right] d\mathbf{k}\, d\omega$$

Then using this representation in (6.10), we see that the beamformer output can be written as

$$bf(t) = \frac{1}{N} \sum_{i=0}^{N-1} w_i \frac{1}{(2\pi)^4} \int_{-\infty}^{\infty} \int_{-\infty}^{\infty} S(\mathbf{k}, \omega) \exp\left[-j(\mathbf{k} - \omega\boldsymbol{\alpha}_0)'\mathbf{x}_i\right] \exp\left(j\omega t\right) d\mathbf{k}\, d\omega$$

(6.15)

Finally, using the definition of the array pattern (6.14), we can rearrange (6.15) to give

$$bf(t) = \frac{1}{(2\pi)^4} \int_{-\infty}^{\infty} \int_{-\infty}^{\infty} S(\mathbf{k}, \omega) W(\mathbf{k} - \omega\boldsymbol{\alpha}_0) \exp\left(j\omega t\right) d\mathbf{k}\, d\omega \qquad (6.16)$$

This equation represents the wideband response [4] of the weighted delay-and-sum beamformer. It shows how the various plane-wave components are attenuated before being combined to form the beamformer output signal.

Let us consider the special case where all the components of $s(\mathbf{x}, t)$ are propagating in the same direction. In this case

$$s(\mathbf{x}, t) = v(t - \boldsymbol{\alpha}'\mathbf{x}) \qquad (6.17)$$

for some waveform $v(t)$. The wavenumber–frequency spectrum is given by

$$S(\mathbf{k}, \omega) = V(\omega)\delta(\mathbf{k} - \omega\boldsymbol{\alpha}) \qquad (6.18)$$

where $V(\omega)$ is the Fourier transform of $v(t)$ and $\delta(\mathbf{k})$ is the 3-D impulse function. Substituting equation (6.18) into (6.16), we see that

$$\begin{aligned} bf(t) &= \frac{1}{(2\pi)^4} \int_{-\infty}^{\infty} \int_{-\infty}^{\infty} V(\omega)\delta(\mathbf{k} - \omega\boldsymbol{\alpha}) W(\mathbf{k} - \omega\boldsymbol{\alpha}_0) \exp\left(j\omega t\right) d\mathbf{k}\, d\omega \\ &= \frac{1}{2\pi} \int_{-\infty}^{\infty} V(\omega) W(\omega(\boldsymbol{\alpha} - \boldsymbol{\alpha}_0)) \exp\left(j\omega t\right) d\omega \end{aligned}$$

(6.19)

If the propagating wave $v(t - \boldsymbol{\alpha}'\mathbf{x})$ crosses the array with precisely the orientation for which the beamformer was designed, then $\boldsymbol{\alpha} = \boldsymbol{\alpha}_0$ and

$$\begin{aligned} bf(t) &= \frac{1}{2\pi} W(0) \int_{-\infty}^{\infty} V(\omega) \exp\left(j\omega t\right) d\omega \\ &= W(0)v(t) \end{aligned}$$

(6.20)

In this case, the beamformer does not distort the signal waveform.

If $\boldsymbol{\alpha} \neq \boldsymbol{\alpha}_0$, the magnitude of the argument of $W(\cdot)$ in (6.19) grows linearly with frequency. In this case, the higher-frequency components in the wave will generally be attenuated more than the lower-frequency components. We can see why this happens by referring to Figure 6.3, which is a view of the $(k_x, k_y)$-plane from above.

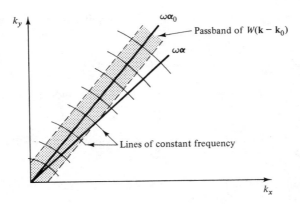

**Figure 6.3** Projection of $(\mathbf{k}, \omega)$-space onto the $(k_x, k_y)$-plane to illustrate the lowpass filtering effect of a misaligned beam.

(For the purposes of illustration, we are again neglecting the $k_z$ variable.) Since all the signal components are presumed to be propagating at the same speed $c$, they lie on the cone in $(\mathbf{k}, \omega)$-space defined by $\mathbf{k} = \omega\boldsymbol{\alpha}$. All signal components at the same frequency on this cone will project onto a circle in the $(k_x, k_y)$-plane. Thus in this figure frequency is measured by the distance from the origin multiplied by the speed $c$. If the array pattern $W(\mathbf{k})$ is such that its value is close to unity only when the magnitude of its argument is small and is close to zero otherwise, the main lobe of the array pattern is given by the shaded region in the figure bounded by lines parallel to the line $\mathbf{k} = \omega\boldsymbol{\alpha}_0$. If the incoming signal is aligned with the beam ($\boldsymbol{\alpha} = \boldsymbol{\alpha}_0$), all frequencies are seen to lie within the main lobe of the beam. But if $\mathbf{k} = \omega\boldsymbol{\alpha} \neq \omega\boldsymbol{\alpha}_0$, we see that the higher-frequency components along the $\mathbf{k}$ line will fall outside the passband of the beamformer. Consequently, wideband signals propagating across the array in directions that are different than the steering direction suffer a distortion that is similar to lowpass filtering.

Equation (6.16) can also be interpreted as a multidimensional filtering operation. To show this, we begin with equation (6.7), which is repeated below.

$$f(\mathbf{x}, t) = \int_{-\infty}^{\infty} \int_{-\infty}^{\infty} h(\mathbf{x} - \boldsymbol{\xi}, t - \tau)s(\boldsymbol{\xi}, \tau)d\boldsymbol{\xi}\, d\tau$$

Applying the convolution theorem to this expression gives

$$f(\mathbf{x}, t) = \frac{1}{(2\pi)^4} \int_{-\infty}^{\infty} \int_{-\infty}^{\infty} H(\mathbf{k}, \omega)S(\mathbf{k}, \omega) \exp\left[j(\omega t - \mathbf{k}'\mathbf{x})\right] d\mathbf{k}\, d\omega \qquad (6.21)$$

If we identify the beamformer output as

$$bf(t) = f(\mathbf{0}, t) \qquad (6.22)$$

then by comparing equations (6.16) and (6.21), we see that the effective frequency response $H(\mathbf{k}, \omega)$ is given by

$$H(\mathbf{k}, \omega) = W(\mathbf{k} - \omega\boldsymbol{\alpha}_0) \qquad (6.23)$$

The array pattern evaluated at $\mathbf{k} - \omega\boldsymbol{\alpha}_0$ is the complex amplitude of the frequency response of the filter used to process the space–time signal $s(\mathbf{x}, t)$.

### 6.2.3 Example of an Array Pattern

The array pattern $W(\mathbf{k})$ tells us how selective a particular beamforming system is. Let us examine it in detail for the very simple case of a linear array of uniformity weighted, equally spaced receivers whose geometry is shown in Figure 6.4. The $N$ receivers are located at the positions $\mathbf{x}_i = (iD, 0, 0)'$ for $0 \leq i \leq N - 1$ and $w_i$ is equal to 1 for each $i$. Using the definition of the array pattern which was given in (6.14), we see that for this case

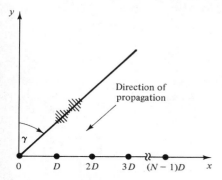

$$W(\mathbf{k}) = \frac{1}{N} \sum_{i=0}^{N-1} \exp(-j\mathbf{k}'\mathbf{x}_i)$$

$$= \frac{1}{N} \sum_{i=0}^{N-1} \exp(-jk_x iD) \qquad (6.24)$$

$$= \frac{\sin(Nk_x D/2)}{N \sin(k_x D/2)} \exp\left[\frac{-j(N-1)k_x D}{2}\right]$$

**Figure 6.4**  Receiver geometry for the beamforming example of Section 6.2.3. Dots represent receiver locations.

The magnitude of this array pattern is shown in Figure 6.5(a). Although the array pattern is ostensibly a function of $\mathbf{k} = (k_x, k_y, k_z)'$, in this case it depends only on $k_x$ because all the receivers are located on the x-axis.

If the receiver signals are delayed to steer the beam according to (6.11), we have

$$\tau_i = -\boldsymbol{\alpha}_0' \mathbf{x}_i = -\boldsymbol{\alpha}_0' \cdot (iD, 0, 0)'$$

$$= -\alpha_{0x} iD \qquad (6.25)$$

and the array response is given by $W(\mathbf{k} - \omega\boldsymbol{\alpha}_0)$, the magnitude of which is shown in Figure 6.5(b). Notice that the array pattern is periodic in $k_x$ with a period of $2\pi/D$. The beamwidth, which is the width of the main lobe of the array pattern, is inversely proportional to the product $ND$. This product can be thought of as the length of the array *aperture*. Because of the periodicity of $W(\mathbf{k} - \omega\boldsymbol{\alpha}_0)$, the main lobe is repeated at intervals of $2\pi/D$. These repeated mainlobes are sometimes called *grating lobes* because they are analogous to the higher diffraction orders caused by an optical diffraction grating.

Figure 6.6 shows one way to visualize the array pattern in wavenumber–frequency space. Since $W(\mathbf{k} - \omega\boldsymbol{\alpha}_0)$ depends only on the x-components of $\mathbf{k}$ and $\boldsymbol{\alpha}_0$, the $k_y$-axis in Figure 6.6 has been oriented into the page so that we can plot the array pattern as a function of $k_x$ and $\omega$. The shaded areas correspond to the main lobe and grating lobes. A signal propagating with a slowness vector $\boldsymbol{\alpha}$ will have components that lie along the line $k_x = \omega\alpha_x$. At low frequencies, those signal components will lie in the main lobe (or passband) of the beam pattern and thus will not be attenuated. As $\omega$ increases, however, the signal components will fall outside the main lobe and be attenuated by the beamformer. At even higher frequencies, the signal components may fall within a grating lobe and be passed on to the beamformer output. To prevent

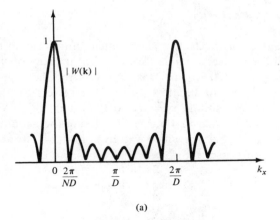

(a)

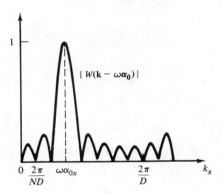

(b)

**Figure 6.5**   (a) Array pattern of
a uniformly weighted, equally spaced
one-dimensional array of receivers.
(b) Response of the array if time delays
are used to steer the beam.

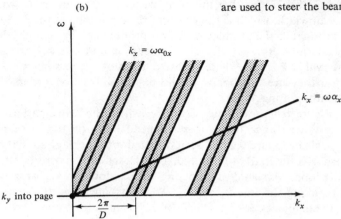

**Figure 6.6**   Effect of grating lobes in wavenumber–frequency space. High-
frequency components of a signal propagating with a slowness vector $\alpha$ will be
unattenuated when they lie in a grating lobe.

this from happening, the receiver signals may be lowpass filtered to eliminate the high-frequency signal components that might otherwise be passed by grating lobes.

Because the receivers are located at discrete points, they in effect sample an incoming plane wave spatially. Consequently, the existence of grating lobes can be interpreted in terms of spatial frequency aliasing resulting from a spatial sampling interval $D$ which is too large. Consider a plane wave with a temporal frequency $\omega$ and wavelength $\lambda = 2\pi/|\mathbf{k}|$ which is propagating past the array of receivers. When observed along the $x$-axis at any instant of time, the plane wave varies sinusoidally with position. The period of this sine wave is $\lambda_x = \lambda/\sin\gamma$, where $\gamma$ is the angle of incidence measured from the perpendicular to the line of receivers. This geometry was depicted in Figure 6.4. The smallest value of $\lambda_x$, corresponding to the highest spatial frequency seen by the array, is observed when $\gamma = \pm 90°$. In this case, $\lambda_x = \lambda$. Consequently, spatial frequency aliasing can result unless

$$D \leq \frac{\lambda}{2} \tag{6.26}$$

If the receiver spacing $D$ satisfies (6.26) for the smallest wavelength that the array receives, grating lobes have no effect on the received signal.

We saw earlier that the beamwidth, and consequently the angular resolution of the beamformer, depends on the size of the array aperture $ND$. When dealing with narrowband plane-wave signals, it is sometimes instructive to measure the aperture size in terms of the wavelength $\lambda$. If the array pattern is plotted as a function of $\sin\gamma$ rather than $k_x$, the width of the main beam is seen to be inversely proportional to the ratio $D/\lambda$. The aperture size of a particular array will be equal to a small number of wavelengths for a low-frequency signal but a large number of wavelengths for a high-frequency signal. Consequently, we would expect the beamwidth to be large for low-frequency signals and small for high-frequency signals.

### 6.2.4 Effect of the Receiver Weighting Function

The choice of the receiver weights $w_l$ has a significant effect on the shape of the array pattern and consequently on the selectivity of the beamformer. If the receiver positions $\mathbf{x}_l$ are equally spaced along the $x$-axis as in Figure 6.4, the problem of choosing the weights is equivalent to the problem of choosing the coefficients of a 1-D finite-length window function. In the absence of other constraints, the use of a standard 1-D window [5] to weight the receiver signals results in a reasonable array pattern, with the usual trade-off between the width of the main lobe and the height of the side lobes. Side lobe levels can be reduced, thus rejecting off-beam signals more thoroughly, only at the expense of increasing the width of the main lobe and reducing the angular resolution.

For two- and three-dimensional arrays that have receivers located on a regularly spaced grid, weighting functions can be designed by the same techniques that we used in Chapter 3 to obtain multidimensional window functions [6, 7]. As an example, consider a 3-D problem in which we want to design the receiver weights $w_l$ for receivers located at $\mathbf{x}_i = (x_i, y_i, z_i)'$. To indicate explicitly the 3-D nature of the weighting function, let us change notation slightly and denote the weighting function by $w(x_i, y_i, z_i)$. We can obtain a separable weighting function by writing

$$w(x_i, y_i, z_i) = f(x_i)g(y_i)h(z_i) \tag{6.27}$$

where $f$, $g$, and $h$ are 1-D window functions chosen to have the appropriate main lobe width and side lobe levels in each wavenumber variable. Separable weighting functions such as (6.27) generally lead to array patterns with a roughly rectanguloid mainlobe. Alternatively, nearly spherically symmetric array patterns can be obtained by using

$$w(x_i, y_i, z_i) = f(\sqrt{x_i^2 + y_i^2 + z_i^2}) \tag{6.28}$$

where, again, $f$ is an appropriately chosen continuous 1-D window function.

Weighting functions can also be designed by using window functions when the array geometry consists of receivers placed on other periodic grids. For example, a weighting function for a hexagonal array can be obtained by using the hexagonal window functions in Chapter 3. When the receivers are not located in a regular fashion, it becomes much more difficult to design the receiver weighting function. The problem is analogous to that of designing a multidimensional FIR filter whose impulse response consists of unequally spaced nonzero samples.

### 6.2.5 Filter-and-Sum Beamforming

The weighted delay-and-sum beamformer can be generalized. For example, in some applications, it is desirable to make the receiver weighting function depend on frequency. A signal component received at one frequency is processed by a different set of weights than a signal component received at another frequency. For the $i$th receiver, the signal component at the frequency $\omega = \omega_0$ can be written as $R_i(\omega_0)\exp(j\omega_0 t)$, where $R_i(\omega)$ is the Fourier transform of the receiver signal $r_i(t)$. Using a frequency-dependent weighting function $W_i(\omega)$, we can compute the output of a delay-and-sum beamformer for the signal component at $\omega = \omega_0$ by writing

$$bf(t, \omega_0) = \frac{1}{N}\sum_{i=0}^{N-1} W_i(\omega_0)R_i(\omega_0)\exp[j\omega_0(t - \tau_i)] \tag{6.29}$$

Now, to apply the frequency-dependent beamforming operation over all frequencies simultaneously, we replace the parameter $\omega_0$ by the variable $\omega$ and integrate. This gives us the filter-and-sum beamformer output

$$fs(t) \triangleq \frac{1}{2\pi}\int_{-\infty}^{\infty} bf(t, \omega)\,d\omega$$
$$= \frac{1}{N}\sum_{i=0}^{N-1}\frac{1}{2\pi}\int_{-\infty}^{\infty} W_i(\omega)R_i(\omega)\exp[j\omega(t - \tau_i)]\,d\omega \tag{6.30}$$

To simplify the notation, we can define the signal

$$q_i(t) \triangleq \frac{1}{2\pi}\int_{-\infty}^{\infty} W_i(\omega)R_i(\omega)\exp(j\omega t)\,d\omega \tag{6.31}$$

so that

$$fs(t) = \frac{1}{N} \sum_{i=0}^{N-1} q_i(t - \tau_i) \tag{6.32}$$

By applying the convolution theorem to equation (6.31), we see that $q_i(t)$ can be written as the convolution

$$q_i(t) = w_i(t) * r_i(t) \tag{6.33}$$

where $w_i(t)$ is the inverse Fourier transform of the frequency-dependent weighting function $W_i(\omega)$. The function $w_i(t)$ can be interpreted as the impulse response of a filter that operates on the receiver signal $r_i(t)$ to yield $q_i(t)$. This filtered signal is then used in the formation of the beam (6.32); hence the name filter-and-sum beamformer.

The output of the filter-and-sum beamformer can also be written in terms of a four-dimensional convolution

$$fs(t) = h(\mathbf{x}, t) * s(\mathbf{x}, t)\Big|_{\mathbf{x}=0} \tag{6.34}$$

It is straightforward to verify that the effective wavenumber–frequency response $H(\mathbf{k}, \omega)$ for the filter-and-sum beamformer is given by

$$H(\mathbf{k}, \omega) = \frac{1}{N} \sum_{i=0}^{N-1} W_i(\omega) \exp\left[-j(\mathbf{k} - \omega\boldsymbol{\alpha}_0)'\mathbf{x}_i\right] \tag{6.35}$$

where the slowness vector $\boldsymbol{\alpha}_0$, which is indicative of the direction in which the beam is steered, is again related to the time delays $\tau_i$ by

$$\tau_i = -\boldsymbol{\alpha}_0' \mathbf{x}_i \tag{6.36}$$

For plane waves propagating with slowness vector $\boldsymbol{\alpha}_0$, we have $\mathbf{k} = \omega\boldsymbol{\alpha}_0$, so the exponential term drops out. In this case, the wavenumber–frequency response $H(\omega\boldsymbol{\alpha}_0, \omega)$ is simply the average of the frequency-dependent weights $W_i(\omega)$.

### 6.2.6 Frequency-Domain Beamforming

The two beamformers that we have already studied—the weighted delay-and-sum beamformer and the filter-and-sum beamformer—both represent time-domain operations. The individual receiver signals are physically delayed, filtered, and added together. A frequency-domain beamformer, on the other hand, creates a beam pattern by performing the necessary delaying, filtering, and summing operations in the frequency domain, through the use of Fourier transforms. For example, if we let $R_i(\omega)$ denote the Fourier transform of the receiver signal $r_i(t)$, then delaying $r_i(t)$ by $\tau_i$ seconds can be implemented in the frequency domain by multiplying $R_i(\omega)$ with the phasor $\exp(-j\omega\tau_i)$. Similarly, weighting $r_i(t)$ with $w_i$ can be accomplished by forming the product $w_i R_i(\omega)$, or more generally, filtering $r_i(t)$ with the impulse response $w_i(t)$ can be accomplished by the multiplication $W_i(\omega)R_i(\omega)$. Using these frequency-domain operations, we can form, conceptually at least, the Fourier transform of the weighted delay-and-sum beamformer output as

$$BF(\omega) = \frac{1}{N} \sum_{i=0}^{N-1} w_i R_i(\omega) \exp\left(-j\omega\tau_i\right) \tag{6.37}$$

and the Fourier transform of the filter-and-sum beamformer output as

$$FS(\omega) = \frac{1}{N} \sum_{i=0}^{N-1} W_i(\omega) R_i(\omega) \exp\left(-j\omega\tau_i\right) \tag{6.38}$$

The component of the beamformer output at the frequency $\omega$ is given by $BF(\omega) \exp(j\omega t)$, and the entire beamformer output is obtained by integrating all the components over frequency

$$bf(t) = \frac{1}{2\pi} \int_{-\infty}^{\infty} BF(\omega) \exp\left(j\omega t\right) d\omega \tag{6.39}$$

Similar reasoning can also be applied to equation (6.38) to give

$$fs(t) = \frac{1}{2\pi} \int_{-\infty}^{\infty} FS(\omega) \exp\left(j\omega t\right) d\omega \tag{6.40}$$

In practice, we do not have access to the spectrum $R_i(\omega)$; to compute it would require integrating over the entire time axis. However, we can compute the Fourier transform of a segment of $r_i(t)$ by applying a finite-extent window $v(t)$ to $r_i(t)$. Let us define the short-time Fourier transform [8] as

$$R_i(t, \omega) \triangleq \int_{-\infty}^{\infty} v(t - \tau) r_i(\tau) \exp\left(-j\omega\tau\right) d\tau \tag{6.41}$$

The limits of integration here are really finite since the window function $v(t - \tau)$ vanishes except in a finite interval. The function $R_i(t, \omega)$ will be a reasonable approximation to the spectrum $R_i(\omega)$ as long as the Fourier transform $V(\omega)$ of the window function is relatively narrow.

The frequency-domain beamformer output is defined as

$$fd(t, \omega) \triangleq \frac{1}{N} \sum_{i=0}^{N-1} w_i R_i(t, \omega) \exp\left[j\omega(t - \tau_i)\right] \tag{6.42}$$

It can be interpreted as an approximation to the beamformer component $BF(\omega) \exp(j\omega t)$. Thus an approximation to the beamformer output signal may be formed by integrating $fd(t, \omega)$ over frequency. [A more general frequency-domain beamformer output could be defined by using equation (6.38), but we shall not pursue that derivation here.] If we consider $\omega$ to be a fixed parameter for the moment, we can think of $fd(t, \omega)$ as the output of a beamforming operation applied to receiver signal components at the frequency $\omega$. In this regard, the frequency-domain beamformer is similar to the filter-and-sum beamformer for a particular choice of narrow bandpass filters. Later, in Section 6.3.3, we shall see that the frequency-domain implementation offers computational advantages when forming several beams simultaneously [9].

With a bit of algebra, we can write $fd(t, \omega)$ in terms of the wavenumber–frequency spectrum $S(\mathbf{k}, \omega)$. Using the relations

$$S(\mathbf{k}, \omega) \triangleq \int_{-\infty}^{\infty} \int_{-\infty}^{\infty} s(\mathbf{x}, t) \exp(j\mathbf{k}'\mathbf{x} - j\omega t) \, d\mathbf{x} \, dt \tag{6.43}$$

$$r_i(t) \triangleq s(\mathbf{x}_i, t) \tag{6.44}$$

$$\tau_i \triangleq -\boldsymbol{\alpha}_0' \mathbf{x}_i \tag{6.45}$$

$$W(\mathbf{k}) \triangleq \frac{1}{N} \sum_{i=0}^{N-1} w_i \exp(-j\mathbf{k}'\mathbf{x}_i) \tag{6.46}$$

and

$$V(\omega) \triangleq \int_{-\infty}^{\infty} v(t) \exp(-j\omega t) \, dt \tag{6.47}$$

we can derive

$$fd(t, \omega) = \frac{1}{(2\pi)^4} \int_{-\infty}^{\infty} \int_{-\infty}^{\infty} W(\mathbf{k} - \omega\boldsymbol{\alpha}_0) V(\theta - \omega) S(\mathbf{k}, \theta) \exp(j\theta t) \, d\mathbf{k} \, d\theta \tag{6.48}$$

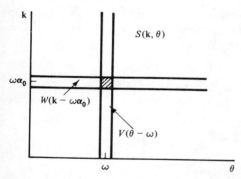

If we consider the array pattern $W(\mathbf{k})$ and the window spectrum $V(\omega)$ to be zero except in a small region where their respective arguments are near zero, the factor $W(\mathbf{k} - \omega\boldsymbol{\alpha}_0)V(\theta - \omega)$ in equation (6.48) will be zero except in the region of $(\mathbf{k}, \theta)$-space where $\mathbf{k} \simeq \omega\boldsymbol{\alpha}_0$ and $\theta \simeq \omega$. This is illustrated conceptually in Figure 6.7, where the horizontal strip represents the nonzero region of the factor $W(\mathbf{k} - \omega\boldsymbol{\alpha}_0)$ and the vertical strip represents the nonzero region of the factor $V(\theta - \omega)$. The intersection of the two strips, indicated by the shaded rectangle, is the only region of $(\mathbf{k}, \theta)$-space where components of $s(\mathbf{x}, t)$ are not significantly attenuated by the beamforming operation. Components within this rectangle contribute to the beamformer output $fd(t, \omega)$.

**Figure 6.7** Contributions to the output of the frequency-domain beamformer come from the small region in $(\mathbf{k}, \theta)$-space indicated by the shaded rectangle.

## 6.3 DISCRETE-TIME BEAMFORMING

So far in this chapter we have considered time to be a continuous variable in our derivations. However, when beamforming systems are to be implemented with digital hardware, it is necessary to sample the received signals in time. Consequently, we must examine the implications of performing the beamforming operation on discrete-time signals. The notation we shall use will be similar to that used in the first part of this chapter. For example, the signal received by the $i$th receiver will be denoted $r_i(t)$. The difference is that the variable $t$ will be regarded as an integer multiple of the sampling period, $t = nT$, so that $r_i(t)$ is a discrete-time signal. In some cases, we shall define $T \triangleq 1$ for simplicity. Other variables that depend on time, such as the propagation speed $c$, will also be redefined to use the sampling period $T$ as the basic unit of time, rather than the second.

Naturally, the usual caveats applying to discrete-time signals and systems will be in force. The sampling rate must be high enough to avoid aliasing. The spectra of the discrete time signals will be periodic, and so forth.

### 6.3.1 Time-Domain Beamforming for Discrete-Time Signals

When the receiver signals $r_i(t)$ are sampled in time, the steering delays $\tau_i$ used in a time-domain beamformer must be integer multiples of the sampling period $T$. Thus the discrete-time version of a weighted delay-and-sum beamformer is given by

$$bf(nT) = \frac{1}{N} \sum_{i=0}^{N-1} w_i r_i(nT - n_i T) \tag{6.49}$$

where $n_i T$ is the steering delay for the $i$th sensor.

Ideally, of course, the steering delays would be set to $-\alpha_0' \mathbf{x}_i$ as dictated by equation (6.11). However, because $n_i$ is constrained to be an integer, the ideal steering delays will generally be quantized. Letting $\{\tau_i\}$ represent the ideal delays, we can define the errors introduced by the time quantization as

$$\Delta \tau_i \triangleq n_i T - \tau_i \tag{6.50}$$

These errors perturb the array pattern. Because of the delay quantization, the response of the weighted delay-and-sum beamformer to a plane wave with slowness vector $\alpha_0$, which was given in (6.23), becomes

$$H(\mathbf{k}, \omega) = \frac{1}{N} \sum_{i=0}^{N-1} w_i \exp\left[-j(\mathbf{k} - \omega\alpha_0)'\mathbf{x}_i\right] \exp\left(-j\omega \, \Delta\tau_i\right) \tag{6.51}$$

A simplified example will help us to understand the effect of quantizing the steering delays. As before, assume that receivers are equally spaced along the x-axis so that the position vector $\mathbf{x}_i$ is given by the ordered triplet $(iD, 0, 0)'$ and that $w_i = 1$ for all $N$ receivers.

If the beam is steered perpendicularly to the array ($\alpha_{0x} = 0$), both $\tau_i$ and $n_i$ will be zero for all the receivers. Consequently, $\Delta\tau_i$ is also zero and $H(\mathbf{k}, \omega)$ is unperturbed. Similarly, when the beam is steered so that $\alpha_{0x} = T/D$, we have

$$\tau_i = -\alpha_0' \mathbf{x}_i = -\alpha_{0x} iD = -iT \tag{6.52}$$

so that $n_i = -i$ and $\Delta\tau_i$ is again zero for all the receivers.

However, if the beam is steered so that $\alpha_{0x} = T/2D$, then $\tau_i = -iT/2$ and $n_i$ is equal to the greatest integer less than or equal to $i/2$. In this case, the steering error becomes

$$\Delta\tau_i = \begin{cases} \dfrac{T}{2} & \text{for } i \text{ odd} \\ 0 & \text{for } i \text{ even} \end{cases} \tag{6.53}$$

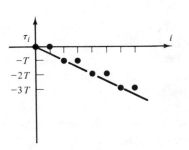

**Figure 6.8** Quantized steering delays. The solid line represents ideal steering delay; the dots represent quantized steering delay.

as pictured in Figure 6.8.

For this case, the sum in equation (6.51) can be split into odd terms and even terms so that

$$H(\mathbf{k}, \omega) = \frac{1}{N} \sum_{i=0}^{N/2-1} \exp\left[-j\left(k_x - \frac{\omega T}{2D}\right) 2iD\right]$$

$$+ \frac{1}{N} \sum_{i=0}^{N/2-1} \exp\left[-j\left(k_x - \frac{\omega T}{2D}\right)(2i+1)D - j\omega \frac{T}{2}\right] \quad (6.54)$$

$$= \left\{\frac{1}{N} \sum_{i=0}^{N/2-1} \exp\left[-ji(2k_x D - \omega T)\right]\right\}[1 + \exp(-jk_x D)]$$

Finally, with some algebraic manipulation, we can evaluate the summation and write

$$H(\mathbf{k}, \omega) = H_1(\mathbf{k}, \omega) H_2(\mathbf{k}, \omega) \quad (6.55)$$

where

$$H_1(\mathbf{k}, \omega) = \frac{2}{N} \exp\left[-j\left(\frac{N}{2} - 1\right)\left(k_x D - \frac{\omega T}{2}\right)\right] \frac{\sin\left[(N/2)(k_x D - \omega T/2)\right]}{\sin(k_x D - \omega T/2)} \quad (6.56a)$$

$$H_2(\mathbf{k}, \omega) = \exp\left(-j\frac{k_x D}{2}\right) \cos\left(\frac{k_x D}{2}\right) \quad (6.56b)$$

The magnitude $|H(\mathbf{k}, \omega)|$ is plotted as a function of $k_x$ in Figure 6.9(a) for the parameter choices $N = 8$, $D = 1$, $T = 1$ and $\omega = \pi/2$. Figure 6.9(b) shows the

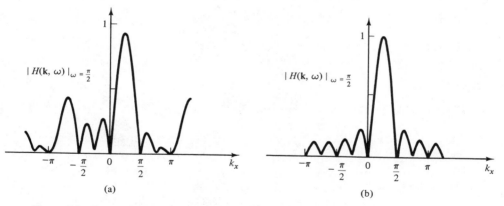

**Figure 6.9**  Example of the effect of steering delay quantization on the wavenumber–frequency response of a beamformer. (a) Quantized beamformer. (b) Ideal beamformer.

magnitude of the ideal frequency response with unquantized steering delays. The ideal response is given by

$$H(\mathbf{k}, \omega) = \frac{1}{N} \exp\left[-j\left(\frac{N-1}{2}\right)\left(k_x D - \frac{\omega T}{2}\right)\right] \frac{\sin\left[(N/2)(k_x D - \omega T/2)\right]}{\sin\frac{1}{2}(k_x D - \omega T/2)} \quad (6.57)$$

For each value of $\omega$, the response changes. Since $H_1(\mathbf{k}, \omega)$ depends on $k_x D - \omega T/2$, it will shift to the right or left as $\omega$ increases or decreases. $H_2(\mathbf{k}, \omega)$, however, does not shift as $\omega$ varies. Consequently, the array pattern $H(\mathbf{k}, \omega)$

$= H_1(\mathbf{k}, \omega) H_2(\mathbf{k}, \omega)$, as a function of $k_x$, can vary substantially from the ideal for different values of $\omega$.

This example serves to illustrate another point. Because of the use of discrete steering delays, sensors with the same steering delay can be paired up. Sensors 0 and 1 have no delay; sensors 2 and 3 are delayed one sample ($T$ seconds); sensors 4 and 5 are delayed two samples; and so forth. Consequently, we can think of the array as being composed of $N/2$ receivers located at positions $(0, 0, 0)'$, $(2, 0, 0)'$, $(4, 0, 0)'$, and so on. Each receiver consists of two omnidirectional sensors whose signals are added with no relative time delay.

The array of $N/2$ receivers has an array pattern given by

$$W_1(\mathbf{k} - \omega\boldsymbol{\alpha}_0) = \frac{2}{N} \sum_{i=0}^{N/2-1} w_i \exp\left[-j(k_x - \omega\alpha_{0x})2iD\right] \tag{6.58}$$

Recall that in this example $w_i = 1$ and $\alpha_{0x} = T/2D$. Thus

$$\begin{aligned} W_1(\mathbf{k} - \omega\boldsymbol{\alpha}_0) &= \frac{2}{N} \sum_{i=0}^{N/2-1} \exp\left[-j(2k_x iD - \omega iT)\right] \\ &= \frac{2}{N} \exp\left[-j\left(\frac{N}{2} - 1\right)\left(k_x D - \frac{\omega T}{2}\right)\right] \frac{\sin\left[(N/2)(k_x D - \omega T/2)\right]}{\sin\left(k_x D - \omega T/2\right)} \\ &= H_1(\mathbf{k}, \omega) \end{aligned} \tag{6.59}$$

Each receiver in turn is a subarray of two sensors whose outputs are added without any relative time delay. Thus each receiver has an array pattern given by

$$\begin{aligned} W_2(\mathbf{k} - \omega\boldsymbol{\alpha}_0) &= \tfrac{1}{2} \sum_{i=0}^{1} w_i \exp\left(-jk_x iD\right) \\ &= \tfrac{1}{2}\left[1 + \exp\left(-jk_x D\right)\right] \\ &= H_2(\mathbf{k}, \omega) \end{aligned} \tag{6.60}$$

Thus the total array pattern (6.54) is the product of the $N/2$-receiver array pattern (6.59) and the subarray pattern (6.60).

This property has practical importance since real sonar and seismic arrays are often made up of subarrays of sensors wired together directly. In addition, real sensors are not necessarily perfectly omnidirectional; their frequency response in $(\mathbf{k}, \omega)$-space helps to shape the overall array pattern.

In computing the beamformer signal $bf(t)$ earlier, we assumed that the receiver signal $r_i(t)$ was equal to $s(\mathbf{x}_i, t)$. In the example above, however, and in general, each receiver signal is given by

$$r_i(t) = \int_{-\infty}^{\infty} g(\mathbf{x}_i, \tau) s(\mathbf{x}_i, t - \tau) \, d\tau \tag{6.61}$$

where $g(\mathbf{x}, t)$ is the inverse Fourier transform of the sensor wavenumber–frequency response $G(\mathbf{k}, \omega)$. Applying the multidimensional convolution theorem to this equation yields

$$r_i(t) = \frac{1}{(2\pi)^4} \int_{-\infty}^{\infty} \int_{-\infty}^{\infty} S(\mathbf{k}, \omega) G(\mathbf{k}, \omega) \exp\left(-j\mathbf{k}'\mathbf{x}_i\right) \exp\left(j\omega t\right) d\mathbf{k} \, d\omega \tag{6.62}$$

The weighted delay-and-sum beamformer output is still given by

$$bf(t) = \frac{1}{N} \sum_{i=0}^{N-1} w_i r_i(t + \boldsymbol{\alpha}_0' \mathbf{x}_i)$$  (6.63)

After combining these two equations, we get

$$bf(t) = \frac{1}{(2\pi)^4} \int_{-\infty}^{\infty} \int_{-\infty}^{\infty} S(\mathbf{k}, \omega) G(\mathbf{k}, \omega) \left\{ \frac{1}{N} \sum_{i=0}^{N-1} w_i \exp\left[-j(\mathbf{k} - \omega\boldsymbol{\alpha}_0)' \mathbf{x}_i\right] \right\}$$
$$\cdot \exp(j\omega t) \, d\mathbf{k} \, d\omega$$  (6.64)

The term in brackets is the array pattern $H(\mathbf{k}, \omega)$ that would have been obtained had the receivers been omnidirectional, that is, if $G(\mathbf{k}, \omega)$ had been equal to 1. Thus the beamformer output can be written as

$$bf(t) = \frac{1}{(2\pi)^4} \int_{-\infty}^{\infty} \int_{-\infty}^{\infty} S(\mathbf{k}, \omega) G(\mathbf{k}, \omega) H(\mathbf{k}, \omega) \exp(j\omega t) \, d\mathbf{k} \, d\omega$$  (6.65)

comparing this result to equation (6.16), we see that the overall array pattern is given by the product $H(\mathbf{k}, \omega)G(\mathbf{k}, \omega)$.

### 6.3.2 Interpolation Beamforming

One way to circumvent the problem of steering delay quantization is through interpolation of the receiver signals $\{r_i(nT)\}$. Methods for digital signal interpolation, which are well understood [10, 11], can be used to evaluate the values of the receiver signals at times between the sampling instants, thus reducing the quantization error by as much as desired. There is a cost, though, in the form of additional computation. While interpolation beamforming has been studied in great detail [12, 13], we shall restrict ourselves to a brief discussion of baseband pre-beamforming interpolation.

Pre-beamforming interpolation, as the name implies, requires that the individual receiver signals be resampled at a higher rate. This results in a lower value for $T$ and more precise control over the steering delays. This resampling is performed digitally from the original sampled receiver signal. Let $T$ denote the clock rate of the original system before interpolation and $\tilde{T}$ denote the effective clock rate of the interpolated signal. For simplicity, we will assume that $I \triangleq T/\tilde{T}$ is an integer. (This assumption also results in a simpler implementation.) The beamformer output is given by

$$bf(n\tilde{T}) = \frac{1}{N} \sum_{i=0}^{N-1} w_i \tilde{r}_i((n - n_i)\tilde{T})$$  (6.66)

where the steering delay for the $i$th receiver is $n_i \tilde{T}$. The interpolator produces the samples $\tilde{r}_i(m\tilde{T})$ from the samples $r_i(nT)$ according to the rule

$$\tilde{r}_i(m\tilde{T}) = \sum_n r_i(nT) g((m - nI)\tilde{T})$$  (6.67)

The function $g(m\tilde{T})$ represents the impulse response of the interpolating filter.

To design the filter $g(m\tilde{T})$, it is helpful to visualize the interpolator as two subsystems in cascade, as illustrated in Figure 6.10. The first subsystem, which is called

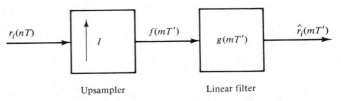

Upsampler                    Linear filter

**Figure 6.10**  Decomposition of a digital interpolator into a cascade of an up-sampler and a linear filter.

an upsampler, stretches out the input sequence by interspersing $I - 1$ samples of value zero. The second subsystem is an LSI filter with impulse response $g(m\tilde{T})$. In the frequency domain, the effect of the upsampler is to compress the spectrum so that $I$ periods of the spectrum occur between $-\pi$ and $\pi$, as shown in Figure 6.11.

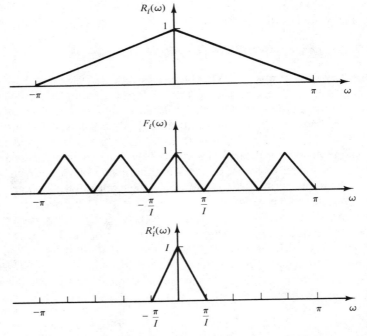

**Figure 6.11**  Fourier transform of an assumed receiver signal $R_i(\omega)$, unsampler output $F(\omega)$, and ideally interpolated receiver signal $R_i'(\omega)$.

Since the spectrum of the interpolated signal is ideally a compressed version of the original one, $g(m\tilde{T})$ should represent the impulse response of a perfect lowpass filter with a normalized cutoff frequency of $\pi/I$ and a gain of $I$. In practice, this lowpass filter will not be perfect, the baseband portion of $\tilde{r}_i(m\tilde{T})$ will be distorted slightly, and small amounts of energy from other frequencies will leak through the stopband.

The beamformer signal $bf(m\tilde{T})$ is formed by adding together the weighted, delayed, interpolated receiver signals $\tilde{r}_i(n\tilde{T})$ giving

$$bf(m\tilde{T}) = \frac{1}{N} \sum_{i=0}^{N-1} w_i \tilde{r}_i(m\tilde{T} - n_i\tilde{T})$$

$$= \frac{1}{N} \sum_{i=0}^{N-1} w_i \sum_p r_i(pT)g((m - pI - n_i)\tilde{T}) \tag{6.68}$$

At this point, the beamformed signal $bf(m\tilde{T})$ is sampled at a rate higher than is necessary to preserve the highest frequency present. Consequently, we can consider reducing the sampling rate without losing any signal components. For simplicity, we shall assume that the new sampling interval for the beamformed signal will be equal to the original receiver sampling interval $T$. If we let $m = nI$, equation (6.68) becomes

$$bf(nI\tilde{T}) = bf(nT) = \frac{1}{N} \sum_{i=0}^{N-1} w_i \sum_p r_i(pT)g((n - p)T - n_i\tilde{T})$$

$$= \frac{1}{N} \sum_{i=0}^{N-1} w_i \left[ \sum_p r_i(pT)g_i((n - p)T) \right] \tag{6.69}$$

where we have defined

$$g_i(nT) \triangleq g(nT - n_i\tilde{T}) = g((nI - n_i)\tilde{T}) \tag{6.70}$$

The expression in brackets can be interpreted in terms of subfilters of the interpolating filter $g(n\tilde{T})$ [13]. Thus $g_i(nT)$ represents every $I$th sample of the impulse response $g(n\tilde{T})$ offset by $n_i$ samples from the origin. Equation (6.69) can be interpreted as a filter-and-sum beamformer; each receiver signal is convolved with the impulse response $w_i g_i(nT)$ before being added to the sum.

Post-beamforming interpolating is very similar except that the beamforming and filtering operations are interchanged. Beamforming is done directly on the upsampled signals $\{f_i(m\tilde{T})\}$ and then the beamformer output is filtered. The performance is similar but there are significant computational savings if the number of receivers is significantly greater than $I$.

### 6.3.3 Frequency-Domain Beamforming for Discrete-Time Signals

It is fairly straightforward to develop the equations for a frequency-domain beamformer using discrete-time receiver signals. As before, let $r_i(nT)$ represent the $n$th sample of the signal from the $i$th receiver, but for brevity of notation, we shall set $T = 1$. The discrete-time output for the frequency-domain beamformer is given by

$$fd(n, \omega) = \frac{1}{N} \sum_{i=0}^{N-1} w_i R_i(n, \omega) \exp[j\omega(n - \tau_i)] \tag{6.71}$$

which is analogous to equation (6.42). The discrete short-time Fourier transform $R_i(n, \omega)$ is given by [8]

$$R_i(n, \omega) = \sum_m r_i(m)v(n - m) \exp(-j\omega m) \tag{6.72}$$

where $v(n)$ represents the impulse response of a narrowband lowpass digital filter. We shall assume that $v(n)$ is equal to zero for values of $n$ outside the range $0 \leq n < M$.

The factor $R_i(n, \omega) \exp(j\omega n)$ in equation (6.71) can be computed efficiently for many values of $\omega$ by using a 1-D FFT algorithm [8]. For example, if we let $\omega = 2\pi l/M$ for $0 \leq l < M$, we can write

$$R_i\left(n, \frac{2\pi l}{M}\right) \exp\left(j\frac{2\pi l}{M}n\right) = \sum_m r_i(m)v(n-m) \exp\left[j\frac{2\pi l}{M}(n-m)\right]$$
$$= \sum_{p=0}^{M-1} r_i(n-p)v(p) \exp\left(j\frac{2\pi l}{M}p\right) \tag{6.73}$$

Except for the sign in the exponent, this equation has the form of an $M$-point DFT computation which can be evaluated by an FFT algorithm when $M$ is a power of 2.

The discrete frequency-domain beamformer output can then be computed using equation (6.71). Since the beam has been formed in the frequency domain, the steering delays $\tau_i$ do *not* have to be quantized. However, in the special case of a linear array with equally spaced receivers where the steering delays have the form

$$\tau_i = -\frac{Mq}{Nl}i \tag{6.74}$$

equation (6.71) becomes

$$fd\left(n, \frac{2\pi l}{M}\right) = \left[\frac{1}{N}\sum_{i=0}^{N-1} w_i R_i\left(n, \frac{2\pi l}{M}\right) \exp\left(j\frac{2\pi q}{N}i\right)\right] \exp\left(j\frac{2\pi l}{M}n\right) \tag{6.75}$$

Except for the sign in the exponent, the term in brackets has the form of a 1-D DFT in the receiver index $i$.

For this special case, we can compute the discrete frequency-domain beamformer output $fd(n, 2\pi l/M)$ by applying a 2-D DFT directly to the receiver signals $r_i(n)$ [9]. Substituting equation (6.73) into equation (6.75) gives us

$$fd\left(n, \frac{2\pi l}{M}\right) = \frac{1}{N}\sum_{i=0}^{N-1}\sum_{p=0}^{M-1} w_i v(p) r_i(n-p) \exp\left(j\frac{2\pi l}{M}p + j\frac{2\pi q}{N}i\right) \tag{6.76}$$

Now, if we define the 2-D sequence

$$x_n(p, i) \triangleq w_i v(p) r_i(n-p) \tag{6.77}$$

where $n$ is taken as a constant for the moment, and we let the $(M \times N)$-point DFT of $x_n(p, i)$ be represented by $X_n(l, q)$, we see that

$$fd\left(n, \frac{2\pi l}{M}\right) = \frac{1}{N} X_n(M - l, N - q) \tag{6.78}$$

The notation $fd(n, 2\pi l/M)$ does not explicitly indicate the direction in which the beam has been steered. The restriction that the steering delays $\tau_i$ satisfy equation (6.74) means that the $x$-component of the desired slowness vector is given by

$$\alpha_{0x} = \frac{Mq}{NlD} \tag{6.79}$$

The values of $M$, $N$ and the receiver separation $D$ are fixed, but the frequency index

$l$ can vary between 0 and $M - 1$, depending on the frequency band of interest. The use of the DFT restricts us to $N$ possible steering directions, corresponding to the values $q$ can assume between 0 and $N - 1$.

By using the 2-D DFT, we are simultaneously computing the discrete frequency-domain beamformer outputs for $M$ values of the frequency index $l$ and $N$ values of the steering direction index $q$. Our attempts to extract signal components by filtering $S(\mathbf{k}, \omega)$ with a frequency-domain beamformer have led us to a formulation that essentially segregates signal components by using a 2-D Fourier transform. In Section 6.5 we examine the techniques for estimating the multidimensional power spectrum $|S(\mathbf{k}, \omega)|^2$. Equation (6.78) thus provides one bridge between the beamforming approach and the spectral estimation approach to segregating the energy carried by propagating waves.

## 6.4 FURTHER CONSIDERATIONS FOR ARRAY PROCESSING APPLICATIONS

In this chapter we use the processing of signals received from an array of sensors as a framework for discussing beamforming, multidimensional digital filtering, and multidimensional power spectrum estimation. Our discussions are fairly general and could be applied to other applications as well. In this section we briefly describe how some of the techniques of this chapter can be tailored more specifically to the array processing application.

We have discussed how the multidimensional space–time signal $s(\mathbf{x}, t)$ is sampled in time and space to give the receiver signals

$$r_i(n) = s(\mathbf{x}_i, n) \tag{6.80}$$

This notation emphasizes the difference in treatment accorded the spatial variables in many practical array processing problems. There are generally many more time samples available than there are space samples. Additional samples in time may be obtained simply by waiting a little longer (assuming the signals under investigation are stationary in time). To obtain additional samples in space, however, we must build longer sensor arrays or position additional sensors in other locations in space. Cost and other engineering constraints usually limit this approach.

In many practical problems, then, we need to process array signals with a wealth of time-series information and a paucity of spatial information. Sometimes the variance of the limited spatial estimates can be reduced by time averaging as, for example, in the case of estimating an autocorrelation coefficient with a large spatial lag. Often, conventional spectral estimation techniques can be applied in the time dimension to obtain the desired temporal resolution while high-resolution techniques must be applied in the spatial dimensions. Also, in most sonar and radar applications, the signals of interest have relatively narrow temporal bandwidths compared to their center frequencies. Consequently, the temporal frequency variable $\omega$ and the spatial wavenumber variable $\mathbf{k}$ are handled differently.

### 6.4.1 Analysis of a Narrowband Beamformer

Let us consider once again the filter-and-sum beamformer of Section 6.2.5,

$$fs(n) = \frac{1}{N} \sum_{i=0}^{N-1} q_i(n - \tau_i) \tag{6.81}$$

where

$$q_i(n) = \frac{1}{2\pi} \int_{-\pi}^{\pi} W_i(\omega) R_i(\omega) \exp(j\omega n)\, d\omega \tag{6.82}$$

As before, $R_i(\omega)$ is the Fourier transform of the $i$th receiver signal $r_i(n)$, $W_i(\omega)$ is the frequency response of the filter applied to $r_i(n)$, and $\{\tau_i\}$ are the steering delays. We can take the Fourier transform of both sides of equation (6.81) to obtain

$$FS(\omega) = \frac{1}{N} \sum_{i=0}^{N-1} W_i(\omega) R_i(\omega) \exp(-j\omega\tau_i) \tag{6.83}$$

To simplify things slightly, let us define the filter response as

$$F_i(\omega) \triangleq \frac{1}{N} W_i(\omega) \exp(-j\omega\tau_i) \tag{6.84}$$

so that (6.83) may be rewritten as

$$FS(\omega) = \sum_{i=0}^{N-1} F_i(\omega) R_i(\omega) \tag{6.85}$$

Because of the relatively high resolution in temporal frequency and because many applications involve only narrowband signals, we can view this equation as representing the beamforming operation at a single frequency $\omega$.

For some applications, it is desirable to compute the power in the output of the filter-and-sum beamformer at the frequency $\omega$, which we shall denote $P_{FS}(\omega)$. To do this, we can obtain the autocorrelation function $E\{fs(n)fs^*(n-m)\}$ and take its Fourier transform. First, we shall define $\phi_{ij}(m)$ to be the cross-correlation function between the $i$th and $j$th receiver signals; it is given by

$$\phi_{ij}(m) \triangleq E\{r_i(n)r_j^*(n-m)\} \tag{6.86}$$

which has a Fourier transform that we shall denote $\Phi_{ij}(\omega)$. ($E\{\cdot\}$ denotes the expected value.) Now it is straightforward to show that

$$P_{FS}(\omega) = \sum_{i=0}^{N-1} \sum_{j=0}^{N-1} F_i^*(\omega) \Phi_{ij}(\omega) F_j(\omega) \tag{6.87}$$

Note how the spatial and temporal variables have been handled differently. Because of the large number of time samples, the expected value operator used in equation (6.86) to obtain $\phi_{ij}(m)$ can be approximated quite accurately by using time averages if the signal is ergodic. Because of the limited number of sensors, however, a separate cross-correlation function is computed for each pair. Equation (6.87) can be written in matrix form as

$$P_{FS}(\omega) = \mathbf{f}^{\dagger} \mathbf{\Phi} \mathbf{f} \tag{6.88}$$

where

$$\boldsymbol{\Phi} \triangleq \begin{bmatrix} \Phi_{00}(\omega) & \cdots & \Phi_{0,N-1}(\omega) \\ \cdot & & \\ \cdot & & \\ \cdot & & \\ \Phi_{N-1,0}(\omega) & \cdots & \Phi_{N-1,N-1}(\omega) \end{bmatrix} \tag{6.89}$$

$$\mathbf{f} = [F_0(\omega), \ldots, F_{N-1}(\omega)]' \tag{6.90}$$

and † denotes the conjugate transpose. (For convenience, we have dropped any explicit reference to the $\omega$ dependence of $\boldsymbol{\Phi}$ and $\mathbf{f}$.)

Up until now, we have not really considered the problem of noise corrupting the signals received by the sensors. In practical applications, of course, noise can come from the medium or the sensors themselves; it can take the form of random, uncorrelated fluctuations or unwanted, interferring signals. Suppose that the receiver signals consist of an ideal signal component and an additive noise component that is uncorrelated with the signal. Then we can write the correlation matrix $\boldsymbol{\Phi}$ in the form

$$\boldsymbol{\Phi} = \sigma_s^2 \mathbf{s}\mathbf{s}^\dagger + \sigma_\epsilon^2 \mathbf{E} \tag{6.91}$$

The vector $\mathbf{s}$ represents the ideal signal component at frequency $\omega$ seen by each receiver. Thus

$$\mathbf{s} = [S_0(\omega), \ldots, S_{N-1}(\omega)]' \tag{6.92}$$

where $S_i(\omega)$ is the 1-D Fourier transform over time:

$$S_i(\omega) = \mathcal{F}[s(\mathbf{x}_i, n)] \tag{6.93}$$

and $s(\mathbf{x}_i, n)$ represents samples of the ideal signal. The vector $\mathbf{s}$ is normalized by the constraint

$$\mathbf{s}^\dagger \mathbf{s} = N \tag{6.94}$$

so that $\sigma_s^2$ in equation (6.91) represents the average signal power at the frequency $\omega$ across the $N$ sensors [14].

The matrix $\mathbf{E}$ represents the noise correlation matrix, which is normalized so that its trace is equal to $N$. Then $\sigma_\epsilon^2$ represents the average noise power at the frequency $\omega$. Thus the $ij$th entry in the matrix $\mathbf{E}$ is given by

$$E_{ij}(\omega) = \frac{\mathcal{F}[E\{\epsilon_i(n)\epsilon_j^*(n-m)\}]}{\sigma_\epsilon^2} \tag{6.95}$$

where $\epsilon_i(n)$ represents the noise in the $i$th receiver. Using (6.87), the power spectrum $P_{FS}(\omega)$ can be written as

$$P_{FS}(\omega) = \sigma_\epsilon^2 \mathbf{f}^\dagger \mathbf{E}\mathbf{f} + \sigma_s^2 |\mathbf{f}^\dagger \mathbf{s}|^2 \tag{6.96}$$

The ratio

$$G \triangleq \frac{|\mathbf{f}^\dagger \mathbf{s}|^2}{\mathbf{f}^\dagger \mathbf{E}\mathbf{f}} \tag{6.97}$$

may be taken as the processing gain of the narrowband filter-and-sum beamformer [14], which together with resolution is an important measure of the performance of a beamformer.

Let us briefly look at the processing gain and output power spectrum for three different filter vectors $\mathbf{f}$, which we shall call $\mathbf{f}_1$, $\mathbf{f}_2$, and $\mathbf{f}_3$. First, however, let us define the steering vector $\mathbf{v}$ needed to determine $\mathbf{f}$ when looking for a particular ideal signal. If $\mathbf{s}$ is the signal vector defined in equation (6.92), $\mathbf{v}$ should be equal to $\mathbf{s}$. Consequently, $\mathbf{v}$ satisfies

$$\mathbf{v}^\dagger \mathbf{v} = N \tag{6.98}$$

When $\mathbf{v} = \mathbf{s}$, the beamformer is said to be matched to the ideal signal. (Cox [14] has studied the performance of several optimum beamformers under conditions of mismatch, when the actual signal differs from the ideal signal.) For the case of a single plane wave, the steering vector $\mathbf{v}$ is given by

$$\mathbf{v} = [\exp(j\omega\tau_0), \dots, \exp(j\omega\tau_{N-1})]' \tag{6.99}$$

where the $\{\tau_i\}$ are the familiar steering delays. The ideal signal can, of course, be more complicated than a single plane wave; for example, it could be a spherical wave emanating from a point in the near-field of the array.

We can now look at the case where the filter vector is given by

$$\mathbf{f}_1 = \frac{\mathbf{v}}{N} \tag{6.100}$$

which corresponds to the delay-and-sum beamformer when the ideal signal is a plane wave. The processing gain from equation (6.97) for this case is given by

$$G_1 = \frac{N^2}{\mathbf{v}^\dagger \mathbf{E} \mathbf{v}} \tag{6.101}$$

When the noise is spatially uncorrelated so that $\mathbf{E} = \mathbf{I}$, the gain $G_1$ becomes equal to $N$, the number of sensors. In general, however, the power spectrum $P_{FS}(\omega)$ is given by

$$P_{FS}(\omega)\Big|_{t=f_1} = \sigma_s^2 + \sigma_\epsilon^2 \frac{\mathbf{v}^\dagger \mathbf{E} \mathbf{v}}{N^2} \tag{6.102}$$

(Although not explicitly indicated here, $\mathbf{E}$, $\mathbf{v}$, $\sigma_s^2$, and $\sigma_\epsilon^2$ are functions of $\omega$.)

Another interesting case occurs when the filter vector is given by

$$\mathbf{f}_2 = \frac{\mathbf{E}^{-1}\mathbf{v}}{\mathbf{v}^\dagger \mathbf{E}^{-1}\mathbf{v}} \tag{6.103}$$

where $\mathbf{E}^{-1}$ is the inverse of the noise correlation matrix. This filter arises as the optimum solution in several detection and estimation problems [14], including a true maximum likelihood estimate of the desired filter-and-sum output signal, the maximization of the output signal-to-noise ratio, and the minimization of the output noise variance [15]. When the noise is spatially uncorrelated so that $\mathbf{E} = \mathbf{E}^{-1} = \mathbf{I}$, the filter vector $\mathbf{f}_2$ reduces to $\mathbf{f}_1$. The processing gain for $\mathbf{f}_2$ is given by

$$G_2 = \mathbf{v}^\dagger \mathbf{E}^{-1}\mathbf{v} \tag{6.104}$$

and the power spectrum is given by

$$P_{FS}(\omega)\Big|_{t=f_2} = \sigma_s^2 + \sigma_\epsilon^2 \frac{1}{\mathbf{v}^\dagger \mathbf{E}^{-1}\mathbf{v}} \tag{6.105}$$

The Capon high-resolution spectral estimate [21], when put into the proper notation [14], results in a filter vector of the form

$$\mathbf{f}_3 = \frac{\mathbf{\Phi}^{-1}\mathbf{v}}{\mathbf{v}^\dagger\mathbf{\Phi}^{-1}\mathbf{v}} \tag{6.106}$$

Note that similarity between $\mathbf{f}_3$ and $\mathbf{f}_2$; the difference is the use of the inverse signal-plus-noise correlation matrix in place of the inverse noise correlation matrix. The gain for $\mathbf{f}_3$ is given by

$$G_3 = \frac{(\mathbf{v}^\dagger\mathbf{\Phi}^{-1}\mathbf{v})^2}{\mathbf{v}^\dagger\mathbf{\Phi}^{-1}\mathbf{E}\mathbf{\Phi}\mathbf{v}} \tag{6.107}$$

and the power spectrum is given by

$$P_{FS}(\omega)\Big|_{\mathbf{f}=\mathbf{f}_3} = \sigma_s^2 + \sigma_\epsilon^2 \frac{\mathbf{v}^\dagger\mathbf{\Phi}^{-1}\mathbf{E}\mathbf{\Phi}^{-1}\mathbf{v}}{(\mathbf{v}^\dagger\mathbf{\Phi}^{-1}\mathbf{v})^2} \tag{6.108}$$

This last expression can be simplified greatly by writing $P_{FS}(\omega)$ in the form given by equation (6.88):

$$\begin{aligned} P_{FS}(\omega)\Big|_{\mathbf{f}=\mathbf{f}_3} &= \mathbf{f}_3^\dagger\mathbf{\Phi}\mathbf{f}_3 \\ &= \frac{1}{\mathbf{v}^\dagger\mathbf{\Phi}^{-1}\mathbf{v}} \end{aligned} \tag{6.109}$$

Although we shall not discuss them here, other optimum narrowband beamformers may be derived, including one based on the Wiener filtering approach for minimizing the mean-squared error between the actual beamformer output and the desired beamformer output [15]. In addition, many array processing applications require beamforming systems that can adapt to slowly varying changes in the properties or statistics of the input signals and noise. In these cases, the filter vector is periodically updated by an adaptation algorithm to reflect the changes in the environment [15].

## 6.5 MULTIDIMENSIONAL SPECTRAL ESTIMATION

If we were confronted with the one-dimensional task of detecting the presence of a sinusoid of unknown frequency in a background of noise, there are several approaches that we might choose to follow. One would be to build a bank of narrow bandpass filters each tuned to a different frequency. By measuring the energy in the output waveforms from these filters, we could detect the presence or absence of a sinusoid in that band of frequencies. Another approach would be to capture a segment of the waveform, evaluate its Fourier transform, and use that to estimate the power spectrum of the entire signal. Again, the presence of a sinusoid could be detected by an energy concentration. Still another approach would be to assume the presence of a sinusoid and attempt to estimate its amplitude and frequency from the observed time waveform.

If we are confronted with the multidimensional counterpart of this problem—detecting the presence of plane waves in a background of noise—we have similar choices. The equivalent to the bandpass filter bank is a bank of beamformers such as we have discussed in the earlier sections of this chapter. The other two approaches involve estimating the parameters of a stationary random process and fall into the collection of techniques known as multidimensional spectral estimation. In the remainder of this chapter, we examine four approaches to the problem of multidimensional spectral estimation: classical spectral estimation, high-resolution spectral estimation, all-pole spectral modeling, and maximum entropy spectral estimation. The first two methods are straightforward extensions of 1-D techniques, but the all-pole modeling and maximum entropy methods, as we shall see, are somewhat different in the multidimensional case.

### 6.5.1 Classical Spectral Estimation

The classical approaches to 1-D spectral estimation, as discussed in [16–19], have been known for well over two decades and are summarized in Chapter 11 of Oppenheim and Schafer [20]. The extension of these techniques to two or more dimensions is straightforward.

The spectral estimation problem, simply stated, is to estimate the power spectrum $P(\mathbf{k}, \omega)$ of $s(\mathbf{x}, t)$ from the measurements $r_i(nT)$. There are a number of definitions for the power spectrum of a random process. Although we would like to think of it as the squared magnitude of the Fourier transform of the process, such Fourier transforms usually do not exist. As an alternative, we will define it as the Fourier transform of the autocorrelation function. (For the deterministic case these definitions are equivalent.) Thus

$$P(\mathbf{k}, \omega) \triangleq \int_{-\infty}^{\infty} \int_{-\infty}^{\infty} \phi_{ss}(\mathbf{x}, t) \exp\left[-j(\omega t - \mathbf{k}'\mathbf{x})\right] d\mathbf{x}\, dt \qquad (6.110)$$

where

$$\phi_{ss}(\mathbf{x}, t) = E[s(\boldsymbol{\xi}, \tau)s^*(\boldsymbol{\xi} - \mathbf{x}, \tau - t)] \qquad (6.111)$$

and $E[\cdot]$ denotes the expectation over the ensemble of the random process.

A number of factors conspire to limit the amount of data available to estimate $P(\mathbf{k}, \omega)$. First, the statistics of $s(\mathbf{x}, t)$, while generally assumed to be wide-sense stationary for the purposes of analysis, may actually evolve with time. Second, the number of receivers is limited, and their outputs will be sampled. Finally, we have available only a limited length record of one instance of the random process which makes it difficult to evaluate an ensemble average. For these reasons, in what is to follow we will assume that the measurements $r_i(nT)$ to be used in estimating $P(\mathbf{k}, \omega)$ are available only for $0 \leq i < N; 0 \leq n < M$ and we will assume that these measurements are noise-free and that $T = 1$, so that

$$r_i(n) = s(\mathbf{x}_i, n) \qquad (6.112)$$

One of the simplest estimates for $P(\mathbf{k}, \omega)$ can be obtained by taking the multidimensional Fourier transform of $r_i(n)$ and squaring its magnitude to obtain

$$\hat{P}(\mathbf{k}, \omega) = \frac{1}{MN} \left| \sum_{i=0}^{N-1} \sum_{n=0}^{M-1} r_i(n) \exp\left(j\mathbf{k}'\mathbf{x}_i - j\omega n\right) \right|^2 \tag{6.113}$$

This is called the *periodogram*. (We will use the carat over $P$ to denote an estimate.) The normalizing factor $1/MN$ arises in the discretization of (6.111). In the case of a linear array with $\mathbf{x}_i = (i, 0, 0)'$, this equation becomes

$$\hat{P}(k_x, \omega) = \frac{1}{MN} \left| \sum_{i=0}^{N-1} \sum_{n=0}^{M-1} r_i(n) \exp\left(jk_x i - j\omega n\right) \right|^2 \tag{6.114}$$

a sampled version of which can be obtained by using a 2-D DFT.

We can generalize the periodogram estimate slightly to form the *modified periodogram* $\hat{P}_M(k_x, \omega)$ if we first multiply $r_i(n)$ by a 2-D window function $g_i(n)$. Let us examine the particular case where $g_i(n)$ is chosen to be

$$g_i(n) = \frac{1}{N} w_i v(M - 1 - n) \tag{6.115}$$

where $w_i$ is nonzero only over the interval $0 \le i < N$ and $v(n)$ is nonzero only over the interval $0 \le n < M$. For this particular 2-D window, the spectral estimate becomes

$$\begin{aligned}\hat{P}_M(k_x, \omega) &= \frac{1}{MN} \left| \sum_{i=0}^{N-1} \sum_{n=0}^{M-1} g_i(n) r_i(n) \exp\left(jk_x i - j\omega n\right) \right|^2 \\ &= \frac{1}{MN} \left| \frac{1}{N} \sum_{i=0}^{N-1} w_i \exp\left(jk_x i\right) \sum_{n=0}^{M-1} r_i(n) v(M - 1 - n) \exp\left(-j\omega n\right) \right|^2 \end{aligned} \tag{6.116}$$

In this case, the modified periodogram can be related to the output of a frequency-domain beamformer, given by equation (6.76).

$$\hat{P}_M(k_x, \omega) = \frac{1}{MN} |fd(M - 1, \omega)|^2 \tag{6.117}$$

There are a number of limitations to the periodogram as a spectral estimator, but the most serious is that the spectral estimates have a large statistical variance and that they are "busy" (i.e., not smooth). Samples of the periodogram separated in frequency by $2\pi/M$ or in wavenumber by $2\pi/N$ are uncorrelated, and as $M$ and $N$ grow larger, these uncorrelated spectral samples are drawn closer together, allowing large changes in amplitude for small changes in frequency or wavenumber. This phenomenon results in a spectral estimate with a high degree of variability from one spectral sample to the next.

This variability can be reduced in a variety of ways. We may consider adjusting the data window $g_i(n)$, smoothing the periodogram estimate in frequency and wavenumber, averaging several spectral estimates together, or applying more than one of these three techniques. A trade-off between spectral resolution and the "busyness" of the spectral estimate occurs when any of these techniques is used.

Let us first consider reducing the variance of the spectral estimate by averaging several such estimates together. This is an attempt to realize the expectation operation that appeared in (6.111). Although we have available only a single instance of the random process, if it is ergodic, we can partition the set of measurements $\{r_i(n)\}$ into several segments each of which can provide a spectral estimate. For example, let us define the periodogram of the $l$th segment of the measurements to be

$$\hat{P}_l(k_x, \omega) \triangleq \frac{1}{\tilde{M}N} \left| \sum_{i=0}^{N-1} \sum_{n=0}^{\tilde{M}-1} r_i(n + l\tilde{M}) \exp\left(jk_x i - j\omega n\right) \right|^2 \tag{6.118}$$

In all there are $L \triangleq M/\tilde{M}$ such periodograms which may be averaged together to form

$$\hat{P}_B(k_x, \omega) \triangleq \frac{1}{L} \sum_{l=0}^{L-1} \hat{P}_l(k_x, \omega)$$

$$= \frac{1}{MN} \sum_{l=0}^{L-1} \left| \sum_{i=0}^{N-1} \sum_{n=0}^{\tilde{M}-1} r_i(n + l\tilde{M}) \exp\left(jk_x i - j\omega n\right) \right|^2 \tag{6.119}$$

The subscript $B$ stands for Bartlett, who first suggested this approach to spectral estimation for the 1-D case [16]. Because the measurements have been divided into segments of length $\tilde{M}$ in the time variable, the resolution in temporal frequency has been reduced by the factor $L$. In exchange for this reduction in resolution, we have obtained a statistically more stable spectral estimate, that is, an estimate with a smaller variance.

We could also consider reducing the wavenumber resolution by partitioning $r_i(n)$ with respect to the receiver index $i$. This would generate even more spectral estimates to be averaged together, reducing the variance of the final spectral estimate even further. In many applications, however, $N$ is much smaller than $M$, and the resulting further loss in wavenumber resolution that would be caused by partitioning with respect to the index $i$ cannot be tolerated.

The method studied by Welch [19] for spectral estimation uses the same data partitioning as the Bartlett procedure but applies a data window $g_i(n)$ to each data segment before its Fourier transform is computed. The data segments are also allowed to overlap. The amount of overlap, the length and number of the data segments, and the shape of the data window can all be adjusted to reduce the variance of the spectral estimate, usually at the expense of some spectral resolution. (The reader is directed to [5] for a comprehensive review of data windows for 1-D spectral estimation.) Just as the modified periodogram is closely related to the output of a frequency-domain beamformer, the Welch estimate can be interpreted as the average power contained in the output of a frequency-domain beamformer.

We can directly trade spectral resolution for a more stable estimate of the power spectrum by forming a *smoothed periodogram* of the form

$$\hat{P}_S(k_x, \omega) = \frac{1}{2\pi} \int_{-\pi}^{\pi} \int_{-\pi}^{\pi} \hat{P}(\kappa_x, \theta) C(k_x - \kappa_x, \omega - \theta) \, d\kappa_x \, d\theta \tag{6.120}$$

Here the raw periodogram is convolved with a smoothing function $C(k_x, \omega)$ to form the smoothed estimate. This is equivalent to multiplying an estimate of the auto-correlation function by a window sequence $c(l, m)$ whose Fourier transform is $C(k_x, \omega)$. The wider the main lobe of $C(k_x, \omega)$ the greater the degree of smoothing, the more stable the spectral estimate, and the poorer the resolution. Note also that if $C(k_x, \omega)$

is not chosen to be always positive, it is possible for the smoothed periodogram to be negative for some values of $(k_x, \omega)$.

To demonstrate what is happening in the space–time domain, let us change notation slightly and write our array of measurements as

$$r(i, n) \triangleq r_i(n) \tag{6.121}$$

Since these measurements are samples of a zero-mean 2-D discrete random process, the autocorrelation sequence can be found as the ensemble average

$$\phi(l, m) = E[r(i, n)r^*(i - l, n - m)] \tag{6.122}$$

If the process $r(i, n)$ is wide-sense stationary, $\phi(l, m)$ will be independent of the indices $(i, n)$. We can then compute an estimate of the autocorrelation sequence by forming the product $r(i, n)r^*(i - l, n - m)$ and averaging it over all possible values of the ordered pair $(i, n)$ to obtain

$$\hat{\phi}(l, m) \triangleq \frac{1}{MN} \sum_{i=l}^{N-1} \sum_{n=m}^{M-1} r(i, n)r^*(i - l, n - m) \tag{6.123}$$

(The limits shown on the sums are the ones valid for $l > 0$ and $m > 0$. For other values of $l$ and $m$, other limits would apply.)

Because we have measurements $r(i, n)$ available only in the region

$$0 \leq i < N; \qquad 0 \leq n < M$$

the estimate of the autocorrelation function $\hat{\phi}(l, m)$ will be nonzero only for $(l, m)$ in the region

$$-N < l < N; \qquad -M < m < M \tag{6.124}$$

Furthermore, as the magnitude of $l$ or $m$ is increased, the estimate of the autocorrelation function is based on fewer samples of $r(i, n)$, and is thus statistically less reliable than for values of $l$ and $m$ closer to the origin.

The periodogram estimate given by

$$\hat{P}(k_x, \omega) = \frac{1}{MN} \left| \sum_{i=0}^{N-1} \sum_{n=0}^{M-1} r_i(n) \exp(jk_x i - j\omega n) \right|^2 \tag{6.125}$$

can be written as the Fourier transform of the estimate of the autocorrelation function

$$\hat{P}(k_x, \omega) = \sum_{l=-N+1}^{N-1} \sum_{m=-M+1}^{M-1} \hat{\phi}(l, m) \exp(jk_x l - j\omega m) \tag{6.126}$$

and the smoothed spectral estimate, $\hat{P}_s(k_x, \omega)$, may be written as

$$\hat{P}_s(k_x, \omega) = \sum_{l=-N+1}^{N-1} \sum_{m=-M+1}^{M-1} c(l, m)\hat{\phi}(l, m) \exp(jk_x l - j\omega m) \tag{6.127}$$

where the window function $c(l, m)$ is the inverse Fourier transform of $C(k_x, \omega)$ which we encountered in equation (6.120).

$$c(l, m) \triangleq \frac{1}{(2\pi)^2} \int_{-\pi}^{\pi} \int_{-\pi}^{\pi} C(k_x, \omega) \exp(-jk_x l + j\omega m) \, dk_x \, d\omega \tag{6.128}$$

The use of a correlation window (as distinguished from the data window encountered in the modified periodogram and Welch estimates) can be interpreted as reducing the weight given to the estimated autocorrelation coefficients $\hat{\phi}(l, m)$ for which $|l| \simeq N$ or $|m| \simeq M$. These are the same samples of $\hat{\phi}(l, m)$ which have a large variance because of the small number of measurements entering their calculation. The multiplicative window in equation (6.127) serves to reduce the contribution of these unreliable values to the spectral estimate.

We have used the problem of processing signals received by an array of distributed sensors as motivation for a discussion of classical techniques for estimating power spectra. Because of this, the notation in this chapter has deviated somewhat from that established in earlier chapters. For this reason, let us pause at this point to summarize the results of this section using the notation of earlier chapters.

Let us consider the multidimensional random process $x(\mathbf{n})$ which has the autocorrelation function

$$\phi(\mathbf{m}) = E\{x(\mathbf{n})x^*(\mathbf{n} - \mathbf{m})\} \tag{6.129}$$

We seek to estimate the power spectrum $P(\omega)$ given formally by

$$P(\omega) = \sum_{\mathbf{m}} \phi(\mathbf{m}) \exp(-j\omega'\mathbf{m}) \tag{6.130}$$

but we are limited by a finite number of measurements. In particular, let us assume that we have values of $x(\mathbf{n})$ in the region

$$0 \leq n_1 < N_1; \quad 0 \leq n_2 < N_2; \quad \ldots; \quad 0 \leq n_M < N_M \tag{6.131}$$

or, more compactly,

$$\mathbf{0} \leq \mathbf{n} < \mathbf{N} \tag{6.132}$$

One way to estimate $P(\omega)$ is to compute the Fourier transform of the measurements

$$X(\omega) = \sum_{\mathbf{n}} x(\mathbf{n}) \exp(-j\omega'\mathbf{n}) \tag{6.133}$$

and then form the periodogram estimate

$$\hat{P}(\omega) = \frac{1}{\det \mathbf{N}} |X(\omega)|^2 \tag{6.134}$$

where $\mathbf{N} = \text{diag}(N_1, N_2, \ldots, N_M)$. The statistical stability of the spectral estimate may be improved, with a reduction in spectral resolution, by partitioning the set of measurements, computing the periodogram estimate of each segment, and averaging over the estimates. This estimate, called the Bartlett estimate, is given by

$$\hat{P}_B(\omega) = \frac{1}{\det \mathbf{N}} \sum_{\mathbf{l}} \left| \sum_{\mathbf{n}} x(\mathbf{n} + \tilde{\mathbf{M}}\mathbf{l}) \exp(-j\omega'\mathbf{n}) \right|^2 \tag{6.135}$$

The modified periodogram uses a data window $g(\mathbf{n})$ to multiply the measurements before computing the Fourier transform. Thus

$$\hat{P}_M(\omega) = \frac{1}{\det \mathbf{N}} \left| \sum_{\mathbf{n}} g(\mathbf{n}) x(\mathbf{n}) \exp(-j\omega'\mathbf{n}) \right|^2 \tag{6.136}$$

The Welch estimate can be viewed as a combination of the Bartlett procedure and the modified periodogram.

$$\hat{P}_W(\omega) = \frac{1}{\det N} \sum_{l} \left| \sum_{n} g(n)x(n + \tilde{M}l) \exp(-j\omega'n) \right|^2 \qquad (6.137)$$

We can form an estimate of the autocorrelation function $\hat{\phi}(m)$ by computing

$$\hat{\phi}(m) = \frac{1}{\det N} \sum_{n} x(n)x^*(n - m) \qquad (6.138)$$

Then the periodogram estimate given by equation (6.134) can be written as

$$\hat{P}(\omega) = \sum_{m} \hat{\phi}(m) \exp(-j\omega'm) \qquad (6.139)$$

The smoothed periodogram estimate may be computed by applying a window to the autocorrelation estimate before computing the Fourier transform. This helps to reduce the contribution of those estimated autocorrelation coefficients with high variances. The smoothed periodogram may be written as

$$\hat{P}_s(\omega) = \sum_{m} c(m)\hat{\phi}(m) \exp(-j\omega'm) \qquad (6.140)$$

### 6.5.2 High-Resolution Spectral Estimation

The term "high-resolution spectral estimation" can be applied to any of a variety of techniques for obtaining a spectral estimate whose frequency resolution is higher than that of the classical estimators discussed in the preceding section. In this section we examine one of these techniques, which is due to Capon [14,21]. This procedure is sometimes mistakenly referred to as the "maximum-likelihood method" because the form of the optimum filter used in the procedure is similar to that found in the maximum-likelihood estimation of the amplitude of a sine wave of known frequency in Gaussian random noise [14,15,22].

To begin, let us assume that the autocorrelation coefficients $\phi(l, m)$ of the 2-D discrete random process $r(i, n)$ are known for $-N < l < N$, $-M < m < M$. One way to estimate the spectral power $P(k_x, \omega)$ at a particular wavenumber–frequency pair $(k_{0x}, \omega_0)$ would be to filter the process with a narrowband bandpass filter whose gain is unity for $(k_x, \omega) = (k_{0x}, \omega_0)$ and is small otherwise. This filter can be chosen to have minimum output power, subject to the constraint that it have unity gain at $(k_{0x}, \omega_0)$. The average power in the output of this filter can be used as an estimate of $P(k_{0x}, \omega_0)$, which we shall denote as $\hat{P}_C(k_{0x}, \omega_0)$.

Instead of using a bandpass filter to estimate $P(k_{0x}, \omega_0)$, we could demodulate $r(i, n)$ by multiplying it by a complex exponential and then pass the result through a narrowband lowpass filter. Let

$$r_e(i, n) \triangleq r(i, n) \exp(jk_x i - j\omega n) \qquad (6.141)$$

denote the demodulated signal and let $g(i, n)$ denote the impulse response of a lowpass filter whose gain at zero frequency is equal to 1. We shall assume that $g(i, n)$ is a real finite-extent impulse response which is zero outside the region $0 \leq i < N$;

$0 \leq n < M$. The constraint on the gain of $g(i, n)$ can then be expressed as

$$\sum_{i=0}^{N-1} \sum_{n=0}^{M-1} g(i, n) = 1 \tag{6.142}$$

If $r_e(i, n)$ is passed through this lowpass filter, the output signal $y(i, n)$ is given by

$$y(i, n) = \sum_{l=0}^{N-1} \sum_{m=0}^{M-1} g(l, m) r_e(i - l, n - m) \tag{6.143}$$

and the average power in the output signal is

$$E\{|y(i, n)|^2\} = E\left\{ \sum_{l=0}^{N-1} \sum_{m=0}^{M-1} g(l, m) r_e(i - l, n - m) \sum_{p=0}^{N-1} \sum_{q=0}^{M-1} g(p, q) r_e^*(i - p, n - q) \right\}$$

$$= \sum_{l=0}^{N-1} \sum_{m=0}^{M-1} \sum_{p=0}^{N-1} \sum_{q=0}^{M-1} g(l, m) g(p, q) E\{r_e(i - l, n - m) r_e^*(i - p, n - q)\} \tag{6.144}$$

$$= \sum_{l=0}^{N-1} \sum_{m=0}^{M-1} \sum_{p=0}^{N-1} \sum_{q=0}^{M-1} g(l, m) g(p, q) \phi_e(p - l, q - m)$$

where

$$\phi_e(l, m) \triangleq E\{r_e(i, n) r_e^*(i - l, n - m)\} \tag{6.145}$$
$$= \phi(l, m) \exp(jk_x l - j\omega m)$$

We seek to minimize $E\{|y(i, n)|^2\}$ by varying the impulse response coefficients $g(i, n)$ subject to the constraint given in equation (6.142). By doing this, we shall pass the desired wavenumber–frequency component with unity gain but minimize the contributions to the output power from all other components.

The constrained minimization can be performed using a Lagrange multiplier. First, the quadratic form

$$Q \triangleq \tfrac{1}{2} E\{|y(i, n)|^2\} + \lambda \left[ 1 - \sum_{l=0}^{N-1} \sum_{m=0}^{M-1} g(l, m) \right] \tag{6.146}$$

is constructed with the constant but as yet unknown Lagrange multiplier $\lambda$. Using equation (6.144), we can differentiate $Q$ with respect to the coefficient $g(l, m)$ and set the result equal to zero to obtain

$$\frac{\partial Q}{\partial g(l, m)} = \sum_{p=0}^{N-1} \sum_{q=0}^{M-1} g(p, q) \phi_e(p - l, q - m) - \lambda = 0 \tag{6.147}$$

or, equivalently,

$$\sum_{p=0}^{N-1} \sum_{q=0}^{M-1} g(p, q) \phi_e(p - l, q - m) = \lambda \tag{6.148}$$

In order to solve for the optimum coefficients $g(p, q)$, we postulate the existence of an inverse function $\psi_e(l, \alpha; m, \beta)$ which satisfies

$$\sum_{l=0}^{N-1} \sum_{m=0}^{M-1} \phi_e(p - l, q - m) \psi_e(l, \alpha; m, \beta) \triangleq \delta(p - \alpha, q - \beta) \tag{6.149}$$

Then, multiplying both sides of (6.148) by $\psi_e$, summing over $l$ and $m$, and applying equation (6.149), we obtain

$$g(\alpha, \beta) = \lambda \sum_{l=0}^{N-1} \sum_{m=0}^{M-1} \psi_e(l, \alpha; m, \beta) \tag{6.150}$$

The unknown parameter $\lambda$ can be easily determined by applying the constraint (6.142) to the expression for $g(\alpha, \beta)$. This gives

$$\lambda = \frac{1}{\displaystyle\sum_{\alpha=0}^{N-1}\sum_{\beta=0}^{M-1}\sum_{l=0}^{N-1}\sum_{m=0}^{M-1}\psi_e(l,\alpha;m,\beta)} \tag{6.151}$$

Combining this result with equation (6.150) yields the optimum impulse response coefficients

$$g(\alpha, \beta) = \frac{\displaystyle\sum_{l=0}^{N-1}\sum_{m=0}^{M-1}\psi_e(l,\alpha;m,\beta)}{\displaystyle\sum_{\alpha=0}^{N-1}\sum_{\beta=0}^{M-1}\sum_{l=0}^{N-1}\sum_{m=0}^{M-1}\psi_e(l,\alpha;m,\beta)} \tag{6.152}$$

By substituting this result into equation (6.144), the reader can show that the average power in the filter output, which we take as the high-resolution spectral estimate $\hat{P}_C(k_{0x}, \omega_0)$, is given by

$$\hat{P}_C(k_{0x}, \omega_0) = E\{|y(i,n)|^2\}$$

$$= \frac{1}{\displaystyle\sum_{\alpha=0}^{N-1}\sum_{\beta=0}^{M-1}\sum_{l=0}^{N-1}\sum_{m=0}^{M-1}\psi_e(l,\alpha;m,\beta)} \tag{6.153}$$

Since $\psi_e$ depends implicitly on the selected point $(k_{0x}, \omega_0)$ for which the spectral estimate is being generated, it may appear at first that a new $\psi_e$ must be computed to generate the spectral estimate at any other point. Fortunately, this is not the case. We can define a single inverse function $\psi(l,\alpha;m,\beta)$ of the form

$$\psi(l,\alpha;m,\beta) \triangleq \psi_e(l,\alpha;m,\beta)\exp[-jk_x(l-\alpha)+j\omega(m-\beta)] \tag{6.154}$$

so that

$$\sum_{l=0}^{N-1}\sum_{m=0}^{M-1}\phi(p-l,q-m)\psi(l,\alpha;m,\beta) = \delta(p-\alpha, m-\beta) \tag{6.155}$$

Then $\psi$ can be computed once and used in equations (6.154) and (6.153) to produce the high-resolution spectral estimate at any point $(k_x, \omega)$ in wavenumber–frequency space.

$$\hat{P}_C(k_x, \omega) = \frac{1}{\displaystyle\sum_{\alpha=0}^{N-1}\sum_{\beta=0}^{M-1}\sum_{l=0}^{N-1}\sum_{m=0}^{M-1}\psi(l,\alpha;m,\beta)\exp[jk_x(l-\alpha)-j\omega(m-\beta)]} \tag{6.156}$$

This expression can be simplified somewhat by summing over those values of $\psi(l,\alpha;m,\beta)$ for which the differences $l-\alpha$ and $m-\beta$ are constant. Let us define the array $\gamma(p,q)$, whose elements consist of sums of $\psi(l,\alpha;m,\beta)$ when $l-\alpha=p$ and $m-\beta=q$.

$$\gamma(p,q) \triangleq \sum_{\substack{l=0 \\ l-\alpha=p}}^{N-1}\sum_{\alpha=0}^{N-1}\sum_{\substack{m=0 \\ m-\beta=q}}^{M-1}\sum_{\beta=0}^{M-1}\psi(l,\alpha;m,\beta)$$

$$= \sum_{l=\max(0,p)}^{\min(N-1,N-1+p)}\sum_{m=\max(0,q)}^{\min(M-1,M-1+q)}\psi(l,l-p;m,m-q) \tag{6.157}$$

Since $0 \le l$, $\alpha < N$ and $0 \le m$, $\beta < M$, the variables $p$ and $q$ range from $-N < p < N$ and $-M < q < M$, respectively. Using equation (6.157), we can rewrite the high-resolution spectral estimate as

$$\hat{P}_c(k_x, \omega) = \frac{1}{\displaystyle\sum_{p=-N+1}^{N-1} \sum_{q=-M+1}^{M-1} \gamma(p, q) \exp(jk_x p - j\omega q)} \tag{6.158}$$

We can interpret the denominator of this expression as an estimate of the inverse power spectrum $Q(k_x, \omega) = 1/P(k_x, \omega)$. The coefficients $\gamma(p, q)$ play the role of the estimates of the autocorrelation coefficients in a periodogram estimate of $Q(k_x, \omega)$. Once the values of $\gamma(p, q)$ are known for $-N < p < N$ and $-M < q < M$, equation (6.158) quickly yields the estimate $\hat{P}_c(k_x, \omega)$.

Now, we must address the problem of determining $\gamma(p, q)$ from the autocorrelation coefficients $\phi(l, m)$, which are assumed to be known for $-N < l < N$, $-M < m < M$. This is equivalent to finding the inverse of an $NM \times NM$ block Toeplitz matrix, for which efficient algorithms exist [23].

To show how our problem reduces to the inversion of a block Toeplitz matrix, we can first form the composite matrix

$$\mathbf{\Phi} \triangleq \begin{bmatrix} \mathbf{\Phi}_0 & \mathbf{\Phi}_{-1} & \cdots & \mathbf{\Phi}_{-M+1} \\ \mathbf{\Phi}_1 & \mathbf{\Phi}_0 & \cdots & \mathbf{\Phi}_{-M+2} \\ \cdot & \cdot & & \cdot \\ \cdot & \cdot & & \cdot \\ \cdot & \cdot & & \cdot \\ \mathbf{\Phi}_{M-1} & \mathbf{\Phi}_{M-2} & \cdots & \mathbf{\Phi}_0 \end{bmatrix} \tag{6.159}$$

where each of the submatrices is $N \times N$. The submatrices along any diagonal are identical. Furthermore, each of the submatrices is a Toeplitz matrix of the form

$$\mathbf{\Phi}_m \triangleq \begin{bmatrix} \phi(0, m) & \phi(-1, m) & \cdots & \phi(-N+1, m) \\ \phi(1, m) & \phi(0, m) & \cdots & \phi(-N+2, m) \\ \cdot & \cdot & & \cdot \\ \cdot & \cdot & & \cdot \\ \cdot & \cdot & & \cdot \\ \phi(N-1, m) & \phi(N-2, m) & \cdots & \phi(0, m) \end{bmatrix} \tag{6.160}$$

Such a Toeplitz arrangement of Toeplitz submatrices is called a *block Toeplitz* matrix. In (6.160), the numbers $\phi(l, m)$ are the known autocorrelation coefficients.

Let us denote the inverse of $\mathbf{\Phi}$ as $\mathbf{\Phi}^{-1}$ and write it in the following form:

$$\mathbf{\Phi}^{-1} \triangleq \begin{bmatrix} \mathbf{\Psi}_{00} & \mathbf{\Psi}_{01} & \cdots & \mathbf{\Psi}_{0,M-1} \\ \mathbf{\Psi}_{10} & \mathbf{\Psi}_{11} & \cdots & \mathbf{\Psi}_{1,M-1} \\ \cdot & \cdot & & \cdot \\ \cdot & \cdot & & \cdot \\ \cdot & \cdot & & \cdot \\ \mathbf{\Psi}_{M-1,0} & \mathbf{\Psi}_{M-1,1} & \cdots & \mathbf{\Psi}_{M-1,M-1} \end{bmatrix} \tag{6.161}$$

where the submatrices are given by

$$
\Psi_{m\beta} \triangleq
\begin{bmatrix}
\psi(0,0;m,\beta) & \cdots & \psi(0,N-1;m,\beta) \\
\vdots & & \vdots \\
\psi(N-1,0;m,\beta) & \cdots & \psi(N-1,N-1;m,\beta)
\end{bmatrix}
\tag{6.162}
$$

Now, it is straightforward to show that the matrix identity

$$
\Phi\Phi^{-1} = I \tag{6.163}
$$

is equivalent to the relationship between $\phi(l,m)$ and $\psi(l,\alpha;m,\beta)$ given by equation (6.155). Thus the values of $\psi$ can be computed from the values of $\phi$ by forming the matrix $\Phi$ and inverting it. Then $\gamma(p,q)$ is easily computed from equation (6.157) and $\hat{P}_C(k_x,\omega)$ from (6.158).

In practice, the autocorrelation coefficients $\phi(l,m)$ are not known and must be estimated from the set of measurements $\{r(i,n)\}$. For the high-resolution spectral estimate, it is interesting to speculate on methods for directly estimating the coefficients $\gamma(p,q)$ used in equation (6.158) from the set of measurements.

### 6.5.3 All-Pole Spectral Modeling

This approach to multidimensional spectral estimation is a straightforward extension of the 1-D technique known as autoregressive (AR) spectral estimation. It makes the assumption that the random process being analyzed is the output of a filter which has a spectrally flat (white noise) input. Because it greatly simplifies the resulting algorithms, this filter is usually assumed to be a recursively computable IIR filter with a transfer function which has a constant numerator, as illustrated in Figure 6.12. To the extent that the model is valid, the power spectrum of $r(i,n)$, $P_A(k_x,\omega)$, is given by

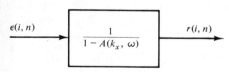

**Figure 6.12**  2-D random process $r(i,n)$ can be modeled as the output of an all-pole IIR filter excited by a white noise input sequence $\epsilon(i,n)$.

$$
P_A(k_x,\omega) = P_\epsilon(k_x,\omega)\left|\frac{1}{1-A(k_x,\omega)}\right|^2 \tag{6.164}
$$

where

$$
A(k_x,\omega) \triangleq \sum_{\substack{p=0 \\ (p,q)\neq(0,0)}}^{N-1}\sum_{q=0}^{M-1} a(p,q)\exp\left(jk_x p - j\omega q\right) \tag{6.165}
$$

and $P_\epsilon(k_x,\omega)$ is the power spectrum of the random process $\epsilon(i,n)$. If $\epsilon(i,n)$ is a white noise process, we have

$$
P_\epsilon(k_x,\omega) = \sigma_\epsilon^2 \tag{6.166}
$$

and

$$
P_A(k_x,\omega) = \frac{\sigma_\epsilon^2}{|1-A(k_x,\omega)|^2} \tag{6.167}
$$

Estimating $P_A(k_x, \omega)$ reduces to estimating the coefficients $\{a(p, q)\}$ from the autocorrelation coefficients of the random process, $\phi(l, m)$, which are assumed known for $-N < l < N, -M < m < M$. There are, however, some important differences in the properties of the 1-D and 2-D all-pole spectral estimates, caused by fundamental differences in the mathematics of functions of one and two variables. We shall examine these differences at the end of this subsection.

We shall begin by assuming that the coefficient array $a(p, q)$ is zero outside the region

$$0 \leq p < N; \qquad 0 \leq q < M \tag{6.168}$$

Then, $r(i, n)$ must satisfy the following:

$$r(i, n) = \sum_{\substack{p=0 \\ (p,q) \neq (0,0)}}^{N-1} \sum_{q=0}^{M-1} a(p, q)r(i - p, n - q) + \epsilon(i, n) \tag{6.169}$$

We can express this relation in terms of the known autocorrelation coefficients, $\phi(l, m)$, by multiplying both sides by $r^*(i - l, n - m)$ and taking the expected value. This gives us

$$\phi(l, m) = \sum_{\substack{p=0 \\ (p,q) \neq (0,0)}}^{N-1} \sum_{q=0}^{M-1} a(p, q)\phi(l - p, m - q) + E\{r^*(i - l, n - m)\epsilon(i, n)\} \tag{6.170}$$

The second term on the right may be expanded by writing $r^*(i - l, n - m)$ in terms of the impulse response $h(i, n)$ of the IIR filter, giving

$$r^*(i - l, n - m) = \sum_{p=-\infty}^{\infty} \sum_{q=-\infty}^{\infty} h^*(p, q)\epsilon^*(i - l - p, n - m - q) \tag{6.171}$$

from which we see that

$$E\{r^*(i - l, n - m)\epsilon(i, n)\} = \sum_{p=-\infty}^{\infty} \sum_{q=-\infty}^{\infty} h^*(p, q)E\{\epsilon(i, n)\epsilon^*(i - l - p, n - m - q)\} \tag{6.172}$$

Again, we can make use of the fact that $\epsilon(i, n)$ is a white noise process. This implies that

$$E\{\epsilon(i, n)\epsilon^*(i - l, n - m)\} = \sigma_\epsilon^2 \delta(l, m) \tag{6.173}$$

which allows (6.172) to be simplified to

$$\begin{aligned} E\{r^*(i - l, n - m)\epsilon(i, n)\} &= \sum_{p=-\infty}^{\infty} \sum_{q=-\infty}^{\infty} h^*(p, q)\sigma_\epsilon^2 \delta(l + p, m + q) \\ &= \sigma_\epsilon^2 h^*(-l, -m) \end{aligned} \tag{6.174}$$

and (6.170) to be written as

$$\phi(l, m) = \sum_{\substack{p=0 \\ (p,q) \neq (0,0)}}^{N-1} \sum_{q=0}^{M-1} a(p, q)\phi(l - p, m - q) + \sigma_\epsilon^2 h^*(-l, -m) \tag{6.175}$$

Since the filter coefficients $\{a(p, q)\}$ are zero outside the first quadrant, let us assume for the moment that $h(i, n)$ is also zero outside the first quadrant. If we restrict our attention to the case $0 \leq l < N; 0 \leq m < M$, the second term on the right side of

(6.175) will be zero except for $l = m = 0$. Since for this filter $h(0, 0) = 1$, we get finally

$$\phi(l, m) = \sum_{\substack{p=0 \\ (p,q) \neq (0,0)}}^{N-1} \sum_{q=0}^{M-1} a(p, q)\phi(l - p, m - q) + \sigma_\epsilon^2 \delta(l, m) \tag{6.176}$$

$$\text{for } 0 \leq l < N \text{ and } 0 \leq m < M$$

After the double sum is subtracted from both sides, this relation may be written in matrix form using the matrix $\boldsymbol{\Phi}$ defined in equation (6.159). This gives us

$$\boldsymbol{\Phi}\mathbf{a} = \mathbf{p} \tag{6.177}$$

where the vectors are given by

$$\mathbf{a} \triangleq [1, -a(1, 0), \ldots, -a(N - 1, 0), -a(0, 1), \ldots, -a(N - 1, M - 1)]' \tag{6.178}$$

and

$$\mathbf{p} \triangleq [\sigma_\epsilon^2, 0, 0, \ldots, 0]' \tag{6.179}$$

As before, $\boldsymbol{\Phi}$ is an $NM \times NM$ matrix, and $\mathbf{a}$ and $\mathbf{p}$ each contain $NM$ elements.

The solution to (6.177) is

$$\mathbf{a} = \boldsymbol{\Phi}^{-1}\mathbf{p} \tag{6.180}$$

where $\boldsymbol{\Phi}^{-1}$ is the matrix given earlier in (6.161). Because of the simple structure of $\mathbf{p}$, we can write the solution as

$$\mathbf{a} = [\sigma_\epsilon^2 \psi(0, 0; 0, 0), \sigma_\epsilon^2 \psi(1, 0; 0, 0), \ldots, \sigma_\epsilon^2 \psi(N - 1, 0; M - 1, 0)]'$$

Thus the filter coefficients are given by

$$a(p, q) = -\sigma_\epsilon^2 \psi(p, 0; q, 0) \tag{6.181}$$

where

$$\sigma_\epsilon^2 = \frac{1}{\psi(0, 0; 0, 0)} \tag{6.182}$$

Now that we have the filter coefficients, we can finally use equation (6.167) as the estimate of the power spectrum.

$$\hat{P}_A(k_x, \omega) \triangleq \frac{\sigma_\epsilon^2}{\left| 1 - \sum_{\substack{p=0 \\ (p,q) \neq (0,0)}}^{N-1} \sum_{q=0}^{M-1} a(p, q) \exp\left(jk_x p - j\omega q\right) \right|^2} \tag{6.183}$$

Equation (6.177) for the filter coefficient vector $\mathbf{a}$ also results from a linear prediction formulation. In this formulation, we seek to minimize the mean-squared error between the actual value of $r(i, n)$ and the predicted value $\hat{r}(i, n)$ obtained from a linear combination of previous values of $r(i, n)$. The predicted value is given by

$$\hat{r}(i, n) = \sum_{\substack{p=0 \\ (p,q) \neq (0,0)}}^{N-1} \sum_{q=0}^{M-1} a(p, q)r(i - p, n - q) \tag{6.184}$$

The predictor coefficients $a(p, q)$ are chosen to minimize the quantity

$$\sigma_\epsilon^2 = E\{| r(i, n) - \hat{r}(i, n) |^2\} \tag{6.185}$$

We can interpret the linear prediction formulation as trying to design an FIR filter to remove the predictable part of the signal. This filter attempts to whiten the spectrum. If $r(i, n)$ were truly generated by passing white noise through an all-pole filter, as in Figure 6.12, then it would be possible to recover the white noise by applying the inverse filter. This is exactly what the linear prediction formulation tries to do.

The all-pole spectral modeling approach is a reasonable one for estimating the power spectrum of a random process $r(i, n)$ which was created by passing white noise through an all-pole causal IIR filter. What happens if this method is applied to a set of autocorrelation measurements $\phi(l, m)$ that do not satisfy the model?

In the 1-D case, it is possible to generate a set of filter coefficients leading to a power spectrum estimate whose inverse Fourier transform equals the known autocorrelation samples even when the random process does not satisfy the all-pole model. (This statement is made assuming that the given autocorrelation points, when written in matrix notation, form a positive-definite matrix.) This property is called the *autocorrelation matching property*. The resulting filter coefficients can be used to extend the autocorrelation function while maintaining the positive-definite property. In the 1-D linear prediction formulation, there are $2N - 1$ known autocorrelation coefficients, only $N$ of which are independent, and $N$ parameters to be found. This leads to $N$ linear equations in $N$ unknowns, which can always be solved because the autocorrelation matrix is positive definite.

In the 2-D case, these statements are not true. Autocorrelation matching is not always possible. The number of known autocorrelation points is $(2M - 1) \times (2N - 1)$, only $2MN - M - N + 1$ of which are independent, but the number of parameters to be found is only $NM$. The set of $NM$ parameters does not contain enough degrees of freedom to match the $2MN - M - N + 1$ independent autocorrelation coefficients.

Furthermore, even though the matrix $\Phi$ is positive definite, a positive-definite extension may not exist [24,25]. This means that an arbitrary set of coefficients $\phi(l, m)$ which satisfy $\phi(l, m) = \phi^*(-l, -m)$ and lead to a positive-definite $\Phi$ matrix is *not*, in general, part of a valid autocorrelation function. (This is in marked contrast to the 1-D case.) According to [24–26], the lack of a positive-definite extension can be attributed to the inability of an arbitrary 2-D positive polynomial to be written as a sum of squares of polynomials. One-dimensional positive polynomials, on the other hand, can always be written in this form.

In the 1-D formulation of the linear prediction problem, the inverse filter has the minimum-phase property, implying that the corresponding all-pole filter in the modeling problem is always stable. (Again, we are assuming that the matrix of correlation measurements is positive-definite.) However, this is not always the case in two dimensions. It is possible to find examples where the 2-D all-pole filter

$$H_z(z_1, z_2) = \cfrac{1}{1 - \sum_{\substack{p=0 \\ (p,q) \neq (0,0)}}^{N-1} \sum_{q=0}^{M-1} a(p, q) z_1^{-p} z_2^{-q}} \tag{6.186}$$

with first-quadrant support is unstable even when $\Phi$ is positive definite [27].

Although this problem may not be serious for power spectral estimation, it is potentially destructive in signal modeling and synthesis applications.

At the beginning of this subsection we derived the all-pole spectral model for the case where the filter coefficients $a(p, q)$ had first quadrant support. Naturally, it is possible to formulate the all-pole model assuming other regions of support for the filter coefficients. For example, we could assume support in the third quadrant; that is,

$$a_3(p, q) \triangleq 0 \quad \text{unless } -N < p \leq 0, \quad -M < q \leq 0 \tag{6.187}$$

[The subscript "3" distinguishes these filter coefficients from the first-quadrant coefficients $a(p,q)$.] Because of the conjugate symmetry in the autocorrelation function

$$\phi(l, m) = \phi^*(-l, -m) \tag{6.188}$$

we can deduce that

$$a_3(p, q) = a^*(-p, -q) \tag{6.189}$$

In this case, the resulting power spectral estimate

$$\hat{P}_{A_3}(k_x, \omega) = \frac{\sigma_\epsilon^2}{\left| 1 - \sum_{\substack{p=-N+1 \\ (p,q) \neq (0,0)}}^{0} \sum_{q=-M+1}^{0} a_3(p, q) \exp\left(jk_x p - j\omega q\right) \right|^2} \tag{6.190}$$

will be the same as $\hat{P}_A(k_x, \omega)$ given by equation (6.183).

If we set up the problem assuming that the second quadrant is the region of support for the filter coefficients, we get a different answer. We can still set up a matrix equation similar to (6.177), but now the vector $\mathbf{a}_2$ reflects the second-quadrant support.

$$\mathbf{a}_2 = [-a_2(-N + 1, 0), \ldots, -a_2(-1, 0), 1,$$
$$-a_2(-N + 1, 1), \ldots, -a_2(0, M - 1)] \tag{6.191}$$

The vector $\mathbf{p}$ must also be rearranged so that its $N$th element is equal to $\sigma_\epsilon^2$ while the rest equal zero. Continuing, we can show that

$$\mathbf{a}_2 = \mathbf{\Phi}^{-1}\mathbf{p}_2 \tag{6.192}$$

or

$$a_2(p, q) = -\sigma_\epsilon^2 \psi(p + N - 1, N - 1; q, 0)$$
$$\text{for } -N < p \leq 0 \text{ and } 0 \leq q < M, \quad (p, q) \neq (0, 0) \tag{6.193}$$

where

$$\sigma_\epsilon^2 = \frac{1}{\psi(N - 1, N - 1; 0, 0)} \tag{6.194}$$

The resulting spectral estimate, $\hat{P}_{A_2}(k_x, \omega)$, is *not* equal to $\hat{P}_A(k_x, \omega)$ in general, as the following examples show.

Figure 6.13(a) shows the contour plot of a first-quadrant all-pole estimate of the power spectrum of two plane waves in white noise. (The crosses indicate the true wavenumber and frequency.) For this example, the autocorrelation coefficients $\phi(l, m)$ can be computed explicitly and used to obtain $\hat{P}_A(k_x, \omega)$. To generate these plots, we used $M = N = 3$. The all-pole estimate produced two peaks which are narrow

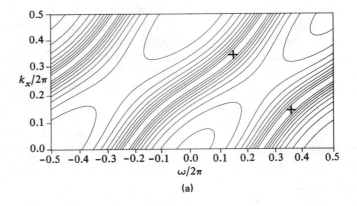

(a)

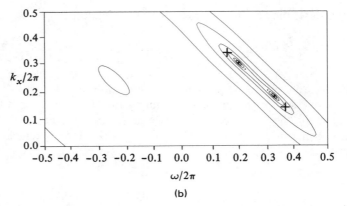

(b)

**Figure 6.13** (a) First-quadrant all-pole spectral estimate of two plane waves in white noise. The crosses "+" represent the locations of the true spectral peaks. The contours are equally spaced in decibels. (b) Second-quadrant all-pole spectral estimate.

along the line connecting the true spectral maxima but very broad perpendicular to this line. In fact, for this example, the peaks are so broad along 45° lines that they are virtually ridges. The periodicity of the wavenumber–frequency spectrum introduces a third ridge in the figure.

Figure 6.13(b) shows the corresponding plot for the second-quadrant all-pole estimate $\hat{P}_{A_2}(k_x, \omega)$. It is significantly different than the first-quadrant estimate. Two spectral peaks are resolved, but the resolution along lines at 45° is much better than the resolution along the line connecting the true peaks. However, the locations of the peaks in the spectral estimate are not correct. They appear to be "pulled in" somewhat along the line connecting the true spectral peaks.

It is possible to combine $\hat{P}_A(k_x, \omega)$ and $\hat{P}_{A_2}(k_x, \omega)$ to reduce the dependence of resolution on direction. Jackson and Chien [28] have proposed that the inverse spectral estimates be averaged to produce a new inverse spectral estimate. Thus,

$$\frac{1}{\hat{P}_J(k_x, \omega)} \triangleq \frac{1}{2}\left[\frac{1}{\hat{P}_A(k_x, \omega)} + \frac{1}{\hat{P}_{A_2}(k_x, \omega)}\right] \tag{6.195}$$

Figure 6.14 shows $\hat{P}_J$ for the same example of two plane waves in noise when $M = N = 3$. The resolution of the spectral estimate, at least in this simple example, is more nearly uniform and the peak locations are correct.

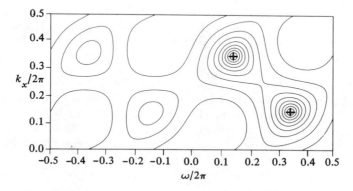

**Figure 6.14**   Combined all-pole spectral estimate of Jackson and Chien.

It is also possible to use other regions of support, such as wedges, to formulate the all-pole model. The basic constraint seems to be that the resulting all-pole filter be recursively computable so that the term $E\{r^*(i - l, n - m)\epsilon(i, n)\}$ in equation (6.170) can be reduced to $\sigma_\epsilon^2 \delta(l, m)$. Pendrell [29] has examined the all-pole modeling approach to spectral estimation for filter masks of various shapes.

### *6.5.4 Maximum Entropy Spectral Estimation

The technique of 1-D maximum entropy spectral estimation has been studied by many researchers. Its success and popularity in applications such as speech analysis [30] and geophysical signal processing [22] have led to its extension to two dimensions [31–36]. Again, however, there are some fundamental differences in the 1-D and 2-D solutions to this problem.

The concept of maximum entropy spectral estimation is relatively simple. We assume, as before, that we know the values of the autocorrelation coefficients $\phi(l, m)$ in the region $-N < l < N$, $-M < m < M$. We seek the spectral estimate whose inverse Fourier transform matches the known autocorrelation coefficients and, at the same time, assumes as little as possible about the structure of the unknown autocorrelation coefficients. Consequently, we want to maximize the randomness or entropy of the spectral estimate subject to the constraint of matching the known autocorrelation coefficients. Following [31, 32, 34, 35], we will define the entropy, $H$, of a random process with a power spectrum $P(k_x, \omega)$ as

$$H \triangleq \frac{1}{(2\pi)^2} \int_{-\pi}^{\pi} \int_{-\pi}^{\pi} \log P(k_x, \omega) \, dk_x \, d\omega \tag{6.196}$$

[Often the scaling constant $1/(2\pi)^2$ is omitted; we retain it for consistency.]

In terms of optimization theory [37], we can state this constrained minimization as a primal problem: Choose $P(k_x, \omega)$ so as to minimize $H$ subject to the constraint that

$$\phi(l, m) = \frac{1}{4\pi^2} \int_{-\pi}^{\pi} \int_{-\pi}^{\pi} P(k_x, \omega) \exp\left(-jk_x l + j\omega m\right) dk_x \, d\omega \tag{6.197}$$
$$\text{for } |l| < N, |m| < M$$

This problem can be reformulated by writing $P(k_x, \omega)$ as a Fourier expansion and then attempting to minimize $H$ by differentiating it with respect to the unconstrained coefficients in the Fourier expansion. We begin by expressing the power spectrum $P(k_x, \omega)$ as the sum of two terms, one over the known autocorrelation coefficients and the other over the unknown autocorrelation coefficients.

$$P(k_x, \omega) = \sum_{l=-N+1}^{N-1} \sum_{m=-M+1}^{M-1} \phi(l, m) \exp\left(jk_x l - j\omega m\right)$$
$$+ \sum_{p} \sum_{q} \phi(p, q) \exp\left(jk_x p - j\omega q\right) \tag{6.198}$$
$$\substack{|p| \geq N \\ \text{or } |q| \geq M}$$

By adjusting the values of $\phi(p, q)$ for which $|p| \geq N$ or $|q| \geq M$, we can maximize the entropy $H$. If we differentiate $H$ with respect to these $\phi(p, q)$ and set the resulting partial derivatives equal to zero, we see that $H$ is maximized when [32]

$$\int_{-\pi}^{\pi} \int_{-\pi}^{\pi} P^{-1}(k_x, \omega) \exp\left(-jk_x p + j\omega q\right) dk_x \, d\omega = 0 \tag{6.199}$$
$$\text{for } |p| \geq N, \quad |q| \geq M$$

This says that the inverse Fourier transform of $1/P(k_x, \omega)$ must be a sequence of finite extent. This, in turn, implies that the maximum entropy estimate $\hat{P}_{\mathrm{ME}}(k_x, \omega)$ must have the following form:

$$\hat{P}_{\mathrm{ME}}(k_x, \omega) = \frac{1}{\displaystyle\sum_{l=-N+1}^{N-1} \sum_{m=-M+1}^{M-1} \lambda(l, m) \exp\left(jk_x l - j\omega m\right)} \tag{6.200}$$
$$\triangleq \frac{1}{Q(k_x, \omega)}$$

where $\lambda(l, m) = \lambda^*(-l, -m)$. The parameters $\{\lambda(l, m)\}$ must be chosen so that the known autocorrelation coefficients are matched; that is,

$$\hat{\phi}(l, m) = \phi(l, m) \qquad \text{for } |l| < N, \quad |m| < M$$

where

$$\hat{\phi}(l, m) = \frac{1}{(2\pi)^2} \int_{-\pi}^{\pi} \int_{-\pi}^{\pi} \hat{P}_{\mathrm{ME}}(k_x, \omega) \exp\left(-jk_x l + j\omega m\right) dk_x \, d\omega \qquad (6.201)$$

The form of the maximum entropy spectral estimate is independent of dimensionality [32]; the 1-D estimate and the $M$-D estimate both consist of the reciprocal of a truncated Fourier expansion, analogous to (6.200). The determination of the $\lambda$ parameters, however, is much more difficult in two dimensions than it is in one dimension.

This same formulation of the maximum entropy problem may be obtained by using the method of Lagrange multipliers to turn the constrained optimization into an unconstrained optimization [35]. The Lagrange multipliers turn out to be the same as the parameters $\{\lambda(l, m)\}$ given in equation (6.200). Consequently, any approach to solving the maximum entropy problem by determining $\{\lambda(l, m)\}$ is actually solving the dual optimization problem [37].

In one dimension, the maximum entropy spectral estimate is identical to the all-pole spectral estimate [30]. Because of the ability to factor 1-D polynomials, the maximum entropy spectral estimate can always be written as the product of a minimum-phase all-pole frequency response and its complex conjugate. This form is identical to that of the 1-D all-pole spectral model. The computationally efficient algorithm for computing the 1-D all-pole spectral model also yields the maximum entropy spectral estimate with its autocorrelation matching property.

In two dimensions, the maximum entropy spectral estimate and the all-pole spectral estimate are not the same. As discussed in the preceding section, the all-pole spectral model does not possess enough degrees of freedom to match the known autocorrelation coefficients. Since this matching is a requirement for the maximum entropy spectral estimate, it is clear that the all-pole spectral estimate does not qualify as the maximum entropy solution.

Woods [33] has proven that the maximum entropy spectral estimate exists and is unique provided that the known autocorrelation coefficients $\phi(l, m)$ are actually part of a valid autocorrelation function. In one dimension, if the matrix $\boldsymbol{\Phi}$ constructed from the known autocorrelation coefficients is positive definite, a valid autocorrelation function exists which contains the known autocorrelation coefficients. In two dimensions, however, the positive definiteness of $\boldsymbol{\Phi}$ is not sufficient to guarantee the existence of a valid autocorrelation function containing the known autocorrelation coefficients [24,25,26], as discussed in the preceding subsection. Consequently, in the existence and uniqueness proofs [33] mentioned above, it is necessary to assume that the known autocorrelation coefficients $\phi(l, m)$ are, in fact, part of a valid 2-D autocorrelation function.

This result can be somewhat disconcerting for practical applications where the autocorrelation coefficients are measured or estimated from measurements of the random process. Because of measurement or estimation errors, the set of numbers taken as the "known" autocorrelation coefficients may not correspond to any valid

2-D autocorrelation function, meaning that a maximum entropy spectral estimate matching these numbers may not exist.

One conceptually simple approach to finding the maximum entropy spectral estimate is to set up the equation relating the entropy $H$ to $P(k_x, \omega)$ and then to apply a nonlinear optimization algorithm to maximize $H$ while simultaneously minimizing the mismatch between the known autocorrelation coefficients and the ones corresponding to the spectral estimate $\hat{P}_{ME}(k_x, \omega)$ [31]. [In practice, $P(k_x, \omega)$ is sampled on a dense frequency grid and the sample values are varied to minimize a discrete approximation to $H$. Consequently, the solution yields samples of $\hat{P}_{ME}(k_x, \omega)$.] This approach has the advantage that errors in measuring or estimating the "known" autocorrelation coefficients can be taken into account, since an exact match is not required. A disadvantage of this approach is the amount of computation required to perform the nonlinear optimization, especially when the number of frequency samples and the number of autocorrelation coefficients to be matched are large.

The maximum entropy problem may be formulated in the dual space as a minimization problem [35]. For example, Burg [38] proposed the following optimization problem to find the 2-D maximum entropy spectral estimate:

$$\min_{\lambda} \left\{ H(\lambda) + \sum_{l=-N+1}^{N-1} \sum_{m=-M+1}^{M-1} \lambda(l, m)\phi^*(l, m) \right\} \tag{6.202}$$

The vector $\lambda$ is simply a notational convenience to represent the parameters $\{\lambda(l, m)\}$ and $H(\lambda)$ reminds us that the entropy functional $H$ depends on these parameters through the relation

$$H(\lambda) = \frac{-1}{4\pi^2} \int_{-\pi}^{\pi} \int_{-\pi}^{\pi} \ell n \, [Q(k_x, \omega)] \, dk_x \, d\omega \tag{6.203}$$

where $Q(k_x, \omega)$ is given by equation (6.200). Because of duality, the problem of maximizing $H(\lambda)$ subject to the correlation matching constraints is turned into a minimization problem. If we differentiate the expression in brackets in (6.202) with respect to $\{\lambda(l, m)\}$ and set the resulting partial derivatives equal to zero, we get the equation

$$\phi(l, m) = \frac{1}{4\pi^2} \int_{-\pi}^{\pi} \int_{-\pi}^{\pi} \frac{\exp(-jk_x l + j\omega m) \, dk_x \, d\omega}{Q(k_x, \omega)}$$
$$= \hat{\phi}(l, m) \quad \text{for } |l| < N, \; |m| < M \tag{6.204}$$

which of course is the correlation matching constraint. Optimization algorithms such as Newton's method may be applied to the minimization problem (6.202) to obtain a spectral estimate that maximizes entropy and satisfies the correlation matching constraint.

The maximum entropy spectral estimation problem may also be cast in the form of a constrained minimization problem [36]. Let us define the vector $\phi$ comprising the known autocorrelation values $\{\phi(l, m)\}$ and form the inner product

$$\langle \phi, \lambda \rangle \triangleq \sum_{l=-N+1}^{N-1} \sum_{m=-M+1}^{M-1} \phi^*(l, m)\lambda(l, m) \tag{6.205}$$

When the parameters $\{\lambda(l, m)\}$ correspond to the desired maximum entropy spectral estimate, equation (6.204) will hold. Then the inner product $\langle \phi, \lambda \rangle$ becomes

$$
\langle \phi, \lambda \rangle = \sum_{l=-N+1}^{N-1} \sum_{m=-M+1}^{M-1} \phi^*(l, m)\lambda(l, m)
$$

$$
= \sum_{l=-N+1}^{N-1} \sum_{m=-M+1}^{M-1} \left[ \frac{1}{4\pi^2} \int_{-\pi}^{\pi} \int_{-\pi}^{\pi} \frac{\exp(jk_x l - j\omega m)\, dk_x\, d\omega}{Q(k_x, \omega)} \right] \lambda(l, m) \qquad (6.206)
$$

$$
= \frac{1}{4\pi^2} \int_{-\pi}^{\pi} \int_{-\pi}^{\pi} \frac{\displaystyle\sum_{l=-N+1}^{N-1} \sum_{m=-M+1}^{M-1} \lambda(l, m) \exp(jk_x l - j\omega m)}{Q(k_x, \omega)} dk_x\, d\omega
$$

Since the numerator of the integrand is simply the definition of $Q(k_x, \omega)$ implied by equation (6.200), the integrand becomes unity and we see that

$$
\langle \phi, \lambda \rangle = 1 \qquad (6.207)
$$

holds for the maximum entropy spectral estimate. This equation can be interpreted as a constraint that the parameters $\{\lambda(l, m)\}$ must obey. The objective then is to *minimize* $H(\lambda)$ subject to the linear constraint $\langle \phi, \lambda \rangle = 1$. Again, because we are solving the dual problem, minimizing $H(\lambda)$ will give us the maximum entropy spectral estimate.

This constrained minimization problem can be reformulated as an unconstrained minimization by using the linear constraint (6.207) to eliminate the parameter $\lambda(0, 0)$ [35,36]. Using the definition of the inner product, we can solve (6.207) for $\lambda(0, 0)$ to obtain

$$
\lambda(0, 0) = \frac{1}{\phi(0, 0)} \left[ 1 - \sum_{\substack{l=-N+1 \\ (l,m) \neq (0,0)}}^{N-1} \sum_{m=-M+1}^{M-1} \phi^*(l, m)\lambda(l, m) \right] \qquad (6.208)
$$

The minimization problem is then solved for the parameters $\{\lambda(l, m): |l| < N, |m| < M, (l, m) \neq (0, 0)\}$.

The computation of the maximum entropy spectral estimate may also be formulated as an iterative procedure [34]. The form of $\hat{P}_{ME}(k_x, \omega)$ given by equation (6.200) requires that

$$
\lambda(l, m) = 0 \qquad \text{for } |l| \geq N \text{ or } |m| \geq M \qquad (6.209)
$$

Similarly, the autocorrelation matching property (6.201) requires that

$$
\hat{\phi}(l, m) = \phi(l, m) \qquad \text{for } |l| < N \text{ and } |m| < M \qquad (6.210)
$$

These two constraints can be applied alternately until convergence is obtained.

**Example 1**

In Figure 6.15, we show a comparison of the results of several of the spectral estimation techniques discussed in this chapter. As input, we used 25 correlation coefficients of the form

$$
\phi(l, m) = \delta(l, m) + \cos(2\pi\alpha_1 l + 2\pi\beta_1 m) + \cos(2\pi\alpha_2 l + 2\pi\beta_2 m)
$$

$$
\text{for } -2 \leq l \leq 2, \quad -2 \leq m \leq 2
$$

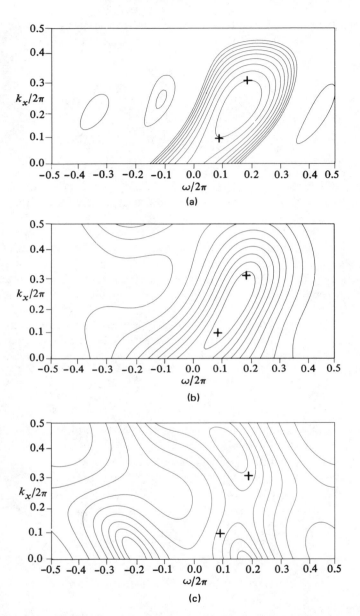

**Figure 6.15** (a) Periodogram spectral estimate. The crosses "+" represent the locations of the true spectral peaks. The contours are equally spaced in decibels. (b) High-resolution spectral estimate. (c) First-quadrant all-pole spectral estimate. (d) Second-quadrant all-pole spectral estimate. (e) Combined all-pole spectral estimate. (f) Maximum entropy spectral estimate computed by the iterative algorithm. (Part (f) courtesy of Jae S. Lim and Naveed A. Malik, *IEEE Trans. Acoustics, Speech, and Signal Processing*, © 1981 IEEE.)

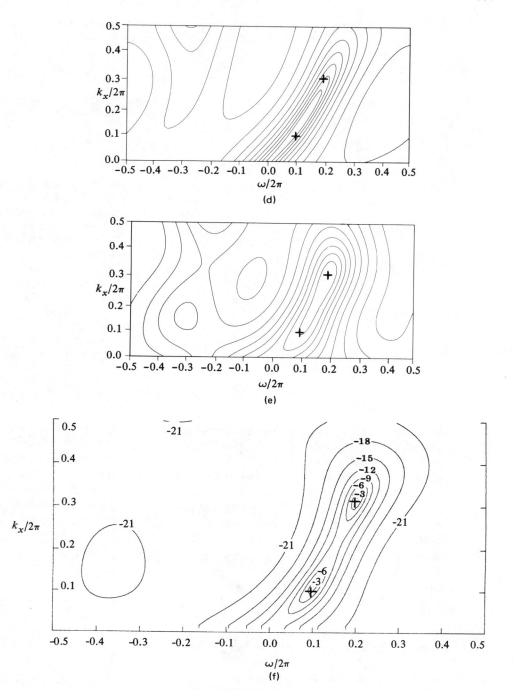

**Figure 6.15**   (*Continued*)

This correlation function corresponds to two plane waves in white noise. (Because of symmetry, only 13 of the 25 coefficients used are independent.) For this example,

$$(\alpha_1, \beta_1) = (0.1, 0.1)$$

$$(\alpha_2, \beta_2) = (0.2, 0.3125)$$

Figure 6.15(a) shows the result of the periodogram estimate of the power spectrum. Because so few correlation samples were used, the periodogram fails to resolve the two peaks. (The true peak locations are indicated by a "$+$.") Figure 6.15(b) shows the high-resolution spectral estimate for this example. Although the two peaks are not resolved by this method, a ridge is formed along the line connecting the two peaks.

In Figure 6.15(c), we show the result for a first-quadrant all-pole spectral estimate. The coefficient array $a(p, q)$ was nonzero for $0 \leq p \leq 2$ and $0 \leq q \leq 2$. The spectral estimate exhibits two clear peaks, but they are not close to the locations of the true peaks. The second-quadrant all-pole spectral estimate is shown in Figure 6.15(d). It also fails to resolve the peaks but it does exhibit a narrow ridge in the general direction of the line connecting the true peak locations. Figure 6.15(e) shows the result of combining the first-quadrant and second-quadrant all-pole estimates according to equation (6.195). The combined estimate also fails to resolve the two peaks, but one can sense the improvement in using a combined estimate compared to either of the single-quadrant estimates.

Finally, in Figure 6.15(f), we see the maximum entropy power spectral estimate. The two peaks are clearly resolved in this case. For this example, 43 iterations were required for convergence and $(64 \times 64)$-point FFTs were used in the computation [34].

## *6.5.5 Extendibility

In Section 6.5.4 we discussed the problem of determining whether or not a set of values $\{\phi(l, m)\}$ corresponds to samples of a valid 2-D autocorrelation sequence. This problem has been called the extendibility problem, and in one dimension it can be approached by setting up a Toeplitz matrix of the known correlation values and checking for positive definiteness. Unfortunately, this approach is not applicable in two dimensions. The problem of extending a finite set of multidimensional autocorrelation samples has been attacked in a more general framework by Lang and McClellan [36]. Here we briefly outline their approach for the 2-D case where the samples $\{\phi(l, m)\}$ are given on the rectangle $|l| < N, |m| < M$.

A set of samples $\{\phi(l, m)\}$ is extendible to a full 2-D autocorrelation function if the samples can be written in the form:

$$\phi(l, m) = \frac{1}{4\pi^2} \int_{-\pi}^{\pi} \int_{-\pi}^{\pi} \Phi(k_x, \omega) \exp(-jk_x l + j\omega m) \, dk_x \, d\omega \qquad (6.211)$$

where $\Phi(k_x, \omega)$ is a nonnegative real function.

Although we shall not go into the details here, the set $E$ of extendible correlation samples can be characterized by an inner product formulation. Let us introduce the vector $\mathbf{p} = \{p(l, m): |l| < N, |m| < M\}$, where $p(-l, -m) = p^*(l, m)$. For each vector

**p** we can define a real-valued trigonometric polynomial

$$P(k_x, \omega) = \sum_{l=-N+1}^{N-1} \sum_{m=-M+1}^{M-1} p(l, m) \exp\left(jk_x l - j\omega m\right) \tag{6.212}$$

We shall use the somewhat misleading but concise terminology that **p** is a "positive" vector if and only if $P(k_x, \omega)$ is nonnegative [36]. Then it can be shown that the vector $\boldsymbol{\phi} = \{\phi(l, m)\}$ will be extendible if and only if

$$\langle \boldsymbol{\phi}, \mathbf{p} \rangle \triangleq \sum_{l=-N+1}^{N-1} \sum_{m=-M+1}^{M-1} \phi^*(l, m) p(l, m) \geq 0 \tag{6.213}$$

for all "positive" vectors **p** [36].

A test for extendibility has been derived by Lang and McClellan [36] from this inner product characterization of the set of extendible functions. Basically, the test consists of solving the following constrained optimization problem. Let

$$\alpha = \min_{\mathbf{p}} \langle \boldsymbol{\phi}, \mathbf{p} \rangle \tag{6.214}$$

where **p** is a "positive" vector that satisfies the linear constraint

$$\langle \boldsymbol{\psi}, \mathbf{p} \rangle = 1 \tag{6.215}$$

The vector $\boldsymbol{\psi}$ consists of a set of correlation values known to be extendible. In particular, we can let $\boldsymbol{\psi}$ be the "white noise" vector

$$\psi(l, m) = \delta(l)\delta(m) \tag{6.216}$$

so that (6.215) becomes simply $p(0, 0) = 1$. If $\alpha \geq 0$, the correlation samples $\{\phi(l, m)\}$ are extendible.


## PROBLEMS

**6.1.** Suppose that we are given the signal

$$s(x, t) = \cos\left(k_x x - \omega t\right)$$

  **(a)** Sketch $s(x, t)$ along the $x$-axis for $t = 0$.
  **(b)** What is the distance $\lambda_x$ along the $x$-axis between successive positive peaks of $s(x, t)$?
  **(c)** Sketch $s(x, t)$ along the $x$-axis for $t = \Delta t$. Let $\Delta x$ represent the amount that this sketch must be shifted so that it coincides with the sketch in part (a). What is the value of $c_x \triangleq \Delta x / \Delta t$ in terms of $k_x$ and $\omega$? In terms of $\lambda_x$ and $\omega$?

**6.2.** Let $s(\mathbf{x}, t)$ be the propagating signal

$$s(\mathbf{x}, t) = \cos\left(\mathbf{k}'\mathbf{x} - \omega t\right)$$

$$= \cos\left(k_x x + k_y y + k_z z - \omega t\right)$$

  **(a)** For a fixed value of $t$, say $t = 0$, what is the value of the vector $\Delta \mathbf{x}$ with the smallest positive magnitude such that $s(\mathbf{x}, t) = s(\mathbf{x} + \Delta \mathbf{x}, t)$? The magnitude of $\Delta \mathbf{x}$ is known as the *wavelength* $\lambda$.
  **(b)** Find an expression for $\lambda$ in terms of $\lambda_x$, $\lambda_y$, and $\lambda_z$. $\lambda_x$ was defined in Problem 6.1(b); $\lambda_y$ and $\lambda_z$ are defined similarly.

(c) The speed of propagation $c$ may be defined as the wavelength $\lambda$ divided by the time required for one period of the wave to propagate past a fixed position, say $\mathbf{x} = \mathbf{0}$. What is the minimum positive value $\Delta t$ for which $s(\mathbf{x}, t) = s(\mathbf{x}, t + \Delta t)$? What is $c \triangleq \lambda/\Delta t$ in terms of $\lambda$ and $\omega$? In terms of $|\mathbf{k}|$ and $\omega$?

(d) The value of $c_x$ that was derived in Problem 6.1(c) is the phase velocity in the $x$ direction. Derive an expression for the speed of propagation $c$ in terms of $c_x$, $c_y$, and $c_z$.

**6.3.** Suppose that a plane wave is passed through a 4-D linear shift-invariant filter with an impulse response $h(\mathbf{x}, t)$. Show that the output is a plane wave propagating in the same direction at the same frequency.

**6.4.** Consider the sensor arrays shown in Figure P6.4. All sensor weights are unity.

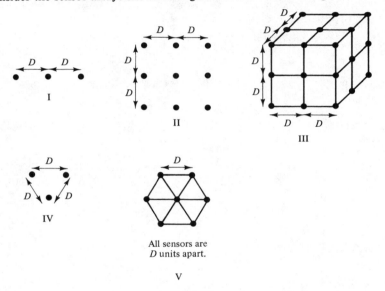

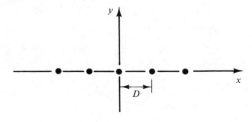

All sensors are
$D$ units apart.

V

**Figure P6.4**

(a) For each array, find the array pattern. (Judicious choice of the origin for the spatial coordinate system will simplify your expressions.)

(b) If each array is to be steered in the direction $\boldsymbol{\alpha}_0 = (\alpha_{0x}, \alpha_{0y}, \alpha_{0z})$, determine the delays for each element.

**6.5.** Consider the sensor array shown in Figure P6.5. There are five sensors along the $x$-axis, separated by a distance $D$.

**Figure P6.5**

(a) Derive the array pattern for this array assuming unity sensor weighting.

(b) Can the array distinguish between two propagating signals whose directions of propagation are given by the unit vectors $\mathbf{u} = (\cos\theta, \sin\theta, 0)$ and $\mathbf{v} = (\cos\theta, -\sin\theta, 0)$? Explain.

(c) Now suppose that a sixth sensor is added at the position $(x, y, z) = (0, -D, 0)$. What is the new array pattern?

(d) If the six-sensor array is steered to pass elemental signals of the form $\exp(j\omega t - j\mathbf{k}'\mathbf{x})$ with the wavenumber vector $(0, \omega/c, 0)'$, what is the response to the signal $\exp(j\omega t - jk_y y)$? To the signal $\exp(j\omega t + jk_y y)$?

**6.6.** There are often several ways to interperet how sensors are spaced to form an array. Consider the array pictured in Figure P6.6. Sensors are placed at $x = \pm 3D, \pm 4D, \pm 5D, \pm 6D$, and $\pm 7D$.

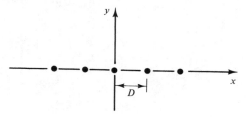

**Figure P6.6**

(a) Compute the array pattern for this array assuming that each sensor has unity weighting.

(b) The array pictured in Figure P6.6 is one example of a "sparse array," a regular array that has some sensors missing. Consequently, we can consider the array to be a regular weighted array with $w(i) = 0$ for $|i| \leq 2$ and $w(i) = 1$ for $3 \leq |i| \leq 7$. Alternatively, we can consider the array pattern to be the difference of two array patterns, one with $w(i) = 1$ only for $|i| \leq 7$ and the other with $w(i) = 1$ only for $|i| \leq 2$. Derive the overall array pattern using this alternative interpretation.

(c) We may also interpret the sparse array as an array of two subarrays of five sensors each. Write down the array pattern of one of the subarrays. What would the overall array pattern be if the subarrays were each replaced by a single omnidirectional sensor? What is the overall array pattern with the subarrays in place? Sketch the overall array pattern.

**6.7.** Assume that we have a planar area of sensors that lie on a regular rectangular grid; for example, see Figure P6.7. What conditions must the sensor weights obey for the weighted delay-and-sum beamformer to have a separable array pattern?

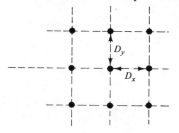

**Figure P6.7**

**6.8.** Suppose that we are given an array of $N$ sensors located at $\mathbf{x}_i = (x_i, y_i, z_i)'$ for $i = 0,$ $\ldots, N - 1$. The array pattern is given by

$$W(\mathbf{k}) = \frac{1}{N} \sum_{i=0}^{N-1} w_i \exp(-j\mathbf{k}'\mathbf{x}_i)$$

(a) Now suppose that the array has been displaced; that is, the location of the $i$th sensor is now $(x_i + d_x, y_i + d_y, z_i + d_z)$. What is the new array pattern? How is it related to the formula given above?

(b) Now suppose that the array size is expanded by a factor $D$ so that the sensor locations become $\{(Dx_i, Dy_i, Dz_i)\}$. What is the new array pattern in terms of $W(\mathbf{k})$?

(c) If the expansion factor is different in each dimension [e.g., sensor locations at $(D_x x_i, D_y y_i, D_z z_i)$], what is the new array pattern?

**6.9.** Show that the effective wavenumber–frequency response $H(\mathbf{k}, \omega)$ for a filter-and-sum beamformer is given by

$$H(\mathbf{k}, \omega) = \frac{1}{N} \sum_{i=0}^{N-1} W_i(\omega) \exp[-j(\mathbf{k} - \omega\boldsymbol{\alpha}_0)'\mathbf{x}_i]$$

**6.10.** In Section 6.2.6 the Fourier transform of the filter-and-sum beamformer was derived. The component of the beamformer output at frequency $\omega$ is given by

$$FS(\omega)e^{j\omega t} = \frac{1}{N} \sum_{i=0}^{N-1} W_i(\omega)R_i(\omega) \exp[j\omega(t - \tau_i)]$$

We can derive a frequency-domain version of the filter-and-sum beamformer if we approximate $R_i(\omega)$ in the equation above by the short-time Fourier transform $R_i(t, \omega)$.

$$fdfs(t, \omega) \triangleq \frac{1}{N} \sum_{i=0}^{N-1} W_i(\omega)R_i(t, \omega) \exp[j\omega(t - \tau_i)]$$

where

$$R_i(t, \omega) = \int_{-\infty}^{\infty} v(t - \tau)r_i(\tau) \exp(-j\omega\tau) \, d\tau$$

(a) Write the discrete-time versions of the equations for $R_i(t, \omega)$ and $fdfs(t, \omega)$.

(b) Assuming that the steering delays have the form

$$\tau_i = -\frac{Mq}{Nl}i$$

derive the formula for $fdfs(n, 2\pi l/M)$ in terms of $W_i(2\pi l/M)$, $v(n)$, and $r_i(n)$.

(c) Explain how $fdfs(n, 2\pi l/M)$ may be efficiently implemented with 1-D FFTs. How is this implementation different from the one for $fd(n, 2\pi l/M)$ derived in Section 6.3.3?

**6.11.** Show that the output of a frequency-domain beamformer

$$fd(t, \omega) \triangleq \frac{1}{N} \sum_{i=0}^{N-1} w_i R_i(t, \omega) \exp[j\omega(t - \tau_i)]$$

can be represented in terms of the wavenumber–frequency spectrum as

$$fd(t, \omega) = \frac{1}{(2\pi)^4} \int_{-\infty}^{\infty} \int_{-\infty}^{\infty} W(\mathbf{k} - \omega\boldsymbol{\alpha}_0)V(\theta - \omega)S(\mathbf{k}, \theta) \exp(j\omega t) \, d\mathbf{k} \, d\theta$$

**6.12.** Consider the $N \times N$ two-dimensional discrete-time beamformer array which is sketched in Figure P6.12. All the receivers have a weight of unity. The beam is steered

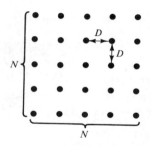

**Figure P6.12**

so that $\alpha_{0x} = \alpha_{0y} = T/(2D)$, where $T$ is the sampling interval. What is the array pattern of this array?

**6.13.** (a) Sketch a block diagram of a discrete-time beamformer which uses pre-beamforming interpolation.

(b) Sketch a similar diagram of a discrete-time beamformer which uses post-beamforming interpolation.

(c) Assume that the up-sampling and down-sampling rates are both $I$ and that the impulse responses of the filters used in conjunction with the interpolation are $M$ samples long. If there are $N$ receivers in the array, how will the computational and storage requirements of these two beamformers compare?

**6.14.** Suppose that we have the following samples of a 2-D autocorrelation function:

$$\phi(0, 0) = 1$$
$$\phi(1, 0) = \phi(-1, 0) = a$$
$$\phi(0, 1) = \phi(0, -1) = b$$
$$\phi(1, 1) = \phi(1, -1) = \phi(-1, 1) = \phi(-1, -1) = 0$$

where $a$ and $b$ are real-valued parameters. Compute the periodogram estimate of the power spectrum $\Phi(k_x, \omega)$.

**6.15.** In this problem we want to get a high-resolution spectral estimate from the same autocorrelation values which were used in Problem 6.14. In particular, let

$$\phi(0, 0) = 1$$
$$\phi(1, 0) = \phi(-1, 0) = a$$
$$\phi(0, 1) = \phi(0, -1) = b$$
$$\phi(1, 1) = \phi(-1, 1) = \phi(1, -1) = \phi(-1, 1) = 0$$

where $a$ and $b$ are real-valued parameters.

(a) Form the matrix $\mathbf{\Phi}$ defined by equation (6.159).

(b) Compute the inverse matrix $\mathbf{\psi} = \mathbf{\Phi}^{-1}$. It has the form

$$\mathbf{\psi} = \begin{bmatrix} f & g & h & i \\ g & f & i & h \\ h & i & f & g \\ i & h & g & f \end{bmatrix}$$

where $f$, $g$, $h$, and $i$ are real-valued variables.

(c) Form the high-resolution spectral estimate

$$\hat{P}_C(k_x, \omega) = \frac{1}{\sum_p \sum_q \gamma(p, q) \exp(jk_x p - j\omega q)}$$

where $\gamma(p, q)$ is given by equation (6.157).

**6.16.** In this problem we want to get an all-pole spectral estimate from the same autocorrelation values that were used in Problems 6.14 and 6.15. Let

$$\phi(0, 0) = 1$$
$$\phi(1, 0) = \phi(-1, 0) = a$$
$$\phi(0, 1) = \phi(0, -1) = b$$
$$\phi(1, 1) = \phi(-1, 1) = \phi(1, -1) = \phi(-1, -1) = 0$$

where $a$ and $b$ are real-valued parameters.

(a) Compute the first-quadrant all-pole spectral estimate $\hat{P}_A(k_x, \omega)$.
(b) Compute the second-quadrant all-pole spectral estimate $\hat{P}_{A_2}(k_x, \omega)$. How does it differ from $\hat{P}_A(k_x, \omega)$?
(c) Compute $\hat{P}_J(k_x, \omega)$ as defined by equation (6.195).
  (*Hint:* Use the matrices $\boldsymbol{\Phi}$ and $\boldsymbol{\Psi}$ derived in Problem 6.15.)

**\*6.17.** In this problem we consider the question of extendibility. Suppose that we have the following samples of a 2-D function:

$$\phi(0, 0) = 1$$
$$\phi(1, 0) = \phi(-1, 0) = \phi(0, 1) = \phi(0, -1) = a$$
$$\phi(1, 1) = \phi(-1, -1) = \phi(1, -1) = \phi(-1, 1) = 0$$

where $a$ is a real-valued parameter. We shall address the question: What values of $a$ are consistent with the existence of a true 2-D autocorrelation function?

(a) Set up the minimization problem

$$\alpha = \min_{\mathbf{p}} \langle \boldsymbol{\phi}, \mathbf{p} \rangle$$

under the constraint $p(0, 0) = 1$, where $\mathbf{p}$ is a "positive" vector. For simplicity, assume that the coefficients $\{p(l, m)\}$ are real and symmetric. [Note that because of the extent of the known values of $\phi$ and the assumed symmetry, we need only consider the coefficients $p(1, 0)$, $p(0, 1)$, $p(1, 1)$, and $p(1, -1)$.]

(b) The positive polynomial $P(\omega_1, \omega_2)$ has the form

$$P(\omega_1, \omega_2) = 1 + 2p(0, 1) \cos \omega_2 + 2p(1, 0) \cos \omega_1 + 2p(1, 1) \cos(\omega_1 + \omega_2)$$
$$+ 2p(1, -1) \cos(\omega_1 - \omega_2) \geq 0$$

Verify that $P(\omega_1, \omega_2) \geq 0$ for the two cases
(1) $p(0, 1) = p(1, 0) = \frac{1}{2}, p(1, 1) = p(1, -1) = \frac{1}{4}$
(2) $p(0, 1) = p(1, 0) = -\frac{1}{2}, p(1, 1) = p(1, -1) = \frac{1}{4}$
[*Hint:* Is $p(l, m)$ separable?]

(c) What conditions on $\{p(1, 0), p(0, 1), p(1, 1), p(1, -1)\}$ are necessary to ensure that $P(\pi, \pi/2) \geq 0$? To ensure that $P(\pi/2, \pi) \geq 0$, $P(0, \pi/2) \geq 0$, and $P(\pi/2, 0) \geq 0$?

(d) Considering parts (b) and (c), you should be able to perform the minimization that you set up in part (a), keeping in mind that the coefficients $\{p(l, m)\}$ must correspond to a nonnegative $P(\omega_1, \omega_2)$. What are the restrictions on the parameter $a$ so that the minimum value $\alpha$ is nonnegative? (Consider both positive and negative values for $a$.)

**\*6.18.** Suppose that the known samples are given by

$$\phi(0, 0) = 1$$

$$\phi(0, 1) = \phi(1, 0) = \phi(0, -1) = \phi(-1, 0) = a$$

where $a$ is a real-valued parameter. (In contrast to Problem 6.17, where nine samples were known, here only five are known.)

**(a)** As in Problem 6.17(a), set up the minimization

$$\alpha = \min_{\mathbf{p}} \langle \boldsymbol{\phi}, \mathbf{p} \rangle$$

assuming that $p(0, 0) = 1$.

**(b)** Write an expression for the positive polynomial $P(\omega_1, \omega_2)$ assuming that $p(0, 1) = p(0, -1)$ and $p(1, 0) = p(-1, 0)$. Is it the same as $P(\omega_1, \omega_2)$ in Problem 6.17(b)?

**(c)** What restrictions must be placed on $p(0, 1)$ and $p(1, 0)$ to ensure that $P(\omega_1, \omega_2) \geq 0$?

**(d)** What restrictions must be placed on the parameter $a$ so that the minimum value $\alpha$ is nonnegative?

# REFERENCES

1. Alan V. Oppenheim, ed., *Applications of Digital Signal Processing* (Englewood Cliffs, N.J.: Prentice-Hall, Inc., 1978).

2. E. J. Kelly, "The Representation of Seismic Waves in Frequency–Wavenumber Space," M.I.T. Lincoln Laboratory Technical Note 1964-15 (Mar. 6, 1964).

3. O. S. Halpeny and Donald G. Childers, "Composite Wavefront Decomposition via Multidimensional Digital Filtering of Array Data," *IEEE Trans. Circuits and Systems,* CAS-22, no. 5 (June 1975), 552–63.

4. E. J. Kelly, "Response of Seismic Arrays to Wide-Band Signals," M.I.T. Lincoln Laboratory Technical Note 1967-30 (June 29, 1967).

5. Fredric J. Harris, "On the Use of Windows for Harmonic Analysis with the Discrete Fourier Transform," *Proc. IEEE,* 66, no. 1 (Jan. 1978), 51–83.

6. Thomas S. Huang, "Two-Dimensional Windows," *IEEE Trans. Audio and Electroacoustics,* AU-20, no. 1 (Mar. 1972), 80–90.

7. Theresa C. Speake and Russell M. Mersereau, "A Note on the Use of Windows for Two-Dimensional FIR Filter Design," *IEEE Trans. Acoustics, Speech, and Signal Processing,* ASSP-29, no. 1 (Feb. 1981), 125–27.

8. Michael R. Portnoff, "Time–Frequency Representation of Digital Signals and Systems," *IEEE Trans. Acoustics, Speech, and Signal Processing,* ASSP-29, no. 1 (Feb. 1980), 55–69.

9. Dan E. Dudgeon, "Fundamentals of Digital Array Processing," *Proc. IEEE,* 65, no. 6 (June 1977), 898–904.

10. Ronald W. Schafer and Lawrence R. Rabiner, "A Digital Signal Processing Approach to Interpolation," *Proc. IEEE*, 61, no. 6 (June 1973), 692–720.

11. Geerd Oetken, Thomas W. Parks, and Hans W. Schüssler, "New Results in the Design of Digital Interpolators," *IEEE Trans. Acoustics, Speech, and Signal Processing*, ASSP-23, no. 3 (June 1975), 301–9.

12. Roger G. Pridham and Ronald A. Mucci, "A Novel Approach to Digital Beamforming," *J. Acoustical Society of America*, 63(3) (Feb. 1978), 425–34.

13. Roger G. Pridham and Ronald A. Mucci, "Digital Interpolation Beamforming for Low-Pass and Bandpass Signals," *Proc. IEEE*, 67, no. 6 (June 1979), 904–19.

14. Henry Cox, "Resolving Power and Sensitivity to Mismatch of Optimum Array Processors," *J. Acoustical Society of America*, 54, no. 3 (Mar. 1973), 771–85.

15. Robert A. Monzingo and Thomas W. Miller, *Introduction to Adaptive Arrays* (New York: John Wiley & Sons, Inc., 1980).

16. M. S. Bartlett, *An Introduction to Stochastic Processes with Special Reference to Methods and Applications* (New York: Cambridge University Press, 1953).

17. R. B. Blackman and J. W. Tukey, *The Measurement of Power Spectra* (New York: Dover Publications, Inc., 1958).

18. G. M. Jenkins and D. G. Watts, *Spectral Analysis and Its Applications* (San Francisco: Holden-Day, Inc., 1968).

19. P. D. Welch, "The Use of Fast Fourier Transform for the Estimation of Power Spectra: A Method Based on Time Averaging over Short, Modified Periodograms," *IEEE Trans. Audio and Electroacoustics*, AU-15, no. 2 (June 1967), 70–73.

20. Alan V. Oppenheim and Ronald W. Schafer, *Digital Signal Processing* (Englewood Cliffs, N.J.: Prentice-Hall, Inc., 1975).

21. J. Capon, "High-Resolution Frequency–Wavenumber Spectrum Analysis," *Proc. IEEE*, 57, no. 8 (Aug. 1969), 1408–18.

22. Richard T. Lacoss, "Data Adaptive Spectral Analysis Methods," *Geophysics*, 36, no. 4 (Aug. 1971), 661–75.

23. James H. Justice, "A Levinson-Type Algorithm for Two-Dimensional Wiener Filtering Using Bivariate Szegö Polynomials," *Proc. IEEE*, 65, no. 6 (June 1977), 882–86.

24. Bradley W. Dickinson, "Two-Dimensional Markov Spectrum Estimates Need Not Exist," *IEEE Trans. Information Theory*, IT-26, no. 1 (Jan. 1980), 120–21.

25. W. Rudin, "The Extension Problem for Positive Definite Functions," *Illinois J. Mathematics*, 7 (1963), 532–39.

26. William I. Newman, "Notes on Multidimensional Power Spectral Analysis," *Astronomy and Astrophysics* (Germany), 70, no. 3, pt. 1 (Nov. 1978), 409–10.

27. Yves Genin and Yves Kamp, "Counterexample in the Least-Squares Inverse Stabilisation of 2D Recursive Filters," *Electronics Letters*, 11, no. 15 (July 24, 1975), 330.

28. Leland B. Jackson and H. C. Chien, "Frequency and Bearing Estimation by Two-Dimensional Linear Prediction," *Proc. Int. Conf. Acoustics, Speech, and Signal Processing*, Washington, D.C. (Apr. 1979), 665–68.

29. J. V. Pendrell, "The Maximum Entropy Principle in Two Dimensional Spectral Analysis," Ph.D. thesis, York University, Ontario, Canada (Nov. 1979).

30. John Makhoul, "Linear Prediction: A Tutorial Review," *Proc. IEEE*, 63, no. 4 (Apr. 1975), 561–80.

31. Stephen J. Wernecke and Larry R. D'Addario, "Maximum Entropy Image Reconstruction," *IEEE Trans. Computers*, C-26, no. 4 (Apr. 1977), 351–64.

32. William I. Newman, "A New Method of Multidimensional Power Spectral Analysis," *Astronomy and Astrophysics* (Germany), 54, no. 2, pt. 2 (Jan. 1977), 369–80.

33. John W. Woods, "Two-Dimensional Markov Spectral Estimation," *IEEE Trans. Information Theory*, IT-22, no. 5 (Sept. 1976), 552–59.

34. Jae S. Lim and Naveed A. Malik, "A New Algorithm for Two Dimensional Maximum Entropy Power Spectrum Estimation," *IEEE Trans. Acoustics, Speech and Signal Processing*, ASSP-29, no. 3 (June 1981), 401–13.

35. James H. McClellan, "Multi-dimensional Spectral Estimation," *Proc. IEEE*, 70, no. 9 (Sept. 1982), 1029–39.

36. Stephen W. Lang and James H. McClellan, "Spectral Estimation for Sensor Arrays," *Proc. 1st ASSP Workshop on Spectral Estimation*, 1 (Aug. 1981), 3.2.1–3.2.7.

37. David G. Luenberger, *Introduction to Linear and Nonlinear Programming* (Reading, Mass.: Addison-Wesley Publishing Company, 1973).

38. John P. Burg, "Maximum Entropy Spectral Analysis," Ph.D. thesis, Stanford University (May 1975).

# 7

# INVERSE PROBLEMS

A number of physical problems can be modeled by an expression of the form

$$y(n_1, n_2) = D[x(n_1, n_2)]$$

(7.1)

where $D[\cdot]$ is a distortion operator that acts on the input sequence $x(n_1, n_2)$ to produce the output sequence $y(n_1, n_2)$. These problems can assume quite different characteristics depending upon what is known and what is to be determined. If the input and the distortion operator are both known and the output is to be determined, the problem is one of *system realization*. System realization has been our primary concern in the first six chapters. If the input and the output are both known and the distortion operator is to be determined, the problem is one of *system identification*. A problem of the remaining type, in which we must determine the input which produced a known distorted output, is known as an *inverse problem*. Common examples of inverse problems include noise removal, deconvolution, and signal extrapolation.

We look at several examples of inverse problems in this chapter. Our treatment will not be exhaustive; indeed, no such treatment could be. Rather our intent is to suggest possible approaches that have been found to be useful for certain multidimensional inverse problems (and for one-dimensional problems as well). This treatment will also provide an opportunity to use some of the material presented in the earlier chapters.

We begin by looking at iterative methods for constrained signal restoration. Because many inverse problems do not have unique solutions, the incorporation of constraints is often necessary to reduce the number of possible solutions. With a number of iterative procedures, this is conveniently done. Then we look at the seismic wave migration problem, in which we attempt to remove distortion introduced into a propagating wavefront by the process of propagation itself. One approach is based on mathematically propagating the wave backward in both time and space using a multi-dimensional all-pass filter. Finally, we look at the recovery of multidimensional signals from their projections. This general problem arises in the area of computer-aided tomography as well as in some other applications.

## 7.1 CONSTRAINED ITERATIVE SIGNAL RESTORATION

We can attempt to solve equation (7.1) for $x(n_1, n_2)$ by trying to find the inverse operator $D^{-1}$ such that

$$x(n_1, n_2) = D^{-1}[y(n_1, n_2)] \qquad (7.2)$$

However, in many cases of practical interest, it may be difficult, or indeed impossible, to determine and implement such an inverse operator. In other cases we may know the operator $D$ only approximately and the implementation of an inverse operator based on an incomplete knowledge of $D$ might be quite unsatisfactory. Even if $D^{-1}$ could be approximated and implemented, the result of applying it to the signal $y(n_1, n_2)$ may cause large errors if $y(n_1, n_2)$ is known imprecisely due to such uncertainties as measurement or quantization noise.

For these reasons alternatives to finding an inverse operator for signal restoration are of interest. One alternative, which is particularly attractive for computer implementations, is the method of successive approximations. It is based on an iteration equation of the form

$$x_{k+1}(n_1, n_2) = F[x_k(n_1, n_2)] \qquad (7.3)$$

where $F$ is an appropriately determined operator. It is not necessary, however, that $F$ depend only upon $D$. In some cases, limited prior knowledge of the properties of $x(n_1, n_2)$ may be incorporated into the iteration in the form of signal constraints. The operator $F$ is usually not unique; many different iterative equations can be derived for a given distortion operator and set of signal constraints.

A convenient way to express prior knowledge or known constraints about a signal $x(n_1, n_2)$ is to define a constraint operator $C$ such that

$$x(n_1, n_2) = C[x(n_1, n_2)] \qquad (7.4)$$

if and only if $x(n_1, n_2)$ satisfies the constraint. For example, if $x(n_1, n_2)$ is a signal that is known to be nonnegative, $C$ could be defined by the positivity operator $P$

$$C[x(n_1, n_2)] \triangleq P[x(n_1, n_2)] = \begin{cases} x(n_1, n_2), & \text{if } x(n_1, n_2) \geq 0 \\ 0, & \text{otherwise} \end{cases}$$

The constraint operator should have the property that signals with the prescribed properties are not changed by the operator, and signals which do not have those properties are converted into signals that do.

Using a constraint operator, equation (7.1) can be expressed as

$$y(n_1, n_2) = DC[x(n_1, n_2)] \qquad (7.5)$$

Combining (7.4) and (7.5) leads to

$$x(n_1, n_2) = C[x(n_1, n_2)] + \lambda\{y(n_1, n_2) - DC[x(n_1, n_2)]\} \qquad (7.6)$$

where $\lambda$ can be a constant, a function of $(n_1, n_2)$, or a function of $x(n_1, n_2)$. Applying the method of successive approximations leads to the iteration

$$\begin{aligned} x_0(n_1, n_2) &= \lambda y(n_1, n_2) \\ x_{k+1}(n_1, n_2) &= \lambda y(n_1, n_2) + (C - \lambda DC)[x_k(n_1, n_2)] \end{aligned} \qquad (7.7)$$

The parameter $\lambda$ can be chosen to assure convergence and improve the rate of convergence. The convergence properties and the uniqueness of the final result of such an iterative procedure are a major concern in practical applications. These issues are discussed in some detail in [1].

A signal $x(n_1, n_2)$ which emerges as a solution of (7.7) is called a *fixed point* of the iteration. If an $x(n_1, n_2)$ exists which satisfies (7.1) and (7.4), it will be a fixed point of the iteration in (7.7).

### 7.1.1 Iterative Procedures for Constrained Deconvolution

If the distortion operator is linear and shift invariant we can write

$$y(n_1, n_2) = x(n_1, n_2) ** h(n_1, n_2) \qquad (7.8)$$

In this case the restoration problem becomes a deconvolution problem and the iterative equation may be written in the following form:

$$x_0(n_1, n_2) = \lambda y(n_1, n_2) \qquad (7.9a)$$

$$x_{k+1}(n_1, n_2) = \lambda y(n_1, n_2) + q(n_1, n_2) ** C[x_k(n_1, n_2)] \qquad (7.9b)$$

where

$$q(n_1, n_2) = \delta(n_1, n_2) - \lambda h(n_1, n_2) \qquad (7.9c)$$

To find the fixed points of equation (7.9b), consider the unconstrained system where $C[\cdot]$ is simply an identity operator. Taking Fourier transforms leads to

$$X_{k+1}(\omega_1, \omega_2) = \lambda Y(\omega_1, \omega_2) + X_k(\omega_1, \omega_2) - \lambda X_k(\omega_1, \omega_2)H(\omega_1, \omega_2) \qquad (7.10)$$

This is a first-order 1-D difference equation in the index $k$, with coefficients that are functions of $(\omega_1, \omega_2)$. With this interpretation the input to the difference equation is the "sequence" $\lambda Y(\omega_1, \omega_2)u(k)$, the impulse response is $[1 - \lambda H(\omega_1, \omega_2)]^k u(k)$, and the output is

$$X_k(\omega_1, \omega_2) = \frac{Y(\omega_1, \omega_2)}{H(\omega_1, \omega_2)}\{1 - [1 - \lambda H(\omega_1, \omega_2)]^{k+1}\}u(k) \qquad (7.11)$$

where $u(k)$ is the 1-D unit step sequence defined in Chapter 1. In the limit as $k \rightarrow \infty$ we get

$$X_\infty(\omega_1, \omega_2) = \frac{Y(\omega_1, \omega_2)}{H(\omega_1, \omega_2)} \tag{7.12}$$

provided that

$$|1 - \lambda H(\omega_1, \omega_2)| < 1 \tag{7.13}$$

This output is the same as the result we would obtain by using an inverse filter, but it does not actually require that the inverse filter be realized. One advantage of the iterative procedure is that it can be stopped after a finite number of iterations, at which point the output may be subjectively preferable to the actual inverse filter output.

The use of this iteration for deconvolution was proposed by Van Cittert in 1931 [2]. An examination of the convergence condition in (7.13) reveals that the procedure will not converge if $H(\omega_1, \omega_2) = 0$ for any $(\omega_1, \omega_2)$. In fact, the convergence criterion in (7.13) is equivalent to

$$\text{Re}\,[H(\omega_1, \omega_2)] > 0 \tag{7.14}$$

If $H(\omega_1, \omega_2)$ satisfies (7.14), a $\lambda$ can be found such that (7.13) will also be satisfied. There are two means by which this convergence criterion can be broadened. The first is through the imposition of a nontrivial constraint operator $C[\cdot]$, and the second is through a modification of the iteration. Let us briefly discuss the latter first.

If we convolve both sides of (7.8) with $h^*(-n_1, -n_2)$, we get

$$y(n_1, n_2) ** h^*(-n_1, -n_2) = x(n_1, n_2) ** h(n_1, n_2) ** h^*(-n_1, -n_2) \tag{7.15}$$

or

$$y'(n_1, n_2) = x(n_1, n_2) ** h'(n_1, n_2) \tag{7.16}$$

where

$$H'(\omega_1, \omega_2) = |H(\omega_1, \omega_2)|^2$$

The signal $y'(n_1, n_2)$ is a distorted version of the input signal, but now the frequency response of the distortion operator satisfies (7.14) as long as $H(\omega_1, \omega_2) \neq 0$. Thus an iteration based on (7.16) is guaranteed to converge if $H(\omega_1, \omega_2) \neq 0$. This is very similar to the modification that was made to the iterative implementation for recursive filters in Chapter 5.

Since the basic unconstrained iteration converges to the inverse filter solution, it possesses most of the undesirable properties that we associate with inverse filtering. In particular, the result of the deconvolution operation is not unique if $H(\omega_1, \omega_2)$ goes to zero. The constraint operator often circumvents this difficulty. For some applications, for example, it may be appropriate to assume that $x(n_1, n_2)$ must have finite support and be positive over its support. A positivity constraint and a finite support constraint can both be incorporated into the algorithm by defining the constraint operator

$$C[x_k(n_1, n_2)] = \begin{cases} x_k(n_1, n_2), & \text{if } P_1 \le n_1 \le Q_1, \quad P_2 \le n_2 \le Q_2, \\ & \text{and} \quad x_k(n_1, n_2) \ge 0 \\ 0, & \text{otherwise} \end{cases} \tag{7.17}$$

The implementation of the iteration of (7.9b) is relatively simple in either one or two dimensions. In both cases the convolution required can be implemented using standard discrete Fourier transform methods.

An example of this technique is shown in Figure 7.1 [1]. In this case a 2-D Gaussian blurring sequence

$$h(n_1, n_2) = \exp\left[-\left(\frac{n_1^2 + n_2^2}{100}\right)\right] \qquad (7.18)$$

is convolved with the sequence

$$x(n_1, n_2) = [\delta(n_1 - 24) + \delta(n_1 - 34)]\delta(n_2 - 32) \qquad (7.19)$$

to produce the 2-D sequence $y(n_1, n_2)$ of Figure 7.1(b). Figure 7.1(c) shows the deconvolution result after 65 iterations, using the iteration in equation (7.9) with $\lambda = 2$ and a positivity constraint.

In many practical applications, the appropriate distortion model is shift varying. In the 1-D case such a distortion operator is described by the superposition sum

$$y(n) = \sum_{m=-\infty}^{\infty} x(m)h(n, m) \qquad (7.20)$$

where $h(n, m)$ is the response of the system to a unit impulse at index $m$. As an example $h(n, m)$ might be

$$h(n, m) = \exp\left[-\left(\frac{n - m}{\sigma(m)}\right)^2\right] \qquad (7.21)$$

In this case, the response of the distorting system to an impulse $\delta(n - m)$ is a Gaussian pulse with standard deviation $\sigma(m)$ centered at $n = m$. For example, if

$$x(n) = \sum_{k=0}^{15} \delta(n - 50 - 10k) \qquad (7.22)$$

is taken to be the input [Figure 7.2(a)] to the shift-varying system with

$$\sigma(m) = 2 + \frac{m - 50}{100} \qquad (7.23)$$

we get the output signal shown in Figure 7.2(b).

When $y(n)$ is processed using the iterative algorithm with positivity and finite support constraints under the (incorrect) assumption that the distortion is *shift invariant* with impulse response

$$h(n) = \exp\left[-\left(\frac{n}{2}\right)^2\right] \qquad (7.24)$$

we get the result shown in Figure 7.2(c) after 500 iterations. Note that the first four or five impulses are well restored; however, the later impulses, where the blurring is much greater than implied by (7.24), are not recovered. When the true shift-varying distortion operator is used, the result after 500 iterations is as shown in Figure 7.2(d). It can be noted that for those impulses where the degree of blur is the greatest, the restoration is less accurate [1, 3]. Additional iterations are necessary to restore these pulses completely.

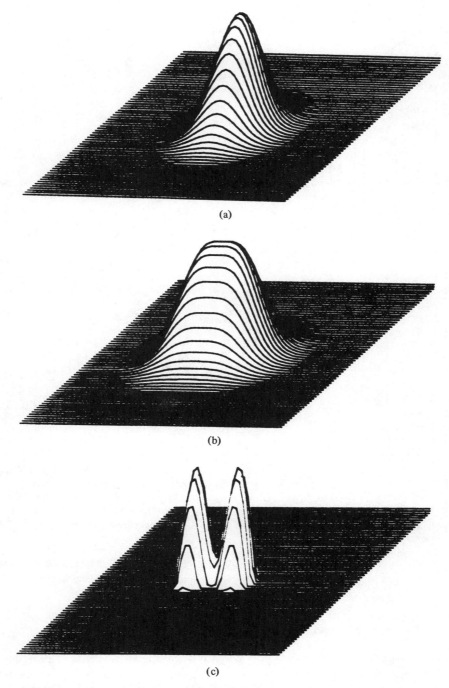

(a)

(b)

(c)

**Figure 7.1** Two-dimensional positive-constrained deconvolution of synthetic data. (a) Gaussian blurring function $h(n_1, n_2)$. (b) Sequence $y(n_1, n_2)$ obtained by convolving $h(n_1, n_2)$ with an impulse pair. (c) Estimate of the impulse pair obtained after 65 iterations with $\lambda = 2$. (Courtesy of Ronald W. Schafer, Russell M. Mersereau, and Mark A. Richards, *Proc. IEEE,* © 1981 IEEE.)

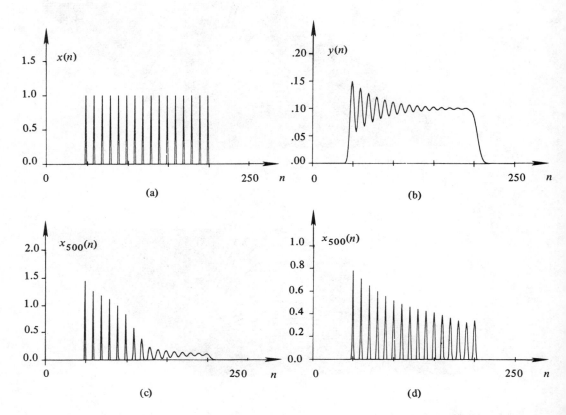

**Figure 7.2**  Positive-constrained reconstruction from a shift-varying blur. (a) Original impulse train $x(n)$. (b) Blurred sequence $y(n)$. (c) Result obtained after 500 iterations assuming a shift-invariant blur. (d) Result obtained after 500 iterations using the correct shift-varying blur. (Courtesy of Ronald W. Schafer, Russell M. Mersereau, and Mark A. Richards, *Proc. IEEE*, © 1981 IEEE. Figure originally appeared in Marucci [3].)

A related iterative restoration procedure, which is based on a stochastic formulation of the problem, has been published by Trussell [4]. His procedure gives equally impressive results.

### 7.1.2 Iterative Procedures for Signal Extrapolation

The general iteration in (7.7) can also be applied to the problem of signal extrapolation. In this case $D[\cdot]$ corresponds to restricting the support of a signal by multiplying it by a window function. A typical constraint might be to assume that the original signal is bandlimited. Algorithms based on the general iteration (7.7) for this problem have been published by Gerchberg [5] and Papoulis [6].

Huddleston [7] has used signal extrapolation techniques to measure the electromagnetic fields associated with directional antennas. The far-field pattern of an antenna is proportional to the (vector) Fourier transform of the electric field measured in the near-field (i.e., very close to the antenna). For large antennas at high frequencies where the distances involved are too great to make a direct measurement of the far-field possible, the indirect approach of measuring the near-field distribution and then computing Fourier transforms can yield satisfactory results. Conversely, in problems associated with radome analysis, the far-field may be measurable, but the near-field may be inaccessible due to the presence of a radome.

In either problem practical considerations often dictate that the observed fields be measured over only a limited portion of a plane which limits the resolution of the calculated fields. If, as an example, the far-field is measured over a limited area and the near-field distribution is assumed to possess finite support, an iterative signal restoration algorithm can be used to extrapolate the far-field measurements outside the measurement area before inverse Fourier transformation.

In Figure 7.3 we show the results of applying this technique to a radome analysis problem in which a near-field distribution is calculated from a truncated far-field distribution. The true near-field pattern should be flat out to a distance of 0.75 wavelength and be zero beyond. The solid curve in that figure used no extrapolation and the dashed one used 55 iterations of the iterative procedure.

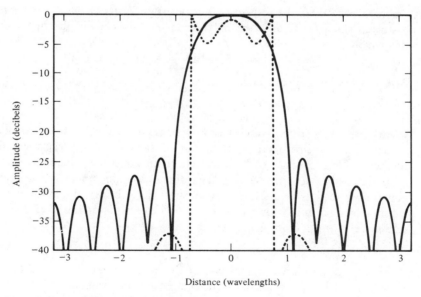

**Figure 7.3**   Near-field antenna patterns computed from truncated far-field patterns, with no extrapolation (solid) and 55 iterations of the extrapolation iteration (dashed). (Courtesy of G. K. Huddleston.)

### 7.1.3 Reconstructions from Phase or Magnitude

In a variety of problems (particularly in optics) the physics of signal generation implies certain constraints on the observed signal and its Fourier transform. For example, in optical systems it is possible to measure the magnitude of a signal (wavefront) and its Fourier transform, but measurement of the phase of either is very difficult. In such a situation the distortion and constraints are intimately related. For example, consistency with a known spectral magnitude function is a constraint, while unknown phase can be thought of as a signal-dependent distortion. Gerchberg and Saxton [8] proposed an iterative algorithm for the reconstruction of a complex or bipolar signal from its magnitude and the magnitude of its Fourier transform. Fienup [9, 10] has considered iterative algorithms for reconstruction from the magnitude of the Fourier transform of a signal under the constraint that the signal is positive. Hayes [11, 12] has discussed algorithms for the reconstruction of signals with finite support from either the phase or magnitude of the Fourier transform.

To illustrate this class of algorithms, suppose that either the magnitude or the phase of the Fourier transform of the signal $x$ is known but not both. Also assume that prior knowledge of the properties of the signal can be expressed as a constraint operator (e.g., $x$ may be known to be positive or to have finite support or both). The distortion can thus be represented in the Fourier domain as

$$Y = DX \tag{7.25}$$

where $X$ and $Y$ are the respective Fourier transforms of the desired signal $x$ and the distorted signal $y$. If only the magnitude of $X$ is known, the operator $D$ could be defined by

$$Y = DX = |X| = X \exp(-j \arg[X]) \tag{7.26a}$$

or if only the phase is known

$$Y = DX = \exp(j \arg[X]) = \frac{X}{|X|} \tag{7.26b}$$

Clearly, the distortion is nonlinear and signal dependent in both cases.

The constraint operator $C$ can be most conveniently represented (and implemented) as the cascade of a Fourier-domain constraint operator $C_F$ and a signal-domain constraint operator $C_S$. The signal-domain constraint operator embodies constraints such as finite support or positivity (or both). When the magnitude of the Fourier transform is known [corresponding to the distortion of (7.26a)], the Fourier-domain constraint is implemented as

$$V_k = C_F U_k = |X| \exp(j \arg[U_k]) \tag{7.27a}$$

where $|X|$ is the known magnitude of the Fourier transform and $U_k$ is the Fourier transform of the input to the Fourier-domain constraint operator. When the phase is known, the Fourier-domain constraint operator can be implemented as

$$V_k = C_F U_k = |U_k| \exp(j \arg[X]) \tag{7.27b}$$

where $\arg[X]$ is the known phase.

If we set $C = \mathfrak{F}^{-1}C_F\mathfrak{F}C_S$, where $\mathfrak{F}$ is the Fourier transform operator, the general iteration can be represented in the frequency domain as

$$X_0 = \lambda Y \tag{7.28a}$$

$$X_{k+1} = \lambda Y + \tilde{X}_k - \lambda D\tilde{X}_k \tag{7.28b}$$

where $\tilde{X}_k$ is the Fourier transform of the output of the constraint operator, that is,

$$\tilde{X}_k = \mathfrak{F}Cx_k = C_F\mathfrak{F}C_Sx_k \tag{7.28c}$$

In both the known-magnitude and known-phase cases, no matter how the iteration is started, and independent of the iteration number, the term $\lambda D\tilde{X}_k$ in (7.28b) can be expressed as

$$\lambda D\tilde{X}_k = \lambda DC_F\mathfrak{F}C_Sx_k = \lambda Y \tag{7.29}$$

because the distortion and the Fourier-domain constraint are so intimately related for both types of distortions. For example, if the Fourier-domain constraint is known phase, the combined operations $DC_F$ will always produce a Fourier transform with unity magnitude and phase equal to the known phase.

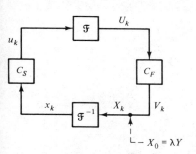

Substituting the result of (7.29) into (7.28b) and expressing the iterative equation in terms of signal-domain quantities, we obtain

$$x_0 = \lambda y \tag{7.30a}$$

$$x_{k+1} = Cx_k = \mathfrak{F}^{-1}C_F\mathfrak{F}C_Sx_k \tag{7.30b}$$

The operations of (7.30) are depicted in Figure 7.4.

In implementing either (7.29) or (7.30), we must implement the Fourier operators $\mathfrak{F}$ and $\mathfrak{F}^{-1}$. For discrete signals, these operators can often be approximated adequately by discrete Fourier transforms. Indeed, if the signal is known to have finite support, it has been shown that it is theoretically possible to recover the signal exactly using the discrete Fourier transform [11].

It is possible to apply the Fourier-domain and signal-domain constraint operators in the reverse order by letting $C = C_S\mathfrak{F}^{-1}C_F$ in the general iteration. This leads to a different iterative equation which requires additional computation and seems to have no offsetting advantages [1].

**Figure 7.4** Block diagram of an algorithm for constrained reconstruction from the magnitude or phase of the Fourier transform. (Courtesy of Ronald W. Schafer, Russell M. Mersereau, and Mark A. Richards, *Proc. IEEE*, © 1981 IEEE.)

In problems of this type where constraints are imposed independently in both the signal domain and the Fourier domain, it is essential that these constraints be consistent and that there be a unique signal that satisfies both constraints. For example, consider the case of known phase with a finite support constraint in the signal domain. Clearly, there are an infinite number of signals whose Fourier transform has a given phase and likewise there are an infinite number of signals that are nonzero in a given region of support. The constraints are consistent if there is at least one signal that satisfies both constraints. On the other hand, the constraints may be consistent

(a)

(b)

(c)

(d)

**Figure 7.5** Examples of signed-magnitude-only and phase-only reconstructions. (a) Signed-magnitude-only image formed by combining the correct Fourier transform magnitude with 1 bit of phase information. (b) Image reconstructed after 20 iterations. (c) Phase-only image (constant Fourier transform magnitude). (d) Image reconstructed after 20 iterations. (Courtesy of M. H. Hayes, *IEEE Trans. Acoustics, Speech, and Signal Processing*, © 1982 IEEE.)

but they may not uniquely define the signal. In this example, if the known phase function has a linear component corresponding to one or more sets of four reciprocal zeros in its $z$-transform, it will be impossible to reconstruct $x$ from knowledge of the phase and finite support region alone, since any set of four reciprocal zeros will produce the same linear phase component. This issue is treated in detail by Hayes [11, 12], who gives conditions for unique reconstruction from phase. A unique reconstruction will occur if and only if none of these reciprocal zeros are present in the $z$-transform of the signal.

In Figure 7.5 we show the result of these iterative procedures for image restoration from Fourier transform magnitude and phase information. In Figure 7.5(a) we show an image with the correct Fourier transform magnitude and a 1-bit quantized phase. In Figure 7.5(b) the result after 20 iterations is shown. In Figure 7.5(c) we show the result of a phase-only image with a constant Fourier transform magnitude and in Figure 7.5(d) we again show the restoration after 20 iterations. An attempt at a pure magnitude-only reconstruction with no estimate of the phase whatsoever did not lead to a convergent iteration. In all cases the finite support of the images and their positivity were used as constraints.

## 7.2 SEISMIC-WAVE MIGRATION [13, 14]

The term "migration" refers to a processing technique which is applied to seismic data sections to compensate for some undesirable geometric effects of wave propagation. Looked at another way, however, it may be viewed as a clever way of using multidimensional digital filters to solve partial differential equations explicitly. The idea is due to Claerbout [15] but the presentation in this section is more closely related to the work of Garibotto [13] and Harris [14]. We will present this problem in its mathematically abstracted form. For a very readable discussion of where this problem fits into the context of seismic exploration, the reader is referred to the appendix of the thesis by Harris [14].

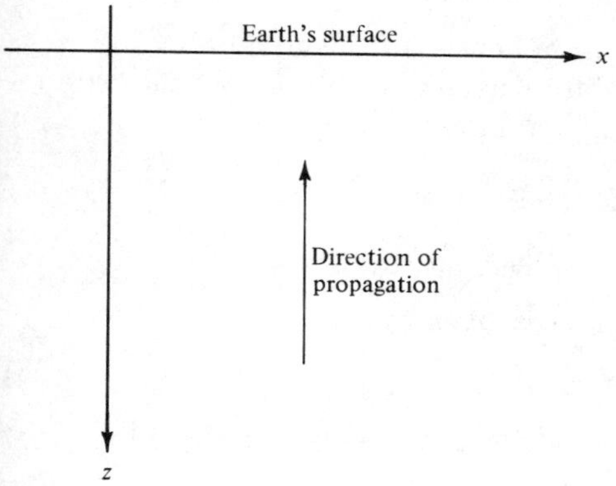

The geometry of the problem is illustrated in Figure 7.6, which is an idealization of a two-dimensional earth. The variable $x$ represents position on the earth, which is assumed to be flat; the variable $z$ represents depth. An acoustic seismic wave, $s(x, z, t)$ is assumed to be propagating upward through the earth with a uniform speed $c$. Its propagation is governed by the two-dimensional hyperbolic wave equation

$$\frac{\partial^2 s}{\partial x^2} + \frac{\partial^2 s}{\partial z^2} = \frac{1}{c^2} \frac{\partial^2 s}{\partial t^2} \qquad (7.31)$$

An array of seismometers measures $s(x, 0, t)$ at

**Figure 7.6**  Geometry of the migration problem.

the earth's surface, which serves as a boundary condition for the partial differential equation. We want to know $s(x, z_0, t)$, the wave profile, at depth $z_0$.

The backward propagation of the acoustic wave from depth 0 to depth $z_0$ is known as *migration*. Typically, the wave field is computed recursively for a large number of discrete depths $z = l \Delta z$ from the field at the preceding depth. The operation of extrapolating the wave field $s(x, z_0, t)$ to $s(x, z_0 + \Delta z, t)$ is a linear filtering operation.

Following Harris [14] and Garibotto [13], let us define the 2-D Fourier transform of the wave field at a depth $z_0$ as

$$S(k_x, z_0, \Omega) = \int \int s(x, z_0, t) \exp\left[-j(\Omega t - k_x x)\right] dx\, dt \qquad (7.32)$$

Taking the Fourier transform of both sides of the wave equation (7.31) results in

$$\frac{d^2 S(k_x, z, \Omega)}{dz^2} = \left(\frac{\Omega^2}{c^2} - k_x^2\right) S(k_x, z, \Omega) \qquad (7.33)$$

which is an ordinary second-order differential equation in the variable $z$. This equation may be solved for the field at $z = z_0 + \Delta z$ given initial conditions at $z = z_0$:

$$S(k_x, z_0 + \Delta z, \Omega) = [A \exp(j \Delta z Q) + B \exp(-j \Delta z Q)] S(k_x, z_0, \Omega) \qquad (7.34)$$

where

$$Q = \sqrt{\frac{\Omega^2}{c^2} - k_x^2} \qquad (7.35)$$

The positive exponent corresponds to an upwardly propagating wave and the negative exponent corresponds to a downwardly propagating wave. Since we are assuming that the wave propagates up, we can set $B = 0$. We can further note that (7.34) is in the form

$$S(k_x, z_0 + \Delta z, \Omega) = A H(k_x, \Omega) S(k_x, z_0, \Omega) \qquad (7.36)$$

where

$$H(k_x, \Omega) = \exp\left(j \Delta z \sqrt{\frac{\Omega^2}{c^2} - k_x^2}\right) \qquad (7.37)$$

Thus, as claimed, the extrapolation operator is a linear shift-invariant filter.

If the wave field is presented in sampled form as $s(n_1 \Delta x, l\Delta z, n_2 \Delta t)$, the extrapolation operation can be performed using a digital filter. Letting

$$x(n_1, n_2) = s(n_1 \Delta x, z_0, n_2 \Delta t)$$

and

$$y(n_1, n_2) = s(n_1 \Delta x, z_0 + \Delta z, n_2 \Delta t)$$

the frequency response of the ideal digital filter is given by

$$H(\omega_1, \omega_2) = \exp\left(j\sqrt{\alpha^2 \omega_2^2 - \omega_1^2}\right) \qquad (7.38)$$

where $\omega_1$ is really the wavenumber $k_x$, $\omega_2$ is the temporal frequency $\Omega$, and

$$\alpha = \frac{1}{c} \frac{\Delta z}{\Delta t}$$

In the region $|\alpha\omega_2| > |\omega_1|$, the transfer function has unit magnitude and is thus completely characterized by its phase function

$$\phi(\omega_1, \omega_2) = \sqrt{\alpha^2\omega_2^2 - \omega_1^2} \tag{7.39}$$

Physically, this region corresponds to those frequencies and wavenumbers for which the waves are propagating. In the region $|\alpha\omega_2| < |\omega_1|$, which is known as the evanescent region, waves do not propagate but are attenuated. These regions are illustrated in Figure 7.7. The objective of the migration filter design is to approximate (7.39) in the

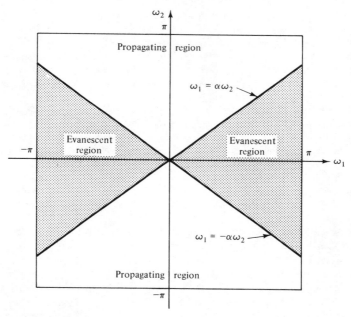

**Figure 7.7** Evanescent and propagating regions of $H(\omega_1, \omega_2)$. (Courtesy of David B. Harris.)

propagating region. This is the only region in which the approximation need be accurate, since the data are presumed to have little or no energy in the evanescent region.

Since the desired transfer function has unit magnitude in the propagation region and may be chosen arbitrarily in the evanescent region, it is convenient to choose an all-pass structure for $H_A(\omega_1, \omega_2)$, a realizable approximation to $H(\omega_1, \omega_2)$. Thus

$$H_A(z_1, z_2) = Gz_2^{N_0}\frac{z_2^{-N}A_N(z_1, z_2^{-1})}{A_N(z_1, z_2)}$$

We have now reduced the problem to a filter design problem and will go no further. Both Garibotto [13] and Harris [14] have considered the design of multi-dimensional all-pass filters for this application. An example of the performance of a migration filter is shown in Figure 7.8.[14]. The earth model is shown in part (a)

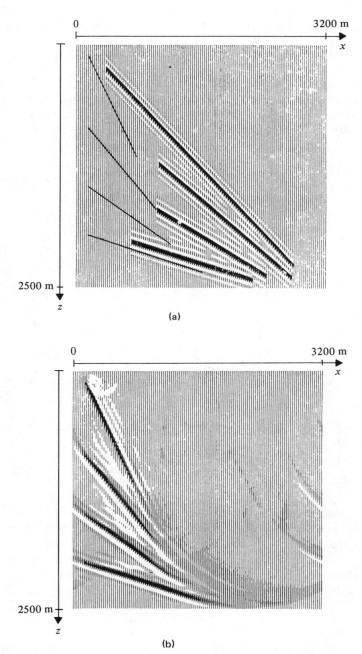

**Figure 7.8** Example of seismic-wave migration of synthetic data. (a) Earth model (solid lines) and synthetic seismic section. (b) Result after migration. (Courtesy of David B. Harris.)

as solid slanting lines superimposed on a synthetic seismic section measured at the surface. In part (b) we see the result after migration. It should be noted that the migrated seismic section depicts the earth model accurately in the case.

## 7.3 RECONSTRUCTION OF SIGNALS FROM THEIR PROJECTIONS

The problem of reconstructing a multidimensional signal from its projections is uniquely multidimensional, having no 1-D counterpart. It has applications that range from computer-aided tomography to geophysical signal processing. It is a problem which can be explored from several points of view—as a deconvolution problem, a modeling problem, an estimation problem, or an interpolation problem. Each viewpoint provides extra information concerning the problem solution.

The problem originates as a continuous rather than a discrete one and since sampling itself is an issue, we begin by formulating the problem in continuous terms and then present several algorithms for its approximate solution. Finally, we consider some generalizations of the problem.

### 7.3.1 Projections

A projection is a mathematical operation that is similar to the physical operation of taking an x-ray photograph with a collimated beam of radiation. The result is a shadow in which the three-dimensional structure of the unknown is reduced to a two-dimensional image that can be measured. Mathematically, we can consider projections of any dimensionality, but our discussion will be made simpler if we consider 1-D projections of 2-D objects. The geometry for modeling the projection operation is shown in Figure 7.9. Let the object being irradiated be described by the

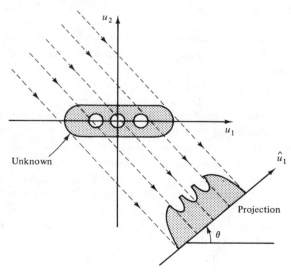

**Figure 7.9**   Collimated beam scanning geometry.

(unknown) density function $x(u_1, u_2)$ and assume that the beam propagation is in a direction which is normal to a line that forms an angle $\theta$ with the $u_1$-axis. By defining a rotated coordinate system, the $(\hat{u}_1, \hat{u}_2)$ system, as

$$\hat{u}_1 = u_1 \cos \theta + u_2 \sin \theta \qquad (7.40a)$$

$$\hat{u}_2 = -u_1 \sin \theta + u_2 \cos \theta \qquad (7.40b)$$

the projecting beam is seen to be parallel to the $\hat{u}_2$-axis and normal to the $\hat{u}_1$-axis.

The combined effects of scattering and absorption result in an exponential attenuation of a beam of photons as it passes through a material. A monoenergetic beam with an input intensity of $I_0$ photons per second, passing through a length $l$ of material, has an output intensity of

$$I = I_0 e^{-\rho l} \qquad (7.41)$$

The absorption parameter $\rho$ depends on the energy of the x-rays and the material itself. If the material is heterogeneous, the simple product $\rho l$ is replaced by a line integral and the output intensity can be expressed as

$$I(\hat{u}_1) = I_0(\hat{u}_1) \exp\left[ -\int_{-\infty}^{\infty} x(u_1, u_2) \, d\hat{u}_2 \right] \qquad (7.42)$$

The quantity

$$p_\theta(\hat{u}_1) \triangleq -\log \frac{I(\hat{u}_1)}{I_0(\hat{u}_1)} = \int_{-\infty}^{\infty} x(u_1, u_2) \, d\hat{u}_2 \qquad (7.43)$$

$$= \int_{-\infty}^{\infty} x(\hat{u}_1 \cos \theta - \hat{u}_2 \sin \theta, \, \hat{u}_1 \sin \theta + \hat{u}_2 \cos \theta) \, d\hat{u}_2 \qquad (7.44)$$

is termed the *projection* of $x$ at an angle $\theta$. It corresponds to a family of line integrals taken along a series of lines which are parallel to the beam and each other. Each different angle $\theta$ in the range $0 \leq \theta < \pi$ results in a different orientation for the beam and a different projection.

The reconstruction problem is the problem of inverting a finite number of equations of the form of (7.44) for different values of $\theta$ to yield an estimate of $x(u_1, u_2)$. This procedure may or may not incorporate *a priori* information about $x(u_1, u_2)$. The problem of determining $x(u_1, u_2)$ exactly given the continuum of projections for all angles between 0 and $\pi$ was solved early in this century by Radon [16].

This problem arises naturally in a variety of contexts. Computer-aided tomography is concerned with the reconstruction of a section of the human body by means of x-rays taken at different orientations. Although originally used only for detecting tumors and similar abnormalities in the head, where patient motion could be easily controlled, as faster algorithms and faster processors were developed, reconstructions of the lower torso have also been performed. Figure 7.10 shows a typical cross-sectional scan of a human torso made with a modern CAT scanner. The current state of the art permits cinematic reconstructions of the beating heart to be made. For those conditions where the x-ray absorption function does not differentiate between healthy and unhealthy tissue, other sources of radiation including positrons and gamma rays have

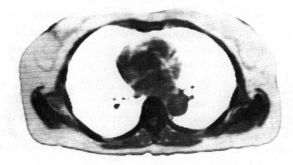

**Figure 7.10**  Reconstruction (computer-aided tomograph) of a cross-sectional scan through the human chest. (Courtesy of the Department of Radiology, Brigham and Women's Hospital, Boston.)

also been used. In a nonmedical context these techniques can be used in the nondestructive testing of materials.

Projections can also result from geologic measurements made from boreholes [17], as shown in Figure 7.11. Two boreholes are drilled, one on either side of a region

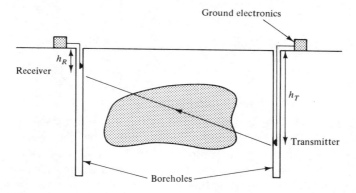

**Figure 7.11**  Use of boreholes to acquire information about underground geologic formations using acoustic beams.

on interest. Down one borehole an acoustic or microwave transmitter is lowered to a depth of $h_T$ and down the other a receiver is positioned at a depth of $h_R$. Under the assumption that the only signal received is the one that propagates in a straight line from the transmitter, we get one sample of a projection of either the acoustic (or microwave) delay function or of the acoustic (or microwave) attenuation function. By varying $h_R$ and $h_T$, different line integrals across the unknown region can be obtained. This method has been proposed for mapping hydrocarbon deposits, measuring the burn front of an *in situ* coal gasification process, and probing underground caverns for nuclear waste disposal.

Other applications of reconstruction algorithms are suggested by considering those situations in which projections are produced. Electron microscopes produce projections of their specimens, radio telescopes measure projections of interstellar space, marginal probability densities are projections of multivariate probability densities, and the response of a linear shift-invariant optical system to a line is a projection of the point spread function (impulse response). All of these problems can be addressed by variations of the algorithms to be presented in this section.

### 7.3.2 Projection-Slice Theorem

Many reconstruction algorithms have been proposed. Some of these are implemented in the space domain and some are implemented in the transform domain. Without regard for how a particular algorithm is implemented, however, it is beneficial to look at the projection operation in both domains.

The projection function defined by (7.44) is a 1-D function. If $x(u_1, u_2)$ has a 2-D Fourier transform, $X(\Omega_1, \Omega_2)$, the 1-D Fourier transform of $p_\theta(\hat{u}_1)$ will exist. Let its Fourier transform be denoted by $S_\theta(\omega)$. It then follows that

$$S_\theta(\omega) = \int_{-\infty}^{\infty} p_\theta(\hat{u}_1) \exp\left(-j\omega\hat{u}_1\right) d\hat{u}_1 \tag{7.45}$$

$$= \int_{-\infty}^{\infty} \int_{-\infty}^{\infty} x(\hat{u}_1 \cos\theta - \hat{u}_2 \sin\theta, \hat{u}_1 \sin\theta + \hat{u}_2 \sin\theta) \\ \cdot \exp\left(-j\omega\hat{u}_1\right) d\hat{u}_2 \, d\hat{u}_1 \tag{7.46}$$

Going back to the unrotated coordinate system, we can write

$$S_\theta(\omega) = \int_{-\infty}^{\infty} \int_{-\infty}^{\infty} x(u_1, u_2) \exp\left[-j\omega(u_1 \cos\theta + u_2 \sin\theta)\right] du_1 \, du_2 \tag{7.47}$$

or

$$S_\theta(\omega) = X(\omega \cos\theta, \omega \sin\theta) \tag{7.48}$$

This result is rather remarkable. It says that the Fourier transform of the projection taken at an angle $\theta$ is the 2-D Fourier transform of the unknown evaluated along a line passing through the origin of the $(\Omega_1, \Omega_2)$-plane and making an angle $\theta$ with the $\Omega_1$-axis. This cross-sectional function will be called the *slice* of $X(\Omega_1, \Omega_2)$ *at angle* $\theta$. Equation (7.48) is known as the projection-slice theorem. The relevant geometry is displayed in Figure 7.12.

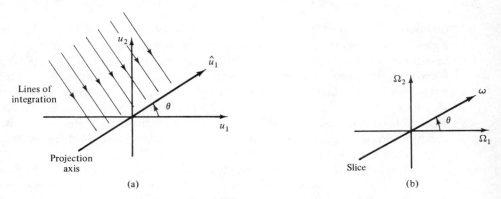

**Figure 7.12**  Relationship between the projection of a two-dimensional function and the slice of its Fourier transform. (Courtesy of Russell M. Mersereau, *Proc. IEEE*, © 1974 IEEE.)

With the projection-slice theorem, we see that knowledge of several projections of an object provides knowledge of the Fourier transform along selected radial lines in the Fourier plane. The problem of reconstructing, or estimating, $x(u_1, u_2)$ is thus equivalent to the problem of interpolating the whole Fourier transform from these radial samples. *A priori* knowledge about $x(u_1, u_2)$ can be used to help in this interpolation process. For example, if it is known that $x(u_1, u_2)$ has finite extent, it follows that $X(\Omega_1, \Omega_2)$ is analytic. This provides one means for performing a Fourier-domain interpolation.

Using the projection-slice theorem it is reasonably straightforward to derive Radon's inversion formula [16] for the special case where all projections are known over the interval $0 \leq \theta < \pi$. The unknown $x(u_1, u_2)$ can be obtained from its Fourier transform $X(\Omega_1, \Omega_2)$ by means of the Fourier transform inversion integral

$$x(u_1, u_2) = \frac{1}{4\pi^2} \int_{-\infty}^{\infty} \int_{-\infty}^{\infty} X(\Omega_1, \Omega_2) \exp\left(j\Omega_1 u_1 + j\Omega_2 u_2\right) d\Omega_1 \, d\Omega_2$$

Describing the 2-D Fourier plane in polar coordinates $(\omega, \theta)$, this integral can be rewritten as

$$\begin{aligned}
x(u_1, u_2) &= \frac{1}{4\pi^2} \int_0^{\pi} \int_{-\infty}^{\infty} X(\omega \cos\theta, \omega \sin\theta) \\
&\quad \cdot \exp\left[j\omega(u_1 \cos\theta + u_2 \sin\theta)\right] |\omega| \, d\omega \, d\theta \\
&= \frac{1}{4\pi^2} \int_0^{\pi} \int_{-\infty}^{\infty} S_\theta(\omega) \exp\left[j\omega(u_1 \cos\theta + u_2 \sin\theta)\right] |\omega| \, d\omega \, d\theta
\end{aligned} \tag{7.49}$$

The inner integral represents the inverse 1-D Fourier transform of the product of $S_\theta(\omega)$ and $|\omega|$. It thus corresponds to a filtered projection function. The implied frequency response, $|\omega|$, corresponds to taking the derivative of the Hilbert transform of $p_\theta(u_1 \cos\theta + u_2 \sin\theta)$. Equation (7.49) can thus be written in the spatial domain as

$$x(u_1, u_2) = \frac{1}{2\pi} \int_0^{\pi} g_\theta(u_1 \cos\theta + u_2 \sin\theta) \, d\theta \tag{7.50}$$

where

$$g_\theta(t) \triangleq \frac{d}{dt} \int_{-\infty}^{\infty} \frac{p_\theta(\tau)}{t - \tau} \, d\tau \tag{7.51}$$

$$\triangleq p_\theta(t) * k(t) \tag{7.52}$$

The function $k(t)$ is the Radon kernel and is the inverse Fourier transform of $|\omega|$ which exists only as a generalized function. We will consider some approximations to it when we consider implementations of reconstruction algorithms in Section 7.3.5.

### 7.3.3 *Discretization of the Reconstruction Problem*

The first step in the development of digital reconstruction algorithms is the discretization of the problem. In any practical context, only a finite number of projections will be available. Furthermore, each projection may be known only over a finite number of sample points, and transforms when evaluated may be known only over a set of discrete points.

If the unknown density, $x(u_1, u_2)$ is bandlimited with a Fourier transform confined to a region $R$, all its projections will also be bandlimited (this follows from the projection-slice theorem). The converse is also true: If *all* the projections of $x(u_1, u_2)$ are bandlimited, $x(u_1, u_2)$ must be bandlimited.

Let us assume that $x(u_1, u_2)$ is known to be bandlimited so that $X(\Omega_1, \Omega_2) \equiv 0$ if $\Omega_1^2 + \Omega_2^2 \geq R_0^2$. This has two important consequences. First, it implies that it is sufficient to reconstruct or estimate samples of $x(u_1, u_2)$. For example, we could estimate the sample values on the Cartesian lattice defined by

$$(u_1, u_2) = \left(\frac{\pi n_1}{R_0}, \frac{\pi n_2}{R_0}\right), \qquad -\infty < n_1, n_2 < \infty \qquad (7.53)$$

Second, it says that each of the projections can be sampled with no loss of information provided that these samples are taken no more than $\pi/R_0$ apart. The fact that the number of samples which must be processed is infinite does not pose any serious difficulties in practice. Although the assumption that $x(u_1, u_2)$ is bandlimited implies that it must have an infinite region of support, this is more a constraint of the mathematical model than one of the real world. As a practical matter, $x(n_1\pi/R_0, n_2\pi/R_0)$ can generally be assumed to have finite extent and, if not, our reconstruction can often be performed only over a finite region.

Given that we can only obtain a finite number of projections $N$, how many do we need? This depends to a great extent upon what is known *a priori* about the signal being reconstructed and how much detail we expect to learn about it. For example, if it is known *a priori* that an unknown signal is circularly symmetric, all its projections are identical and only one of them is necessary to reconstruct $x(u_1, u_2)$. On the other hand, if absolutely nothing is known beforehand about $x(u_1, u_2)$, it follows from the projection-slice theorem that an infinite number of projections will be required. To steer a somewhat more reasonable middle course, let us assume that $x(u_1, u_2)$ has an effective diameter $d$ and that it is sufficient for the reconstruction to resolve details as small as $r$, where $r \ll d$. From the 2-D sampling theorem with the domains reversed, we know that if $x(u_1, u_2)$ has diameter $d$, it is completely specified by samples of its Fourier transform spaced $2\pi/d$ apart in $\Omega_1$ and $\Omega_2$. With this sample spacing no point in the frequency domain will be more than $\pi\sqrt{2}/d$ from one of these sample values. If we demand that the sampling of the Fourier domain which is performed by the projection-slice theorem satisfy the same constraint, we require that

$$\frac{\pi}{N}R_0 < \frac{\pi}{d}$$

Thus the number of projections $N$ must satisfy

$$N > R_0 d \qquad (7.54)$$

where $R_0$ is the highest frequency that we want to estimate in the Fourier transform. From the resolution requirement,

$$R_0 > \frac{\pi}{r} \tag{7.55}$$

which implies that

$$N > \frac{\pi d}{r} \tag{7.56}$$

Although this formula should be interpreted primarily as a rule of thumb, it is a reasonably reliable one. Note that the number of views increases as the desired resolution improves. This is not surprising.

### 7.3.4 Fourier-Domain Reconstruction Algorithms

Let us assume that $N$ projections of $x(u_1, u_2)$ are available at the equispaced angles $\theta_i = \pi i / N$, $i = 0, 1, \ldots, N - 1$, that each projection has been sampled at the same sampling rate, and that the $M$-point DFT of each sampled projection has been computed. These DFT values can be interpreted as samples of the Fourier transform of $x(u_1, u_2)$ on a regular polar raster, such as the one depicted in Figure 7.13(a). If we further assume that $x(u_1, u_2)$ has finite support and that it is approximately bandlimited so that it can be represented adequately by its $(N \times N)$-point discrete Fourier transform, the reconstruction problem becomes one of interpolation of the Fourier transform. We can interpolate from the known transform values on the polar raster to the unknown ones on the DFT raster, perform an inverse DFT, and use the result as an estimate of the samples of $x$.

To perform the necessary interpolation we might consider the use of either zeroth-order or linear interpolation. In the Fourier domain, most points $(\omega_1, \omega_2)$ on the DFT raster are surrounded by four polar samples, as depicted in Figure 7.14. With zeroth-order interpolation, each DFT sample is assigned the value of the nearest polar sample and with linear interpolation it is assigned a weighted average of the four nearest polar samples, the weighting varying inversely with the Euclidean distance between the points.

If we are free to vary the sampling rates of the individual projections, we can alter the form of the polar raster of DFT samples to facilitate the process of interpolation. The raster in Fig. 7.13(b), for example, results if the spacing between samples in the projection at angle $\theta$ is $\pi/W_\theta$, where

$$W_\theta = \frac{R_0}{\max(|\cos \theta|, |\sin \theta|)} \tag{7.57}$$

With this raster, all the interpolation is confined to the rows and columns of a rectangular DFT lattice and is thus 1-D. This results not only in reduced computation but also in less error resulting from the interpolation process [18].

Some results of these algorithms are shown in Figures 7.15 and 7.16. It can be seen that the reconstructions made using linear interpolation are superior to those made with zeroth-order interpolation and that the modified polar raster gives superior reconstructions to the normal polar one. From Figure 7.16 one can discern how the quality of the reconstruction improves with the number of projections. Further details on these algorithms can be found in [18, 19].

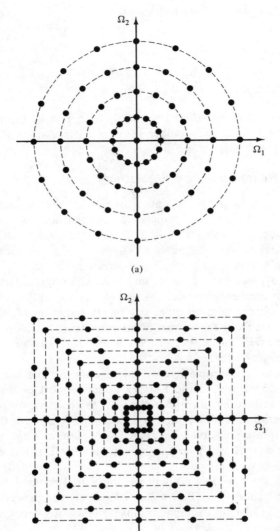

(a)

(b)

Figure 7.13 (a) Polar raster of samples in the Fourier domain, obtained by sampling all projections at the same sampling rate. (b) Concentric squares raster, obtained by varying the sampling rate with the angle of the projection. (Courtesy of Russell M. Mersereau, *Proc. IEEE*, © 1974 IEEE.)

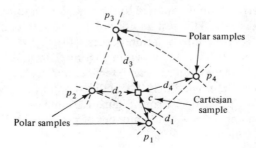

Figure 7.14 Parameters for the definition of zeroth-order and linear interpolation. (Courtesy of Russell M. Mersereau, *Proc. IEEE*, © 1974 IEEE.)

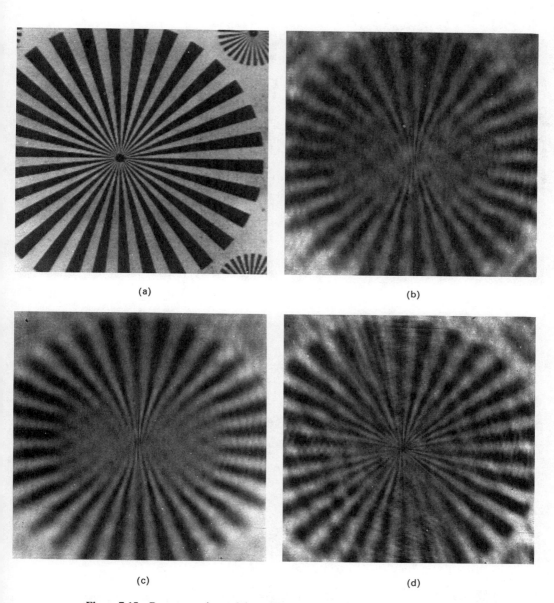

(a)

(b)

(c)

(d)

**Figure 7.15**   Reconstructions of the original image shown in (a) made from 64 equiangular projections using various interpolation algorithms. (b) Zeroth-order interpolation, polar raster. (c) Linear interpolation, polar raster. (d) Linear interpolation, concentric squares raster. (Courtesy of Russell M. Mersereau, *Proc. IEEE*, © 1974 IEEE.)

(a)

(b)

(c)

(d)

**Figure 7.16** Reconstructions made using linear interpolation from a concentric squares raster using: (a) 16 projections; (b) 32 projections; (c) 64 projections; (d) 128 projections. (Courtesy of Russell M. Mersereau, *Proc. IEEE*, © 1974 IEEE.)

### *7.3.5 Convolution/Back-Projection Algorithm*

The interpolation reconstruction algorithms of the preceding section were derived in a rather obvious fashion from the projection-slice theorem. The convolution/back-projection algorithms can similarly be derived by discretizing Radon's inversion formula (7.50). They represent the most widely used class of algorithms for a variety of reasons, not the least of which is their performance. They have the feature that they can be implemented completely in the spatial domain.

Assume that projections are available at angles $\theta_0, \theta_1, \ldots, \theta_{N-1}$ and let

$$\Delta\theta_i \triangleq \theta_i - \theta_{i-1}, i = 1, \ldots, N-1 \tag{7.58a}$$

$$\Delta\theta_0 \triangleq \theta_0 - \theta_{N-1} + \pi \tag{7.58b}$$

From equation (7.50) we see that the unknown signal $x(u_1, u_2)$ can be approximated as

$$x(u_1, u_2) = \frac{1}{2\pi} \sum_{i=0}^{N-1} \Delta\theta_i g_i(u_1 \cos \theta_i + u_2 \sin \theta_i) \tag{7.59}$$

where

$$g_i(t) = p_{\theta_i}(t) * k(t) \tag{7.60}$$

$$\mathcal{F}\{k(t)\} = |\omega| \tag{7.61}$$

These equations have the following interpretation. The 1-D projection at angle $\theta_i$ is passed through a 1-D filter, with impulse response $k(t)$ and frequency response $|\omega|$. Note that all the projections are passed through the same filter. The output of this filter is the function $g_i(t)$. Using the rotated coordinate system $(\hat{u}_1, \hat{u}_2)$, we see that

$$g_i(u_1 \cos \theta + u_2 \sin \theta) = g_i(\hat{u}_1) \tag{7.62}$$

The signal $g_i$ that appears in the sum in (7.59) is thus interpreted as a 2-D signal which is filtered in the $\hat{u}_1$ variable and uniform in the $\hat{u}_2$ variable. Since the operation of computing this function involves starting with a 1-D function to obtain a 2-D one, we will refer to this operation as *back-projection*. The function $g_i(u_1 \cos \theta_i + u_2 \sin \theta_i)$ is obtained from $g_i(t)$ by back-projecting that function in the $\hat{u}_2$ direction (i.e., parallel to the original lines of integration which defined the projection). Since the orientation of the $(\hat{u}_1, \hat{u}_2)$ coordinate system is different for each projection angle, each of the back-projected, filtered projections will have a different orientation.

In implementing this algorithm, there are two additional issues that must be considered. First, the filter given by (7.61) possesses no dc gain. As a result, the mean gray level of the reconstruction is zero. This is undesirable in many applications where the reconstructed density should be nonnegative. This is not a serious difficulty, however, since it simply means that a dc bias should be added to the reconstruction. This bias level can be chosen so that the mean gray level of the reconstruction is equal to the mean gray level of the unknown. These mean gray levels can be measured from the projections themselves.

A second implementation issue concerns the choice of the projection filter $k(t)$. Ideally, this filter should have the frequency response $|\omega|$, but if all the projections are bandlimited, the high-frequency behavior of the filter becomes immaterial. Since the filter gain increases with increasing frequency, high-frequency noise will be amplified. Thus to minimize the deterioration that can result from such noise, the filter $k(t)$ is typically chosen to have an approximately linear response out to some cutoff frequency beyond which the response goes to zero. The exact shape of the frequency response is also governed by computational convenience [20, 21].

Some reconstructions obtained using this algorithm are shown in Figures 7.17 and 7.18. The resolution here is noticeably better than for the reconstructions

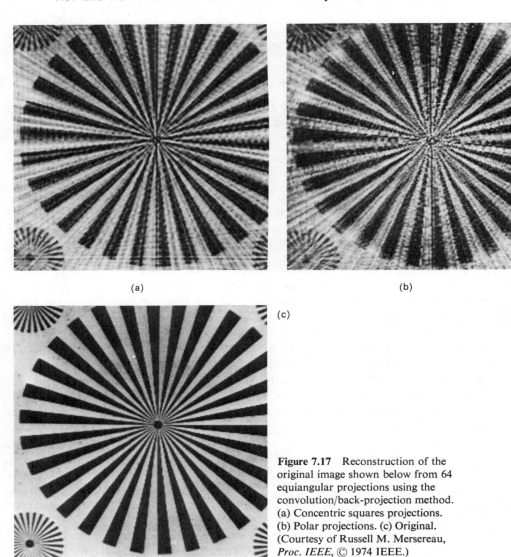

(a)                                                            (b)

(c)

**Figure 7.17** Reconstruction of the original image shown below from 64 equiangular projections using the convolution/back-projection method. (a) Concentric squares projections. (b) Polar projections. (c) Original. (Courtesy of Russell M. Mersereau, *Proc. IEEE*, © 1974 IEEE.)

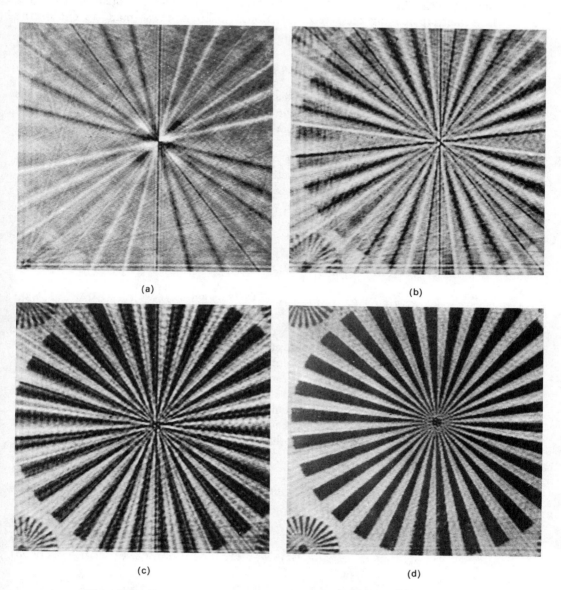

(a)                                                    (b)

(c)                                                    (d)

**Figure 7.18**  Reconstructions made using the convolution/back-projection method applied to concentric squares projections. (a) 16 projections. (b) 32 projections. (c) 64 projections. (d) 128 projections. (Courtesy of Russell M. Mersereau, *Proc. IEEE,* © 1974 IEEE.)

obtained using linear interpolation. However, it should also be noted that the noise is greater. In making these reconstructions, no attempt was made to choose the reconstruction filter $k(t)$ optimally.

### 7.3.6 Iterative Reconstruction Algorithms [22]

Iterative algorithms form a third class of reconstruction algorithms. The iterations are similar to those we saw in Section 7.1. These algorithms are by far the most computationally intensive, but they are the only ones that can incorporate *a priori* information about $x(u_1, u_2)$.

Let the distortion operator $D_i$ correspond to the combined operations of evaluating the projection of an object at an angle $\theta_i$ followed by a back-projection which is also taken with respect to the angle $\theta_i$. There must be one distortion operator for each projection angle. As before, we can let $N$ equal the number of projections. Since the procedure is iterative, let $x_k(u_1, u_2)$ denote the estimate of $x(u_1, u_2)$ from the $k$th iteration. One can then define the iterative procedure by

$$x_0(u_1, u_0) \triangleq \sum_{i=1}^{N} \lambda_i p_{\theta_i}(\hat{u}_1) \tag{7.63a}$$

$$x_k(u_1, u_2) \triangleq x_{k-1}(u_1, u_2) + \sum_{i=1}^{N} \lambda_i[p_{\theta_i}(\hat{u}_1) - D_i x_{k-1}(u_1, u_2)] \tag{7.63b}$$

where the $\{\lambda_i\}$ are a set of parameters that can be used to guarantee the convergence of the iteration and optimize the convergence rate. If an $x(u_1, u_2)$ exists which satisfies (7.63) for each $i$, it will be a fixed point of the iteration. As with the single distortion case that we studied in Section 7.1, the $\{\lambda_i\}$ can be chosen to be functions of $u_1$ and $u_2$. The signal used for the zeroth iterate is somewhat arbitrary. One choice would be to use the reconstruction obtained from the convolution/back-projection algorithm.

Suppose in addition that we know that $x$ satisfies some *a priori* constraints. Let $C$ be a constraint operator such that

$$x = Cx \tag{7.64}$$

In this case the recursion could be modified to

$$x_0(u_1, u_2) = C\left[\sum_{i=1}^{N} \lambda_i p_{\theta_i}(\hat{u}_1)\right] \tag{7.65a}$$

$$x_k(u_1, u_2) = x_{k-1}(u_1, u_2) + \sum_{i=1}^{N} \lambda_i[p_{\theta_i}(\hat{u}_1) - D_i C x_{k-1}(u_1, u_2)] \tag{7.65b}$$

### *7.3.7 Fan-Beam Reconstructions

Signal reconstruction from projections, particularly in the context of computer-aided tomography, is an active research area for which we have only tried to outline possibilities rather than to explore the state of the art. More sophisticated versions of all three algorithmic types exist which will outperform the simple algorithms we have presented. Before leaving this general area, however, we should say a few words about a generalization of the reconstruction problem which is particularly important in computer-aided tomography—reconstruction from fan beams.

To this point it has been assumed that our projections have been made by means of a collimated beam of radiation. In the mathematical model, this meant that the projections were made by integrating over a series of parallel straight lines. Collimated x-ray beams, however, are difficult to obtain without allowing extra time for data gathering. Long collection times result in motion artifacts for reconstructions which are to be performed in the vicinity of the heart and lungs. An alternative is to use noncollimated fan beams, such as the ones illustrated in Figure 7.19. As before,

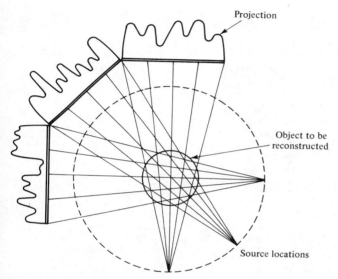

**Figure 7.19**   Fan-beam projections.

the projection can be modeled as a series of line integrals over straight lines, but those lines are not parallel.

The use of fan beams instead of collimated beams complicates the reconstruction problem in several ways. First, fan-beam projections must be taken over a complete 360° range of angles. With collimated beams a 180° range is sufficient since two projections which differ in projection angle by 180° are redundant. Fan-beam reconstruction is further complicated by the fact that the projection-slice theorem does not hold for fan-beam projections. Thus there is no generalization of the interpolation algorithms for the fan-beam case. The other two approaches, the iterative procedure and the convolution/back-projection procedure can be generalized.

The iterative equations in (7.65) that were used for the collimated beam case can be used for fan beams if we simply redefine the operator $D_i$. That operator should be chosen to compute the fan-beam projection at the angle $\theta_i$ and then perform a fan-beam back-projection.

Herman *et al.* [23] have generalized the convolution/back-projection algorithm by deriving a version of Radon's inversion formula which is appropriate to the diverging beam case. We shall simply summarize their procedure below. It may be helpful to refer to the geometrical arrangement shown in Figure 7.20. The object to

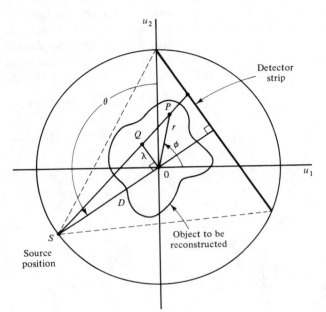

**Figure 7.20**  Illustration of the parameters of the fan-beam algorithm. [23]

be recovered is centered at point $O$. The x-ray source is located at position $S$, which always lies on a circle of radius $D$ around $O$. As in the parallel-beam case, we will let $\theta$ denote the position or orientation of the source. Any sample of the projection at angle $\theta$ can be described by the parameter $\lambda$, which gives the distance $OQ$. The line $OQ$ is parallel to the detector array and it passes through the center of the object to be reconstructed. The object $x$ will be described in terms of the polar coordinates $r$ and $\phi$. $P$ denotes a generic point in the reconstruction.

As in the parallel-beam case the reconstruction is performed in two steps. First the projections are filtered, then they are back-projected, weighted, and summed. To perform the filtering, the original projection data, $p_\theta(n\,\Delta\lambda)$ is first weighted by a function $J(n)$, then convolved with a kernel sequence $k(n)$. Thus

$$g_\theta(n) = \sum_{m=-\infty}^{\infty} J(m)p_\theta(m\,\Delta\lambda)k(n-m) \qquad (7.66)$$

where

$$J(n) = \frac{\Delta\lambda}{\sqrt{1 + (n\,\Delta\lambda/D)^2}} \qquad (7.67)$$

and

$$k(n) = \begin{cases} \dfrac{1}{8\,\Delta\lambda^2}, & n = 0 \\[2mm] \dfrac{1}{2\pi^2\,\Delta\lambda^2 n^2}, & n \text{ is odd} \\[2mm] 0, & \text{otherwise} \end{cases} \qquad (7.68)$$

The parameter $\Delta\lambda$ is the sample spacing in the variable $\lambda$. The weighting sequence $J(n)$ is to correct for the fact that intensities are naturally reduced as the beam diverges. The kernel $k(n)$ is similar to the discretized Radon kernel from the parallel-beam case. It should be noted that both the weighting function $J(n)$ and the impulse response of the filter $k(n)$ are independent of the projection orientation $\theta$.

The unknown density can be recovered by back-projecting and weighting $g_{\theta_i}$.

$$x(r, \phi) = \sum_{i=1}^{N} w(i, r, \phi) g_{\theta_i}(f(\lambda)) \qquad (7.69)$$

(The function $f(\lambda)$ represents an interpolation operation that must be applied to $\lambda$.) The back-projection must be done in a manner that is consistent with the original diverging beam. The abscissa of $g_{\theta_i}$ must be linearly stretched or compressed, and in actual computations, the filtered projections must be interpolated. The weighting function $w$ is given by

$$w(i, r, \phi) = \frac{\Delta\theta_i}{2U^2} \qquad (7.70)$$

where

$$U = 1 + \frac{r}{d}\sin(\theta_i - \phi) \qquad (7.71)$$

This weights the back-projected convolution at any point with the inverse square of the relative distance of the point from the virtual x-ray source.

## 7.4 PROJECTION OF DISCRETE SIGNALS

The preceding section dealt with projections of continuous multidimensional signals. We can also define projections of discrete arrays. In some cases such a projection can amount to a reversible rearrangement of the samples from a finite extent $N$-dimensional sequence into an $M$-dimensional sequence. The primary motivation is one of implementation convenience.

We can define a projection operator for a 2-D array as a 2-D LSI digital filter with the impulse response

$$h(n_1, n_2) = \sum_{p=-\infty}^{\infty} \delta(n_1 - pm_1, n_2 - pm_2) \qquad (7.72)$$

which is drawn in Figure 7.21 for the special case $m_1 = 3$, $m_2 = 2$. The integer pair $(m_1, m_2)$ defines the orientation of the projection, which is also given by the angle

$$\theta = \tan^{-1}\left(-\frac{m_1}{m_2}\right) \qquad (7.73)$$

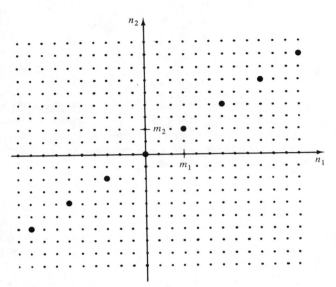

**Figure 7.21** Impulse response corresponding to a projection operator with $m_1 = 3$, $m_2 = 2$.

Unlike the continuous case, the discrete projection is not defined for all angles since $m_1$ and $m_2$ must be integers.

A projection of a 2-D array is really a 1-D sequence. Let $g(n_1, n_2)$ denote the result of projecting the 2-D signal $x(n_1, n_2)$ at the angle $\theta$ given by (7.73). Thus

$$g(n_1, n_2) = x(n_1, n_2) ** h(n_1, n_2) \tag{7.74}$$

$$= \sum_{p=-\infty}^{\infty} \sum_{k_1=-\infty}^{\infty} \sum_{k_2=-\infty}^{\infty} x(n_1 - k_1, n_2 - k_2)\delta(k_1 - pm_1, k_2 - pm_2) \tag{7.75}$$

$$= \sum_{p=-\infty}^{\infty} x(n_1 - pm_1, n_2 - pm_2) \tag{7.76}$$

From (7.76) we see that

$$g(n_1 - qm_1, n_2 - qm_2) = g(n_1, n_2) \tag{7.77}$$

for any integer $q$. Thus

$$g(n_1, n_2) = g(0, m_2n_1 - m_1n_2) \tag{7.78}$$

$$\triangleq g(m_2n_1 - m_1n_2) \tag{7.79}$$

In the last equation we have written the projection $g$ as a function of a single variable.

Projections of arrays with finite support are interesting because they are frequently invertible. Let $x(n_1, n_2)$ be an array with its support limited to the finite region $\{(n_1, n_2): 0 \le n_1 < N_1, 0 \le n_2 < N_2\}$. For this array consider the projection $(m_1, m_2) = (-1, N_2)$. Then

$$g(N_2n_1 + n_2) = x(n_1, n_2), \qquad 0 \le n_1 \le N_1 - 1, \quad 0 \le n_2 \le N_2 - 1 \tag{7.80}$$

The sequence $g(n)$ is simply the concatenation of the columns of the array $x$; the invertibility of $g$ is thus straightforward. More generally, a projection may cause several samples of $x$ to be mapped to the same sample of $g$. When this occurs, the projection is not invertible.

Let us now consider the $z$-transform of the projection mapping (7.80).

$$
\begin{aligned}
G_z(z) &= \sum_{p=0}^{N_1 N_2 - 1} g(p)z^{-p} \\
&= \sum_{n_1=0}^{N_1-1} \sum_{n_2=0}^{N_2-1} g(N_2 n_1 + n_2)z^{-(N_2 n_1 + n_2)} \\
&= \sum_{n_1=0}^{N_1-1} \sum_{n_2=0}^{N_2-1} x(n_1, n_2)z^{-N_2 n_1} z^{-n_2} \\
&= X_z(z^{N_2}, z)
\end{aligned}
\tag{7.81}
$$

The $z$-transform of a projection is thus the multidimensional $z$-transform evaluated on a particular contour. In the more general case of a projection defined by a pair of relatively prime integers $(m_1, m_2)$, it can be shown that

$$
G_z(z) = X_z(z^{m_2}, z^{-m_1}) \tag{7.82}
$$

This is the projection-slice theorem for discrete projections.

If instead of looking in the $z$-plane we restrict our attention to the Fourier plane $z_1 = \exp(j\omega_1)$, $z_2 = \exp(j\omega_2)$, then (7.82) becomes

$$
G(\omega) = X(m_2\omega, -m_1\omega) \tag{7.83}
$$

Because $X(\omega_1, \omega_2)$ is doubly periodic in $(\omega_1, \omega_2)$, $G(\omega)$ corresponds to the evaluation of $X$ along a series of parallel lines, each of which forms an angle $\theta = \tan^{-1}(-m_1/m_2)$ with the $\omega_1$-axis. This is shown in Figure 7.22 [24] for the invertible projection $m_1 = -1$, $m_2 = N_2$.

For the case of an invertible projection, the specification of $g$ is equivalent to the specification of $x$. Thus $G$ and $X$ are equivalent Fourier representations for $x$, but $X$ is a 2-D Fourier transform and $G$ is a 1-D Fourier transform, from which $x$ can be computed.

If $x$ is an $N_1 \times N_2$ array then the $N_1 N_2$-point sequence $g(p)$, obtained by concatenating the columns of $x$, has a DFT from which it can be recovered. This DFT consists of $N_1 N_2$ evenly spaced samples of $G(\omega)$ and from the projection-slice theorem it also consists of samples of $X(\omega_1, \omega_2)$. This set of samples, shown in Figure 7.23 [24], represents an alternative DFT for the 2-D array, which can be expressed as

$$
X_M(k_1, k_2) = \sum_{n_1=0}^{N_1-1} \sum_{n_2=0}^{N_2-1} x(n_1, n_2) \exp\left[-j\frac{2\pi}{N_1 N_2}(N_2 n_1 + n_2)(k_1 + k_2 N_1)\right] \tag{7.84}
$$

$$
0 \le k_1 \le N_1 - 1, \quad 0 \le k_2 \le N_2 - 1
$$

The alternative DFT is essentially the same as the slice DFT discussed in Section 2.5.1. It is a generalized DFT with the periodicity matrix

$$
\mathbf{N} = \begin{bmatrix} N_1 & -1 \\ 0 & N_2 \end{bmatrix} \tag{7.85}
$$

and it can be evaluated using a single $N_1 N_2$-point 1-D DFT.

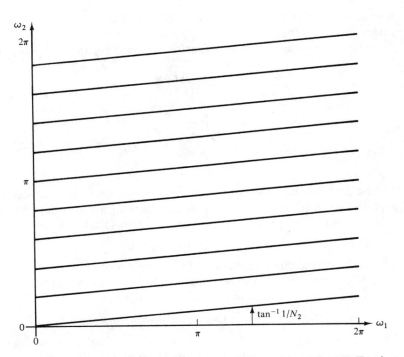

**Figure 7.22**  Lines in the 2-D Fourier plane corresponding to the 1-D Fourier transform of the projection at angle $\theta = \tan^{-1} (1/N_2)$. (Courtesy of Russell M. Mersereau and Dan E. Dudgeon, *IEEE Trans. Acoustics, Speech, and Signal Processing*, © 1974 IEEE.)

Projections can also be defined for multidimensional arrays which are not rectangularly sampled. This is done formally in [25].

Discrete projections can also be used for linear filtering using the configuration depicted in Figure 7.24. Referring to that figure, we see that

$$\hat{y}(n) = \hat{x}(n) * \hat{h}(n) \tag{7.86}$$

In the Fourier domain, this becomes

$$\hat{Y}(\omega) = \hat{X}(\omega)\hat{H}(\omega) \tag{7.87}$$

or

$$Y(m_2\omega, -m_1\omega) = X(m_2\omega, -m_1\omega)H(m_2\omega, -m_1\omega) \tag{7.88}$$

where $h$ is the inverse projection of $\hat{h}$ and $H$ is the Fourier transform of $h$. To realize the 2-D system we need

$$Y(\omega_1, \omega_2) = X(\omega_1, \omega_2)H(\omega_1, \omega_2) \tag{7.89}$$

for all $(\omega_1, \omega_2)$. Equation (7.89) will be true if the projection at orientation $(m_1, m_2)$ is invertible for the sequence $\hat{y}$. This will permit the 2-D sequence $y(n_1, n_2)$ to be constructed from the 1-D sequence $\hat{y}(n)$ without loss of information. An appropriate orientation can always be found when the three arrays have finite extent.

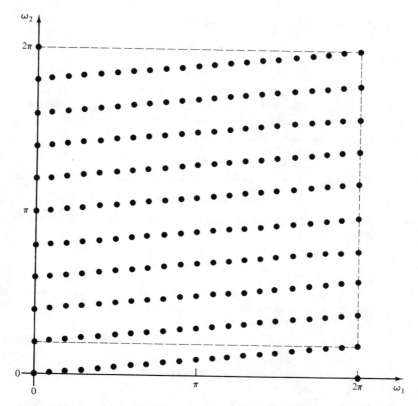

**Figure 7.23** Raster of samples in the 2-D Fourier plane which corresponds to the (20 × 10)-point modified 2-D DFT. (Courtesy of Russell M. Mersereau and Dan E. Dudgeon, *IEEE Trans. Acoustics, Speech, and Signal Processing,* © 1974 IEEE.)

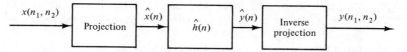

**Figure 7.24**    2-D LSI system implemented using projections.

## PROBLEMS

**7.1.** Find a constraint operator $C$ that is appropriate for each of the following sets of sequences.

(a) The set of sequences with finite extent whose support is limited to some region $R$.

(b) The set of bandlimited sequences whose Fourier transform is nonzero only in some region $W$.

(c) The set of bounded sequences that satisfy $L \leq x(n_1, n_2) \leq U$ for all $(n_1, n_2)$.

**7.2.** Consider a simple iteration of the form

$$x_{k+1}(n_1, n_2) = F[x_k(n_1, n_2)]$$

for the simple case

$$F[x] = ax + b$$

(a) If $x_0(n_1, n_2) = b$, work out expressions for $x_1(n_1, n_2)$, $x_2(n_1, n_2)$, and $x_3(n_1, n_2)$.
(b) What conditions on $a$ and $b$ must be imposed if the iteration is to converge?
(c) To what sequence $x_\infty(n_1, n_2)$ will the iteration converge?

**7.3.** Let $F$ be an operator that is used in the iteration

$$x_{k+1}(n_1, n_2) = F[x_k(n_1, n_2)] \qquad\qquad \text{(P7.3a)}$$

The operator $F$ is said to be a *contraction mapping* (or simply a *contraction*) if

$$\| F[x_i] - F[x_j] \| \le r \| x_i - x_j \| \qquad \text{for } 0 \le r < 1$$

for any $x_i$ and $x_j$ for which $\| x_i \|$ and $\| x_j \|$ are finite. The $l_2$ norm of a sequence can be defined as

$$\| x(n_1, n_2) \| = \left[ \sum_{n_1} \sum_{n_2} x^2(n_1, n_2) \right]^{1/2}$$

If the inequality is true only for $r = 1$, the operator is called *nonexpansive*. If $F$ is a contraction, the iteration in equation (P7.3a) is guaranteed to converge to a unique solution for any choice of the starting sequence which has a finite $l_2$ norm.
(a) Consider the iteration

$$x_{k+1}(n_1, n_2) = C[x_k(n_1, n_2)] + \lambda\{y(n_1, n_2) - DC[x_k(n_1, n_2)]\} \qquad \text{(P7.3b)}$$

Show that this iteration converges if the operator $G = (I - \lambda D)C$ is a contraction, where $I$ is the identify operator. (*Hint:* Find the operator $F$ and require that it be a contraction.)
(b) Using the result from part (a), show that the iteration in equation (P7.3b) converges if:
(1) $C$ is a contraction and $(I - \lambda D)$ is nonexpansive.
(2) $C$ is nonexpansive and $(I - \lambda D)$ is a contraction.
(3) Both $C$ and $I - \lambda D$ are contractions.

**7.4.** We saw in Section 7.1.1 that the Van Cittert iteration for unconstrained deconvolution could be interpreted as a first-order difference equation in $k$ whose coefficients are functions of $(\omega_1, \omega_2)$.
(a) Where is the "pole" of this difference equation? If $0 < H(\omega_1, \omega_2) \le 1$, what value of $\lambda$ maximizes the rate of convergence?
(b) Using the difference equation analogy we can generate a second-order iteration which produces a new estimate of the undistorted signal using the *two* previous estimates. This iteration will have the form

$$X_{k+1}(\omega_1, \omega_2) = A Y(\omega_1, \omega_2) + B X_k(\omega_1, \omega_2) + C X_{k-1}(\omega_1, \omega_2)$$

What constraints must exist between $A$, $B$, and $C$ if we require

$$X_{-1}(\omega_1, \omega_2) = 0$$

$$X_0(\omega_1, \omega_2) = A Y(\omega_1, \omega_2)$$

$$X_\infty(\omega_1, \omega_2) = \frac{Y(\omega_1, \omega_2)}{H(\omega_1, \omega_2)}?$$

**7.5.** The Van Cittert iteration converges if

$$|1 - \lambda H(\omega_1, \omega_2)| < 1 \tag{P7.5}$$

This condition places severe constraints on the possible values assumable by $H(\omega_1, \omega_2)$.

**(a)** If $\hat{z} = H(\omega_1, \omega_2)$, sketch the region of the complex $\hat{z}$-plane defined by equation (P7.5).

**(b)** The parameter $\lambda$ can be chosen by the user. If

$$\text{Re}\,[H(\omega_1, \omega_2)] > 0$$

show that a value of $\lambda$ can always be found so that (P7.5) is satisfied.

**7.6.** A nonexpansive operator was defined in Problem 7.3.

**(a)** Show that a finite-support constraint operator is nonexpansive.

**(b)** Show that the positivity operator is nonexpansive.

**(c)** An alternative to the positivity operator defined in the text is the operator *ABS* defined by

$$ABS[x(n_1, n_2)] = |x(n_1, n_2)|$$

Show that this operator is also nonexpansive.

**7.7.** As still another approach to the problem of finding the input $x(n_1, n_2)$ which produces a given output $y(n_1, n_2)$ (see Figure P7.7), we might consider solving for the sequence

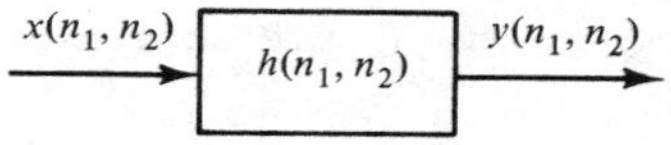

**Figure P7.7**

$x(n_1, n_2)$ which minimizes the functional

$$J = \sum_{n_1} \sum_{n_2} [y(n_1, n_2) - (x ** h)(n_1, n_2)]^2$$

This can be done iteratively using the method of steepest descent. With this procedure we set

$$x_{k+1}(n_1, n_2) = x_k(n_1, n_2) - \lambda_k \frac{\partial J}{\partial x(n_1, n_2)}\bigg|_{x(n_1, n_2) = x_k(n_1, n_2)}$$

**(a)** By explicitly evaluating the gradient, write out the equations that must be implemented at each iteration.

**(b)** How is this iteration different from equation (7.9)?

***(c)** Working backward, equation (7.9) can be seen to be equivalent to minimizing a functional by the method of steepest descent. What is the functional? [Assume that $h(n_1, n_2) = h(-n_1, -n_2)$.]

**7.8.** The input–output relation for a 1-D linear shift-varying system is given by the superposition sum

$$y(n) = \sum_{m=-\infty}^{\infty} h(n, m)x(m)$$

If $x$ and $y$ are finite-extent sequences, they can be written as vectors and the superposition sum assumes the form

$$\mathbf{y} = \mathbf{Hx}$$

If the extent of **x** equals the extent of **y**, the signal $x$ can be recovered using the matrix iteration

$$\mathbf{x}_{k+1} = \lambda\mathbf{y} + (\mathbf{I} - \lambda\mathbf{H})\mathbf{x}_k$$

**(a)** If $\mathbf{x}_0 = \lambda\mathbf{y}$, find expressions for $\mathbf{x}_1$, $\mathbf{x}_2$, and $\mathbf{x}_3$.
**\*(b)** How must $\lambda$ be chosen to guarantee convergence? Can such a $\lambda$ always be found? (Consider an eigenvector expansion of **H**.)
**\*(c)** In working part (b) you should have found that there were some **H** for which the iteration did not converge. Modify the iteration so that it will converge for any **H** that has nonzero eigenvalues.

**7.9.** An iteration for reconstructing a signal $x(n_1, n_2)$ from its phase was given in Section 7.1.3 as

$$x_0(n_1, n_2) = \lambda\mathcal{F}^{-1}\{\exp\{j\arg[X(\omega_1, \omega_2)]\}\}$$

$$x_{k+1}(n_1, n_2) = Cx_k(n_1, n_2) = \mathcal{F}^{-1}C_F\mathcal{F}C_S x_k(n_1, n_2)$$

where $\mathcal{F}$ and $\mathcal{F}^{-1}$ are the Fourier transform and inverse Fourier transform operators, $C_S$ is a spatial domain constraint operator, and

$$C_F[X_k(\omega_1, \omega_2)] = |X_k(\omega_1, \omega_2)|\exp\{j\arg[X(\omega_1, \omega_2)]\}$$

**(a)** Show that the operator $C$ is nonexpansive if $C_S$ is nonexpansive. (A nonexpansive operator was defined in Problem 7.3.)
**(b)** If, instead, we wish to reconstruct $x(n_1, n_2)$ from its magnitude, we would use the operator

$$C_F[X_k(\omega_1, \omega_2)] = |X(\omega_1, \omega_2)|\exp\{j\arg[X_k(\omega_1, \omega_2)]\}$$

Show that, in this case, we cannot guarantee that $C$ will be nonexpansive.

**7.10.** Let $s(x, y, z, t)$ be an acoustic seismic wave which is propagating upward through the earth with a uniform velocity $c$. Define a partial Fourier transform of the wavefront as

$$S(k_x, k_y, z, \Omega) \triangleq \int_{-\infty}^{\infty}\int_{-\infty}^{\infty}\int_{-\infty}^{\infty} s(x, y, z, t)\exp[-j(\Omega t - k_x x - k_z y)]\,dx\,dy\,dt$$

**(a)** If $S(k_x, k_y, z, \Omega)$ is the partial Fourier transform of $s(x, y, z, t)$, determine an expression for the partial Fourier transform of $\partial s(x, y, z, t)/\partial t$ in terms of $S(k_x, k_y, z, \Omega)$. Your expression should involve no integrals.
**(b)** Determine a similar expression for $\partial s(x, y, z, t)/\partial x$.
**(c)** The hyperbolic wave equation requires that

$$\frac{\partial^2 s}{\partial x^2} + \frac{\partial^2 s}{\partial y^2} + \frac{\partial^2 s}{\partial z^2} = \frac{1}{c^2}\frac{\partial^2 s}{\partial t^2}$$

By taking the partial Fourier transform of both sides, convert this partial differential equation into an ordinary differential equation.

**7.11.** Show that the following transfer functions correspond to all-pass systems (i.e., the magnitude of their Fourier transform is constant).
**(a)** $H_{1z}(z_1, z_2) = \dfrac{z_1^{-N_1}z_2^{-N_2}A_z(z_1^{-1}, z_2^{-1})}{A_z(z_1, z_2)}$ where $A_z(z_1, z_2) = \sum_{n_1=0}^{N_1-1}\sum_{n_2=0}^{N_2-1} a(n_1, n_2)z_1^{-n_1}z_2^{-n_2}$
with $\{a(n_1, n_2)\}$ real-valued.

**(b)** $H_{2z}(z_1, z_2) = \dfrac{z_1^{-N_1} A_z(z_1^{-1}, z_2)}{A_z(z_1, z_2)}$ where $A_z(z_1, z_2) = \sum\limits_{n_1=0}^{N_1-1} \sum\limits_{n_2=-N}^{N} a(n_1, n_2) z_1^{-n_1} z_2^{-n_2}$ with $\{a(n_1, n_2)\}$ real-valued and $a(n_1, n_2) = a(n_1, -n_2)$.

**7.12.** **(a)** Determine the 1-D projection of a 2-D uniform circular disk of radius $R$.
**(b)** What is the (continuous) Fourier transform of this projection?

**7.13.** Derive the following property of the 2-D continuous Fourier transform: If an object is rotated through an angle $\theta_0$, its Fourier transform is rotated through the same angle $\theta_0$.

**7.14.** A 2-D circularly symmetric function is completely specified by either its cross section or by its projection. In fact, either of these one-dimensional functions can be determined from the other. Bracewell [26] refers to this relationship as the Abel transform.
**(a)** Derive an integral equation that expresses the projection in terms of the cross section.
**(b)** Derive an integral equation that expresses the cross section in terms of the projection.

**7.15.** Show that an unknown but separable signal

$$x(u_1, u_2) = f(u_1)g(u_2)$$

can be reconstructed exactly from two projections. How should the projection angles be chosen?

**7.16.** By varying the sampling rates of individual projections, the locations at which samples of a multidimensional Fourier transform are taken from those projections can be controlled. How should the projections be sampled to product the elliptical raster shown in Figure P7.16?

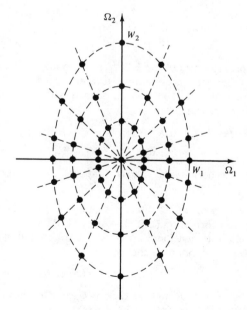

**Figure P7.16**

**7.17.** The filtering of the projections in a discrete implementation of the convolution/back-projection algorithm loses information about the average dc level of the signal being reconstructed. One means of restoring this average level is to simply add a constant signal to the reconstruction so that the original signal and the reconstruction have the same average value. How can the average dc level of the unknown be determined from the projections of that unknown?

**7.18.** Consider a 2-D array that has finite extent $N_1 \times N_2$, where $N_1 = 3$, $N_2 = 4$:

$$
\begin{array}{ccc}
j & k & l \\
g & h & i \\
d & e & f \\
a & b & c
\end{array}
$$

(*i.e.*, the sample $(n_1, n_2) = (0, 0)$ has value $a$, the sample $(n_1, n_2) = (1, 0)$ has value $b$, and so forth.)

Determine the 1-D projections at the following orientations:
**(a)** $m_1 = -1$, $m_2 = 4$
**(b)** $m_1 = -4$, $m_2 = 1$
**(c)** $m_1 = 3$, $m_2 = 4$
and give the resulting order of the labeled samples $(a - l)$.

**7.19.** Consider the projection of the $(N_1 \times N_2 \times N_3)$-point 3-D array onto a 1-D array defined by

$$g(N_2 N_3 n_1 + N_3 n_2 + n_3) = x(n_1, n_2, n_3),$$
$$0 \leq n_1 \leq N_1 - 1, \quad 0 \leq n_2 \leq N_2 - 1, \quad 0 \leq n_3 \leq N_3 - 1$$

**(a)** Interpret samples of the 1-D $N_1 N_2 N_3$-point DFT of $g(n)$ as samples of $X(\omega_1, \omega_2, \omega_3)$. Where are those samples located?
**(b)** A modified 3-D DFT can be defined using the samples of $G(k)$ if that array is mapped according to

$$X_M(k_1, k_2, k_3) \triangleq G(k_1 + k_2 N_1 + k_3 N_1 N_2),$$
$$0 \leq k_1 \leq N_1 - 1, \quad 0 \leq k_2 \leq N_2 - 1, \quad 0 \leq k_3 \leq N_3 - 1$$

Find the periodicity matrix that defines the DFT relating $x(n_1, n_2, n_3)$ and $X_M(k_1, k_2, k_3)$.

**7.20.** The projection of a 2-D discrete signal is a special case of a linear dimensionality-reducing transformation (DRT). A linear DRT is a rearrangement of the samples of a finite-extent $R$-dimensional array $f_R(\mathbf{n})$ onto an $S$-dimensional array $g_S(\mathbf{m})$, $(R \geq S)$, by means of a mapping of the form

$$g_S(\mathbf{Tn}) = f_R(\mathbf{n})$$

where $\mathbf{T}$ is an $S \times R$ matrix. How is $G_S(\boldsymbol{\omega})$, the Fourier transform of $g_S(\mathbf{m})$, related to $F_R(\boldsymbol{\Omega})$, the Fourier transform of $f_R(\mathbf{n})$?

# REFERENCES

1. Ronald W. Schafer, Russell M. Mersereau, and Mark A. Richards, "Constrained Iterative Restoration Algorithms," *Proc. IEEE*, 69 (Apr. 1981), 432–50.

2. P. H. Van Cittert, "Zum Einfluss der Splatbreite auf die Intensitatswerteilung in Spektrallinien II," *Z. für Physik*, 69 (1931), 298–308.

3. R. Marucci, "Signal Recovery from the Effects of a Non-invertible Distortion Operator," M.S. thesis, School of Electrical Engineering, Georgia Institute of Technology (1981).

4. H. J. Trussell, "Maximum Power Signal Restoration," *IEEE Trans. Acoustics, Speech, and Signal Processing*, ASSP-29, no. 5 (Oct. 1981), 1059–61.

5. R. W. Gerchberg, "Super-resolution through Error Energy Reduction," *Optica Acta*, 21 (1974), 709–720.

6. Athanasios Papoulis, "A New Algorithm in Spectral Analysis and Bandlimited Extrapolation," *IEEE Trans. Circuits and Systems*, CAS-22, no. 9 (Sept. 1975), 735–42.

7. G. K. Huddleston, "Aperture Synthesis of Monopulse Antenna for Radome Analysis Using Limited Measured Pattern Data," *Proc. IEEE Southeastcon '81 Conf.*, 350–54.

8. R. W. Gerchberg and W. O. Saxton, "A Practical Algorithm for the Determination of Phase from Image and Diffraction Plane Pictures," *Optik*, 35 (1972), 237–46.

9. J. R. Fienup, "Reconstruction of an Object from the Modulus of Its Fourier Transform," *Optics Letters*, 3 (1978), 27–29.

10. J. R. Fienup, "Iterative Method Applied to Image Reconstruction and to Computer-Generated Holograms," in *Proc. Soc. Photo-optical Instrumentation Engineers*, Session 207–02, *Applications of Digital Image Processing III* (1979).

11. Monson H. Hayes, "The Reconstruction of a Multidimensional Sequence from the Phase or Magnitude of Its Fourier Transform," *IEEE Trans. Acoustics, Speech, and Signal Processing*, ASSP-30, no. 2 (Apr. 1982), 140–54.

12. Monson H. Hayes III, "Signal Reconstruction from Phase or Magnitude," Sc.D. thesis, Department of Electrical Engineering and Computer Science, Massachusetts Institute of Technology (June 1981).

13. Giovanni Garibotto, "2-D Recursive Phase Filters for the Solution of Two-Dimensional Wave Equations," *IEEE Trans. Acoustics, Speech, and Signal Processing*, ASSP-27, no. 4 (Aug. 1979), 367–72.

14. David B. Harris, "Design and Implementation of Rational 2-D Digital Filters," Ph.D. thesis, Department of Electrical Engineering and Computer Science, Massachusetts Institute of Technology (Nov. 1979).

15. J. Claerbout, *Fundamentals of Geophysical Data Processing* (New York: McGraw-Hill Book Company, 1976).

16. J. Radon, "[On the Determination of Functions from Their Integrals along Certain Manifolds]," *Ber. Saechs. Akad. Wiss. Leipzig, Math. Phys. Kl.*, 69 (1917), 262–77 (in German).

17. Kris A. Dines and R. Jeffrey Lytle, "Computerized Geophysical Tomography," *Proc. IEEE*, 67, no. 7 (July 1979), 1065–73.

18. Russell M. Mersereau and Alan V. Oppenheim, "Digital Reconstruction of Multidimensional Signals from Their Projections," *Proc. IEEE*, 62, no. 10 (Oct. 1974), 1319–38.

19. Russell M. Mersereau, "Direct Fourier Transform Techniques in 3-D Image Reconstruction," *Computers in Biology and Medicine*, 6 (1976), 247–58.

20. L. A. Shepp and B. F. Logan, "The Fourier Reconstruction of a Head Section," *IEEE Trans. Nuclear Science*, 21 (1974), 21–43.

21. Berthold K. P. Horn, "Density Reconstruction Using Arbitrary Ray-Sampling Schemes," *Proc. IEEE*, 66, no. 5 (May 1978), 551–62.

22. G. T. Herman and A. Lent, "Iterative Reconstruction Algorithms," *Computers in Biology and Medicine*, 6 (1976), 273–94.

23. G. T. Herman, A. V. Lakshminarayanan, and A. Naparstek, "Convolution Reconstruction Techniques for Divergent Beams," *Computers in Biology and Medicine*, 6 (1976), 259–71.

24. Russell M. Mersereau and Dan E. Dudgeon, "The Representation of Two-Dimensional Sequences as One-Dimensional Sequences," *IEEE Trans. Acoustics, Speech, and Signal Processing*, ASSP-22, no. 5 (Oct. 1974), 320–5.

25. Russell M. Mersereau, "Dimensionality-Changing Transformations with Non-rectangular Sampling Strategies," *Transformations in Optics*, W. T. Rhodes et al., eds. (Seattle: Society of Photo-optical Instrumentation Engineers, 1983.)

26. Ronald N. Bracewell, *The Fourier Transform and Its Applications* (New York: McGraw-Hill Book Company, 1978).

# INDEX